Whole Organ Approaches to Cellular Metabolism

Springer

New York
Berlin
Heidelberg
Barcelona
Budapest
Hong Kong
London
Milan
Paris
Santa Clara
Singapore
Tokyo

Authors and discussants in the planning conference, *Whole Organ Approaches to Cellular Metabolism,* held at the Montreal General Hospital, July 14–16, 1995. *Left to right,* using informal names: *Back row, standing:* Nicole Siauve, Hans van Beek, Sasha Popel, Andi Deussen, Moise Bendayan, Jan Schnitzer, Eugenio Rasio, Tom Harris, Mel Silverman, Rick Haselton, Said Audi, Chris Dawson, Colin Rose, and Dick Effros. *Front row, sitting:* Fernando Vargas, Sandy Pang, Jim Bassingthwaighte, Francis Chinard, Carl Goresky, Jack Linehan, Andreas Schwab, Dick Weisiger, Harry Goldsmith. (Absent: Keith Kroll.)

James B. Bassingthwaighte
Carl A. Goresky
John H. Linehan

Editors

Whole Organ Approaches to Cellular Metabolism

Permeation, Cellular Uptake, and
Product Formation

With 190 Illustrations

 Springer

James B. Bassingthwaighte
Department of Bioengineering
University of Washington
Seattle, WA 98195, USA

John H. Linehan
Biomedical Engineering Department
Marquette University
Milwaukee, WI 53233-1881,USA

Carl A. Goresky (deceased)
formerly, Division of Gastroenterology
Department of Medicine
McGill University School of Medicine
Montreal, Quebec H3G
Canada

Library of Congress Cataloging-in-Publication Data
Bassingthwaighte, James.
    Whole organ approaches to cellular metabolism : permeation, cellular uptake, and product
formation / James B. Bassingthwaighte, Carl A. Goresky, John H. Linehan.
        p.   cm.
    Includes bibliographical references and index.
    ISBN 0-387-94975-5 (alk. paper)
    1. Metabolism.   2. Cell metabolism.   3. Endothelium.   4. Capillaries.   I. Goresky,
Carl A., 1932–1996.   II. Linehan, John H.   III. Title.
    QP171.B37   1998
    572'.4—dc21   97-19015

Printed on acid-free paper.

Production managed by Terry Kornak; manufacturing supervised by Joe Quatela.
Typeset by Princeton Editorial Associates, Scottsdale, AZ, and Roosevelt, NJ.
Printed and bound by Maple-Vail Book Manufacturing Group, York, PA.
Printed in the United States of America.

9 8 7 6 5 4 3 2 1

ISBN 0-387-94975-5 Springer-Verlag New York Berlin Heidelberg   SPIN 10556859

Carl Arthur Goresky
August 25th, 1932 to March 21st, 1996

Carl Goresky was the epitome of the physician–scientist, and even more. Two dozen scientists gathered at the Montreal General Hospital in July 1995 to give tribute to Carl's scientific contributions; they met in admiration, respect, and love for the man, rather than the symbol of science. They met to plan this book on the methods and approaches to making discoveries about cellular metabolism in the intact organ. This is part of the issue of carrying forward the information from genomics, proteomics, and molecular and cellular biology into physiological phenotyping and an understanding of the behavior of an intact organ and organism. Such research can be undertaken only by studying intact systems, an approach Carl pioneered and promoted.

Carl grew up in Castlegar, in the mountains of British Columbia, where his father was the town physician. Carl played the piano so well that he could have made a career of it; he climbed mountains, hunted, collected minerals, and worked as a stevedore on the Columbia River barges. At 16 he went to McGill, and by 22 had completed a B.Sc. and his M.D. As a part of a medical residency at Johns Hopkins Medical School he spent 2 years with Dr. Francis Chinard. Francis had pioneered the multiple indicator dilution technique for estimating solute transport and volumes of distribution (Chinard et al., 1955). Carl brought the technology, including a sample collecting system and many ideas, back to McGill, where he completed a Ph.D. His advisor was an encouraging, brilliant man, Arnold Burgen, whose policy was to give free reign to such a "student," which was just as well because Dr. Burgen left for Oxford before the thesis was complete.

The first part of the thesis was the hallmark 1963 paper (Goresky, 1963). It demonstrated that a set of solutes passing through the liver following simultaneous bolus injection into the portal vein emerged into the hepatic vein in a characteristic way. The shapes of their outflow dilution curves were identical, relative to their mean transit time, and could be superimposed upon each other by scaling the time axis by their individual mean transit times. The observation that the curves superimposed defined all the solutes to be flow-limited in their exchange between blood and tissue: RBC, plasma protein, sucrose, sodium, and water. This conceptual step was based on the deeper idea that the capillary–tissue exchange unit was axially distributed, not a lumped compartment or mixing chamber. These two ideas, coupled with Christian Crone's demonstration that the bolus injection technique could be used to measure capillary permeability (Crone, 1963), set the stage for the use of the multiple indicator dilution technique to elucidate substrate transmembrane transport and intracellular metabolism. Carl's paper on sulfobromophthalein published in 1964, the remainder of the thesis, did exactly that. A refinement of the analysis to correct for catheter delay was published the same year with Carl's first student Mel Silverman, who worked later with Francis Chinard.

Kenneth Zierler, Chinard's compatriot as an undergraduate and colleague as a faculty member at Hopkins, had watched Carl's development in Francis' laboratory in 1958–59, and his excellent performance as chief medical resident the next year. As a reviewer of the 1964 papers for *Circulation Research* he saw the brilliance of these: "There was so much meat in it, so creative." Of the 1963 work he said, "Carl made at least three very important points in this paper, which was obviously technically meticulous." The first point concerned the axially distributed geometry of the capillary, which Carl called a "linear two-compartment system," but which Ken preferred to call a linear two-*component* system to distinguish it from the mixing chamber idea associated with the word *compartment*. His second point was Carl's simple diagram of the system of partial, rather than ordinary, differential equations. The third was the flow-limited behavior described above.

By "technically meticulous" I think Ken was referring not only to the experimental methods but also the methods of analysis. From his first paper onward, Carl used mathematical phrasing, and characterized the biology in terms of the parameters of a precisely hypothesized physiological system. The wealth of papers that followed over 34 years had his mathematical mark upon them. Each advanced the field another step. The flow-limited transport idea applied to gasses carried by erythrocytes, the "red cell carriage effect" (Goresky et al., 1975). The use of Michaelis–Menten expressions for saturable transformation appeared in the 1964 papers. Crone demonstrated this for transport across the brain capillary membrane barrier for glucose a year later (Crone, 1965).

The general, model-free mass balance expressions were laid out by Zierler (Meier and Zierler, 1954; Zierler, 1962a, 1962b), but Carl had developed the next stages through *model-dependent* analyses of the observations: (1) passive barrier limitation (Goresky et al., 1970); (2) concentrative transport (Goresky et al., 1973); (3) carrier-mediated transport (Silverman and Goresky, 1965); (4) intra-tissue diffusion (Goresky and Goldsmith, 1973); (5) intraorgan flow heterogeneity (Rose and Goresky, 1976); (6) transport limitations by two barriers in series (Rose et al., 1977; Rose and Goresky, 1977); (7) reaction via intracellular enzymes (Goresky et al., 1983); (8) receptor binding (Cousineau et al., 1986); and (9) oxygen transport (Rose and Goresky, 1985).

As Carl unraveled the mysteries of increasingly complex systems, he maintained the purity, even if not the simplicity, of the mathematics he used. He believed in finding the analytical solutions to the partial differential equations, and while getting advice from Glen Bach of the Department of Mechanical Engineering, fought his way through each new method of solution. He didn't really trust the accuracy of numerical methods, I suspect, or didn't feel that they offered so much benefit that mathematical elegance could be sacrificed. I like numerical methods for the freedom of concept that they offer, and for speed of solution, but these were secondary issues for him. Carl was strongly principled.

Carl maintained close relationships with many colleagues inside and outside of McGill over his career. Foremost among these were Francis Chinard, his early mentor, and Ken Zierler, Mel Silverman, Arnold Burgen, and others. My relationship with Carl began in 1960 when Carl came to the Mayo Clinic to see his classmate Andy Engel; Carl and I were both beginning our independent studies using indicator dilution methods. Thereafter we met regularly not only at scientific meetings but also at each other's homes and institutions, sharing our efforts to sort out what we didn't understand. Carl made everyone feel a partner in these explorations; while the average guru tells one how it is, Carl helped everyone to reason their way toward an answer.

Carl's qualities as a teacher were seldom equalled. He was patient, careful, and kind, and led the residents and fellows through a topic. The GI residents loved him; when he died in the Montreal General, they all came as a group to his bedside to pay their respects. But when presenting a new topic at a scientific meeting he didn't always think of himself as a teacher but as the presenter of the information, in all its glory. Some presentations were difficult for the general audience, though great for the cogniscenti; Carl was modest to a fault, in the sense that he seemed to think that everyone was as smart and quick as he was. At McGill and on many occasions elsewhere he was a magnificent teacher. One of the best lectures I have ever heard, Carl gave out of the blue; he was asked to explain indicator dilution methods to an evening meeting of the National Academy of Engineering in Washington, D.C. Knowing that the biology was unknown to his audience, but that quantitative approaches were known, he gave a most erudite comprehensive review of the concepts and applications in a half hour, with just chalk and blackboard.

Carl provided leadership in the medical sciences. He edited the journal *Clinical and Investigative Medicine* throughout his last 12 years. He headed the Division of Gastroenterology at the two McGill hospitals, the Royal Victoria and the Montreal General, having brought their two gastroenterology divisions into the first merger between the two hospitals. His efforts in science and medicine were recognized for the impact he had on both. He received the Landis Award of the Microcirculatory Society, the Gold Medal of the Canadian Liver Foundation, the Distinguished Achievement Award of the American Association for the Study of Liver Diseases, and many others. In 1995 he was named officer of the Order of Canada, equivalent to a knighthood in the United Kingdom.

Behind him he leaves many colleagues who will carry on his efforts. Harry Goldsmith and Andreas Schwab, his close friends and colleagues in the research unit, Colin Rose in Cardiology, Phil Gold and Doug Kinnear in Medicine, all at the Montreal General, Eugenio Rasio and Moise Bendayan at the University of Montreal, Jocelyn Dupuis at the Montreal Heart Institute, Mel Silverman and Sandy Pang at the University of Toronto, and others scattered around the globe, continue, like myself, to learn from him and to build upon his ideas. Gone he may be, but never to be forgotten.

James B. Bassingthwaighte

References

Chinard, F. P., G. J. Vosburgh, and T. Enns. Transcapillary exchange of water and of other substances in certain organs of the dog. *Am. J. Physiol.* 183:221–234, 1955.

Crone, C. The permeability of capillaries in various organs as determined by the use of the "indicator diffusion" method. *Acta Physiol. Scand.* 58:292–305, 1963.

Crone, C. Facilitated transfer of glucose from blood into brain tissue. *J. Physiol.* 181:103–113, 1965.

Meier, P., and K. L. Zierler. On the theory of the indicator-dilution method for measurement of blood flow and volume. *J. Appl. Physiol.* 6:731–744, 1954.

Zierler, K. L. Circulation times and the theory of indicator–dilution methods for determining blood flow and volume. In: *Handbook of Physiology,* Sect. 2: *Circulation,* Washington, D.C.: American Physiological Society, 1962, pp. 585–615.

Zierler, K. L. Theoretical basis of indicator-dilution methods for measuring flow and volume. *Circ. Res.* 10:393–407, 1962.

Preface

The field of capillary–tissue exchange physiology has been galvanized twice in the past 25 years. A 1969 conference at the National Academy of Sciences in Copenhagen resulted in the book *Capillary Permeability: The Transfer of Molecules and Ions Between the Capillary Blood and the Tissue* (Crone and Lassen, 1970). It focused on the physiochemical aspects of transcapillary water and solute transport. The field has matured considerably since. This volume was designed as the successor to the 1970 book, and was created at a gathering of the authors at McGill University. It too captures the breadth of a field that has been dramatically enriched by numerous technical and conceptual advances. In 1970 it was already known that the capillary wall was not merely a "cellophane bag" exerting steric hindrances on solute particles. Instead, the endothelial surface was recognized as the site of binding reactions and permeation by passive or carrier-mediated transport. Furthermore, the cells of the blood could traverse evanescent wide openings in the "zippered" clefts. Today, research priorities have turned more to cell–cell interactions, toward understanding the utility of the gap junctional connections between endothelial cells and neighboring smooth muscle cells, neuronal twigs, and the parenchymal cells of organs. New discoveries in the past few years have revealed the critical importance of the close relationships between the endothelial cells and the parenchymal cells. Endothelial cell transporters, enzymes, and receptors play critical roles in substrate transport to the parenchymal cells of the organ, and in receptor-mediated responses related both to vasoregulation and to the functions of the parenchymal cells of the organ. Thus the focus has shifted away from permeation mechanisms and toward cellular metabolism.

This book brings together contributions from prominent researchers in the kinetics of blood–tissue exchange processes, in endothelial biochemistry and metabolism, and in cellular to whole body imaging, around the central theme of endothelial and parenchymal cellular function. The planning meeting "Whole Organ Approaches to Cellular Metabolism" was sponsored by the Commission on Bioengineering in Physiology of the International Union of Physiological Sciences, and supported generously by the Whitaker Foundation. Harry Goldsmith organized a setting conducive to group discussion at the Montreal General Hospital. There was a focus on the interpretation of high-resolution data which provide insight into cellular function using simulation analysis applied to physiological systems. This is the only workable approach for whole animal and human studies using nuclear magnetic resonance, positron emission tomography, and X-ray computed tomography—imaging modes that are well suited for acquisition of data in situations where modeling is essential to understanding of cellular function. Examples are studies of cancerous growth processes, myocardial and cerebral ischemia, and the stages of recovery from injury. Positron emission tomography is particularly useful for examining the distribution of receptors or the dynamics of changing states of flow and metabolism. Noninvasive imaging methods are the key to the identification of the local densities of receptors and the assessment of their normal functions. The whole organ analytical approach provides the mechanism for integrating knowledge from all of these areas and relating them to a common set of underlying processes.

As this book was being brought together Carl Goresky died of renal adenocarcinoma. He worked strenuously to the end, and on his last day worked on Chapter 1, the principles. The book is dedicated to his memory, to the many ideas he pioneered, and to the leadership he provided in science and medicine.

Another colleague has been lost just as his career was blossoming. Keith Kroll, who was born on December 9, 1948 and died on July 15, 1997, had the same spirit of perseverance and dedication as did Carl as he struggled with a devastatingly rapid progression of gastric adenocarcinoma. His last two years saw him emerge as a leader in the understanding of cellular energy balance in the heart.

Carl Goresky and Keith Kroll were determined, brilliant scholars, kindly teachers, and wonderful colleagues. While we try to follow in their footsteps, we cannot do what they would have done.

James B. Bassingthwaighte
John H. Linehan

Contents

4.   Metabolism in the Liver

5.   Metabolism in the Lung

Contributors

Susan B. Ahlf, Department of Veteran Affairs Medical Center, Milwaukee, WI 53295, USA

Said H. Audi, Research Service, Physiology, Veterans Administration Medical Center, Milwaukee, WI 53295, USA

Glen G. Bach, Department of Mechanical Engineering, McGill University, Montreal, Quebec, Canada

James B. Bassingthwaighte, Center for Bioengineering, University of Washington, Seattle, WA 98195, USA

Moïse Bendayan, Department of Anatomy, University of Montreal, Montreal, Quebec, Canada

Julie Biller, Department of Pulmonary and Critical Care, Medical College of Wisconsin, Milwaukee, WI 53226, USA

Robert D. Bongard, Department of Physiology, Medical College of Wisconsin, Milwaukee, WI 53226, USA

Soledad Calvo, National Institute of Child Health and Human Development, National Institutes of Health, Bethesda, MD 20892

Francis P. Chinard, Department of Medicine, New Jersey Medical School, Newark, NJ 07103-2714, USA

Christopher A. Dawson, Department of Physiology, Medical College of Wisconsin, and Veterans Administration Medical Center, Milwaukee, WI 53295-1000, USA

Andreas Deussen, Institute für Physiologie, Medizinische Fakultät Karl Gustav Karus, Technische Universität Dresden, Dresden, D-01307, Germany

Abhijit Dutta, Fluent, Inc., Centerra Resource Park, Lebanon, NH 03766-1442, USA

Richard M. Effros, Department of Pulmonary and Critical Care Medicine, MCW Clinic at Froedtert, Milwaukee, WI 53226, USA

Wanping Geng, Bioavail Corporation International, Toronto, Ontario M1L 4S4, Canada

Carl A. Goresky, formerly Division of Gastroenterology, Department of Medicine, McGill University School of Medicine, Montreal, Quebec H3G, Canada

Thomas R. Harris, Biomedical Engineering Department, Vanderbilt University, Nashville, TN 37203, USA

Frederick R. Haselton, Department of Biomedical Engineering, Vanderbilt University, Nashville, TN 37235, USA

Elizabeth Jacobs, Pulmonary Division, Medical College of Wisconsin, Milwaukee, WI 53226, USA

Richard B. King, Department of Bioengineering, University of Washington, Seattle, WA 98195-7962, USA

Gary S. Krenz, Department of Mathematics, Statistics, and Computer Science, Marquette University, Milwaukee, WI 53201-1881, USA

Keith Kroll, formerly Center for Bioengineering, University of Washington, Seattle, WA 98195, USA

John H. Linehan, Biomedical Engineering Department, Marquette University, Milwaukee, WI 53233-1881, USA

Marilyn P. Merker, Departments of Anesthesiology and Pharmacology, Medical College of Wisconsin, Milwaukee, WI 53226, USA

Yoshiyuki Okamoto, Department of Chemistry, Polytechnic University, Brooklyn, NY 11201, USA

Lars E. Olson, Department of Biomedical Engineering, Marquette University, Milwaukee, WI 53201-1881, USA

K. Sandy Pang, Faculty of Pharmacy, University of Toronto, Toronto, Ontario M5S 2S2, Canada

Aleksander S. Popel, Department of Biomedical Engineering, School of Medicine, Johns Hopkins University, Baltimore, MD 21205, USA

Eugenio A. Rasio, Department of Nutrition, Notre Dame Hospital, Montreal, Quebec H2L, Canada

Gary M. Raymond, Department of Bioengineering, University of Washington, Seattle, WA 98195-7962, USA

David L. Roerig, Departments of Anesthesiology and Pharmacology, Medical College of Wisconsin, Milwaukee, WI 53226, USA

Eduardo Rojas, National Institute of Diabetes, Digestive and Kidney Diseases, National Institutes of Health, Bethesda, MD 20892, USA

Colin P. Rose, University Medical Clinic, Montreal General Hospital, Montreal, Quebec H3G 1A4, Canada

Tukin K. Roy, Departments of Anesthesiology and Critical Care Medicine, Johns Hopkins Medical Institutions, Baltimore, MD 21205, USA

Jan E. Schnitzer, Department of Pathology, Harvard University Medical School, and Beth Israel Hospital, Boston, MA 02215, USA

Andreas J. Schwab, Department of Medicine, McGill University School of Medicine, and Montreal General Hospital, Montreal, Quebec H3G 1A4, Canada

Lisa M. Schwartz, Department of Pathology, Duke University Medical Center, Durham, NC 27710, USA

Melvin Silverman, Department of Medicine, University of Toronto School of Medicine, Toronto, Ontario M5S 1A8, Canada

Johannes H.G.M. van Beek, Laboratorium voor Fysiologie, Vrije Universiteit, 1081 BT Amsterdam, The Netherlands

Fernando F. Vargas, Department of Human Physiology, University of California Davis School of Medicine, Davis, CA 95616-8644, USA

Raul Vinet, Department of Pharmacology, Faculty of Medicine, University of Chile, Santiago, Chile

Richard A. Weisiger, Department of Medicine, University of California, San Francisco, San Francisco, CA 94143-0538, USA

Introduction

Whole Organ Approaches to Cellular Metabolism are based on making observations via a variety of techniques at varied resolution. Whole organ data are interpreted in terms of the structures and behavior of tissues: cells of different types, subcellular structures and processes, and the physical chemistry of molecular motions, reactions, and surface phenomena. Often one obtains some data at the suborgan level to aid in the process. This book is structured so as to give some insight first into the general theory of mass balance and conservation principles, then into the more biophysical and molecular aspects of the field, and finally into a succession of applications to various organs. There is no attempt to provide complete coverage of the organs of the body or of full ranges of solutes, substrates, hormones, or pharmaceuticals. The principles developed and illustrated should be adaptable to the study of any organ.

The book is divided into five sections, the first two of which cover the basic fundamentals and the general mechanisms involved in transport. The last three sections are focused on particular organ systems.

Section 1 provides a general background of the principles and practice of indicator dilution methods in the study of cellular metabolism. They are mainly based on mass balance: the expectation that what goes in is retained or comes out.

Section 2 concerns the physical chemistry of transport mechanisms: the interaction between convection, diffusion, permeation, and reaction. These five chapters provide background for the physiological behavior of endothelial cells in their interactions with cells in the blood, with smooth muscle cells, and with the organ's parenchymal cells. Research initiated in the 1940s and still vital includes influences of the glycocalyx, of pH, surface charges, and of zeta potentials on the interactions of solutes and ions to surfaces, on the apparent affinity of receptors, and on the asymmetries of transport rates. Electrophysiology of membrane channels and cell-to-cell conductance of small solutes are topics related to broader phenomena such as calcium cycling. Shear-dependent channel activation of NO release illustrates how endothelial cells can communicate with others.

Section 3 focuses in the first five chapters on the role of endothelial cellular biochemistry in cardiac metabolism. Intraendothelial reactions can have remarkable influences on solute fluxes. A classic example is the finding that after an isolated heart is perfused with solution containing tracer adenosine for 30 minutes, more than 90% is to be found in endothelial cells, not in the myocytes (Sparks et al., 1984). Why is the endothelial capacity for purine so high, when presumably it is the myocytes that have the high ATP turnover? Although this section uses purine handling as an example of the kind of interactions that will be found for other solutes and other signalling pathways, the relationship between purine and energy production from oxygen utilization is of prime importance in cardiac research. The final chapter of this section (Chapter 12) details the theory and experimental results on oxygen transport and metabolism, and although it emphasizes events in skeletal muscle, elucidates the processes occurring in all organs.

Section 4, on the liver, exemplifies how one may examine cellular metabolism in vivo. The absence of a hindering endothelial barrier in the sinusoid facilitates the interpretation of metabolic transformations inside hepatocytes. These chapters range over normal metabolic and pharmacokinetic processes, and into the intracellular diffusional processes that must play a role in the liver's excretory functions. The multiple indicator dilution technique has been the key technology leading to enhanced understanding of hepatic function at the whole organ level. Weisiger's studies (Chapter 16), using optical methods to examine solute concentrations at the cellular level, illustrate that the techniques of cell biology are essential in interpreting whole organ data to a more refined level.

Section 5, on the metabolic functions of the lung, illustrates the power of the multiple tracer indicator dilution approach to dissect events occurring along the pulmonary capillary endothelium, regions a fraction of a micrometer thick. New insight is provided into the complex processes of water transport, which underlie all those processes concerning solutes. Molecular interactions at the surfaces and composite processes occurring within the blood-to-air barrier are all explored to create new insight into barrier function.

It is not fortuitous that the applications are mainly in the heart, lung, and liver, for these are the organs studied most extensively. However, Chinard's pioneering contributions (e.g., Chinard et al., 1955, 1997) on the kidney, Crone's in the brain (Crone, 1963, 1965), Renkin's (Renkin, 1959a, 1959b) and Zierler's (Andres et al., 1954; Meier and Zierler, 1954) in skeletal muscle, and Yudilevich's (Yudilevich and Martin de Julian, 1965; Yudilevich et al., 1979) in the salivary gland demonstrated that the techniques of experimentation and analysis are general. Although the multiple indicator dilution techniques are most easily applied to organs with a single inflow and single outflow, they can be used in more complex organs with multiple inflows and outflows, as suggested by theory (Perl et al., 1969), and applied to the interpretation of brain image sequences (Raichle et al., 1978). For those who wish to determine the metabolic status of intact tissues and organs, this book provides a take-off point for the future.

James B. Bassingthwaighte

References

1. Andres, R., K. L. Zierler, H. M. Anderson, W. N. Stainsby, G. Cader, A. S. Ghrayyib, and J. L. Lilienthal, Jr. Measurement of blood flow and volume in the forearm of man; with notes on the theory of indicator-dilution and on production of turbulence, hemolysis, and vasodilation by intra-vascular injection. *J. Clin. Invest.* 33:482–504, 1954.

2. Chinard, F. P., G. J. Vosburgh, and T. Enns. Transcapillary exchange of water and of other substances in certain organs of the dog. *Am. J. Physiol.* 183:221–234, 1955.

3. Chinard, F. P., G. Basset, W. O. Cua, G. Saumon, F. Bouchonnet, R. A. Garrick, and V. Bower. Pulmonary distribution of iodoantipyrine: Temperature and lipid solubility effects. *Am. J. Physiol.* 272 (*Heart Circ. Physio.* 41):H2250–H2263, 1997.

4. Crone, C. The permeability of capillaries in various organs as determined by the use of the "indicator diffusion" method. *Acta Physiol. Scand.* 58:292–305. 1963.

5. Crone, C. Facilitated transfer of glucose from blood into brain tissue. *J. Physiol.* 181:103–113, 1965.

6. Crone, C. and N. A. Lassen. *Capillary Permeability: The Transfer of Molecules and Ions Between Capillary Blood and Tissues.* Copenhagen: Munksgaard, 1970, 681 pp.

7. Meier, P., and K. L. Zierler. On the theory of the indicator-dilution method for measurement of blood flow and volume. *J. Appl. Physiol.* 6:731–744, 1954.

8. Perl, W., R. M. Effros, and F. P. Chinard. Indicator equivalence theorem for input rates and regional masses in multi-inlet steady-state systems with partially labeled input. *J. Theor. Biol.* 25:297–316, 1969.

9. Raichle, M. E., M. J. Welch, R. L. Grubb, C. S., Higgins, M. M. Ter-Pogossian, and K. B. Larson. Measurement of regional substrate utilization rates by emission tomography. *Science* 199:986–987, 1978.

10. Renkin, E. M. Transport of potassium-42 from blood to tissue in isolated mammalian skeletal muscles. *Am. J. Physiol.* 197:1205–1210, 1959a.

11. Renkin, E. M. Exchangeability of tissue potassium in skeletal muscle. *Am. J. Physiol.* 197:1211–1215, 1959b.

12. Sparks, H. V. Jr., M. W. Gorman, F. L. Belloni, and B. Fuchs. Endothelial uptake of adenosine: implications for vascular control. *The Peripheral Circulation,* edited by S. Hunyor, J. Ludbrook, J. Shaw, and M. McGrath. Amsterdam: North Holland, 1984, pp. 23–32.

13. Yudilevich, D. L., and P. Martin de Julian. Potassium, sodium, and iodide transcapillary exchange in dog heart. *Am. J. Physiol.* 205:959–967, 1965.

14. Yudilevich, D. L., F. V. Sepulveda, J. C. Bustamante, and G. E. Mann. A comparison of amino acid transport and ouabain binding in brain endothelium and salivary epithelium studied by rapid paired-tracer dilution. *J. Neural Transm.* 15:15–27, 1979.

1
Introduction

1
Modeling in the Analysis of the Processes of Uptake and Metabolism in the Whole Organ

James B. Bassingthwaighte, Carl A. Goresky, and John H. Linehan

Introduction

In the whole organ approach to cellular metabolism, the processes of capillary permeation, cellular entry, and intracellular reaction kinetics need to be examined in detail. The area is complex and varies from organ to organ, but there is a set of principles unifying the approaches to studies of these processes. The general approach to endogenous metabolism has been to carry out tracer studies within a variety of concentration steady states, and, for xenobiotics, to study the disposition of tracer within a variety of developed and maintained steady-state bulk concentrations.

To study processes in vivo, a nondestructive approach is needed. The multiple-indicator dilution technique, which was introduced by Chinard et al. (1955) and is based on the use of multiple simultaneous controls, has been the approach of choice. Generally, the tracer substance under study and a reference tracer that does not escape the capillaries are introduced simultaneously into the inflowing blood stream, and their outflow dilution curves are recorded. From a comparison of the outflow patterns for the two substances, information concerning the transcapillary passage of the study substance can be deduced. If the study substance enters tissue cells, a second reference is also usually added to the injection mixture, one that enters the interstitial space but does not enter the tissue cells. The reference substances are chosen so that, as closely as possible, they are carried in the blood stream in the same way as the substance of interest.

The anatomical structure within which the events of exchange and metabolism take place is a tissue or organ. The microcirculatory pathways in this structure consist of small-in-diameter long capillaries that are more or less anastomosing and situated between tissue cells, which are surrounded by an interstitial space. In highly metabolically active organs, the microcirculation tends to be densely packed and the intercapillary distances are small; in poorly metabolizing tissues, intercapillary distances can be much larger. In the former case, diffusion gradients can be essentially flat; in the latter case, gradients can be well developed.

Two general approaches have been taken to obtaining information from tracer studies: a stochastic one, in which model-independent parameters are derived, describing the data; and a model-dependent one, in which the detailed events underlying the processing of the study substance are included in the description.

Model-Independent Descriptions of Transport Functions

Probability Density Function of Transit Times

The underlying conditions necessary for the application of this approach are that the system be linear and stationary. In a linear system, if two inputs are given together, the output is the sum of the individual responses. Stationarity implies that the distribution of transit times does not change over the period of observation.

The description to be developed is a stochastic one. It applies to a tracer that, when added to the circulation, enters an organ but is not transferred or lost, and so emerges completely at the outflow. The conservation of matter applies; what is introduced into the circulation equals what leaves via the circulation. Thus,

$$q_0 = F \int_0^\infty C(t)\, dt, \tag{1.1}$$

where q_0 is the amount of tracer injected, $C(t)$ is the outflow concentration of tracer, and F is the steady flow in the system. If the concentration is normalized in terms of the amount of tracer injected (this automatically defines the injected tracer as a unit amount), we find

$$1.0 = F \int_0^\infty [C(t)/q_0]\, dt, \tag{1.2}$$

$$= \int_0^\infty h(t)\, dt, \tag{1.3}$$

where $h(t)$ is the transport function of the circulation. The function $h(t)$ provides a description of the frequency function of transit times or probability density function of transit times in the circulation of the organ. There are two underlying assumptions: There is no recirculation of tracer; and the tracer has been introduced at the entrance at time zero as an impulse function or Dirac delta function $\delta(t)$ (a spike of infinitely narrow width unit area, which means the area equals 1.0). The transfer function for a vascular substance and for a tracer leaving and reentering the circulation will be quite different. In this development we use the notation originally defined by Zierler (Meier and Zierler, 1954; Zierler, 1962, 1965) and formalized by international agreement (Bassingthwaighte et al., 1986).

Calculation of Flow

When reorganized, Eq. (1.1) provides the key to calculating flow,

$$F = \frac{q_0}{\int_0^\infty C(t)\, dt}. \tag{1.4}$$

Flow is calculated as the inverse of the area under the concentration curve for a substance completely recovered at the outflow, normalized in terms of the amount of tracer injected.

Related Functions

Functions related to the probability density function of transit times $h(t)$, the transport function, are illustrated in Figure 1.1. The function $H(t)$ represents the fraction of tracer that has exited from the system since $t = 0$. The residue function $R(t)$ represents the fraction of the material in the system that has not yet left. Thus,

$$R(t) = 1 - H(t). \tag{1.5}$$

$R(t)$ is the complement of $H(t)$. It is the kind of function observed with positron emission tomography (PET). The rate of exit, $h(t)$, is the rate of diminution of $R(t)$:

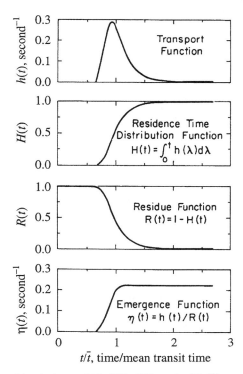

FIGURE 1.1. Relationships between $h(t)$, $H(t)$, $R(t)$, and $\eta(t)$. The curve of $h(t)$ is in this instance given by a unimodal density function having a relative dispersion of 0.33 and a skewness of 1.5. However, the theory is general and applies to $h(t)$s of all shapes. The tail of this $h(t)$ curve becomes monoexponential and hence $\eta(t)$ becomes constant. (From Bassingthwaighte et al., 1986.)

$$\frac{dR(t)}{dt} = -h(t). \tag{1.6}$$

From the point of view of the contents, the fractional rate of loss is the rate of exit $h(t)$ divided by the contents itself. This defines the fractional escape rate from the system, the emergence function $\eta(t)$,

$$\eta(t) = -h(t)/R(t) \tag{1.7}$$

$$= \frac{-dR(t)/dt}{R(t)} \tag{1.8}$$

$$= -d[\ln R(t)]/dt. \tag{1.9}$$

This is the negative of the slope of the residue function, which is seen most commonly as a washout curve recorded by external detection.

Input Functions May Be of Different Form

Since the response of the system $h(t)$ to an impulse or Dirac delta function $\delta(t)$ is, by definition, the transfer function of the system, the outflow response $C_{out}(t)$ to any input to the system $C_{in}(t)$ can be calculated from the convolution of the input function $C_{in}(t)$ with the transfer function $h(t)$:

$$C_{out}(t) = C_{in}(t) * h(t)$$

$$= \int_0^t C_{in}(\tau)h(t - \tau) \, d\tau, \tag{1.10}$$

where τ is a dummy variable of integration that takes values from 0 to t. The input function often will be quite dispersed in relation to an impulse function. When the input is quite dispersed, the tracer entering early may leave early and the residue function $R(t)$ may never reach a level at which all of the tracer is in the system at one time. Careful experimental evaluation is needed to perceive this.

In any experiment, the injection and collecting systems constitute an additional transfer function. To recover the response of the organ itself, it is necessary to deconvolute the impulse response of the injection and collecting systems.

Solute Extraction During Transcapillary Passage

Transit Times, Volumes, and Moments

The mean transit time for a tracer that is completely recovered in the outflow is the first moment of the probability density function $h(t)$:

$$\overline{t} = \int_0^\infty th(t) \, dt. \tag{1.11}$$

The integration is for infinite time; there can be no recirculation of the tracer. Thus one must either prevent recirculation by surgical maneuvers or guess that the tail of the dilution curve has a specific form. The most common form of extrapolation, which was designed to exclude recirculation of the tracer in experiments in vivo, is that proposed by Hamilton et al. (1932), in which the tail of the curve is predicted to follow a single exponential time course. Bassingthwaighte and Beard (1995) proposed a power law function as a probable improvement. The first moment can also be calculated directly from the residue function:

$$\overline{t} = \int_0^\infty R(t)\, dt. \tag{1.12}$$

The tail of the residue function can also be extrapolated with either a chosen model function such as an exponential, e^{-kt}, or a power law, kt^α.

Alternately, the mean transit time can also be calculated directly from the outflow time concentration curve, $C_v(t)$:

$$\overline{t} = \frac{\int_0^\infty t C_v(t)\, dt}{\int_0^\infty C_v(t)\, dt}. \tag{1.13}$$

When the indicator is consumed or sequestered, the value for $\overline{t}$ will become smaller, because the consumption of molecules remaining in the system for a longer time is greater than that for particles having a shorter exposure to the sites of consumption.

The product of flow F and the mean transit time gives an estimate of the volume of distribution V for the indicator, provided that the indicator is not consumed:

$$V = F \cdot \overline{t}. \tag{1.14}$$

The volume calculated is dictated by the flow of medium used in the calculation. For a vascular volume, F is blood flow, and the volume calculated is a vascular volume. Interstitial substances are confined to the plasma phase of blood. Hence, for these, one would use plasma flow, and the volume calculated then will be the sum of the vascular plasma volume plus the interstitial space, expressed in terms of an equivalent plasma volume. One can also use the vascular water flow. The calculated volume of distribution for labeled water then will be the sum of the vascular water space and the water contents of the interstitial and cellular spaces.

Higher moments can also be used to describe $h(t)$. These provide a quantitative description of shape and are useful for the comparison of curves. These can be expressed in terms of $h(t)$. The useful forms are the moments around the mean, the nth central moment μ_n being

$$\mu_n = \int_0^\infty (t - \overline{t})^n \cdot h(t)\, dt. \tag{1.15}$$

Numerical approaches to this computation are given by Bassingthwaighte (1974). The standard deviation (SD) is $\mu_2^{1/2}$, which is the square root of the variance of $h(t)$. It provides a measure of the temporal spread or dispersion of $h(t)$. The relative dispersion of SD/$\overline{t}$ is especially useful; it is the standard deviation divided by the mean transit time, which gives a measure of relative spread. Within the vascular system a bolus of injectate undergoes spatial spreading, so that the temporal spread increases with passage downstream. On the other hand, the value for SD/$\bar{t}$ will reflect spatial dispersion in a manner more or less independent of flow, so long as the flow characteristics remain constant over a range of flows, since SD and $\bar{t}$ change proportionately.

Two other standard parameters that use the third and fourth moments for their computation are skewness and kurtosis. Each is calculated as the nth central moment divided by the standard deviation raised to the power n:

$$\beta_{n-2} = \mu_n/SD^n. \tag{1.16}$$

The skewness β_1 (i.e., the value of the expression when $n = 3$) is a measure of asymmetry. Right skewness is indicated by a positive β_1 and left skewness by a negative β_1. The skewness of most circulatory transport functions have values of the order of +1.0 (Bassingthwaighte et al., 1966). The value for kurtosis β_2 [Eq. (1.16) with $n = 4$] can be used to evaluate the degree of deviation from a Gaussian probability function. The kurtosis has a value of 3.0 for a Gaussian function, >3.0 for leptokurtotic (sharp-pointed) density functions, and <3.0 for platykurtic (flat-topped) functions.

Flow Distributions and Fractals

The regional flows in all organs are heterogeneous. This was not generally recognized until autoradiographic and microsphere techniques evolved (Yipintsoi et al., 1973) but is seen in the brain (Sokoloff et al., 1977), kidney (Grant and Lumsden, 1994), and lung (Glenny et al., 1991).

The heterogeneity of regional blood flows in the heart has been recognized for a long time (e.g., Bassingthwaighte, 1970; Bassingthwaighte et al., 1972; Buckberg et al., 1971). The observation that the standard deviations were about 25% of the mean flows in isolated blood-perfused hearts turned out to be true also for the distributions in awake animals (King et al., 1985), as shown in Figure 1.2. It was later affirmed that this was not due to an artifact of the method (Bassingthwaighte et al., 1990) and therefore had to be considered as a normal variation. (Why this is so broad in monofunctional organs like the liver or the heart is an unanswered question.)

Because differences in capillary transit times mean that there are different contact times for an indicator to escape from different capillaries, the fraction of solute escaping is lower from high-velocity capillaries than from slower ones. This was recognized in Renkin's (1959a,b) and Crone's (1963) expression:

$$E = 1 - e^{-PS/F}, \tag{1.17}$$

where E is an extraction (calculated from data in different ways in different experiments), PS is a membrane permeability–surface area product, and F is the regional flow. The exponent is dimensionless, and $\exp(-PS/F)$ is the fraction of indicator travelling from the entrance of the exchange region to the exit without escaping, that is, it is the "throughput fraction" and E is the "extracted fraction."

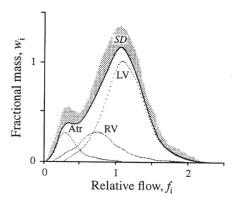

FIGURE 1.2. The probability density function, $w(f)$, of relative flows in the hearts of 13 awake baboons, from 13,114 estimates at four to six times in 2,706 tissue pieces: $w_i(f_i)$ is the fraction of the organ per unit mean flow having a flow f_i, where f_i is dispersionless and is the local flow F_i (in the same units). The solid line represents the distribution for the whole heart, while the dotted lines are its left ventricular (LV), right ventricular (RV), and atrial (Atr) components. The standard deviation (SD) for the whole heart curve is shown by the shaded region. The area under each curve represents its fraction of the total heart weight, and the mean is its flow relative to the total heart flow; these are, by definition, unity for the total heart curve. For the regional curves, the values are LV 0.70 and 1.14, RV 0.20 and 0.81, and Atr 0.10 and 0.41. The relative dispersion of each curve is a measure of the spatial heterogeneity of the flow. The values are, for the whole heart 0.38, for LV 0.30, for RV 0.32, and for Atr 0.17. (Data and figure from King et al., 1985, their Figure 4, with permission. Copyright 1985 American Heart Association.)

Thus E must differ in pathways with different flows, necessitating the use of multicapillary models for the analysis of organs with flow heterogeneity.

Models to account for flow heterogeneity are generally composed of a set of parallel paths with differing flows (see Chapter 7). However, there are many other arrangements to be found in nature, not only with respect to the numbers of generations of arteries and veins represented, but also with respect to collateral connections between arteries of the same generation and likewise between veins. Furthermore, even given an uncomplicated parallel arrangement as in Figure 1.1, there are several ways to approximate the solutions to the equations. We will come back to this in a later section on multipath models.

The probability density function of relative regional myocardial flows, $w(f)$, that is observed depends on the level of resolution used for making the observations. Figure 1.3 shows that dividing the organ into smaller pieces gives a broader dispersion of regional flows, that is, more heterogeneity of flows is revealed by dividing ever more finely so long as each piece is internally heterogeneous. When the pieces are all internally uniform, then further dividing does not increase the variance. The correlation between neighboring units is demonstrated in Figure 1.4 because the slope of the regression between the log voxel size and the log dispersion (log RD) is shallower than -0.5, indicating that the increase in RD with further dividing is less than occurs with random independent values of regional flows. (The fractal D here is $-$[the slope $- 1$] or 1.24; the Hurst coefficient H is $2 - D$ or 0.76. The nearest neighbor correlation is $r_1 = 0.43$.) The conclusion is that neighboring regions tend to be alike statistically and are therefore not so likely to have strikingly different solute concentrations, a situation that will not foster net

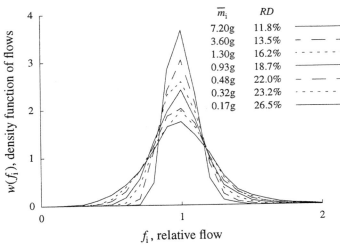

FIGURE 1.3. Probability density functions of regional myocardial blood flows at eight levels of resolution. Data are from the hearts of awake baboons where the measurements were made from the deposition of 15-μm diameter microspheres. Histogram bin widths are 0.1 times the average flow for the heart; m_i is the average volume element mass (or voxel size) for a particular histogram, and RD is the square root of the variance over the mean for that m_i. (Data from King et al., 1985. Figure modified from Bassingthwaighte et al., 1989a, their Figure 4, left panel, with permission. Copyright 1989 American Heart Association.)

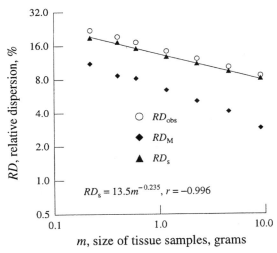

FIGURE 1.4. Fractal regression for spatial flow variation in left ventricular myocardium of a baboon. Plotted are the relative dispersion of the observed density function (RD_{obs}), the methodological dispersion of (RD_M), and the spatial dispersion (RD_s) at each piece mass calculated using $RD_s^2 = RD_{obs}^2 - RD_M^2$. The regression equation is $RD_s(m) = RD_s(m = 1\ g) \cdot m^{1-D}$, where D is the fractal dimension, here 1.235. (Figure reproduced from Bassingthwaighte et al., 1989a, Fig. 5, left panel, with permission. Copyright 1989 American Heart Association.)

exchange of solute between regions. (If neighboring blood-tissue exchange units have similar flows and concentrations, they will not exchange material with one another, which simplifies modeling.)

The fractal description, a reference RD and the fractal exponent, give a two-parameter description of the heterogeneity that is independent of the piece size into which the tissue has been divided. This allows for easy and precise comparisons of the estimates of heterogeneity obtained in different laboratories.

Anatomic and Virtual Volumes

Anatomic observations on capillary densities provide estimates of the blood volume in the exchange region. In well-perfused organs such as the heart, liver, kidney, brain, and lung, the capillary densities are high, as for example in the heart in Figure 1.5. From the capillary densities and diameters one can estimate volumes of the exchange region. Other estimates need to be put together with these to make sure that the overall properties of volume, specific gravity, tissue composition, and water content are all compatible with one another. Such data are shown in Table 1.1.

A more physiological measure of the volume fractions into which specific solutes equilibrate is obtained by determining the volumes of distribution of tracer-labeled solutes of different classes. A comparison of two different extracellular markers in the quick-frozen hearts of rabbits is shown in Figure 1.6. ^{58}CoEDTA is a particularly good marker because it is a gamma-emitter and the cobalt is extremely tightly bound to the EDTA; the compound is polar and inert and remains extracellular (Bridge et al., 1982). The [^{14}C]sucrose and ^{58}CoEDTA were injected intravenously 25 minutes before the animal was sacrificed so that equilibration could occur between blood and extravascular space. The volumes of distribution represent $V'_{ecf} = V_p + V'_{isf}$, where V'_{ecf} (ml/g) is extracellular fluid space, V_p is the total plasma space, milliliters of plasma per gram of tissue, and V'_{isf} is the interstitial space, which is the extravascular, extracellular distribution space. The estimates of V'_{ecf} can be used to derive V'_{isf} when V_p is measured by means of an intravascular plasma marker such as ^{131}I-albumin.

In other experiments, Gonzalez and Bassingthwaighte (1990) determined V_p, V_{rbc} (erythrocyte space), V'_{Na} (sodium space), and total water space. For the tracers, the calculation of volume of distribution V'_{tracer}, ml/g, is

$$V'_{tracer} = \frac{1}{\rho_p} \cdot \frac{C_{tracer} \text{ (heart sample)}}{C_{tracer} \text{ (plasma sample)}}, \tag{1.18}$$

where C_{tracer} has the units of radioactivity level per gram of tissue or plasma and ρ_p is the specific gravity, g/ml, of plasma. The prime on the V' is therefore a statement that the volume of distribution is a virtual volume with a concentration equal to that in the plasma. The observations for the set of tracers are shown in Figure 1.7. In this case, there is no true extracellular marker such as sucrose or CoEDTA, but the sodium space V'_{Na} is only slightly larger than V'_{ecf} since the intracellular Na levels are normally less than one-tenth of the extracellular levels.

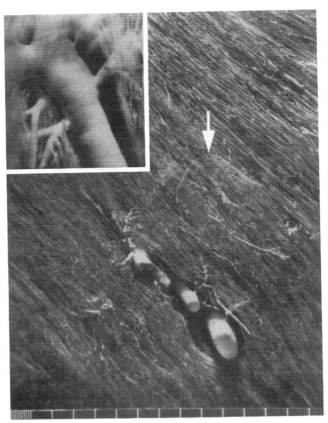

FIGURE 1.5. Subepicardial vasculature of the dog heart, as shown in section parallel to epicardium, 1 mm deep. Scale divisions are 10 and 100 μm. A 26-μm arteriole, accompanied by two venules, gives rise to three 10-μm arterioles, two short and one long (arrow). Inset (same scale) shows a 160-μm vein giving rise to small venules and branching rapidly into the parallel capillaries. (From Bassingthwaighte et al., 1974, with the permission of Academic Press.)

The Multiple-Indicator Dilution (MID) Experiment

Each experiment must be defined to serve a particular purpose. The multiple-indicator dilution (MID) experiment is strongest and most accurately interpretable with respect to physiological events that occur close to the capillary, that is, for capillary permeability and intraendothelial reactions. Reactions that occur in the parenchymal cells of organs are masked to some extent behind the intervening processes of penetration of the capillary wall or endothelial reactions. When endothelial permeabilities are high, and when intraendothelial reactions are negli-

TABLE 1.1. Constituents of the heart.

					Source reference
Total mass as percent of body weight					Diem, 1962
Human	0.45				King et al., 1985
Baboon	0.35				King et al., 1985
Regional myocardial mass and blood flows:					
		Mass fraction,	Fraction of flow,	Mean flow,	
Region	Sp. gr.	%	%	ml min^{-1} g^{-1}	
Left ventricle	1.063	74	80	1.0	Polimeni, 1974
Right ventricle	1.062	20	17	0.8	Bassingthwaighte et al., 1974
Atria		5	3	0.6	
Left ventricle tissue volume fraction					
Cells	70%				
Interstitium	16%				
Capillaries	3.5%				
Arteries and veins	10%				
Capillary dimensions (arrayed in parallel)					Bassingthwaighte et al., 1974
Functional lengths		500–1,000 μm			
Diameters (mean ± SD)		5.0 ± 1.3 μm (SD)			
Capillary density		3,100–3,800/mm^2			
Intercapillary distance		17.5–19 μm			
Chemical composition:					

	Composition, %		
Component	By weight	By volume	
Water	78	82.6	Yipintsoi et al., 1972
Fat	1.5	1.67	Dible, 1934
Protein	17	12.5	Diem, 1962
Carbohydrate	0.7	0.7	Diem, 1962
Mineral ash	1.1	0.5	Diem, 1962

gible, then the estimation of parameters governing parenchymal events is least influenced. A conceptual diagram of the capillary–tissue exchange unit, the basis for a mathematical model, is shown in Figure 1.8. See Chapter 7 for the development of the equations and the application to data analysis.

In order to obtain measures of the influencing processes as directly as possible and to minimize the influences of indeterminacy of these processes on the parameter of particular interest, the MID technique is based on the principle of obtaining multiple simultaneous sets of data that relate to the behavior of the solute under study. Thus, for example, if one wishes to determine the capillary permeability of a solute, then the relevant reference solute is one that does not escape, to any significant extent, from the capillary blood during single transcapillary passage; for example, albumin is a reference solute for determining the capillary permeability to sucrose. In this situation, the albumin transport characteristics within the vascular space may be assumed to be the same as those of the sucrose; thus the shape of the albumin impulse response curve accounts for transport through the convective region, by flow, eddies, and mixing, and by dispersion through a

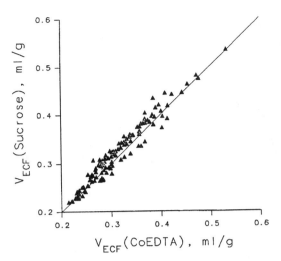

FIGURE 1.6. Estimates of volumes of distribution of extracellular markers. A comparison of extracellular fluid space (V_{ecf}) estimates from [^{14}C]sucrose vs. ^{58}Co-EDTA (from γ-counting rather than β-counting). Regression line, V_{ecf}(Suc) = 0.001 + 1.03 V_{ecf}(^{58}Co-EDTA), with a correlation coefficient of 0.974, is not distinguishable from line of identity. Average V_{ecf} for two tracers is 0.32 $\pm$ 0.06 (N = 130) ml of plasma-equivalent volume/ gram of myocardium. (Figure from Gonzalez and Bassingthwaighte, 1990.)

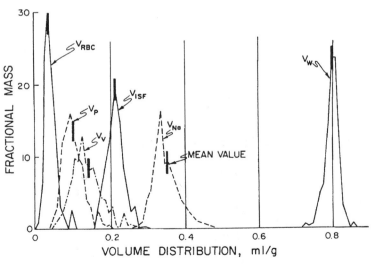

FIGURE 1.7. Probability density functions of volumes of distributions in rabbit left ventricle. Density functions for six animals are combined by superimposing the mean of each individual heart upon the average mean; this provides a correct and realistic representation of spread of the data around the mean. (Figure from Gonzalez and Bassingthwaighte, 1990.)

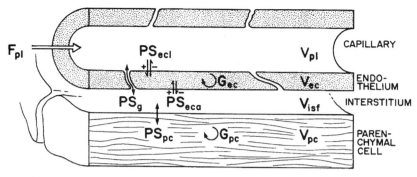

FIGURE 1.8. Representation of model used for analysis of indicator dilution curves. F_{plasma}, plasma (perfusate) flow; PS, permeability-surface areas for adenosine passage through endothelial cell luminal membrane (PS_{ecl}); water-filled channels between endothelial cells (PS_g); endothelial cell abluminal membrane (PS_{eca}); and parenchymal cell membrane (PS_{pc}). G, intracellular consumption (metaoblism) of adenosine by endothelial cells (G_{ec}) or by parenchymal cells (G_{pc}). V, volume of plasma (V_{plasma}), endothelial cell (V_{pc}), interstitial (V_{isf}), and parenchymal cell (V_{pc}) spaces. (Figure from Gorman et al., 1986.)

network of serial/parallel vessels between inflow and outflow. For sucrose, then, which is an extracellular tracer, the same information is used, and the only additional information to be provided from the sucrose independently of the albumin is the capillary PS and the interstitial volume of distribution.

The difference between the intravascular indicator and the permeating one is usefully expressed as an extraction (see Figure 1.9). The instantaneous extraction $E(t)$ is a measure of the fraction of the permeant indicator that escapes, and thereby can provide a measure of the rate of escape across the capillary wall:

$$E(t) = 1 - h_D(t)/h_R(t). \tag{1.19}$$

Its equivalent can be calculated from the slopes of the residue functions:

$$E'(t) = 1 - dR_D(t)/dR_R(t). \tag{1.20}$$

A net extraction that represents the cumulative difference between flux into tissue and return flux from tissue to blood can be taken directly from the residue curves without calculating the derivatives used in Eq. (1.20):

$$E'_{\text{net}}(t) = \frac{R_D(t) - R_R(t)}{1 - R_R(t)}. \tag{1.21}$$

Putting aside the fact that the capillaries are fed by arteries and arterioles and drained by venules and veins, one can diagram capillary–tissue regions as in Figure 1.8. If there is efflux but no return flux from the interstitial space into the capillary lumen, then Bohr's (1909) conceptual model is suitable for analysis of the indicator dilution curves, as proposed by Renkin (1959a,b) for the arteriovenous extraction of a constant infusion of tracer potassium and by Crone

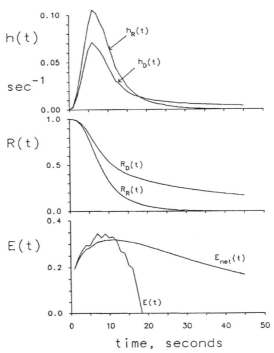

FIGURE 1.9. Multiple-indicator dilution curves. *Top:* Following injection into the left main coronary artery, outflow dilution curves were obtained by sampling at one-second intervals from dog coronary sinus for an intravascular reference indicator, ^{131}I-labeled albumin, providing $h_R(t)$, and a permeating inert hydrophilic molecule that does not enter cells, [14C]sucrose, providing $h_D(t)$. The subscript R is for intravascular reference and D for diffusible (or permeating) indicators. *Middle:* From the $h(t)$ values, the transport functions, one can calculate $R(t)$, residue function, giving $R_D(t)$ and $R_R(t)$. *Bottom:* The difference between $h_D(t)$ and $h_R(t)$ can be expressed as an instantaneous extraction $E(t)$, and the difference between $R_D(t)$ and $R_R(t)$ as a net extraction $E_{net}(t)$. (Figure from Bassingthwaighte and Goresky, 1984.)

(1963) using pulse injection and calculating the instantaneous extraction $E(t)$. The relevant expression considers the loss of the permeant tracer across a single barrier and is that developed by Bohr (1909), which may be expressed in modern terminology as:

$$PS_g = F_S \log_e(1 - E), \tag{1.22}$$

where PS_g is the capillary permeability–surface area product (milliliters per gram per minute), F_s is the flow of solute containing perfusate (milliliters per gram per minute), assuming uniformity of flow throughout the organ and adequacy of single capillary modeling, and E is an extraction relative to the intravascular tracer, assuming that it is due solely to the unidirectional flux of tracer from blood into tissue. The implicit assumption is that the interstitial concentration of the permeating solute remains at zero, which would occur if the interstitial volume was infinite and the interstitial diffusivity high, or if the indicator became bound rapidly at extravascular sites. Crone (1963) recognized that these idealized conditions did not hold and proposed that the E in Eq. (1.19) be taken from the first few seconds of the outflow dilution curves when the interstitial concentration is nearly zero. If it were not for the heterogeneity of regional flows in an organ, this technique would work quite well for solutes of low permeability, when the ratio PS_c/F_s is low, efflux is small, and return flux is slow. Nowadays one can account for the return flux, using fully developed mathematical models of the system diagrammed in Figure 1.8 (e.g., Bassingthwaighte et al., 1989b) and thereby improve the accuracy of the estimates of PS.

A second example: When the goal is to estimate the permeability of the endothelial luminal surface to a solute, the two references are desired—one intravascular as before, and a second one that permeates through the clefts between endothelial cells but does not enter cells, like sucrose. Ideally, this second reference substance should not enter cells, that is, it is an extracellular reference, and should penetrate the interendothelial clefts with exactly the same ease as does the solute under study. An example is to use sucrose as the extracellular reference solute for studies of adenosine uptake by endothelial cells (Schwartz et al., 1997) because sucrose has a molecular size, aqueous diffusion coefficient, degree of hydrophilicity, and charge quite similar to that of adenosine, and therefore is taken to have almost the same cleft PS_g as does adenosine. Thus, if the fraction of adenosine permeating the interendothelial clefts can be inferred accurately by reference to sucrose, then the remainder of its transcapillary extraction must be explained by either binding to endothelial surfaces or by transport across the endothelial luminal surface membrane. [There seems to be no evidence for surface binding. In the dog, virtually all of the adenosine, which is 98% or 99% extracted during single passage through the coronary circulation, is metabolized (Kroll et al., 1997), and therefore must have been transported across cell walls or metabolized extracellularly.] Thus, by use of the sucrose reference, the endothelial PS_{ecl} is inferred accurately by

$$PS_{ecl}(Ado) = PS_{cap}(Ado) - 1.12 \cdot PS_g(sucrose), \qquad (1.23)$$

where the 1.12 is the ratio of the free diffusion coefficient of adenosine divided by that for sucrose, to account for their differences in molecular diffusivities. (Sucrose is not a perfect reference; if the extracellular reference solute had the same diffusion coefficient as adenosine, the factor would be 1.00 instead of 1.12.) The second parameter with a value that should be equal to that for adenosine is the interstitial volume, V'_{isf}.

A third reference tracer is desirable when the solute of interest undergoes facilitated or passive diffusional transport across a membrane and then undergoes intracellular reactions. The third reference solute is one that has the same extracellular behavior (cleft permeation, interstitial volume of distribution, and intravascular transport characteristics) as adenosine, and uses the same transporter across the cell wall (and has the same apparent affinity and transport conduc-

TABLE 1.2. Reference tracers used in determining intracellular reaction.

Solute class	Information provided on solute under study
Intravascular	Convective delay and dispersion in all vessels perfused
Extracellular	Cleft PS, PS_g, and interstitial volume, V'_{isf}
Unreacted but transported analog	Cell PS, PS_{pc}, and intracellular volume of distribution V'_{pc}

tance), but does not react inside the cell. For adenosine there is such a solute, O-methyl adenosine. Dilution curves from O-methyl adenosine should therefore provide evidence on parameter values for PS_{pc} and for V'_{pc}, independent of the data from adenosine but equivalent in value. (This has not yet been firmly established, but is likely.) This third reference therefore supplies PS_{pc} and V'_{pc} for adenosine, given that the premises are verified experimentally. (See Table 1.2.)

Data Acquisition in the MID Experiment

For the study of tracer transport and metabolism in an organ with a single inflow and a single outflow, the most explicit information is obtained with the MID technique, injecting a set of tracer-labeled solutes simultaneously. By this approach it is certain that the intravascular transit times, arterial and venous, are identical for all of the solutes (unless there is entry of the permeating solute into red blood cells, and a consequent "red cell carriage effect," whereby the fraction carried by the red blood cells has a higher velocity than does the fraction carried in the plasma and there is a reduction in the rate of escape from the capillary because of the relative unavailability of the red blood cell fraction). With radioactive tracers, the relative amounts of radioactivity chosen for each solute depend on the methods of analysis of the samples. Often with organic solutes it is easiest and most accurate to use three β-emitters together: [131]I, ^{14}C, and ^{3}H are commonly used since they can be attached to a wide variety of solutes and can be distinguished by liquid scintillation counting without any preceding chemical separation (Bukowski et al., 1992). When four or five tracers are being used simultaneously, then either physicochemical separation will be needed, for example, by HPLC, or combinations of gamma- and beta-emitting tracers should be used (Goresky, 1963).

A diagram of the experimental setup for examining the reactions involving the transformation of hypoxanthine to xanthine to uric acid in the isolated perfused guinea pig heart is shown in Figure 1.10.

The set of tracers are injected simultaneously as a compact bolus into the inflow to the organ, attempting to provide mixing of the bolus of fluid with the perfusate so that it is thorough and complete. If mixing is not complete before the first branchpoints in the arterial system, then different parts of the organ would receive tracer doses out of proportion to the fraction of flow, and therefore would be misrepresented in the outflow. If a disproportionately large fraction of the dose entered a particular region, then the characteristics of this region bias the results of the interpretation of the analysis of the outflow dilution curves. In theory, the

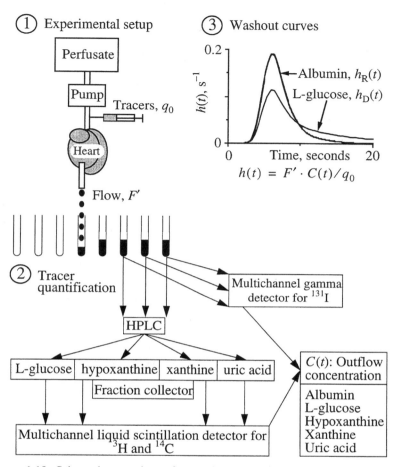

FIGURE 1.10. Schematic overview of procedures underlying the application of the multiple-indicator dilution technique to investigation of multiple metabolites. HPLC is high pressure liquid chromatography.

fraction of the dose entering each element of the fluid at the injection site should be in exact proportion to the fraction of flow through that element. This is flow-proportional labeling (Gonzalez-Fernandez, 1962). It is virtually impossible to achieve this exactly, but the proportion is reasonably well approximated if there is adequate mixing by disturbances in the flow patterns before the branchings occur. (In contrast, cross-sectional mixing is clearly undesirable: Simply labeling a cross-section of the fluid stream as if by inserting a cylinder of tracer-labeled fluid into the column of fluid at the entrance to the system creates equal labeling of the infinitely slow laminae at the wall of the vessel with the highest flow laminae at the center of the stream; this creates a striking bias or weighting of the peripheral laminae. In theory, the layer at the wall, which is not moving, should contain no label.

Data are normally acquired by sampling the outflow. In isolated organ systems the outflow from the whole organ is collected in a series of timed collections. The essential information is the time of beginning each collection, the time of ending, and the volume of the sample. For studies of the heart, samples are collected at 1- or 2-s intervals for the first 30 s and with gradually diminishing frequency thereafter. Nowadays, with computer-controlled sampling, we spread out the length of the collection periods gradually so that where the outflow concentrations change gradually we take fewer samples. The idea, very roughly, is that in order to minimize the work of chemical analysis on the samples, fewer samples are needed when the concentrations of the tracers change slowly. The sample volume per unit time is also a measure of the flow during the period of obtaining the samples.

Normalization of the Indicator–Dilution Curves to Fraction of Dose

The concentration–time curves obtained must be normalized to the "fraction of dose emerging," either per unit volume of outflow or per unit time. We prefer the latter because this puts it into the framework directly defined by the differential equations. To do this, the calibration of the dose injected is critical. Because the tracer concentrations in the injectate are high, precluding isotope counting on accurately measurable volumes of aliquots of the injectate solution, serial dilutions are normally required. We normally determine the activity of 10 to 20 aliquots, at several different dilutions, attempting to obtain something better than 1% accuracy in the estimates of the dose. This allows normalization to the unit response, $h(t)$, the fraction of dose per second emerging with the outflow:

$$h(t) = \frac{F \cdot C_{\text{out}}(t)}{q_0}. \tag{1.24}$$

This normalized outflow response is not identical to the formal impulse response of the system for two reasons: One is that the input function is not usually a Dirac delta function, an infinitely thin spike at $t = 0$, but is more spread out due to the finite duration of the injection itself and to the physical spread of the injectate in the volume of perfusate in the inflow tubing or blood vessels due to the force of the injection, a jet effect. Nevertheless, since the injectate form should be identical for all of the tracers within the injection syringe, direct comparisons can be made between the several tracers using the mathematical models, given that one has a close approximation to the input function.

Dose Calibration

The calibration of the injected dose is not easy. To do this we take 20 aliquots of the injectate solution, using a few different dilutions that are compatible with the radioactivity levels that can be assessed by the gamma or beta scintillation counting methods being used. These samples contain all of the tracers being used, and therefore the samples must be put through the same set of separation and counting processes as are applied to each sample. From these 20 samples we obtain 20 estimates for the dose of each of the tracers injected.

In addition, samples of each of the tracers that were used to make up the dose are assessed separately and individually. This is done to provide an additional check on the amounts put into the dose on samples in which there is only one tracer and no inaccuracy introduced by spillover corrections that enlarge the statistical error (and sometimes even bias) in the assessment of the dose. All of this is to get the estimates of the dose, and thereby the fraction of dose per sample or per unit outflow volume, which sounds pretty tedious, and it is, but the other side of the coin is that the measure of the dose and the volumes of the outflow samples are the two least accurate measures that one makes in the indicator dilution technique.

Recording the Input Function

Since no separation of the various tracers occurs before the bolus enters a region where interactions with elements of the biological systems occur, the recording of any one of the input tracer concentration–time curves provides exact information on the whole set. This has the advantage of reducing sources of error and allows one to get the best possible estimates of the transport functions of the system. Having an optically labeled indicator is a cheap and effective way of getting the input function, by taking a small sampling stream from the inflow tubing through a densitometer. In those conditions where the total inflow is small, as in experiments where one is perfusing a small organ, the whole of the inflow can go through the densitometer to give the input function, as was done by Deussen and Bassingthwaighte (1996) and Harris et al. (1990).

Multiple Data Sets

The most power in the analysis is gained when one has available the maximum amount of simultaneously acquired, self-consistent data. Thus one needs to obtain along with the largest practical set of indicator dilution curves the full set of experimental data. Flow, temperature, pH, pCO_2, pO_2, and so on, will be important in defining the physiological status of the preparations, as are the organ weight, water content, the functional performance of the organ, and the concentrations of solutes and substrates in its venous effluent.

In multiple-indicator dilution experiments the idea is to obtain information on the whole of the set of solutes simultaneously, as described above. When the choices of reference solutes include an intravascular reference, an extracellular solute matched in cleft permeability to the prime permeating solute, and a solute that is transported like the prime permeating solute but which is not metabolized intracellularly, then we have a matched set. This is not a universally defined set, for there will be situations in which there is extracellular metabolism, for example by a phosphatase, so the matching is specific for the prime substrate only.

Approaches to the Analysis

Conservation Principles

The general theory outlined at the beginning of the chapter needs to be applied to each tracer in a multiple-indicator dilution study. Usually, it is not too difficult to determine the fractional recovery of the inert, nonreactive reference intravascular and extracellular indicators. For any tracer-labeled substrate that undergoes transforming reactions, the tests for conservation require finding all of the tracer no matter what chemical form it may have taken. For [^{14}C]glucose, one must find its fractional storage (as glycogen, fat, protein, etc.). Thus, many experiments require chemical and isotopic analyses of the sequence of outflow samples, and also the same analysis of the whole organ. The latter exercise is particularly expensive and tedious, given the known heterogeneity of flows and metabolic functions. See Chapter 9 with respect to metabolic heterogeneity.

Residue Plus Outflow Data

Residue function data are not often available simultaneously with outflow dilution data. The exceptions are imaging studies where one may have residue data by positron emission tomography, single photon emission computed tomography, coincidence detection of positron emitters from the whole organ, single photon residue detection, X-ray computed tomography, or magnetic resonance contrast signals.

Residue data are particularly valuable when the retention of the tracer within the organ is prolonged: In this case, the outflow concentrations are low and difficult to measure, and therefore their integrals are not as accurate as desired. Having both outflow and residue is ideal from the mass balance point of view:

$$\int_0^t h(\lambda) \, d\lambda + R(t) = 1.0, \qquad (1.25)$$

which is to say that at time t the total dose is to be found as the sum of the outflow to that time, plus the residue at that time. An example of this approach applied to the estimation of myocardial oxygen consumption from [^{15}O]oxygen input, residue and output curves is provided by Deussen and Bassingthwaighte (1996).

Using Multiple Data Sets Together

The power of the multiple-indicator dilution method is in the combining of information of several different types, gathered more or less simultaneously, into a common data set for which a complete and self-consistent explanation may be found. The "explanation" is an analysis that accounts for the full set of information, which is quantitative and physiologically realistic, and which provides estimates of parameter values and confidence limits on the estimates. If one has only a single-indicator dilution curve consisting of 30 to 50 points and a complex model to fit to these data, then the fit may not be well determined: There may be too many different sets of parameters that will fit the single curve equally well. The more complex the model the more data are needed to refine and constrain the estimates of the parameters.

The standard multiple-indicator dilution study is ordinarily designed to obtain information on three tracers simultaneously: an intravascular reference tracer, an extracellular reference tracer having an extracellular volume of distribution similar to that of the test solute, and the test solute itself. Can more be done? The answer is clear: The more information that can be gathered on the status of the organ at the time of the experiment the clearer and more definitive will be the results of the analysis. By the same token, information that may not be quite simultaneous, but which can be reasonably considered to be applicable to the status at the time of the indicator dilution curves, is really invaluable.

In an intact organ there is inevitably some considerable heterogeneity of flow and capillary transit times. Various methods for estimating the capillary transit time distribution are discussed by Audi in Chapter 22. Very often, for studies of the myocardial tracer exchanges, we have assumed that information acquired on other hearts at other times is applicable to the particular study. The assumption is based on the often-repeated observation that the heterogeneity of flows in the myocardium is very much the same from heart to heart and is stable in the same heart from one time to another. If this is really true, then we would not have to measure the flow heterogeneity, except occasionally as a check. This is probably true, not only because the heterogeneity does not change very much, but also because small changes in heterogeneity have a negligible influence on the estimates of the parameters of interest, the permeabilities and the reaction rates of the solutes. But one cannot count on this if the physiological conditions are abnormal or are being manipulated. For example, when combinations of drugs are used to change vasomotion or influence metabolism, then the regional flows should be determined using microsphere distributions in the same experiment. Likewise, one may need to determine the volumes of distribution for the vascular and extracellular solutes and for the water content of the tissue samples, all in the interest of reducing the degrees of freedom for the analysis that follows using particular models. When one needs to model local metabolism from tissue samples, then it becomes essential to have a measure of the flow in each of those samples.

The desired information set therefore consists of $w(f)$, the probability density function of regional tissue flows (milliliters per gram per unit time); the regional volumes of distribution for the vascular, extracellular, and water markers (and sometimes others, for example, markers of intracellular binding spaces); the dilution curves $h_R(t)$, $h_E(t)$, and $h_D(t)$; and, in certain cases, a second permeating marker, $h_{D2}(t)$, which permeates the cell but is not metabolized.

For linear systems this is often enough. But the fact is that systems are seldom linear in biology. For tracer levels of a solute, the transporter or the enzyme assisting its reaction is operating in its linear range, far below the levels at which it becomes more than a tiny fraction bound by the available solute. Concentrations are such that a substantial fraction of the transporter or enzyme molecules are occupied. To test for this, one needs to repeat the experiment with different ambient concentrations of the substrate. Normally one can only raise these levels, so that is the experiment to do: Repeat the MID study at two or three different levels of the substrate to determine whether the apparent behavior changes. The expectation is that if it is near saturation of the transporter or enzyme, then raising the concentration will reduce the fraction that is transported or reacted.

A way of testing this idea efficiently was introduced by Linehan, Dawson, and colleagues in a series of studies in the early 1980s (Rickaby et al., 1981; Malcorps et al., 1984). It was so designated the "bolus sweep" technique by Bassingthwaighte and Goresky (1984) in recognition of its power, namely, the power to estimate the transport rate for the substrate in its passage across a membrane at a variety of different concentrations within a few seconds. The technique was simply to add to the tracer in the injectate a known amount of the mother substance (the substance for which the tracer is the marker, e.g., a few millimoles of glucose added to the chemically insignificant few microcuries of radioactive glucose). As the bolus enters the exchange regions, its initial part has a very low concentration of the mother substance, and therefore the tracer has full access to the transporter. A few moments later, as the chemical concentration of the substrate in the bolus rises toward its peak level, it exerts more and more competition for the transport sites, and so reduces the effective fractional transport of the tracer. After the peak, the nontracer chemical mother substance concentration again falls and lessens the competition for the tracer, and the effective transport rate again rises (see Chapter 17). The nice physiological result of this approach is that one can test the behavior of a substance that has marked biological effects if it is given as a steady concentration into the inflow; by giving it so transiently, large pharmacological responses are avoided and the parameter values represent those in the more physiological situation.

Reactants and Products

Even better than having a set of independent tracers is having a set that is intimately related, not only by parallels in molecular behavior, but also by having one give rise to another—a set of substrates and metabolites related by being in a sequence or particular arrangement of reactions. This situation is the focus of Chapter 7, by Bassingthwaighte et al. Such combinations provide for a series of checks and balances, for example, the amount of a substrate reacted should be the same as the amount of products formed in the reaction.

A further independent check is provided by steady-state measurements of chemical arteriovenous differences in the same preparation and at the same time as the tracer transient information is obtained. When tracer glucose extraction is measured by the multiple-indicator dilution technique, the peak instantaneous extractions in the brain (Crone, 1963) and the heart (Yipintsoi et al., 1970; Kuikka et al., 1986) are high, 50% or so, but the arteriovenous differences are small, less than 10% in the brain and only 2% or 3% in the heart. Why should these be so different when the tracer [^{14}C]glucose is chosen to have the same molecular characteristics as the "mother substance," the chemical glucose? The answer is simply that the tracer extraction compared to the reference intravascular tracer reflects the unidirectional fluxes, more or less, across the capillary membrane and provides a measure of the fraction that escapes, transiently, from the blood into the extravascular tissue. In contrast, the arteriovenous difference in chemical concentrations represents the net loss of material, glucose, into some other form (lactate or carbon dioxide); this is a net flux due to a loss or transformation of material. The net fluxes are always much less than the unidirectional tracer fluxes; if glucose consumption were zero, the net arteriovenous difference would be zero, but the tracer transient would still be about 50% extracted.

Summary

The general theory of input–output relationships is applicable to the study of any system in which one has access to signals at the input and can observe the resultant responses at the output. The general theory is derived for linear stationary systems, that is, those with time-invariant responses to a given input, and with responses proportional to the input. While in modern indicator dilution studies we deal with nonlinear systems, it is best done using specific models to serve as analogs of the system; even in these cases we depend on certain aspects of the system to be linear and stationary; in particular, the responses to an intravascular reference tracer and to an extracellular reference marker should fall under the standard theory. This theory therefore provides a basis for the linear and nonlinear studies that follow.

References

Bassingthwaighte, J. B. Blood flow and diffusion through mammalian organs. *Science* 167:1347–1353, 1970.

Bassingthwaighte, J. B., and D. A. Beard. Fractal ^{15}O-water washout from the heart. *Circ. Res.* 77:1212–1221, 1995.

Bassingthwaighte, J. B., and C. A. Goresky. Modeling in the analysis of solute and water exchange in the microvasculature. In: *Handbook of Physiology. Sect. 2, The Cardiovascular System. Vol IV, The Microcirculation,* edited by E. M. Renkin and C. C. Michel. Bethesda, MD: American Physiological Society, pp. 549–626, 1984.

Bassingthwaighte, J. B., F. H. Ackerman, and E. H. Wood. Applications of the lagged normal density curve as a model for arterial dilution curves. *Circ. Res.* 18:398–415, 1966.

Bassingthwaighte, J. B., W. A. Dobbs, and T. Yipintsoi. Heterogeneity of myocardial blood flow. In: *Myocardial Blood Flow in Man: Methods and Significance in Coronary Disease,* edited by A. Maseri. Torino, Italy: Minerva Medica, 1972, pp. 197–205.

Bassingthwaighte, J. B., T. Yipintsoi, and R. B. Harvey. Microvasculature of the dog left ventricular myocardium. *Microvasc. Res.* 7:229–249, 1974.

Bassingthwaighte, J. B., F. P. Chinard, C. Crone, C. A. Goresky, N. A. Lassen, R. S. Reneman, and K. L. Zierler. Terminology for mass transport and exchange. *Am. J. Physiol.* 250 (*Heart Circ. Physiol.* 19):H539–H545, 1986.

Bassingthwaighte, J. B., R. B. King, and S. A. Roger. Fractal nature of regional myocardial blood flow heterogeneity. *Circ. Res.* 65:578–590, 1989a.

Bassingthwaighte, J. B., C. Y. Wang, and I. S. Chan. Blood-tissue exchange via transport and transformation by endothelial cells. *Circ. Res.* 65:997–1020, 1989b.

Bassingthwaighte, J. B., M. A. Malone, T. C. Moffett, R. B. King, I. S. Chan, J. M. Link, and K. A. Krohn. Molecular and particulate depositions for regional myocardial flows in sheep. *Circ. Res.* 66:1328–1344, 1990.

Bohr, C. Über die spezifische Tätigkeit der Lungen bei der respiratorischen Gasaufnahme und ihr Verhalten zu der durch die Alveolarwand stattfindenden Gasdiffusion. *Skand. Arch. Physiol.* 22:221–280, 1909.

Bridge, J. H. B., M. M. Bersohn, F. Gonzalez, and J. B. Bassingthwaighte. Synthesis and use of radiocobaltic EDTA as an extracellular marker in rabbit heart. *Am. J. Physiol.* 242 (*Heart Circ. Physiol.* 11):H671–H676, 1982.

Buckberg, G. D., J. C. Luck, B. D. Payne, J. I. E. Hoffman, J. P. Archie, and D. E. Fixler. Some sources of error in measuring regional blood flow with radioactive microspheres. *J. Appl. Physiol.* 31:598–604, 1971.

Bukowski, T., T. C. Moffett, J. H. Revkin, J. D. Ploger, and J. B. Bassingthwaighte. Triple-label β liquid scintillation counting. *Anal. Biochem.* 204:171–180, 1992.

Chinard, F. P., G. J. Vosburgh, and T. Enns. Transcapillary exchange of water and of other substances in certain organs of the dog. *Am. J. Physiol.* 183:221–234, 1955.

Crone, C. The permeability of capillaries in various organs as determined by the use of the "indicator diffusion" method. *Acta Physiol. Scand.* 58:292–305, 1963.

Deussen, A., and J. B. Bassingthwaighte. Modeling [^{15}O]oxygen tracer data for estimating oxygen consumption. *Am. J. Physiol.* 270 (*Heart Circ. Physiol.* 39):H1115–H1130, 1996.

Dible, J. H. Is fatty degeneration of the heart muscle a phanerosis? *J. Pathol. Bacteriol.* 39:197–207, 1934.

Diem, K. *Documenta Geigy. Scientific Tables.* Ardsley, NY: Geigy Pharmaceuticals, 1962.

Glenny, R., H. T. Robertson, S. Yamashiro, and J. B. Bassingthwaighte. Applications of fractal analysis to physiology. *J. Appl. Physiol.* 70:2351–2367, 1991.

Gonzalez, F., and J. B. Bassingthwaighte. Heterogeneities in regional volumes of distribution and flows in the rabbit heart. *Am. J. Physiol.* 258 (*Heart Circ. Physiol.* 27):H1012–H1024, 1990.

Gonzalez-Fernandez, J. M. Theory of the measurement of the dispersion of an indicator in indicator-dilution studies. *Circ. Res.* 10:409–428, 1962.

Goresky, C. A. A linear method for determining liver sinusoidal and extravascular volumes. *Am. J. Physiol.* 204:626–640, 1963.

Gorman, M. W., J. B. Bassingthwaighte, R. A. Olsson, and H. V. Sparks. Endothelial cell uptake of adenosine in canine skeletal muscle. *Am. J. Physiol.* 250 (*Heart Circ. Physiol.* 19):H482–H489, 1986.

Grant, P. E., and C. J. Lumsden. Fractal analysis of renal cortical perfusion. *Invest. Radiol.* 29:16–23, 1994.

Hamilton, W. F., J. W. Moore, J. M. Kinsman, and R. G. Spurling. Studies on the circulation. IV. Further analysis of the injection method, and of changes in hemodynamics under physiological and pathological conditions. *Am. J. Physiol.* 99:534–551, 1932.

Harris, T. R., G. R. Bernard, K. L. Brigham, S. B. Higgins, J. E. Rinaldo, H. S. Borovetz, W.

J. Sibbald, K. Kariman, and C. L. Sprung. Lung microvascular transport properties measured by multiple indicator dilution methods in patients with adult respiratory distress syndrome. A comparison between patients reversing respiratory failure and those failing to reverse. *Am. Rev. Respir. Dis.* 141:272–280, 1990.

King, R. B. Modeling membrane transport. *Advances in Food and Nutrition Research* 40:243–262, 1996.

King, R. B., J. B. Bassingthwaighte, J. R. S. Hales, and L. B. Rowell. Stability of heterogeneity of myocardial blood flow in normal awake baboons. *Circ. Res.* 57:285–295, 1985.

Kroll, K., D. J. Kinzie, and L. A. Gustafson. Open system kinetics of myocardial phosphoenergetics during coronary underperfusion. *Am. J. Physiol.* 272 (*Heart Circ. Physiol.* 41): H2563–H2576, 1997.

Kuikka, J., M. Levin, and J. B. Bassingthwaighte. Multiple tracer dilution estimates of D- and 2-deoxy-D-glucose uptake by the heart. *Am. J. Physiol.* 250 (*Heart Circ. Physiol.* 19):H29–H42, 1986.

Malcorps, C. M., C. A. Dawson, J. H. Linehan, T. A. Bronikowski, D. A. Rickaby, A. G. Herman, and J. A. Will. Lung serotonin uptake kinetics from indicator-dilution and constant-infusion methods. *J. Appl. Physiol.* 57:720–730, 1984.

Meier, P., and K. L. Zierler. On the theory of the indicator-dilution method for measurement of blood flow and volume. *J. Appl. Physiol.* 6:731–744, 1954.

Polimeni, P. I. Extracellular space and ionic distribution in rat ventricle. *Am. J. Physiol.* 227:676–683, 1974.

Renkin, E. M. Transport of potassium-42 from blood to tissue in isolated mammalian skeletal muscles. *Am. J. Physiol.* 197:1205–1210, 1959a.

Renkin, E. M. Exchangeability of tissue potassium in skeletal muscle. *Am. J. Physiol.* 197:1211–1215, 1959b.

Rickaby, D. A., J. H. Linehan, T. A. Bronikowski, and C. A. Dawson. Kinetics of serotonin uptake in the dog lung. *J. Appl. Physiol.* 51 (*Respirat. Environ. Exercise Physiol.* 2):405–414, 1981.

Schwartz, L. M., T. M. Bukowski, J. H. Revkin, and J. B. Bassingthwaighte. Capillary transport and metabolism of adenosine and inosine in rabbit and guinea pig hearts (unpublished data).

Sokoloff, L., M. Reivich, C. Kennedy, M. H. Des Rosiers, C. S. Patlak, K. D. Pettigrew, O. Sakurada, and M. Shinohara. The [^{14}C]deoxyglucose method for the measurement of local cerebral glucose utilization: Theory, procedure, and normal values in the conscious and anesthetized albino rat. *J. Neurochem.* 28:897–916, 1977.

Yipintsoi, T., R. Tancredi, D. Richmond, and J. B. Bassingthwaighte. Myocardial extractions of sucrose, glucose, and potassium. In: *Capillary Permeability (Alfred Benzon Symp. II),* edited by C. Crone and N. A. Lassen. Copenhagen: Munksgaard, 1970, pp. 153–156.

Yipintsoi, T., P. D. Scanlon, and J. B. Bassingthwaighte. Density and water content of dog ventricular myocardium. *Proc. Soc. Exp. Biol. Med.* 141:1032–1035, 1972.

Yipintsoi, T., W. A. Dobbs, Jr., P. D. Scanlon, T. J. Knopp, and J. B. Bassingthwaighte. Regional distribution of diffusible tracers and carbonized microspheres in the left ventricle of isolated dog hearts. *Circ. Res.* 33:573–587, 1973.

Zierler, K. L. Theoretical basis of indicator-dilution methods for measuring flow and volume. *Circ. Res.* 10:393–407, 1962.

Zierler, K. L. Equations for measuring blood flow by external monitoring of radioisotopes. *Circ. Res.* 16:309–321, 1965.

2
Mechanisms of Endothelial Transport, Exchange, and Regulation

2

Transport Functions of the Glycocalyx, Specific Proteins, and Caveolae in Endothelium

Jan E. Schnitzer

Our current understanding of the molecular, cellular, and physical basis of the barriers and pathways that mediate capillary permeability will be presented in this chapter. Some of the endothelial cell surface proteins involved in the structure and function of specific transendothelial transport pathways will be examined. Many studies utilizing a diversity of experimental approaches including biochemical, morphological, physiological, and theoretical analyses have led to a better understanding of the molecular basis of the interaction of plasma proteins with the endothelium, especially as it relates to those regions that form the critical permselective molecular filters controlling the transendothelial transport of blood molecules. New molecular approaches to investigating capillary permeability are only beginning to identify and characterize the molecular constituents that form or create the main pathways and barriers to molecular transport across vascular endothelium. The emphasis will be on the specific transport proteins of the luminal endothelial cell glycocalyx with a special focus on those proteins associated with noncoated plasmalemmal vesicles or caveolae. Aquaporin and albondin are the two proteins that will be discussed in the greatest detail because of their role in the selective transport of water and albumin, respectively.

Endothelial Cell Heterogeneity

A primary function of the blood circulation or vasculature is to meet the specific nutritional requirements of the tissue cells in each organ. The vascular system is lined with a very attenuated monolayer of highly differentiated endothelial cells that are strategically located in many organs to form a critical barrier that regulates the exchange of water, solutes, macromolecules, and even cells from the circulating blood to the underlying tissue cells. The selectivity of the endothelial cell barrier varies in different vascular beds and is strongly dependent on the structure and type of endothelium lining the microvasculature (100, 157). Endothelial cells are quite responsive to the local tissue environment and may modulate their barrier function in accordance with the needs of the underlying tissue cells in many organs. There is considerable heterogeneity among the endothelia lining the microvascular beds of different organs. The differences are apparent by electron microscopy of tissue specimens and can be grouped readily into three distinct morphological types: continuous (found in tissues such as heart, lung, brain, and skeletal muscle), fenestrated (kidney glomerulus, intestinal mucosa, and endo-crine glands), and discontinuous or sinusoidal (liver, spleen, and bone marrow). Both sinusoidal and fenestrated endothelia have open extramembranous passage-ways for the direct transport of many molecules through either obvious gaps between the cells or circular "windows" of about 100 nm in diameter called fenestrations, respectively. Conversely, continuous endothelia lack such distinct membrane discontinuities, and it is less clear exactly how molecules traverse this barrier.

The extensive phenotypic differentiation of endothelia between organs reflects the influence of the local tissue environment and provides for considerable differences in endothelial cell barrier functions. For instance, the expression of fenestrations in cultured endothelial cells can be modulated by the extracellular matrix used to coat the plastic dishes upon which the cells grow (85). More recently, it was observed that specific growth factors, in this case the vascular endothelial growth factor (VEGF), can induce the formation of fenestrations in vivo and that the normal tissue expression of VEGF correlates quite nicely with those tissues having blood vessels lined with fenestrated endothelium (114). Not surprisingly, VEGF was originally called the vascular permeability factor (VPF) because it increased blood microvessel permeability that was especially evident in the vasculature of tumors where it is highly expressed by the tumor cells (31, 32). In terms of transport behavior, both the fenestrated and sinusoidal endothelia do not present a continuous membrane barrier to the circulating blood molecules, but instead have obvious membrane discontinuities that allow for large hydraulic conductivities and less restricted transport of many blood molecules.

Even endothelia of the continuous type vary considerably in their barrier func-tion. Most endothelia of the continuous type have an abundance of noncoated plasmalemmal vesicles or caveolae that appear to function as vesicular carriers for proteins into and across the cell. These characteristic vesicles can constitute 50–70% of the cell surface plasma membrane and occupy 10–15% of the total cell volume (about 500–600 vesicles/μm^3) (9, 53, 67, 148). These endothelia are endowed with simplified intercellular junctions that are thought not to be as restrictive as those found in epithelia, thereby allowing significant small solute transport. Other continuous endothelia that are located at the blood–brain barrier form a very restrictive barrier by having few, if any, caveolae and elaborating epithelial-like, tight junctions that do not allow the transport of even small solutes. Such brain endothelia must express specific transport proteins to meet the meta-bolic demands of the brain tissue cells even for such small molecules as glucose and amino acids. Lastly, differences even exist between segments of the same microvasculature. For instance, arteriolar, capillary, and venular endothelia vary considerably in their response to various chemical and cellular mediators that affect the transport of blood molecules and cells (18, 20, 96). The "tightness" of the paracellular pathway, especially for venular capillaries, can change significantly in response to thrombin and histamine.

Transport Barriers and Pathways

The luminal cell surface of the endothelium is exposed directly to the circulating blood and represents a key barrier to the movement of molecules and cells from the blood to the interstitium. As with any cellular barrier, the lipid portion of the membrane tends to restrict the direct transmembrane transport of both small and large circulating blood molecules including water, ions, and proteins because of electrostatic, steric, and/or hydrophobic exclusion forces (17, 36, 113, 121–123, 128, 131, 140, 161). Such molecules can cross this cellular barrier via membrane discontinuities observed in fenestrated and sinusoidal endothelia, between the cells via the intercellular junctions, and/or by specialized facilitory pathways that require specific interactions with components of the plasma membrane. This chapter focuses primarily on the later pathways by defining our current understanding of the mechanisms mediating specific transport across endothelia of the continuous type.

Several pathways exist for the transport of plasma molecules across continuous endothelium: (i) Intercellular junctions are highly regulated structures that form the paracellular pathway for the pressure-driven filtration of water and small solutes (16, 62, 75, 78, 102, 109, 152); (ii) noncoated plasmalemmal vesicles (also called caveolae) transcytose plasma macromolecules apparently by shuttling their contents adsorbed from blood from the luminal to antiluminal aspect of the endothelium (48, 49, 94, 99, 125, 160); and (iii) transendothelial channels may form transiently in very attenuated regions of the cell by the fusion of two or more caveolae each located on apposing plasma membranes to provide a direct conduit for the exchange of both small and large plasma molecules (166). Capillary permeabilities are dependent not just on the structure of these transport pathways, but in many vascular beds also on the interaction of plasma proteins such as albumin and orosomucoid with the endothelial glycocalyx (16, 19, 22, 23, 25, 30, 42, 54, 55, 62, 63, 75, 78, 79, 86, 90, 91, 95, 119, 120). This interaction with plasma appears to be required to maintain "normal" capillary permeability.

At present there is little general agreement as to the pathways and mechanisms of transendothelial transport of albumin and of other macromolecules in situ as well as in culture (19, 94, 100, 102, 109, 119, 157). Initially, physiological studies indicated that albumin and other macromolecules cross the endothelium through a hydraulically conductive pathway, modeled simply as pores or filtration slits that usually are assumed to be located along intracellular junctions (100, 102). More recently, it has become apparent that the coupling of macromolecular transport to water flux is reduced significantly in the presence of plasma (63). Therefore, under physiological conditions, it appears that at least some macromolecular transport is mediated through pathways that are not predominantly involved in water transport; this suggests that transcellular pathways such as vesicular carriers may indeed be important elements mediating macromolecular transport. The concept that endothelium can transcytose molecules "in quanta" via shuttling of its abundant population of plasmalemma vesicles is over 40 years old (99). More recent work (49, 70, 124, 134, 135, 160, 165, 168) indicates that these caveolae may even contain specific binding proteins that mediate preferential internalization and in some cases the transcytosis of a select group of ligands by a process conveniently named receptor-mediated transcytosis.

The Endothelial Cell Glycocalyx

The luminal surface of the endothelium is exposed directly to the blood and is covered by an elaborate carbohydrate-containing, proteinaceous, polyanionic surface coat called the glycocalyx, which includes plasmalemmal components such as the glycosylated ectodomains of integral membrane proteins, proteoglycans, and glycolipids, and adsorbed plasma proteins such as albumin (130). The luminal endothelial glycocalyx with its complement of specific proteins plays an important role in determining capillary permeability. The ability of the endothelium to act as both a plasma-modified filtration barrier to transvascular exchange and a specific translocator of molecules [i.e., insulin (70), transferrin (66, 165), LDL (160), glucose (35), fibrinogen (168), and albumin (49, 94, 144, 168)] is dependent on interactions occurring at its luminal surface. Many important normal and pathological vascular functions, such as transendothelial cell migration, blood-born metastasis, thrombosis, inflammation, and coagulation, are dependent on interactions occurring at the level of the glycocalyx (34, 65, 73). Capillary permeability in many vascular beds appears to be dependent on specific molecular interactions occurring at the level of the endothelial glycocalyx. As discussed below, these interactions may increase capillary permselectivity of the endothelial cell barrier through the "serum effect," facilitate transport of select molecules through the actions of specific cell surface transport proteins, or increase microvessel permeability in general to many molecules by altering or disrupting barriers and pathways across the endothelium.

Restriction Through the "Serum Effect"

Proteins circulating through the vasculature have an important role in establishing and maintaining normal capillary permeabilities. Many investigators over the past 70 years have noted that the absence of serum proteins in vascular perfusates increased capillary permeability (16, 19, 22, 23, 25, 30, 42, 54, 55, 62, 63, 75, 78, 79, 86, 90, 91, 95, 119, 120). As little as 10% serum is required in vascular perfusates to maintain normal capillary permeabilities (54). This phenomenon has been called the "serum or plasma effect" (54, 55, 75, 78). Orosomucoid and albumin are the two serum components that mediate the majority of this effect (3, 11–14, 19, 22, 23, 25, 30, 54, 55, 62, 63, 79, 86, 90, 91, 106, 107, 110, 111, 119,

120). Recently, another aspect of the effect of plasma on capillary permeability has been delineated; plasma in perfusates causes a significant reduction in the coupling of macromolecular transport and hydraulic conductivity (63), which suggests that plasma greatly reduces macromolecular transport through water conductive pathways, probably at the intercellular junctions. It is also clear that plasma proteins modify the glycocalyx of the endothelium in microvessels and can increase the apparent thickness of this coat detected by cationic electron dense tracers (2). Moreover, loss of the glycocalyx by partial digestion of the cell surface proteins greatly increases transport across the endothelium (1).

Although the effect is clear and appears to require cell surface binding, the mechanisms responsible for the plasma effect are not. One proposed mechanism involves molecular binding to the endothelial glycocalyx, especially within specialized regions of the glycocalyx (microdomains) that involve specific transendothelial transport pathways. Albumin binds to the luminal glycocalyx of continuous endothelium (94, 119, 124, 130, 168) via distinct binding proteins (45, 46, 124, 127, 129, 134, 135, 143), and its binding within transport pathways, such as plasmalemmal vesicles and the introit of intercellular junctions, may form a molecular filter within these pathways (3) that can electrostatically (122) and sterically (19, 21, 121, 123) restrict the transport of water, small solutes, and macromolecules across the microvascular wall.

More recent work provides an alternative explanation for albumin's role in the "serum effect." Albumin binding at the endothelial cell surface appears to close a receptor-gated, passive conductance channel for Ca^{2+}. In the absence of albumin, Ca^{2+} influx from the extracellular milieu occurs that somehow increases hydraulic conductances without any obvious changes in junctional morphology (58). Thus, albumin's effect could be mimicked by removing extracellular calcium, suggesting that the electrostatic and steric exclusion effects are minimal. This may be true for the transport of small molecules, but it is clear that albumin prevents the entry of large macromolecules such as ferritin in the vesicular pathway (119). Thus, it is likely that the effect of albumin on increasing the overall restrictiveness of the endothelial cell barrier will be a composite of both mechanisms. Comparable information about orosomucoid is lacking and will be required to better understand its effect on capillary permeabilities. Although not a part of the glycocalyx, it is worth noting that many serum proteins under certain conditions can bind to the extracellular matrix underlying the endothelium and also contribute resistance in series to transvascular exchange or net transport.

Selective Facilitation via Receptor-Mediated Processes

It is clear for certain endothelia that specific receptor-mediated processes are necessary for transvascular molecular transport. This is best documented for the continuous endothelium of the brain, which forms a very impermeable barrier and requires specific molecular transport mechanisms for even small solutes. For example, specific glucose transport proteins are present on the surface of brain endothelium that mediate the transport of glucose across this very tight endothelial barrier. Specific receptors on the surface of various endothelia play a significant role in the transendothelial exchange of a variety of larger plasma molecules, including insulin (70), transferrin (66, 165), ceruloplasmin (156), albumin (45, 46, 124, 127, 129, 134, 135, 143), and LDL (160). It has been demonstrated that albumin and orosomucoid interact specifically with the endothelium (124, 130, 140) and that albumin appears to traverse the endothelium via caveolae, apparently via receptor-mediated transcytosis (48, 94, 135, 138). Albumin acts as a carrier for small ligands bound to it, such as fatty acids (41). Several different albumin binding proteins have been identified (45, 46, 124, 129, 135, 143); however, the endothelial proteins mediating orosomucoid binding are unknown and need to be identified. It appears that specific binding to the endothelial glycocalyx can indeed initiate receptor-mediated transport of macromolecules across vascular endothelium via caveolae.

Augmentation of General Leakiness

For many years, the paracellular pathway was thought to be passive and static. Recent data indicate that the binding of serum proteins to the endothelial glycocalyx proteins can cause significant increases in paracellular transport. Specific receptors for these proteins appear to initiate this process. This phenomenon occurs primarily through the interaction of hormones, cytokines, and other ligands with their cognate endothelial cell surface receptors. So far, ligands such as thrombin, histamine, platelet activating factor, and vascular permeability factor or vascular endothelial growth factor (VPF/VEGF) induce significant increases in microvascular permeability. For instance, thrombin increases intracellular calcium and appears to loosen intracellular junctions to effectively increase paracellular transport. Receptor binding causes large junctions to effectively increase paracellular transport. Receptor binding causes large increases in intracellular Ca_{2+} that alter the cytoskeleton and may widen the junctional gap between the endothelial cells (for review, 20). On the other hand, VPF/VEGF appears to increase permeability through another mechanism. It appears to modulate intracellular organelles to increase direct transcellular transport via a circuitous yet patent transendothelial chain of vesicular compartments (71). VPF/VEGF may also induce fenestrations (114). Histamine increases permeability by inducing a large transcellular opening in the venular endothelium (96). Is it possible that the effects of histamine and VPF/VEGF are somehow interrelated? Are these structures related to caveolae? Could all of these internal structures represent altered caveolae, for instance through vesiculation and aggregation? These and other questions need to be addressed in future work.

Other Factors and Signaling Events

Physical forces in the microcirculation are likely to contribute significantly to transport. Hemodynamic forces continuously act on the endothelial cells lining the blood vessels and in many ways it may be endothelial cell surface including the glycocalyx that may "sense" the physical stimuli. Shear stress, blood flow, and intravascular pressures are known to induce endothelial cell secretion of vasoactive substances through G-protein–coupled signal transduction at the cell surface (7). Although the mechanisms require further definition, it is clear that the endothelial glycocalyx with its cell surface proteins must play a role in transducing physical signals from the circulating blood into the cell where activation of internal biochemical machinery results in endothelial cell responses that may even influence other vascular cells. In this regard, it has been proposed that caveolae may be mechanotransduction organelles at the cell surface (126, 154). Interestingly, shear stress at the endothelial cell surface stimulates nitric oxide production via G-proteins (76) and both G-proteins and the enzyme mediating nitric oxide production nitric oxide synthase reside in caveolae (43, 132). Lastly, the endothelial cell glycocalyx has a role in the regulation of capillary hematocrit (29) and the flow of red blood cells through the microcirculation can affect capillary permeability.

Blood Proteins Maintaining Normal Capillary Permeability

Albumin

Through its interactions with the glycocalyx of the endothelium, albumin has several important vascular functions and plays many roles in the transendothelial transport of molecules. The presence of albumin in blood affects transcapillary transport (19, 42, 62, 63, 79, 86, 90, 119) and fluid homeostasis (152). As the major blood protein, it is the principal determinant of the oncotic pressure of the plasma and thereby strongly influences transendothelial fluxes of water and small solutes via Starling forces (152). In addition, its binding within the endothelial glycocalyx mediates a significant part of the "serum effect," apparently by creating a permselective barrier that limits the transendothelial passage of many molecules. Upon removal of albumin from vascular perfusates, for instance, the net flux of solutes and water increases significantly while osmotic reflection coefficients decrease in both individually perfused capillaries (22, 62, 79, 90) and whole organ preparations (42, 86, 119). Chemical modification of arginine (but not lysine) residues within albumin significantly reduces albumin binding to the endothelium (48, 109, 120) and increases transcapillary water fluxes (91, 109, 120).

Albumin also serves as a carrier of many drugs, metabolites, amino acids, heavy metal ions, sterols, bile acids, hormones, and bilirubin and is the main transport molecule for free fatty acids in blood (72, 105). As a result, albumin is considered to be a multifunctional blood transport protein. These carrier activities may involve its own transport across the endothelium. Albumin transcytosis via caveolae appears to occur selectively in certain tissues with vascular beds lined with continuous endothelium (124) and is influenced greatly by the ligands bound to albumin, such as fatty acids (41). Once transported to the interstitial fluid, albumin is actively involved in the presentation of important ligands to various cells (61, 98, 167). For example, the ratio of fatty acids (an important substrate for myocardial energy production) to albumin in vascular perfusates strongly affects myocardial fatty acid oxidation (61). Exposure to free fatty acids increases the transfer of albumin across cultured endothelial monolayers (59). The interaction of fatty acids with albumin may increase both its binding and subsequent transcytosis across continuous endothelium (41); however, more recently, other laboratories have noted increased fatty acid transvascular exchange without a concomitant increase in albumin transport (68). These latter data suggest that uncoupling of fatty acid from albumin occurs and that binding of the carrier molecule, albumin, may occur at the cell surface, but only to release its fatty acid for transport across the endothelium by an undetermined mechanism. Other functions of albumin interacting with the endothelial luminal surface may include the reduction of erythrocyte and platelet adhesion and the restriction of the surface binding of other plasma proteins. Thus, albumin serves to decrease nonspecific transport across the endothelial cell barrier while appearing to facilitate the transfer of itself and select bound ligands as it carries them across the endothelium.

Orosomucoid

Orosomucoid (also called α1-acid glycoprotein) is a highly sialylated, polyanionic serum glycoprotein that is the predominant component of the seromucoid fraction of plasma (for review, 118). It is synthesized mainly by hepatocytes. As a positive acute phase protein, its blood plasma concentration increases during conditions such as inflammation, infections, pregnancy, and cancer (74). Even though its primary structure has been known for many years (118), its specific physiological function is uncertain. Recent studies (23, 55) indicate that orosomucoid has a very important role in contributing to the "serum effect" on capillary permselectivity. Orosomucoid strongly reduces the transvascular transport of polyanionic macromolecules such as albumin (55) and α-lactalbumin (23). Although its plasma concentration (0.3–1.1 mg/ml) is considerably less than that of albumin (25–50 mg/ml), orosomucoid appears to have a disproportionately large effect on capillary charge selectivity to macromolecules (55). In whole organ preparations, orosomucoid reduces the transvascular escape of the polyanionic macromolecule, albumin (55). In individually perfused microvessels of the frog mesentery, circulating orosomucoid reduces the permeability for the polyanionic macromolecule, α-lactalbumin, increased permeability for the polycationic macromolecule, ribonuclease, and did not affect hydraulic conductivities (23). Since ribonuclease and α-lactalbumin have very similar hydraulic radii (~20 Å), these differences reflect charge and not size effects. Since orosomucoid is highly polyanionic with a pI of 2.7 (118), it has been suggested that orosomucoid binding to the endothelial glycocalyx may mediate this effect by increasing the net negative charge of the microvessel wall (23, 55).

Besides this general effect on capillary permselectivity, orosomucoid probably has more specific roles in the circulation that remain unknown. Orosomucoid interacts with high affinity with a number of small ligands, including basic lipophilic drugs and endogenous molecules such as catecholamines, plasminogen activating factor, and steroid hormones (5, 74, 88). As a carrier within the circulation that affects the ligand's pharmacokinetics and body distribution (5, 88), orosomucoid must be transported across microvascular endothelium (5), possibly by receptor-mediated processes. The interaction of orosomucoid with endothelium has been examined in culture (140). Reversible binding of bovine serum orosomucoid has been detected at the surface of bovine lung microvascular endothelial cells (BLMVEC) in culture at 4°C. This binding reaches equilibrium after 10 minutes of incubation, is not calcium-dependent, and is specifically competed with unlabeled orosomucoid but not transferrin, immunoglobulin, gelatin, ovalbumin, mannan, fucoidan, mucin, or asialomucin. In addition, binding is not affected by the presence of galactose, glucose, fucose, mannose, N-acetylgalactosamine, or N-acetylglucosamine, which suggests that this binding is not mediated by carbohydrate recognition via lectin-like molecules such as the asialoglycoprotein receptor. Scatchard analysis of the orosomucoid binding to BLMVEC produces a concave-upward–shaped curve that revealed a higher affinity binding component with a K_D of 8.6 nM, a maximum receptor number of 40,000/cell, and a moderate affinity component with a K_D of 9.1 μM with a maximum binding of 10^7 binding sites per cell (140). Other blood vessel-associated cells including fibroblasts and smooth muscle cells clearly bind less orosomucoid with a lower binding affinity than the BLMVEC monolayers. The total increase in the negative charge of the glycocalyx from orosomucoid binding is about 17 mEq/L (140), which agrees well with previous work estimating the charge necessary to account for the effect of orosomucoid on increasing capillary permselectivity to charged macromolecules (23). These results indicate that orosomucoid binds to BLMVEC specifically in a manner that fulfills many of the requirements of ligand–receptor interactions including cell-type specificity, saturability, reversibility, selective competibility, and dependence on time and ligand concentration. More recently, it has been shown that orosomucoid conjugated to gold particles, when perfused through the vasculature, can bind to the endothelial cell surface for transport across the endothelium via caveolae (111).

Other Proteins

Although albumin and orosomucoid contribute significantly to the "serum effect," they do not constitute the whole effect even when combined, so that it is clear that other unidentified components also contribute. In addition, many other blood proteins are likely to be important carriers for various essential small molecules such as transferrin for iron, ceruloplasmin for copper, and LDL for cholesterol. In each case, interactions with the endothelial cell surface or specific receptors have been observed. Because the vascular endothelium in many organs forms a critical barrier that impedes the diffusive and convective transport of large molecules, selective mechanisms must exist to overcome this hindrance so that specific molecules such as carrier proteins and even hormones released into the circulation can be delivered to their physiologically relevant target tissues. Insulin appears to require binding to its receptor on endothelium for specific transport (70). Recently, human chorionic gonadotropin has been shown to be transcytosed by coated pits and vesicles through direct receptor binding (47). Surprisingly, rat testicular endothelium appear to express the same receptor for human chorionic gonadotropin as the intended target, namely the Leydig cell in the underlying tissue (47).

Cellular and Molecular Analysis

Molecular Mapping of the Endothelial Glycocalyx and Caveolae

Although the pathways and barriers have been identified morphologically and functional evidence supports their role in transport, until recently there has been very little information on the specific molecules that form important structures affecting transport, such as the glycocalyx and plasmalemmal vesicles. In the past, the glycocalyx has been visualized as a whole by various general staining procedures (for review, 100). Its charge characteristics have been examined in situ using different anionic and cationic electron-dense probes, and its carbohydrate composition has been assessed in situ by lectin binding coupled with specific enzymatic digestions (4, 93, 106–108, 149, 163). However, these studies could not identify the individual molecular components involved in these reactions; their identification and characterization are important prerequisites for understanding endothelial surface interactions with the molecular and cellular components of the blood. With new technological advances in selectively identifying endothelial surface proteins, including the recent development of a new membrane isolation technique that provides a highly purified fraction of luminal endothelial cell membrane directly from tissue (see below), it is now possible to identify and purify important molecular components of the glycocalyx and other main barriers and pathways mediating transcapillary exchange.

Culture Versus Native Tissue Conditions

Defining the interactions between the endothelial cell surface and circulating blood molecules, especially with regard to transvascular transport, has been rather difficult. The development of techniques for isolating and growing endothelial cells in culture allowed some important characterizations of these processes. Unfortunately, endothelial cells are quite sensitive to their growing environment. In tissue they are influenced by subendothelial and supraendothelial effectors. Subendothelial factors include the basement membrane, released autocrine and paracrine factors, and perivascular and tissue cells. Supraendothelial effectors relate to the blood circulation and range from specific interaction with blood components such as cells, hormones, and carrier proteins to mechanotransduction of hemodynamic physical forces from pressure, blood flow, and shear stresses. It has become quite evident that endothelial cells change considerably when isolated and grown in culture, creating an apparently "de-differentiated" phenotype that is very different both morphologically and functionally from that found in vivo. For instance, when grown in culture, continuous and fenestrated endothelia lose many characteristic features including nearly all of their caveolae and fenestrations, respectively (15, 92, 142). Such losses (>100-fold) represent a significant and obvious alteration in plasma membrane organization or topology that is likely to reflect major perturbations not only in membrane functions in transport or signal transduction but also in other cellular functions. Hence it becomes necessary to understand the limitations of the cultured systems and is probably prudent to focus more attention on the endothelium as it exists normally in the tissue. At a minimum, it becomes necessary to verify any results derived from cultured systems by appropriate tests performed in situ or in vivo.

With these issues in mind, several laboratories have developed methods of identifying cell surface proteins of the endothelium by in situ labeling procedures (6, 89, 141). In general, the proteins on the luminal cell surface of the endothelium were labeled selectively using a chemical agent reactive with the proteins but kept rather restricted to the intravascular space. These studies gave the first hints at the molecular level that endothelia are indeed heterogeneous, varying considerably in cell surface expression of proteins. Unfortunately, each labeling procedure has its inherent propensity for labeling specific sites that are not equally present in all proteins; therefore, only a small subset may be identified and relative quantification becomes rather difficult to assess. Because the proteins of endothelial cell plasma membranes, especially those found at the luminal cell surface that is critical for transport, can represent less than 0.1% of the total protein in any organ system, it becomes difficult not only to resolve the labeled proteins with great clarity by gel electrophoresis but more importantly to analyze and purify them. Obviously many of these problems can be overcome if the endothelial cell membrane of interest could be purified in sufficient quantity directly from tissue. It is the luminal endothelial cell plasma membrane directly in contact with the circulation that represents the critical interface for transvascular exchange and especially for specific transport processes. Thus, a new procedure is described below that easily overcomes many of these problems by isolating the luminal endothelial cell surface membrane directly from tissue.

Purification of Luminal Plasma Membranes Directly from Endothelium in Tissue

This method requires intravascular perfusion of the tissue and can be performed on any normal or diseased tissue amenable to such perfusion. Most of the work so far has been performed on the rat lung microvasculature primarily because of the availability via direct perfusion of both a large endothelial cell surface area and a rich population of caveolae. Moreover, this tissue has been quite useful in performing functional studies examining transport in situ (24, 69, 135, 138, 139). The lung is perfused in situ at 10–13°C via the pulmonary artery with a solution of polycationic colloidal silica particles that coats the luminal endothelial cell plasma membrane directly adjacent to the circulating blood. This coating creates a stable silica pellicle that specifically marks this membrane of interest and enhances its density sufficiently to facilitate its purification from tissue homogenates by centrifugation (64, 137). The silica-coated membrane pellets have many associated caveolae and display ample enrichment for various endothelial cell surface markers with little contamination from other tissue components (64, 137). A typical membrane sheet has silica coated on one side and caveolae still attached on the other side. Quantified immunoblotting shows enrichments up to 30-fold in the silica-coated membrane pellets relative to the starting tissue homogenate for proteins known to be expressed on the surface of endothelium, such as caveolin and angiotensin converting enzyme (ACE) (64, 137). Conversely, proteins of intracellular organelles and even the plasma membranes of other lung tissue cells are not detected.

Purification of Endothelial Cell Caveolae

A distinctive feature of many endothelia is an abundant population of noncoated plasmalemmal vesicles or caveolae. Little is directly known about their molecular architecture and function. One way to better define their role in transport would be to purify them directly from the isolated luminal endothelial cell plasma membranes so that they can be studied, especially with regard to their molecular composition. Defining the molecules comprising a caveolae provides critical data for a more definitive assessment of function. Such a "molecular fingerprint" provides critical data necessary for creating specific ways to test the function of caveolae and their specific molecules. These studies are only just beginning (see the section on caveolae as transport vesicles).

In order to address the issue of transport across endothelium in situ and whether caveolae do indeed have the molecular machinery for being discrete and specific carriers, this author's laboratory has selectively isolated the luminal plasmalemma with its subtending caveolae from the rat lung microvasculature, purified the caveolae from this plasmalemmal fraction, and begun to dissect it biochemically into structural and functional components. In the purified endothelial cell plasma membranes, as found in situ before homogenization, the caveolae are attached in their normal orientation to the membrane on the opposite side of the silica coating. They are stripped from these endothelial membranes by shearing during homogenization at 4°C in the presence or absence of the nonionic detergent Triton X-100. The caveolae are dissolved by many detergents but appear to be quite resistant to Triton X-100 extraction at low but not higher temperatures. Although it is clear that the Triton X-100 is not necessary for the purification, it does appear to facilitate the shearing of the caveolae from the plasma membrane. The physical basis of this phenomenon is unclear. The caveolae that are physically detached from the plasmalemma by the shearing are then isolated by sucrose density centrifugation to yield a homogeneous population of morphologically distinct small noncoated plasmalemmal vesicles mostly with diameters of 70–90 nm (133, 137). Small membrane-bound, electronlucent openings or stomata, which originally were described many years ago (101) as characteristic of endothelial caveolae (they probably indicate a previous attachment to another vesicle as part of a chain of vesicles), were visible for many vesicles in favorable sections. Higher magnification views of the purified vesicles reveal vesicular structures typical for caveolae in vivo including their characteristic flask-shape with an open neck region. Single plasmalemmal vesicles and chains of membrane bound vesicles were present. Some vesicles still maintained their narrowed necks as in caveolae that are attached to the plasmalemma of endothelium in vivo.

Biochemical analysis shows that caveolae represent specific microdomains of the cell surface with their own unique molecular topography (133, 137). As with caveolae found on the endothelial cell surface in vivo by electron microscopy, the purified caveolae contain almost all at the cell surface of the structural protein caveolin, the plasmalemmal Ca^{2+}-ATPase, the cholera-toxin binding sialoglycolipid GM_1, and the inositol 1,4,5-triphosphate surface receptor (133, 137). By contrast, other cell membrane markers, including angiotensin converting enzyme, B and 4.1, and β-actin, were nearly absent from the purified caveolae in spite of being present amply in the silica-coated membranes. The repelleted silica-coated membrane that had been sheared lacked caveolae as assessed by both electron microscopy and immunoblotting for caveolar proteins. Both electron microscopy and biochemical analysis revealed that the caveolae had been removed from the highly purified, silica-coated endothelial cell membranes, yielding a pure fraction of caveolae.

Molecular Mapping

It is now possible to perform extensive detailed molecular mapping of the endothelial cell plasma membrane and its caveolae. Table 2.1 summarizes the results of these studies to date. Interestingly, many of the discovered molecules are not easily detected in whole tissue lysates or even cultured endothelial cells and require fractionation and membrane purification for easy detection by Western analysis. In tissue, this observation should not be too surprising, because the endothelium represents a very small part of the total cells in most tissues. For instance, VAMP, a protein mediating vesicle docking and fusion shown to be amply present on synaptic vesicles in neurons, turns out to be present on the cell surface of endothelium exclusively in caveolae (132). This finding would be very difficult to get biochemically without purification of the endothelial plasma membrane as it exists in vivo. Two other such proteins mediating specific transport in endothelium are discussed below.

TABLE 2.1. Molecular mapping of endothelial caveolae.

Structural elements
Caveolin
Cholesterol
Glycolipids—GM$_1$
Signaling molecules
GTP-binding proteins (GTP-ases)
Heterotrimeric G proteins
Monomeric GTPases (not Rab5)
Src-like nonreceptor tyrosine kinases
Cross-linked GPI-anchored proteins
Inositol trisphosphate receptor
Ca^{2+} ATPase
PI-3-kinase
Sphingomyelin
Phosphoinositides
Transport/carrier/fusion proteins
Albondin
Albumin
VAMPs (synaptobrevins)
NSF
SNAP
Annexin II & IV
Aquaporin-1

Newly Discovered Endothelial Cell Surface Transport Proteins

Aquaporin and Water Transport in Endothelium

Classically, water transport across endothelium of the continuous type found in the microvessels of many organs such as lung was thought to occur almost completely via the paracellular pathway through intercellular junctions. Direct transmembrane and transcellular transport was considered to be minimal. For both epithelial and endothelial cell barriers, there has been considerable controversy over the specific pathways responsible for transport (17, 80, 101, 113, 161). For many years, water transport across these barriers was professed to occur via the paracellular pathway through intercellular junctions. Although water transport across artificial lipid membranes seemed minimal, biophysical studies on cellular membranes predicted that significant transmembrane transport is possible in some cells (36, 84, 150, 161). Moreover, it was discovered that this transmembrane transport can be inhibited by mercurial alkylating agents such as HgCl$_2$ (84, 150). Yet, the critical molecular mechanism for this transport remained quite elusive until finally several specific water channel proteins now called *aquaporins* were identified in the last few years (3, 33, 159).

Aquaporin-1 (originally known as CHIP-28) is a 28-kDa integral membrane protein first found in erythrocytes (for review, 3, 33, 159). It forms a tetrameric transmembrane channel selective for water transport and has structural similarities with other membrane transport channels. It is sensitive to mercurial alkylation with $HgCl_2$, which modifies Cys-189 of the protein and prevents specific transmembrane water transport. It is the major mercurial-sensitive, water-transporting protein in erythrocytes and certain renal epithelia. In erythrocytes, aquaporin allows rapid cell swelling or shrinkage in response to small changes in osmolality. Reconstitution of purified erythrocyte aquaporin-1 in liposomes increased their water permeabilities while maintaining low proton and urea permeabilities. Expression of mRNA encoding aquaporin-1 in *Xenopus* oocytes and Chinese hamster ovary cells greatly increases their water permeability. Its constitutive presence in renal epithelia of both the proximal tubule and thin descending loop of Henle permits water reabsorption. To date, five members of the aquaporin family have been identified, all with different tissue and cell distributions (3, 33, 159).

Aquaporin-1 forms a transmembrane water channel as a multisubunit oligomer comprised of four 28-kDa protein subunits that are identical except that only one has a large N-linked glycan. It has a rather diverse tissue distribution and can be expressed on the surface of many fluid-transporting cells including various secretory or resorptive epithelia of the colonic crypts, choroid plexus, ciliary body, iris, lung alveolus, sweat gland, and gallbladder (37, 56, 57, 97, 159). Some staining of endothelium by in situ hybridization or antibodies has been noted (37, 56, 97). Water channel proteins embedded in plasma membranes can mediate the transmembrane and transcellular transport of water at physiologically relevant flow rates.

With the development of a methodology for purifying luminal endothelial cell plasma membranes directly from tissue as described above, it was discovered that aquaporin-1 is expressed on the endothelial cell surface in vivo at levels comparable to the plasma membranes of other aquaporin-1 expressing cells such as erythrocyte and renal epithelial cells (136). Although barely detectable in the starting rat lung homogenates, both the nonglycosylated and glycosylated forms of aquaporin-1 are amply present in the purified silica-coated membranes. Surprisingly, aquaporin-1 is found concentrated, but not exclusively, in caveolae. Mass balance calculations show that about 70% of aquaporin on the cell surface is in the caveolae, with the remaining 30% distributed over the rest of the plasmalemma.[1] Dual-labeling immunofluorescence performed on bovine lung microvascular endothelial cells grown in culture provide further confirmation of these

---

[1]These isolated membranes lack contamination from other tissue components, including circulating erythrocytes, which are not found to be coated by the perfused silica as would be expected because of: i) the a priori flushing of the blood from the tissue, ii) the inherent proclivity of the silica to coat patent perfusable blood vessels and not occluded ones, and iii) the demonstrated selectivity of the silica for coating only the luminal cell surface of the vascular endothelium without any evidence of coating of other tissue or blood components as described previously (64, 132, 133, 137).

findings by showing that aquaporin-1 and the caveolar structural protein, caveolin, colocalize on the cell surface. Thus, physiologically relevant levels of aquaporin expression may indeed exist on endothelia, mostly within caveolae.

Endothelia can express aquaporin-1 at levels comparable to other cells shown to mediate selective transmembrane water transport. For erythrocytes and certain renal epithelia, it has been clearly demonstrated that mercurial alkylating agents such as $HgCl_2$ effectively inhibited transmembrane and transcellular transport of water but not other solutes by modifying the Cys-189 residue of aquaporin-1. When the effects of these known inhibitors of aquaporin-1 mediated water transport were tested on the uptake of tritiated water perfused through the rat lung microvasculature in situ, it was found that $HgCl_2$ reduced the water transport significantly (136). This reduction was concentration-dependent, achieving a maximum plateau of about 60% inhibition at about 0.03 mM. The effects of $HgCl_2$ on the uptake of SO_4 and inulin were small to negligible, indicating that $HgCl_2$ was not causing a general redistribution of vascular transport or a general increase in endothelial barrier function by "tightening" intercellular junctions. It should also be noted that, under conditions of constant pressure, no changes were observed in the vascular resistance and flow. $HgCl_2$ appeared to specifically inhibit the tissue uptake of water. This effect of $HgCl_2$ was rapidly reversible (136).

The lipid composition of membranes can affect significantly the ability of water to cross the bilayer. Some membrane lipids such as cholesterol and sphingomyelin prevent such water transport. Interestingly, lipid analysis of the purified caveolae reveals that caveolae contain much of the cholesterol and sphingomyelin found in the plasma membrane (81), suggesting that water and solute transmembrane transport in caveolae should be quite small except through selective pathways such as aquaporin. It is possible that aquaporin may function in the volume regulation of not only the cell but also the caveolae. Could caveolae play a role in the sensing of osmotic pressures/stressors by being osmo-sensors? This may relate to their ability to organize signaling cascades at the cell surface (81) and also sense other physical stimuli such as fluid flow forces (154).

Like certain epithelia, continuous endothelia appear to express aquaporin-1 to facilitate the direct transcellular transport of water. Interestingly, caveolae may play an important role in the transvascular transport of water, not only by containing these specific water channels but also by increasing the total membrane surface area available for this transport. Moreover, quite recently, water transport has been documented across ion and glucose transporters (82), which may apply to many different transport proteins embedded in the plasma membrane including various glucose transport proteins known to be expressed in endothelia. Thus, tissue and/or cell water homeostasis may be strongly influenced by transmembrane pathways that may or may not depend on the presence of significant hydrostatic or osmotic gradients. This new arena of transport research is only just beginning to yield its secrets and in some cases is creating far more questions than answers. But with additional information, a more precise understanding of water transport processes in both endothelium and epithelium may be possible.

In this light, some reevaluation may be necessary of the current dogma that most, if not all, transport of water is paracellular through intercellular junctions. It logically has been assumed in the past that most water transport does not occur directly across the endothelial cell membrane because normally the hydrophobicity of lipid bilayers prevents such transmembrane transport. This logic also prevailed in modeling transepithelial transport for a number of years. For some epithelial cell barriers, the paracellular pathway for water transport is still well accepted, but in the case of certain renal epithelial barriers expressing aquaporins, the transport seems to be selectively transcellular. It becomes apparent that this may also be true for certain endothelia, primarily, if not solely, endothelia of the continuous type expressing aquaporin and not endothelia of the fenestrated or sinusoidal type, where obvious membrane discontinuities exist for less selective transport of water and other solutes. Sinusoidal and fenestrated endothelia have open extramembranous passageways for direct water transport through either obvious gaps between the cells or circular "windows" of about 100 nm in diameter called *fenestrations*. Conversely, continuous endothelia do not have such distinct membrane discontinuities. Consistent with this idea, aquaporin-1 is expressed in rat lung and heart but not cortical brain or liver. These newly discovered pathways deserve future attention in microvascular research.

Because albondin and albumin also reside in caveolae, the possibility exists that the binding of albumin to albondin could influence water transport via aquaporin, possibly through protein phosphorylation. This speculation provides an alternative mechanism for the effect of albumin on water conductivity as part of the "serum effect."

Distinct Albumin Binding Proteins and Albumin Transport

As discussed above, albumin has multiple effects on capillary permeability, especially via its interactions with endothelium. The binding of albumin to cultured microvascular endothelium has been quantified and immunolocalized to the surface of cultured rat microvascular endothelial monolayers (20). Specific albumin binding is saturable, reversible, competitive, and dependent on cell type and cell number, and negative cooperative in nature (130). Albumin binding is sensitive to Pronase but not trypsin digestion of the cell surface and is inhibited significantly by the presence of *Limax flavus* (LFA), *Ricinus communis* (RCA), and *Triticum vulgare* (WGA) agglutinins but not several other lectins (129). Recently, a group of rat endothelial plasmalemmal sialoglycoproteins has been identified both in situ and in culture (141). One of these sialoglycoproteins, with an apparent molecular mass of 60 kDa, named "gp60" (129) and later "albondin" (135), was identified as a putative albumin binding protein because it: i) interacts with albumin conjugated to beads; ii) binds RCA, LFA, and WGA but not other lectins; and iii) is sensitive to Pronase and sialidase but not trypsin digestion (129, 141, 144). Two other laboratories have subsequently confirmed the identity of gp60/albondin as a major albumin binding protein (45, 158). Further characterization of gp60/albondin showed that it: i) is expressed on the surface of endothelium both in

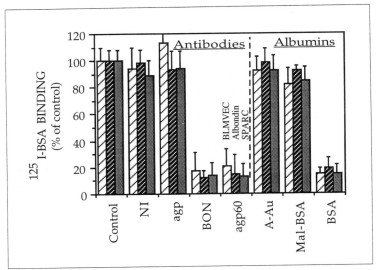

FIGURE 2.1. Inhibition of ^{125}I-BSA binding to cells and ABP.

situ and in culture; ii) is a sialoglycoprotein that contains O-linked oligosaccharides but few, if any, N-linked glycans; and iii) may also be antigenically related to another sialoglycoprotein, namely glycophorin (45, 124, 129, 141, 144, 145, 158).

In the past decade, three other albumin binding proteins (ABPs) have been identified (45, 46, 117). The first one was called SPARC (also called osteonectin) and is not a plasmalemmal protein but a secreted protein that binds albumin as it circulates through the vasculature (117). The other two ABPs are smaller than gp60, with an apparent molecular mass of 30 and 18 kDa (45, 46), and are called gp30 and gp18. Why do so many ABPs exist? Do they have the same functions? Are they interrelated in some way? The structural and functional relationships of these ABP to each other have been unclear until recently.

First, it was found that albondin and SPARC are functionally and immunologically related albumin binding proteins (134). The possibility was investigated that albondin, gp30, and gp18, and the secreted albumin binding protein, SPARC, may have regions of structural homology that might be expected to coincide with their common ability to bind albumin. Polyclonal antiserum raised against purified bovine SPARC (BON) interacted not only with SPARC secreted by cultured bovine endothelial cells but also with albondin in lysates of rat, bovine, and human cultured endothelial cells. In an albumin binding assay using immobilized purified SPARC and albondin, the IgG fraction of BON inhibited albumin binding to both proteins (see Figure 2.1). They also inhibit binding to the surface of bovine lung microvascular endothelial cells (BLMVEC). BON inhibition of binding was lost when BON IgG was preabsorbed to immobilized SPARC (this procedure removed the antibodies recognizing SPARC and resulted in complete loss of recognition of albondin). Although SPARC has many other functions than just binding to albumin (116, 134), it is clear that both albondin and SPARC share at least this one common function and may share a common albumin binding domain.

Gp30 and gp18 were originally identified as ABPs primarily using ligand blotting with albumin–gold complexes (A–Au) (46). At first it was thought that these proteins may be albumin receptors (46); however, subsequent experiments showed that native monomeric bovine serum albumin (BSA) do not compete very effectively with the binding of A–Au to gp30 and gp18 on blots (143), indicating that A–Au is somehow different from BSA. Albumin adsorption to a variety of surfaces frequently causes conformational changes (112). Various conformationally modified forms of albumin were able to compete with A–Au binding to gp30 and gp18 (143). For these albumins, the modification was chemically induced and not caused by surface adsorption. Formaldehyde-treated albumin (Fm-BSA) and maleic anhydride–treated albumin (Mal–BSA) competed much more effectively than BSA for the binding of A–Au to gp30 and gp18. It appears that BSA binds about 1000-fold less avidly to gp30 and gp18 than A–Au. Neither A–Au nor Mal–BSA can be considered equivalent probes for native BSA in studying the interaction of BSA with the endothelial cell surface and its binding proteins. Conversely, A–Au and Mal–BSA appear in many respects to be nearly equivalent probes for their high affinity interaction with gp30 and gp18.

These results suggest that the conformational changes in BSA induced by either chemical modification or surface adsorption to the gold particles can somehow be equivalent at least in terms of their ability to create a ligand with a much greater affinity for gp30 and gp18 than native BSA. Comparative biochemical characterization shows conformational changes for Mal–BSA, Fm-BSA, and A–Au including evidence for sufficient protein unfolding to cause a significant increase in sensitivity to trypsin digestion (143). Such unfolding of albumin may unmask a binding domain(s) that is recognized avidly by gp30 and gp18. Exposure of this recognition domain may be induced by both surface adsorption and chemical modification. In addition, such conformational changes also could release small ligands such as fatty acids or small ions that are normally bound to albumin, which in principle might increase albumin's affinity for gp30 and gp18 while lowering it for albondin. Although the exact mechanism remains unclear, conformational changes in albumin can severely affect its binding to specific proteins apparently located on the endothelial cell surface. Cumulatively, these studies reveal that albondin interacts with native albumin while modified albumins bind rather selectively to gp30 and gp18.

Analysis of Albondin and Albumin Binding

The 60-kDa albumin binding protein, albondin, has been purified from endothelial cell lysates by sequential detergent extractions and lectin affinity chromatography (135). Polyclonal rabbit antiserum (agp60) made against albondin specifically recognizes a 60-kDa protein in human, bovine, and rat endothelial cell lysates by immunoblotting and can immunoprecipitate radiolabeled 60-kDa protein from lysates of endothelial cells whose surface proteins have been radioiodinated (135). As shown in Figure 2.1, the antibodies recognizing albondin inhibit specific BSA binding to endothelium. The IgG fraction from various anti-sera were tested for their ability to inhibit the binding of ^{125}I-labeled BSA (^{125}I-BSA) to the surface of confluent BLMVEC monolayers and directly to immobilized purified albondin and SPARC. The specific binding of ^{125}I-BSA is ablated almost totally by agp60 IgG and native unlabeled BSA in a concentration-dependent manner (38, 158). Both BON and agp60 IgG inhibit greater than 80% of specific ^{125}I-BSA binding to the BLMVEC, immobilized albondin, and SPARC whereas the other antibodies are ineffective. The effects of 20–50 µg/ml of BON and agp60 IgG are comparable with the inhibition achieved with 1 mg/ml of BSA. Specific antibody interaction with albondin significantly reduces endothelial cell surface binding of albumin.

Because both BON and agp60 IgG inhibit BSA interactions with BLMVEC, SPARC, and albondin, and because BON recognizes albondin (134), the ability of agp60 to recognize SPARC was tested and found to bind SPARC (135). This functional and immunological cross-relationship strongly suggests that anti-albondin serum recognizes SPARC, apparently via a common albumin binding domain, and that both proteins share similar albumin binding domains.

It is clear that endothelium can bind not only native albumins (41, 48, 94, 119, 124, 129, 144) but also modified albumins such as A–Au and Mal–BSA (10, 28, 44, 46, 127, 143). Figure 2.1 shows that unlabeled native BSA inhibits ^{125}I-BSA binding to the BLMVEC and directly to SPARC and albondin, whereas A–Au and Mal–BSA did not [even at more than 100-fold molar excess and at levels known to saturate gp30 and gp18 binding (127, 143)]. ^{125}I-labeled modified albumins to BLMVEC and directly to gp30 and gp18 by blotting. All of the antibodies, along with native BSA, were quite ineffective inhibitors whereas Mal–BSA inhibited both ^{125}I–A–Au and ^{125}I–Mal–BSA binding very well (135). Cumulatively, the results to date indicate that different receptors mediate the binding of native and modified albumins to the endothelial cell surface, with gp30 and gp18 recognizing the modified albumins and albondin mediating native albumin binding to the cell surface of endothelium.

Pathways of Albumin Transport

The transendothelial transport pathway for various molecules including albumin varies considerably between tissues and appears to depend greatly on the type of endothelium lining the microvasculature of the tissue. Albumin has been localized immunocytochemically both in situ and in culture to the endothelial glycocalyx of the plasmalemma proper (49, 94, 119, 130), caveolae (48, 94, 119, 162, 168), and in some cases coated pits (28, 44). Albumin crosses the continuous, sinusoidal, and fenestrated endothelial barriers via different transport pathways (for a more complete discussion, see 124). Briefly, many vascular beds lined with continuous endothelium appear to transport albumin via caveolae (48, 94, 119, 162, 168). The only exception so far noted in the literature for continuous endothelium has been that of the brain. This highly restrictive endothelium neither binds nor transports albumin (103, 147), which may be due in part to the lack of caveolae. Albumin crosses sinusoidal endothelium freely via the large gaps between the cells. Sinusoidal endothelia of the liver and bone marrow endocytose albumin–gold complexes (A–Au) via coated vesicles but do not transcytose albumin via caveolae (28, 44). Some fenestrated endothelia internalize albumin via coated vesicles; however, A–Au is transported across diaphragmed fenestrations and not via caveolae (10). It appears that only continuous endothelium have caveolae that apparently transcytose albumin.

Expression of ABP

The available data clearly provide evidence that albumin is transported across continuous endothelia via caveolae whereas it crosses other endothelia via other pathways. If an albumin binding protein is involved in specific transcytosis of albumin, then it should be present in tissues with microvascular beds lined with continuous endothelia that bind and transport albumin via caveolae and is absent in those tissues with other endothelia that do not. Blotting of tissue lysates has revealed that gp30 and gp18 are expressed in all of the tissues examined and appear to be ubiquitously distributed, whereas the expression of albondin in tissues was variable. Albondin was detected most strongly in heart and lung; moderately in diaphragm, skeletal muscle, fat, mesentery, and small intestinal muscularis; mildly in kidney; and very poorly or not at all in cortical brain, liver, adrenal, pancreas, and small intestinal mucosa. Tissues appear to express albondin in accordance with the type of endothelium lining their microvasculature. Albondin was expressed in situ selectively in tissues with continuous endothelium (except for brain) and not exclusively with other types of endothelia. Brain endothelium have few plasmalemmal vesicles and do not interact with albumin in situ or in culture (124). The ubiquity of tissue and cell expression of gp30 and gp18 but apparently not albondin suggests that albondin may indeed have specific endothelial functions that are not ascribed to the other proteins. Because only continuous endothelium with caveolae not only bind and transcytose albumin but also appear to require albumin binding for maintaining normal permeability, it appears that the selective expression of albondin correlates quite favorably with its putative function(s).

A variety of cultured cells associated with the vascular wall have been examined for their expression of albondin, gp30, and gp18 (124). All three albumin binding proteins were expressed by microvascular endothelial cells isolated from rat fat pads (RFC); however, gp30 and gp18 but not albondin were detected in lysates from rat fibroblasts (NRK-F) and smooth muscle cells (A-10). The NRK-F bind about tenfold less albumin than the RFC (130). Microvascular endothelial cells can be isolated using a fluorescence-activated cell sorter from cortical brain (RBE), heart (RHE), and lung (RLE) tissues (124). The RFC, RHE, and RLE grown in culture express albondin and bind albumin whereas the RBE cells do not express albondin and bind 10–15-fold less albumin. It appears that albondin is expressed specifically by vascular endothelium that bind albumin, whereas gp30 and gp18 are expressed by many cell types.

Gp30 and Gp18 as Scavenger Receptors

It is now clear that gp30 and gp18 are expressed by many cells and tissues and interact more avidly with various modified albumins than with native albumin (143). Some of these modified albumins are also known to interact with scavenger receptors (52, 60). Receptor-mediated endocytosis of a variety of modified proteins by many cells including macrophages, fibroblasts, endothelial, and smooth muscle cells has been collectively termed "scavenger function" (8, 146, 153) and may be necessary for normal physiological removal of proteins that are modified by aging or oxidative processes (i.e., inflammation). In the last few years, several different scavenger receptors have been studied extensively in terms of their role in protein catabolism and possible link to atherosclerosis, aging, and diabetes (8, 146, 153). Kinetic binding experiments indicate that cells can express a number of distinguishable scavenger receptors, each with different ligand specificities. Competition studies (143) reveal that gp30 and gp18 bind a variety of molecules that are known to interact with one or more types of scavenger receptors [fucoidan and Mal–BSA but not acetylated-LDL, malondialdehyde-treated LDL, and several polyanionic molecules (dextran sulfate, polyglutamic acid, polyinosinic acid, and heparin)].

If gp30 and gp18 are indeed scavenger receptors for modified albumins, then cells expressing gp30 and gp18 should bind, internalize, and degrade their ligands. Recent work (127, 143) shows that RFC cells degraded A–Au 50 times more than native BSA. NRK-F (fibroblasts) also express gp30 and gp18 and degrade A–Au. Interestingly, the same molecules that compete with A–Au blotting of gp30 and gp18 inhibit A–Au degradation. Electron microscopy confirms that A–Au binds preferentially within caveolae and appears to be endocytosed by caveolae for delivery to endosomes, where it is processed for lysosomal degradation (138). The results to date indicate that gp30 and gp18 mediate the high affinity binding, internalization, and degradation of modified albumins but not native BSA. They are consistent with the observation that A–Au is endocytosed by many different endothelial cell types in various tissues and ultimately accumulates within lysosomes (10, 28, 44, 94, 127, 138, 162). Scavenging of modified proteins may be necessary to remove old/altered proteins that can, under certain conditions, be pathological to local tissue and cells. Gp30 and gp18 may play a role in such a mechanism for the catabolism of old or modified albumins and possibly even other proteins.

Albondin-Mediated Capillary Permeability to Albumin

Although the experiments described above indicate that albondin is the major endothelial plasmalemma protein mediating the cell surface binding of native albumin, its direct role in the transendothelial transport of albumin needed to be established. In order to begin to address this issue, the "single-sample" method (69) has been used to determine the permeability–surface area (*PS*) product for albumin in the rat lung preparations perfused in situ. The *PS* product is directly proportional to the amount of ^{125}I-BSA that accumulates within the tissue after perfusion through the vasculature and provides a direct way to measure the effects of potential inhibitors on capillary permeability to albumin as established in (24, 69). In this assay, unlabeled BSA significantly inhibits ^{125}I-BSA transport into the tissue in a concentration-dependent manner (135). The maximal effect was about a 60% reduction in the *PS* product and was achieved at BSA concentrations equal to or greater than 1 mg/ml. Interestingly, Mal–BSA does not reduce this transport. The albumin-dependent transport may require cell surface binding, whereas the remainder is independent of cold albumin and is likely to be binding-independent and reflect only fluid-phase and/or paracellular transport. It is inherent to any vesicular transport pathway that some nonspecific fluid-phase transport must occur, as, for example, in receptor-mediated endocytosis by clathrin-coated vesicles of LDL where about 50% of the internalization by the cells is through fluid-phase uptake (50, 51). It is also logical to consider that the invasiveness of this in situ procedure (removal of the circulating blood, handling of the tissue, etc.) may contribute to excess nonspecific leakiness that is not normally present in vivo and cannot so far be avoided in this system.

In the same system at concentrations shown to be effective in preventing albumin binding to the BLMVEC surface, agp60 IgG specifically inhibited the transvascular transport of ^{125}I-BSA (135). When only the specific, concentration-dependent transport is examined (by subtracting the transport detected in perfusates containing excess unlabeled BSA), agp60 inhibited greater than 95% of the binding-specific transport. The agp60 is similarly effective in the presence of 10% serum, which represents a more physiological condition for assessing capillary permeability. This inhibitory effect appears to be specific for albumin because agp60 IgG does not hinder the tissue uptake of other molecules such as ferritin and inulin. Ferritin is transported across rat lung endothelium by fluid-phase uptake via caveolae (119). Inulin is a small solute (MW 5000–5500) that is thought to cross the endothelial barrier paracellularly through intercellular junctions. Thus, the effect of agp60 is not a general consequence of altering either vesicular or paracellular transport. These studies indicate that, under a variety of different perfusion conditions, albondin mediates most, if not all, of the specific albumin transport detected. Antibody binding to albondin can greatly reduce capillary permeability to albumin in situ.

Transport in Cultured Endothelium

The transport of native and modified albumins across confluent cultured BLMVEC monolayers has also been examined (135). ^{125}I-BSA transport is inhibited significantly by unlabeled BSA, agp60 IgG, and low temperature (8–10°C) but not Mal–BSA or nonimmune IgG. Native BSA maximally inhibited the total ^{125}I-BSA transport by 50–60% at concentrations ≥1 mg/ml. The agp60 IgG at 50–

100 μg/ml results in nearly equivalent inhibition. Both native BSA and agp60 prevent transport in a concentration-dependent manner that is comparable to their effects on specific ^{125}I-BSA surface binding. These results agree well with the in situ rat lung studies and strongly suggest that 50% of the total ^{125}I-BSA transport is dependent on surface binding to albondin, whereas the other 50% is not. After subtraction of the concentration-independent component, agp60 appears to inhibit about 90% of specific ^{125}I-BSA transport. Low temperature (8–10°C) reduces ^{125}I-BSA transport to about 20% of the total transport observed at 37°C. At 8°C, agp60 does not reduce ^{125}I-BSA transport, indicating that, under these conditions, its binding to albondin does not affect this paracellular pathway nor is it required for paracellular transport.

Unlike native albumin, modified albumin transport across the endothelial cell monolayers is not dependent on cell surface binding. Mal–BSA and agp60 did not reduce the transport of ^{125}I–Mal–BSA across the BLMVEC monolayers at either 8 or 37°C. Interestingly, BSA has a small but statistically significant effect of about 25% at both temperatures, possibly secondary to its role in the "serum effect." Although BLMVEC can bind and internalize ^{125}I–Mal–BSA, its transendothelial transport is obviously independent of this process because an excess concentration of unlabeled Mal–BSA is not inhibitory (unlike the situation for native BSA transport). This finding is consistent with past observations (39, 72) that 90% of ^{125}I–Mal–BSA that is endocytosed by the cells is later released in degraded form (conversely, 90% of ^{125}I-BSA is released undegraded). The binding independence of this transport would suggest further that gp30 and gp18, which bind Mal–BSA avidly, are not involved in its transport across the cell monolayers but rather its internalization and degradation. This transport behavior across the monolayers is consistent with fluid–phase and/or paracellular transport.

Of course, low temperature reduced ^{125}I–Mal–BSA transport, in this case by only twofold, which probably reflects the expected reduction in viscosity and diffusion at 10°C. More importantly, this reduction is much less than the fivefold diminution found for ^{125}I-BSA transport. Because Mal–BSA and BSA are at least physically similar in size, this suggests that a selective process for facilitating native albumin transport exists at the normal permissive temperature of 37°C. In directly paired experiments under equivalent conditions, there is threefold less transport of modified than native BSA (135). On the other hand, at low temperatures, where vesicle- and receptor-mediated transport should be minimal and paracellular transport should prevail, both albumins are transported equally and are essentially equivalent probes. At permissive temperatures for vesicular transport, much more native BSA is transported. Because antibodies recognizing albondin can significantly reduce the binding and transport of native but not modified albumin, it would appear that selective binding of native albumin to albondin on the endothelial cell surface can increase its transendothelial transport two- to threefold over a similarly sized, modified albumin probe that does not interact with albondin. Vesicular transport of albumin in endothelium is enhanced considerably by its specific interaction with the plasmalemmal receptor albondin.

Differential Role of ABP in Transcytosis and Endocytosis

The data presented above help clarify some of the structural and functional inter-relationships between the albumin binding proteins. Albondin and SPARC both interact with albumin, have at least one common immunological epitope, and probably share a common albumin binding domain. SPARC is not a membrane-associated protein, so it is not expected to mediate endothelial surface interactions with albumin. Although antibodies against SPARC and albondin recognize both albondin and SPARC, the observation that neither antiserum recognizes gp30 and gp18 suggests that they are structurally and functionally different from either albondin or SPARC. The differences in cell and tissue expressions for albondin and both gp30 and gp18 also suggest functional differences. Albondin appears to be expressed specifically in those tissues known to have vascular beds lined with continuous endothelium that interacts with albumin not only to initiate its transcytosis via plasmalemmal vesicles but also to increase capillary permselectivity. Conversely, gp30 and gp18 are expressed by a variety of cells and are found in all tissues examined regardless of the type of endothelium lining the vasculature.

Endothelial cells distinctly recognize modified and native albumins via different albumin receptors. The existence of three membrane-associated ABP may be necessary for the differential cellular processing of modified and native albumins by endothelium (125). The work has shown that the conformational change in albumin induced by surface adsorption or chemical means (143) is apparently sufficient to prevent recognition by albondin and SPARC while greatly increasing its affinity for gp30 and gp18. These results bear a striking resemblance to the known interaction of LDL with its receptor, wherein chemical modification of LDL (i.e., acetylation or oxidation) prevents its interaction with the native LDL receptor but induces binding to another set of receptors, namely scavenger receptors such as the acetylated LDL receptor (8, 151). Gp30 and gp18 appear to be scavenger receptors that mediate the internalization and degradation of modified albumins. Gp30 and gp18 may function in endothelial endocytosis of various albumins in many different types of endothelia (and possibly other cells), whereas the role of albondin in the binding and transcytosis of native albumin may be limited to certain continuous endothelia.

Cumulatively, the data indicate that albondin is responsible for the surface binding of native but not modified albumins resulting in cellular uptake and transcytosis, whereas gp30 and gp18 function as scavenger receptors mediating the binding, endocytosis, and degradation of modified but not native albumins. Albondin mediates native albumin binding to vascular endothelium, which affects its capillary permeability and transendothelial transport. More work is needed to complete these studies and provide more evidence that albondin does indeed mediate its effects on capillary permeability through albumin transcytosis via caveolae and that gp30 and gp18 mediate scavenger endocytosis via caveolae. It should be added that, during the writing of this chapter, many of these findings on the function of albondin and its relationship to SPARC were confirmed by another laboratory (158).

New Evidence on Caveolae as Transport Vesicles in Endothelium

Noncoated plasmalemmal vesicles and caveolae are interchangeable terms for the same structures in endothelium. By electron microscopy, they are recognized in many cell types as invaginated pits or vesicles that do not have a thick electron-dense fuzzy coat (distinct for coated vesicles) and that have a rather consistent maximum diameter of ~800 Å. Their cytoplasmic surfaces have bipolar-oriented, thin striations that may be formed from caveolin polymerization (104, 115). Although the existence of caveolae in many cells, especially endothelium, has been known for over 40 years, defining definitively their precise function(s) has been somewhat problematic. The role of caveolae in the transport of molecules into and across endothelium by receptor mediated, adsorptive- and fluid-phase processes has been controversial, especially as it relates to capillary permeability. The select labeling within caveolae has been interpreted as indicative of specific discrete transport by caveolae. Yet, such ultrastructural examination unfortunately cannot prove definitively that caveolae function in transport. Caveolae exist as a racemose structure with many vesicles linked to each other in a chain to form a grape-like branching "cave" that penetrates deep into the cell. Conventional sections of these structures for electron microscopy can be quite misleading. They can show about 50% of the noncoated vesicles apparently free in the cytoplasm while serial ultrathin sectioning reveals that 1% or less of them are actually free and unattached to other membranes (12–14, 38–40). The additional discovery that aldehyde fixation artefactually increases the number of noncoated invaginations creates even more confusion (87, 164). It is therefore not surprising that many investigators have concluded that caveolae are not dynamic vesicular carriers but rather fixed permanent invaginations incapable of budding (12–14, 38, 39). It appears that past methodologies cannot resolve this issue and, therefore, other new approaches will be required to define whether caveolae are dynamic traffickers or static microdomains on the cell surface.

The recent identification of two pharmacological agents that inhibit the apparent endocytosis and transcytosis of select ligands that preferentially bind within caveolae has revived the case for caveolae as carrier vesicles. Cholesterol binding agents such as filipin, which remove cholesterol from the plasma membrane, cause caveolae to disassemble, which reduces the cell surface density of caveolae and thereby produces a significant reduction in the scavenger endocytosis of modified albumins, transcytosis of insulin and native albumin, and even transcapillary permeability of albumin in the rat lung (138). Moreover, such transport by caveolae is also sensitive to alkylation with N-ethylmaleimide (NEM) (110, 139), suggesting by analogy with other vesicular carrier systems a dependence on specific NEM-sensitive factors that mediate caveolae formation, docking, and/or fusion with target membranes.

It is starting to become clear that caveolae contain the molecular machinery required for them to be specific discrete carrier vesicles (132, 133, 139). Transport by discrete vesicular carriers is well established for intracellular membrane trafficking largely because of recent discoveries about the molecular mediators of vesicle formation, docking, and fusion. A general mechanism sensitive to NEM is required for the transport of a divergent group of vesicular carriers ranging from the specialized synaptic vesicles to the more common exocytic and endocytic carriers. Using the method described earlier for purifying endothelial caveolae, it has been shown that caveolae are indeed like other carrier vesicles and utilize similar NEM-sensitive molecular machinery for transport (132, 133, 139). Endothelial caveolae have key proteins that mediate different aspects of vesicle formation, docking, or fusion including the vSNARE, VAMP, or cellubrevin; small and large GTP binding proteins; the calcium-dependent, lipid binding protein annexin II and VI; and the NEM-sensitive factor NSF along with SNAP (132).

The purified caveolae are also likely to be transport vesicles because they contain not only the important blood carrier protein, albumin, which appears to be transcytosed by endothelial caveolae in situ (48, 94, 168), but also albondin, which mediates albumin's specific binding and transport in endothelium (135). Native albumin interacts with the endothelial cell surface and localizes preferentially within caveolae for apparent transport into or across the endothelium. Many continuous endothelia appear to have albumin binding sites for specific transcytosis via an abundant population of caveolae. Ligands bound to albumin, such as fatty acids, may increase its endothelial binding and transcytosis (41) in agreement with albumin's nutritional role as a blood carrier protein. Antibodies raised to albondin inhibited the specific endothelial transport and capillary permeability of albumin and also the binding of albumin both to the endothelial cell surface in culture and to its equivalent in situ, the purified luminal endothelial cell membrane. Normally, in the presence of nonimmune IgG as a control, BSA perfused in situ binds the endothelial cell surface as detected in the luminal membranes purified afterward from the lung tissue using the silica-coating methodology. Moreover, purification of the caveolae from this membrane reveals that about 85% of the bound BSA is found in the purified caveolae. Anti-albondin IgG before and during the in situ BSA perfusion inhibited BSA binding to both the purified cell membrane and caveolae by more than 90% (146a). About 75–80% of albondin is concentrated in the caveolae, whereas 20–25% remains free at the cell surface. Ultimately, in these experiments, the anti-albondin IgG inhibited BSA transport and accumulation in the total lung tissue by nearly 75%. It appears that albondin expressed primarily within caveolae may facilitate the efficient delivery across the endothelium of albumin as a carrier for its ligands to meet the nutritional requirements of the underlying tissue cells.

By purifying caveolae to homogeneity directly from luminal endothelial plasmalemma derived from rat lungs, it has been demonstrated that caveolae represent: i) specific microdomains on the endothelial cell surface that have their own molecular topography consisting of certain preferentially distributed proteins including caveolin, albondin, Ca^{2+}-ATPase, and an IP_3 receptor but not other cell surface proteins; and ii) transport vesicles with the necessary specific molecular machinery for the select binding and transcytosis of albumin. Experiments now must be performed to define the precise role of these proteins in transport. Although the molecular mapping and pharmacological studies indicate that caveolae are necessary for the observed transport and appear to utilize mechanisms similar to other well-defined vesicular carriers, it still remains a challenge of the future to show directly that caveolae work as discrete carriers by budding from the plasma membrane, and to define the transport mechanism and how it is regulated. If caveolae do indeed function in transport similar to other discrete vesicular carriers capable of targeted delivery of their molecular cargo, then the molecules organized in caveolae need to be specifically manipulated to demonstrate directly the budding, docking, and fusion of caveolar vesicles. Until this type of evidence is produced, the controversy will continue. The next decade may provide the answer!

Other Implied Functions

The results discussed here have important implications in elucidating the function of caveolae not only in transport as discussed but also in other areas. The finding that both the plasmalemmal Ca^{2+}-ATPase and inositol trisphosphate receptors are concentrated within caveolae (137) suggests a physiological role for this invaginated microdomain in the regulation of intracellular Ca^{2+} concentrations. Caveolae can associate closely with Ca^{2+} storage compartments in the cells including ER in endothelium and SR in smooth muscle cells (11). Like the SR and ER, caveolae may have the ability to transport Ca^{2+} in both directions with a Ca^{2+} pump for removal and an inositol trisphosphate-activated channel for Ca^{2+} influx. By controlling their activity via the inositol phospholipid signaling pathway (inositol trisphosphate and its metabolites), the plamsalemmal Ca^{2+} channel may be opened concomitantly with inhibition of the Ca^{2+} pump (27, 77, 83), resulting in net Ca^{2+} influx through the caveolae that may be localized in its effects or may alter cytosolic Ca^{2+} levels more globally. Inositol 1,4,5-trisphosphate appears to inhibit Ca^{2+}-ATPase through G proteins (26), which also can be found in caveolae (132). Such microenvironmental changes may, as in the case of synaptic vesicles, regulate the fission–fusion process essential to discrete vesicular transport between the vesicular carrier, in this case the caveolae, and its source and target site, the plasmalemma.

Because calcium levels play an important role in modulating the endothelial barrier function (18, 20), it is of particular interest with regard to caveolae that albumin via its interaction with endothelium prevents transient increases in intracellular calcium levels and increases capillary permselectivity in general (58). The presence within caveolae of albumin and albondin along with Ca^{2+} transporters suggests that caveolae may be the surface microdomains containing the molecular machinery responsible for this behavior through local modulation of Ca^{2+} transport and signalling that ultimately regulates capillary permeability. Lastly, although Ca^{2+} has been found concentrated within caveolae (155), the molecular mechanism for this accumulation is unclear. Interestingly, albondin shares immunological epitopes with another albumin binding protein named SPARC that is secreted and known to bind Ca^{2+} with high and low affinity (117, 134, 135, 155).

Summary

In most organs, the endothelium is the critical barrier preventing the passage of molecules and cells circulating in the blood to the underlying tissue cells. When this barrier is disrupted and the endothelium becomes dysfunctional, tissue homeostasis is disturbed and various pathologies may ensue, including atherosclerosis, cerebral strokes, coagulopathies, ischemia, diabetic microangiopathy, and even hypertension. Conversely, the endothelium must overcome its inherent necessary restrictiveness by utilizing specific transport mechanisms to provide essential nutrients to the underlying tissue cells in accordance with their particular metabolic requirements. Defining the interactions between the endothelial cell surface and circulating blood molecules, especially with regards to transvascular transport, has been rather difficult. The development of techniques for isolating and growing endothelial cells in culture allowed some important characterizations of these processes. Unfortunately, it has become quite evident that endothelial cells change when grown in vitro, creating an apparently "de-differentiated" phenotype that is much different both morphologically and functionally from that found in vivo. For instance, many continuous endothelia lose nearly all of their noncoated plasmalemmal vesicles or caveolae when grown in culture. Hence, it is necessary for us to refocus our attention on the endothelium in tissue. With this in mind, a process has been developed for purifying the luminal endothelial cell surface membrane directly from tissue. This plasma membrane, directly in contact with the circulating blood, represents the critical interface for specific transport and can be isolated using an in situ silica-coating procedure followed by tissue homogenization and centrifugation. Furthermore, the caveolae have been separated from this membrane and purified. Using these membrane fractions, we have begun to carry out "molecular mapping" studies defining the molecules comprising the endothelial cell surface and its transport structures. In combination with functional transport assays, key proteins involved in the specific transport of water, calcium, insulin, and albumin have been identified in certain endothelia. For example, we found that a mercurial-sensitive transmembrane water channel called aquaporin appears to mediate water uptake in the rat lung. Other work has shown that molecules such as glucose and transferrin utilize, at least in the brain, specific transport proteins for uptake. Moreover, specific proteins have been found in the caveolae that appear to function in ligand binding and in the regulated budding, docking, and fusion of this vesicular carrier. Caveolae are rich in albondin, insulin receptor, GTP-binding proteins, and annexins. In situ perfusion of albumin through rat lung shows that antibodies to albondin inhibit its binding within caveolae and its tissue uptake. Caveolae share the basic NEM-sensitive transport machinery operable for a very divergent group of other vesicular carriers ranging from the specialized synaptic vesicles mediating neurotransmission to the more common exocytic and endocytic carriers found in many cells. This includes NSF, SNAP, and VAMPs and permits the caveolae to endocytose or transcytose specific bound ligands. Although we are only beginning to unravel the mysteries of this processing system at the endothelial cell surface, the current data implicate a regulated mechanism for the transport of molecules as ubiquitous and small as water and glucose to the much larger and less plentiful macromolecules. It may be necessary to begin to reevaluate the current "dogma" about the pathways relevant to transport across continuous endothelia and to incorporate the physiologically significant contribution of these newly discovered pathways into the current models of transcapillary exchange.

References

1. Adamson, R. H. Permeability of frog mesenteric capillaries after partial pronase digestion of the endothelial glycocalyx. *J. Physiol. (London)* 428:1–13, 1990.
2. Adamson, R. H. and G. Clough. Plasma proteins modify the endothelial glycocalyx of frog mesenteric microvessels. *J. Physiol. (London)* 455:473–486, 1992.
3. Agre, P., G. M. Preston, B. L. Smith, J. S. Jung, S. Raina, C. Moon, W. B. Guggino, and S. Nielsen. Aquaporin CHIP: the archetypal molecular water channel. *Am. Physiol.* 265:F463–F476, 1993.
4. Bankston, P. W., G. A. Porter, A. J. Milici, and G. E. Palade. Differential and specific labeling of epithelial and vascular endothelial cells of the rat lung by Lycopersicon esculentum and Griffonia simplicifolia I lectins. *Eur. J. Cell Biol.* 54:187–195, 1991.
5. Belaiba, R., P. Riant, S. Urien, F. Bree, E. Albengres, J. Barre, and J. P. Tillement. Blood binding and tissue transfer of drugs: the influence of α_1-acid glycoprotein binding. In: *Progress in Clin. and Biol. Res., Alpha$_1$-Acid Glycoprotein: Genetics, Biochemistry, Physiological Functions, and Pharmacology,* edited by P. Baumann, C. B. Eap, W. E. Müller, and J.-P. Tillement. New York: Alan R. Liss, Inc., 1989, pp. 287–305.
6. Belloni, P. N. and G. L. Nicolson. Differential expression of cell surface glycoproteins on various organ-derived microvascular endothelia and endothelial cell cultures. *J. Cellular Physiol.* 136:398–410, 1988.
7. Berthiaume, F. and J. A. Frangos. Flow-induced prostacyclin production is mediated by a pertussin toxin-sensitive G protein. *FEBS Lett.* 308:277–279, 1992.
8. Brown, M. S. and J. L. Goldstein. Lipoprotein metabolism in the macrophage: Implications for cholesterol deposition in artherosclerosis. *Annu. Rev. Biochem.* 52:223–261, 1983.
9. Bruns, R. R. and G. E. Palade. Studies on blood capillaries. I. General organization of blood capillaries in muscle. *J. Cell Biol.* 37:244–276, 1968.
10. Bumbasirevic, V., G. D. Pappas, and R. P. Becker. Endocytosis of serum albumin-gold conjugates by microvascular endothelial cells in rat adrenal gland: Regional differences between cortex and medulla. *J. Submicrosc. Cytol. Pathol.* 22:135–145, 1990.
11. Bundgaard, M. The three-dimensional organization of smooth endoplasmic reticulum in capillary endothelia: its possible role in the regulation of free cytosolic calcium. *J. Struct. Biol.* 107:76–85, 1991.
12. Bundgaard, M. and J. Frokjaer-Jensen. Functional aspects of the ultrastructure of terminal blood vessels: A quantitative study on conservative segments of the frog mesenteric microvasculature. *Microvasc. Res.* 23:1–30, 1982.
13. Bundgaard, M., J. Frokjaer-Jensen, and C. Crone. Endothelial plasmalemmal vesicles

as elements in a system of branching invaginations from the cell surface. *Proc. Natl. Acad. Sci. USA* 76:6439–6442, 1979.

14. Bundgaard, M., P. Hagman, and C. Crone. The three-dimensional organization of plasmalemmal vesicular profiles in the endothelium of rat heart capillaries. *Microvasc. Res.* 25:358–368, 1983.

15. Carley, W. W., A. J. Milici, and J. A. Madri. Extracellular matrix specificity for the differentiation of capillary endothelial cells. *Exp. Cell Res.* 178:426–434, 1988.

16. Chambers, R. and B. W. Zweifach. Intercellular cement and capillary permeability. *Physiol. Rev.* 27:436–463, 1947.

17. Crone, C. Tight and leaky endothelia. In: *Water Transport Across Epithelia,* edited by H. H. Ussing et al., Copenhagen: Munksgaard, 1981, pp. 259–267.

18. Crone, C. Modulation of solute permeability in microvascular endothelium. *Fed. Proc.* 45:77–83, 1986.

19. Curry, F. E. Effect of albumin on the structure of the molecular filter at the capillary wall. *Fed. Proc.* 44:2610–2613, 1985.

20. Curry, F. E. Modulation of venular microvessel permeability by calcium influx into endothelial cells. *FASEB J.* 6:2456–2466, 1992.

21. Curry, F. E. and C. C. Michel. A fiber matrix model of capillary permeability. *Microvasc. Res.* 20:96–99, 1980.

22. Curry, F. E., C. C. Michel, and M. E. Philips. Effect of albumin on the oncotic pressure exerted by myoglobin across capillary walls in frog mesentery. *J. Physiol.* 387:69–82, 1987.

23. Curry, F. E., J. E. Rutledge, and J. F. Lenz. Modulation of microvessel wall charge by plasma glycoprotein orosomucoid. *Am. J. Physiol.* 257:H1354–1359, 1989.

24. Czartolomna, J., N. F. Voelkel, and S.-W. Chang. Permeability characteristics isolated perfused rat lungs. *J. Appl. Physiol.* 70:1854–1860, 1991.

25. Danielli, J. F. Capillary permeability and oedema in the perfused frog. *J. Physiol.* 98:109–129, 1940.

26. Davis, F. B., P. J. Davis, S. D. Blas, and D. Z. Gombas. Inositol phosphates modulate red blood cell Ca(2+)-adenosine triphosphatase activity in vitro by a guanine nucleotide regulatory protein. *Metabolism* 44:865–868, 1995.

27. Davis, F. B., P. J. Davis, W. D. Lawrence, and S. D. Blas. Specific inositol phosphates inhibit basal and calmodulin-stimulated Ca(2+)-ATPase activity in human erythrocyte membranes in vitro and inhibit binding of calmodulin to membranes. *FASEB J.* 5:2992–2995, 1991.

28. De Bruyn, P. P. H., S. Michelson, and P. W. Bankston. In vivo endocytosis by bristle-coated pits and intracellular transport of endogenous albumin in the endothelium of sinuses of liver and bone marrow. *Cell Tissue Res.* 240:1–7, 1985.

29. Desjardins, C. and B. R. Duling. Heparinase treatment suggests a role for the endothelial glycocalyx in regulation of capillary hematocrit. *Am. J. Physiol.* 258:H247–54, 1990.

30. Drinker, C. K. The permeability and diameter of the capillaries in the web of the brown frog (*R. temporaria*) when perfused with solutions containing pituitary extract and horse serum. *J. Physiol.* 63:249–269, 1927.

31. Dvorak, H. F., L. F. Brown, M. Detmar, and A. M. Dvorak. Vascular permeability factor/vascular endothelial growth factor, microvascular hyperpermeability, and angiogenesis. *Am J. Pathol.* 146:1029–1039, 1995.

32. Dvorak, H. F., J. A. Nagy, J. T. Dvorak, and A. M. Dvorak. Identification and

characterization of the blood vessels of solid tumors that are leaky to circulating macromolecules. *Am. J. Pathol.* 133:95, 1988.

33. Engel, A., T. Walz, and P. Agre. The aquaporin family of membrane water channels. *Curr. Opin. Struct. Bio.* 4:545–553, 1994.

34. Fajardo, L. F. The complexity of endothelial cells. *Am. J. Clin. Pathol.* 92:241–250, 1989.

35. Farrell, C. L. and W. M. Pardridge. Blood-brain barrier glucose transporter is asymmetrically distributed on brain capillary endothelial lumenal and ablumenal membranes: An electron microscopic immunogold study. *Proc. Natl. Acad. Sci. USA* 88:5779–5783, 1991.

36. Finkelstein, A. *Water Movement Through Lipid Bilayers, Pores, and Plasma Membranes. Theory and Reality.* New York: Wiley, 1987.

37. Folkesson, G. H., M. A. Matthay, H. Hasegawa, F. Kheradmand, and A. S. Verkman. Transcellular water transport in lung alveolar epithelium through mercury-sensitive water channels. *Proc. Natl. Acad. Sci. USA* 91:4970–4974, 1994.

38. Frokjaer-Jensen, J. Three-dimensional organization of plasmalemmal vesicles in endothelial cells. An analysis by serial sectioning of frog mesenteric capillaries. *J. Ultrastructure Res.* 73:9–20, 1980.

39. Frokjaer-Jensen, J. The endothelial vesicle system in cryofixed frog mesenteric capillaries analysed by ultrathin serial sectioning. *J. Electron Microscopy Tech.* 19:291–304, 1991.

40. Frokjaer-Jensen, J., R. C. Wagner, S. B. Andrews, P. Hagman, and T. S. Reese. Three-dimensional organization of the plasmalemmal vesicular system in directly frozen capillaries of the rete mirabile. *Cell Tissue Res.* 254:17–24, 1988.

41. Galis, Z., L. Ghitescu, and M. Simionescu. Fatty acid binding to albumin increases its uptake and transcytosis by the lung capillary endothelium. *Eur. J. Cell Biol.* 47:358–365, 1988.

42. Gamble, J. Influence of pH on capillary filtration coefficient of rat mesenteries perfused with solutions containing albumin. *J. Physiol. Lond.* 387:69–82, 1983.

43. Garcia-Cardena, G., P. Oh, J. Liu, J. E. Schnitzer, and W. C. Sessa. Targeting of nitric oxide synthase to endothelial cell caveolae via palmitoylation: Implication for nitric oxide signaling. *Proc. Natl. Acad. Sci. USA* 93:6448–6453, 1996.

44. Geoffrey, J. S. and R. P. Becker. Endocytosis by endothelial phagocytes: Uptake of bovine serum albumin-gold conjugates in bone marrow. *J. Ultrastructural Res.* 89:223–239, 1984.

45. Ghinea, N., M. Eskenasy, M. Simionescu, and N. Simionescu. Endothelial albumin binding proteins are membrane-associated components exposed on the cell surface. *J. Biol. Chem.* 264:4755–4758, 1989.

46. Ghinea, N., A. Fixman, D. Alexandru, D. Popov, M. Hasu, L. Ghitescu, M. Eskenasy, M. Simionescu, and N. Simionescu. Identification of albumin-binding proteins in capillary endothelial cells. *J. Cell Biol.* 107:231–239, 1988.

47. Ghinea, N., M. T. V. Hai, M.-T. Groyer-Picard, and E. Milgrom. How protein hormones reach their target cells. Receptor-mediated transcytosis of hCG through endothelial cells. *J. Cell Biol.* 125:87–97, 1994.

48. Ghitescu, L. and M. Bendayan. Transendothelial transport of albumin: A quantitative immunocytochemical study. *J. Cell Biol.* 117:745–755, 1992.

49. Ghitescu, L., A. Fixman, M. Simionescu, and N. Simionescu. Specific binding sites for albumin restricted to plasmalemmal vesicles of continuous capillary endothelium: Receptor-mediated transcytosis. *J. Cell Biol.* 102:1304–1311, 1986.

50. Goldstein, J. L. and M. S. Brown. Binding and degradation of low density lipoproteins by cultured human fibroblasts: Comparison of cells from a normal subject and from a patient with homozygous familial hypercholesterolemia. *J. Biol. Chem.* 249:5153–5162, 1974.

51. Goldstein, J. L., M. S. Brown, R. G. W. Anderson, D. N. Russel, and W. J. Schneider. Receptor-mediated endocytosis: Concepts emerging from the LDL receptor system. *Ann. Rev. Cell Biol.* 1:1–39, 1985.

52. Haberland, M. E., R. R. Rasmussen, C. L. Olch, and A. M. Fogelman. Two distinct receptors account for recognition of maleyl-albumin in human monocytes during differentiation in vitro. *J. Clin. Invest.* 77:681–689, 1986.

53. Haraldsson, B. Physiological studies of macromolecular transport across capillary walls. *Acta Physiol. Scand. Suppl.* 553:1–40, 1986.

54. Haraldsson, B. and B. Rippe. Serum factors other than albumin are needed for the maintenance of normal capillary permselectivity in rat hindlimb muscle. *Acta Physiol. Scand.* 123:427–436, 1985.

55. Haraldsson, B. and B. Rippe. Orosomucoid as one of the serum components contributing to normal permselectivity in rat skeletal muscle. *Acta Physiol. Scand.* 129:127–135, 1987.

56. Hasegawa, H., S.-C. Lian, W. E. Finkbeiner, and A. S. Verkman. Extrarenal tissue distribution of CHIP28 water channels by in situ hybridization and antibody staining. *Am. J. Physiol.* 266:C893–C903, 1994.

57. Hasegawa, H., T. Ma, W. Skach, M. A. Matthay, and A. S. Verkman. Molecular cloning of a mercurial-insensitive water channel expressed in selected water-transporting tissues. *J. Biol. Chem.* 269:5497–5500, 1994.

58. He, P. and F. E. Curry. Albumin modulation of capillary permeability—Role of endothelial cell [Ca++]. *Am. J. Physiol.* 265:H74–H82, 1993.

59. Hennig, B., D. M. Shasby, A. B. Fulton, and A. A. Spector. Exposure to free fatty acids increases the transfer of albumin across cultured endothelial monolayers. *Arteriosclerosis* 4:489–497, 1984.

60. Horiuchi, S., K. Takata, and Y. Morino. Characterization of a membrane-associated receptor from rat sinusoidal liver cell that binds formaldehyde-treated serum albumin. *J. Biol. Chem.* 260:475–481, 1985.

61. Hutter, J. F., H. M. Piper, and P. G. Spieckermann. Myocardial fatty acid oxidation: Evidence for an albumin-receptor-mediated transfer of fatty acids. *Basic Res. Cardiol.* 79:274–282, 1984.

62. Huxley, V. H. and F. E. Curry. Albumin modulation of capillary permeability: Test of an adsorption mechanism. *Am. J. Physiol.* 248:H264–H273, 1985.

63. Huxley, V. H. and F. E. Curry. Differential actions of albumin and plasma on capillary solute permeability. *Am. J. Physiol.* 260:H1645–H1654, 1991.

64. Jacobson, B. S., J. E. Schnitzer, and G. E. Palade. Isolation and partial characterization of the luminal plasmalemma of microvascular endothelium from rat lungs. *Eur. J. Cell Biol.* 58:296–306, 1992.

65. Jaffe, E. A. Cell biology of endothelial cells. *Hum. Pathol.* 18:234–239, 1987.

66. Jeffries, W. A., M. R. Brandon, S. V. Hunt, A. F. Williams, K. C. Gatter, and D. Y. Mason. Transferrin receptor on endothelium of brain capillaries. *Nature* 312:162–163, 1984.

67. Johansson, B. R. Size and distribution of endothelial plasmalemmal vesicles in consecutive segments of the microvasculature in cat skeletal muscle. *Microvasc. Res.* 17:107–117, 1979.

68. Judd, R. L., M. G. Sarr, and J. M. Miles. Role of albumin in extravascular transport of free fatty acids. *FASEB J.* 6:A1495, 1992.

69. Kern, D. F., D. Levitt, and D. Wangensteen. Endothelial albumin permeability measured with a new technique in perfused rabbit lung. *Am. J. Physiol.* 245:H229–H236, 1983.

70. King, G. L. and S. M. Johnson. Receptor-mediated transport of insulin across endothelial cells. *Science* 227:1583–1586, 1985.

71. Kohn, S., J. Nagy, H. Dvorak, and A. Dvorak. Pathways of macromolecular tracer transport across venules and small veins. Structural basis for the hyperpermeability of tumor blood vessels. *Lab. Invest.* 67:596, 1992.

72. Kragh-Hansen, U. Molecular aspects of ligand binding to serum albumin. *Pharmacol. Res.* 33:17–53, 1981.

73. Kramer, R. H. and G. L. Nicolson. Interactions of tumor cells with vascular endothelial cell monolayers: a model for metastatic invasion. *Proc. Natl. Acad. Sci. USA* 76:5704–5708, 1979.

74. Kremer, J. M. H., J. Wilting, and L. H. M. Janssen. Drug binding to human alpha-1-acid glycoprotein in health and disease. *Pharmacological Reviews* 40:1–47, 1988.

75. Krough, A. and G. A. Harrop. Some observations on stasis and oedema. *J. Physiol.* 54:125–126, 1921.

76. Kuchan, M. J., H. Jo, and J. A. Frangos. Role of G proteins in shear stress-mediated nitric oxide production by endothelial cells. *Am. J. Physiol.* 267:C753–C758, 1994.

77. Kuno, M. and P. Gardner. Ion channels activated by inositol 1,4,5-trisphosphate in plasma membrane of human T-lymphocytes. *Nature* 326:301–304, 1987.

78. Landis, E. M. and J. R. Pappenheimer. Exchange of substances through the capillary walls. In: *Handbook of Physiology, Sect. 2, Circulation II,* Washington, DC: Am. Physiol. Soc., 1963, 961–1034.

79. Levick, J. R. and C. C. Michel. The effect of bovine albumin on the permeability of frog mesenteric capillaries. *Q. J. Exper. Physiol. Cogn. Med. Sci.* 58:87–97, 1973.

80. Levitt, D. G. Routes of membrane water transport: comparative physiology. In: *Water Transport Across Epithelia,* edited by H. H. Ussing et al., Copenhagen: Munksgaard, 1981, pp. 248–257.

81. Liu, J., P. Oh, T. Horner, R. A. Rogers, and J. Schnitzer. Organized endothelial cell surface signal transduction in caveolae distinct from GPI-anchored protein microdomains. *J. Biol. Chem.* 272:7211–7222, 1997.

82. Loo, D. D. F., T. Zeuthen, G. Chandy, and E. M. Wright. Cotransport of water by the Na+/glucose contransporter. *Proc. Natl. Acad. Sc. (USA)* 93:13367–13370, 1996.

83. Luckhof, A. and D. E. Clapham. Inositol 1,3,4,5-tetrakisphosphate activates an endothelial Ca(2+)-permeable channel. *Nature (Lond.)* 355:356–358, 1992.

84. Macey, R. I. Transport of water and urea in red blood cells. *Am. J. Physiol.* 246:C195–C203, 1984.

85. Madri, J. A. and S. K. Williams. Capillary endothelial cell culture: Phenotype modulation by matrix components. *J. Cell Biol.* 97:153–165, 1983.

86. Mann, G. E. Alterations of myocardial capillary permeability by albumin in the isolated, perfused rabbit heart. *J. Physiol. Lond.* 319:311–323, 1981.

87. McGuire, P. G. and T. A. Twietmeyer. Morphology of rapidly frozen endothelial cells. Gluteraldehyde fixation increases the number of caveolae. *Circ. Res.* 53:424–429, 1983.

88. McNamara, P. J., R. C. Jewell, and M. N. Gillespie. The interaction of alpha-1-acid glycoprotein with endogenous autocoids, in particular, platelet activating factor

(PAF). In: *Progress in Clin. and Biol. Res., Alpha₁-Acid Glycoprotein: Genetics, Biochemistry, Physiological Functions, and Pharmacology,* edited by P. Baumann, C. B. Eap, W. E. Müller, and J.-P. Tillement. New York: Alan R. Liss, Inc., 1989, pp. 307–319.

89. Merker, M., W. W. Carley, and G. L. Gillis. In situ iodination of angiotensin-converting enzyme and other pulmonary endothelial membrane proteins. Biochem. Pharm. 38:983–992, 1989.

90. Michel, C. C. Filtration coefficients and osmotic reflexion coefficients of the walls of single frog mesenteric capillaries. *J. Physiol. Lond.* 309:341–355, 1980.

91. Michel, C. C., M. E. Phillips, and M. R. Turner. The effects of native and modified bovine serum albumin on the permeability of frog mesenteric capillaries. *J. Physiol. Lond.* 360:333–346, 1985.

92. Milici, A. J., M. B. Furie, and W. W. Carley. The formation of fenestrations and channels by capillary endothelium in vitro. *Proc. Natl Acad Sci USA* 82:6181–6185, 1985.

93. Milici, A. J. and G. A. Porter. Lectin and immunolabeling of microvascular endothelia. *J. Electron Microsc. Tech.* 19:305–315, 1991.

94. Milici, A. J., N. E. Watrous, H. Stukenbrok, and G. E. Palade. Transcytosis of albumin in capillary endothelium. *J. Cell Biol.* 105:2603–2612, 1987.

95. Myhre, K. and J. B. Steen. The effect of plasma proteins on the capillary permeability in the rete mirabile of the eel (*Anguilla vulgaris* L.). *Acta Physiol. Scand.* 99:98–104, 1977.

96. Neal, C. R. and C. C. Michel. Transcellular openings through microvascular walls in acutely inflamed frog mesentery. *Exper. Physiol.* 77:917–920, 1992.

97. Nielsen, S., B. L. Smith, E. I. Christensen, and P. Agre. Distribution of the aquaporin CHIP in secretory and resportive epithelia and capillary endothelia. *Proc. Natl. Acad. Sci. USA* 90:7275–7279, 1993.

98. Ockner, R. K., R. A. Weisiger, and J. L. Gollan. Hepatic uptake of albumin-bound substances: albumin receptor concept. *Am. J. Physiol.* 245:G13–G18, 1983.

99. Palade, G. E. Transport in quanta across the endothelium of blood capillaries. *Anat. Rec.* 136:254, 1960.

100. Palade, G. E. The microvascular endothelium revisited. In: *Endothelial Cell Biology in Health and Disease,* edited by N. Simionescu and M. Simionescu, New York & London: Plenum Press, 1988, pp. 3–22.

101. Palade, G. E. and R. R. Bruns. Structural modulations of plasmalemmal vesicles. *J. Cell Biol.* 37:633–649, 1968.

102. Pappenheimer, J. R., E. M. Renkin, and L. M. Borrero. Filtration diffusion and molecular seiving through peripheral capillary membranes. A contribution to the pore theory of capillary permeability. *Am. J. Physiol.* 167:13–46, 1951.

103. Pardridge, W. M., J. Eisenberg, and W. T. Cefalu. Absence of albumin receptor on brain capillaries in vivo or in vitro. *Am. J. Physiol.* 249:E264–E267, 1985.

104. Peters, K.-R., C. C. Carley, and G. E. Palade. Endothelial plasmalemmal vesicles have a characteristic stripped bipolar surface structure. *J. Cell Biol.* 101:2233, 1985.

105. Peters, T., Jr. Serum albumin. *Adv. Prot. Chem.* 37:161–245, 1985.

106. Pino, R. M. The cell surface of a restrictive fenstrated endothelium I. Distribution of lectin-receptor monosaccharides on the choriocapillaries. *Cell Tissue Res.* 243:145–155, 1986.

107. Pino, R. M. and C. L. Thouron. Identification of lectin-receptor monosaccharides on the endothelium of retinal capillaries. *Curr. Eye Res.* 5:625–628, 1986.

108. Porter, G. A., G. E. Palade, and A. J. Milici. Differential binding of lectins Griffonia simplicifolia I and Lycopersicon esculentum to microvascular endothelium: organ-specific localization and partial glycoprotein characterization. *Eur. J. Cell Biol.* 51:85–95, 1990.

109. Powers, M. R., F. A. Blumenstock, J. A. Cooper, and A. B. Malik. Role of albumin arginyl sites in albumin-induced reduction of endothelial hydraulic conductivity. *J. Cellular Physiol.* 141:558–564, 1989.

110. Predescu, D., R. Horvat, S. Predescu, and G. E. Palade. Transcytosis in the continuous endothelium of the myocardial microvasculature is inhibited by N-ethylmaleimide. *Proc. Natl. Acad. Sci. USA* 91:3014–3018, 1994.

111. Predescu, D. and G. E. Palade. Plasmalemmal vesicles represent the large pore system of continuous microvascular endothelium. *Am. J. Physiol.* 265:H725–H733, 1993.

112. Reed, R. G. and C. M. Burrington. The albumin receptor effect may be due to a surface-induced conformational change in albumin. *J. Biol. Chem.* 264:9867–9872, 1989.

113. Renkin, E. M. Capillary transport of macromolecules: pores and other pathways. *J. Appl. Physiol.* 134:375–382, 1985.

114. Roberts, W. G. and G. E. Palade. Increased microvascular permeability and endothelial fenestration induced by vascular endothelial growth factor. *J. Cell Sci.* 108:2369–2379, 1995.

115. Rothberg, K. G., J. E. Heuser, W. D. Donzell, Y.-S. Ying, J. R. Glenney, and R. G. W. Anderson. Caveolin, a protein component of caveolae membrane coats. *Cell* 68:673–682, 1992.

116. Sage, E. H. and P. Bornstein. Extracellular proteins that modulate cell-matrix interactions. *J. Biol. Chem.* 266:14831–14834, 1991.

117. Sage, H., C. Johnson, and P. Bornstein. Characterization of a novel serum albumin-binding glycoprotein secreted by endothelial cells in culture. *J. Biol. Chem.* 259:3993–4007, 1984.

118. Schmid, K. Human plasma α_1-acid glycoprotein. In: *Progress in Clin. and Biol. Res., Alpha₁-Acid Glycoprotein: Genetics, Biochemistry, Physiological Functions, and Pharmacology,* edited by P. Baumann, C. B. Eap, W. E. Müller, and J.-P. Tillement. New York: Alan R. Liss, Inc., 1989, pp. 7–22.

119. Schneeberger, E. E. and M. Hamelin. Interaction of serum proteins with lung endothelial glycocalyx: Its effect on endothelial permeability. *Am. J. Physiol.* 247:H206–H217, 1984.

120. Schneeberger, E. E., L. R. D., and B. A. Neary. Interaction of native and chemically modified albumin with pulmonary microvascular endothelium. *Am. J. Physiol.* 258:L89–98, 1990.

121. Schnitzer, J. E. Analysis of steric partition behavior of molecules in membranes using statistical physics: Application to gel chromatography and electrophoresis. *Biophys. J.* 54:1065–1076, 1988.

122. Schnitzer, J. E. Glycocalyx electrostatic potential profile analysis: ion, pH, steric, and charge effects. *Yale J. Biol. Med.* 61:427–446, 1988.

123. Schnitzer, J. E. Fiber matrix model re-analysis: Matrix exclusion limits define effective pore radius describing capillary and glomerular permselectivity. *Microvasc. Res.* 43:342–346, 1992.

124. Schnitzer, J. E. Gp60 is an albumin binding glycoprotein expressed by continuous endothelium involved in albumin transcytosis. *Am. J. Physiol.* 31:H246–H254, 1992.

125. Schnitzer, J. E. Update on the cellular and molecular basis of capillary permeability. *Trends in Cardiovasc. Med.* 3:124–130, 1993.

126. Schnitzer, J. E. Molecular architecture of endothelial caveolae: Possible stress-sensing organelles. *Annals Biomed. Engineering* 23:S34, 1995.

127. Schnitzer, J. E. and J. Bravo. High affinity binding, endocytosis and degradation of conformationally-modified albumins: Potential role of gp30 and gp18 as novel scavenger receptors. *J. Biol. Chem.* 268:7562–7570, 1993.

128. Schnitzer, J. E. and W. W. Carley. Electrostatic and steric partition function of the endothelial glycocalyx. *Fed. Proc.* 45:1152a, 1986.

129. Schnitzer, J. E., W. W. Carley, and G. E. Palade. Albumin interacts specifically with a 60-kDa microvascular endothelial glycoprotein. *Proc. Natl. Acad. Sci. USA* 85:6773–6777, 1988.

130. Schnitzer, J. E., W. W. Carley, and G. E. Palade. Specific albumin binding to microvascular endothelium in culture. *Am. J. Physiol.* 254:H425–H437, 1988.

131. Schnitzer, J. E. and K. Lambrakis. Electrostatic potential and Born energy of charged molecules interacting with phospholipid membranes: Calculation via 3-D numerical solution of the full Poisson equation. *J. Theor. Biol.* 152:203–222, 1991.

132. Schnitzer, J. E., J. Liu, and P. Oh. Endothelial caveolae have the molecular transport machinery for vesicle budding, docking and fusion including VAMP, NSF, SNAP, annexins and GTPases. *J. Biol. Chem.*270:14399–14404, 1995.

133. Schnitzer, J. E., D. P. McIntosh, A. M. Dvorak, J. Liu, and P. Oh. Separation of caveolae from associated microdomains of GPI-anchored proteins. *Science* 269: 1435–1439, 1995.

134. Schnitzer, J. E. and P. Oh. Antibodies to SPARC inhibit albumin binding to SPARC, gp60 and microvascular endothelium. *Am. J. Physiol.* 263:H1872–H1879, 1992.

135. Schnitzer, J. E. and P. Oh. Albondin-mediated capillary permeability to albumin: Differential role of receptors in endothelial transcytosis and endocytosis of native and modified albumins. *J. Biol. Chem.* 269:6072–6082, 1994.

136. Schnitzer, J. E. and P. Oh. Aquaporin-1 in plasma membrane and caveolae provides mercury-sensitive water channels across lung endothelium. *Am. J. Physiol.* 270: H416–H422, 1996.

137. Schnitzer, J. E., P. Oh, B. S. Jacobson, and A. M. Dvorak. Caveolae from luminal plasmalemma for rat lung endothelium: Microdomains enriched in caveolin, Ca2+/ATPase and inositol trisphosphate receptor. *Proc. Natl. Acad. Sci. USA* 92:1759–1763, 1995.

138. Schnitzer, J. E., P. Oh, E. Pinney, and A. Allard. Filipin-sensitive caveolae-mediated transport in endothelium: Reduced transcytosis, scavenger endocytosis, and capillary permeability of select macromolecules. *J. Cell. Biol.* 127:1217–1232, 1994.

139. Schnitzer, J. E., P. Oh, E. Pinney, and J. Allard. NEM inhibits transcytosis, endocytosis and capillary permeability: Implication of caveolae fusion in endothelia. *Am. J. Physiol.* 37:H48–H55, 1995.

140. Schnitzer, J. E. and E. Pinney. Quantitation of specific binding of orosomucoid to cultured microvascular endothelium: Role in capillary permeability. *Am. J. Physiol.* 263:H48–H55, 1992.

141. Schnitzer, J. E., C.-P. J. Shen, and G. E. Palade. Lectin analysis of common glycoproteins detected on the surface of continuous microvascular endothelium in situ and in culture: Identification of sialoglycoproteins. *Eur. J. Cell Biol.* 52:241–251, 1990.

142. Schnitzer, J. E., A. Siflinger-Birnboim, P. J. Del Vecchio, and A. B. Malik. Segmental

differentiation of permeability, protein glycosylation, and morphology of cultured bovine lung vascular endothelium. *Biochem. Biophys. Res. Comm.* 199:11–19, 1994.

143. Schnitzer, J. E., A. Sung, R. Horvat, and J. Bravo. Preferential interaction of albumin binding proteins, gp30 and gp18, with modified albumins: Presence in many cells and tissues with a possible role in catabolism. *J. Biol. Chem.* 264:24544–24553, 1992.

144. Schnitzer, J. E., J. B. Ulmer, and G. E. Palade. A major endothelial plasmalemmal sialoglycoprotein, gp60, is immunologically related to glycophorin. *Proc. Nat. Acad. Sci. USA* 87:6843–6847, 1990.

145. Schnitzer, J. E., J. B. Ulmer, and G. E. Palade. Common peptide epitopes in glycophorin and the endothelial sialoglycoprotein gp60. *Biochem. Biophys. Res. Comm.* 187:1158–1165, 1992.

146. Schwartz, C. J. Pathophysiology of the atherogenic process. *Am. J. Cardiology* 64:23–30, 1989.

146a. Schnitzer, J. E., J. Liu, and P. Oh. Purification of endothelial caveolae reveals the molecular machinery for fusion-dependent, albondin-mediated transcytosis of albumin. *Microcirculation* 2:93, 1995.

147. Simionescu, M., N. Ghinea, A. Fixman, M. Lasser, L. Kukes, N. Simionescu, and G. E. Palade. The cerebral microvasculature of the rat: Structure and luminal surface properties during early development. *J. Submicrosc. Cytol. Pathol.* 20:243–261, 1988.

148. Simionescu, M., N. Simionescu, and G. E. Palade. Morphometric data on the endothelium of blood capillaries. *J. Cell Biol.* 60:128–152, 1974.

149. Simionescu, M., N. Simionescu, and G. E. Palade. Differentiated microdomains on the luminal surface of capillary endothelium: Distribution of lectin receptors. *J. Cell Biol.* 94:406–413, 1982.

150. Solomon, A. K., J. A. Dix, M. F. Lukacovic, M. R. Toon, and A. S. Verkman. The aqueous pore in the red cell membrane: Band 3 as a channel for anions, cations, nonelectrolytes, and water. *Ann. NY Acad. Sci.* 414:97–124, 1983.

151. Sparrow, C. P., S. Parthasarathy, and D. Steinberg. A macrophage receptor that recognizes oxidized low density lipoprotein but not acetylated low density lipoprotein. *J. Biol. Chem.* 264:2599–2604, 1989.

152. Starling, E. H. On the absorption of fluids from the connective tissue spaces. *J. Physiol. (London)* 19:312–326, 1896.

153. Steinberg, D., S. Parthasarathy, T. E. Carew, J. C. Khoo, and J. L. Witztum. Beyond cholesterol; modification of low-density lipoprotein that increase its atherogenicity. *New Engl. J. Med.* 320:915–924, 1989.

154. Sung, A., V. Rizzo, P. Oh, and J. E. Schnitzer. Rapid mechanotransduction occurs in situ at the endothelial cell surface primarily in caveolae. *Mol. Biol. Cell* 7:276a, 1996.

155. Suzuki, S. and H. Sugi. Evidence for extracellular localization of activator calcium in dog coronary artery smooth muscle as studies by the pyroantimonate method. *Cell Tissue Res.* 257:237–246, 1989.

156. Tavassoli, M., T. Kishimoto, and M. Kataoka. Liver endothelium mediates the hepatocyte's uptake of ceruloplasmin. *J. Cell Biol.* 102:1298–1303, 1986.

157. Taylor, A. E. and D. N. Granger. Exchange of macromolecules across the microcirculation. In: *Handbook of Physiology,* edited by E. M. Renkin and C. C. Michel, Bethesda, MD: Am. Physiol. Soc., 1984, pp. 467–520.

158. Tiruppathi, C., A. Finnegan, and A. B. Malik. Isolation and characterization of a cell surface albumin-binding protein from vascular endothelial cells. *Proc. Natl. Acad. Sci. USA* 93:250–254, 1996.

159. van Os, C. H., P. M. T. Deen, and J. A. Dempster. Aquaporins: Water selective channels in biological membranes. Molecular structure and tissue distribution. *Biochim. et Biophys. Acta* 1197:291–309, 1994.

160. Vasile, E., M. Simionescu, and N. Simionescu. Visualization of binding, endocytosis, and transcytosis of low-density lipoprotein in arterial endothelium in situ. *J. Cell Biol.* 96:1677–1689, 1983.

161. Verkman, A. S. Mechanisms and regulation of water permeability in renal epithelia. *Am. Physiol.* 257:C837–C850, 1989.

162. Villaschi, S., L. Johns, M. Cirigliano, and G. G. Pietra. Binding and uptake of native and glycosylated albumin-gold complexes in perfused rat lungs. *Microvasc. Res.* 32:190–199, 1986.

163. Vorbrodt, A. W., A. S. Lossinsky, D. H. Dobrogowska, and H. M. Wisniewski. Distribution of anionic sites and glycoconjugates on the endothelial surfaces of the developing blood-brain barrier. *Dev. Brain Res.* 29:69–79, 1986.

164. Wagner, R. C. and S. B. Andrews. Ultrastructure of the vesicular system in rapidly frozen capillary endothelium of the rete mirabile. *J. Ultrastructure Res.* 90:172–182, 1985.

165. Wagner, R. C., C. S. Robinson, P. J. Cross, and J. J. Devenney. Endocytosis and exocytosis of transferrin by isolated capillary endothelium. *Microvasc. Res.* 25:387–396, 1983.

166. Wagner, R. C. and C. S.-C. Transcapillary transport of solute by the endothelial vesicular system: Evidence from thin serial section analysis. *Microvasc. Res.* 42:139–150, 1991.

167. Weisiger, R. A., J. L. Gollan, and R. K. Ockner. Receptor for albumin on the liver cell surface may mediate uptake of fatty acids and other albumin-bound substances. *Science* 211:1048–1051, 1981.

168. Yokota, S. Immunocytochemical evidence for transendothelial transport of albumin and fibrinogen in rat heart and diaphragm. *Biomed. Res.* 4:577–586, 1983.

3
Study of Blood Capillary Permeability with the Rete Mirabile

Eugenio A. Rasio, Moïse Bendayan, and Carl A. Goresky

Introduction

Some animal species are equipped with a special vascular network, called rete mirabile, that consists of counterflowing parallel arterial and venous vessels (10). The rete provides an efficient transfer system capable of trapping heat or chemical compounds for adaptive purposes. Heat conservation is vital for arctic animals or tropical animals that are hypersensitive to a rapidly cooling milieu. In some fast-swimming fish predators, capillary nets are strategically located within dark muscles, thus increasing their power by conserving the released metabolic heat.

Another function of the countercurrent rete is manifested by the oxygen transfer from the maternal to the fetal blood in the placenta of some animals, or from the blood to the swimbladder of deep-sea fishes (38). In the eel, the number and dimensions of rete blood capillaries, the nonstaggered pattern of contact between arterial and venous capillaries, and the speed of blood flow are best suited for ensnaring oxygen in a most efficient way. At great water depths, the oxygen pressure in the bladder may reach 200 to 300 atmospheres, while it is only approximately one-fifth of an atmosphere in the eel blood and surrounding water. Such an enormous oxygen pressure difference across the swimbladder wall creates a large diffusion of oxygen into the capillary system of the swimbladder: The rate of swimbladder oxygen loss is linear with hydrostatic pressure (14). To avoid the rapid depletion of gas in the swimbladder, and to maintain the fish's neutral buoyancy, oxygen deposition in the swimbladder takes place through the countercurrent system of the rete. Acidification of the venous blood by lactate and CO_2 generated by nonoxidative glucose metabolism of the swimbladder epithelium, and their back-diffusion into the arterial blood, increase the oxygen tension by the Root effect and create high partial pressures of oxygen at the arterial output of the rete (15). Oxygen then is conveyed to the swimbladder microvasculature for diffusion into the cavity. To deflate the bladder, the gas is expelled by the mouth, through a valve that controls communication between the bladder and the esophagus.

The rete mirabile of the eel swimbladder (6) provides an interesting model for the study of the structure, metabolism, and function of the microvasculature and of their interrelationships in normal or pathological conditions. Capillary vessels can be readily incubated in vitro (23), without resorting to lengthy and potentially damaging isolation procedures such as those needed when capillaries are intermingled with contaminant cells of surrounding tissues. Large quantities of basement membrane material can be easily separated from endothelial cells and pericytes (7). For permeability studies (25), the rete offers significant advantages over whole-organ, single-capillary, or endothelial cell monolayer methodologies. Diffusion and osmotic methods used in whole-organ studies of permeability have provided valuable quantitative and qualitative data (2, 12). However, the reliability of whole-organ methods varies with the nature and size of the test solute because errors due to shunting, heterogeneous organization, backflux, or poor mixing are weighted differently in the equations used to measure permeability. That is not the case with the rete, where permeability can be assessed at once, and with accuracy, for sets of compounds spanning a wide range of shapes, sizes, charges, and water or lipid solubilities.

Single-capillary studies (16) enable the quantitative evaluation of permeability because the surface area of diffusion is known and the transcapillary concentration gradient is measurable. Rapid changes of capillary permeability can be detected with electrophysiological methods (18). However, exposure of the single capillary and the invasive procedure required to measure permeability may artifactually induce gaps or functional alterations in the endothelium. The preparative procedures of the rete do not involve handling of the capillary network.

Improved techniques for growing endothelial cells in culture gave rise to a wealth of permeability studies using monolayers of vascular endothelium from a variety of species and vascular sites. These studies have given impetus to the investigation of the effects and mechanisms of the action of a variety of physiological and pathological factors (40). The relevance of some of these in vitro observations to the in vivo situation is controversial. For instance, monolayer permeability to macromolecules is usually 10 to 100 times higher than that of continuous endothelium in vivo, due to the presence of small gaps between the cultured endothelial cells. Furthermore, the successful use of agents that reduce the porosity of the monolayer through changes of the endothelial cell cytoskeleton organization may have more to do with cell spreading and the subsequent reduction of gaps than with true regulatory function at the intercellular junction. Much in the same way as gaps in endothelial monolayers, arterio-venous shunts in the rete capillaries could spuriously increase permeability values and limit interpretation of the functional data. However, such shunts have never been seen, one obvious reason being that they would defeat the functional purpose of the rete.

Morphological Studies

A branch of the mesenteric artery ramifies at the cephalic pole of the rete into thousands of straight, parallel arterial capillaries, which may reach 1 cm in length in a mature silver eel. These capillaries, at the caudal pole of the rete, collect into

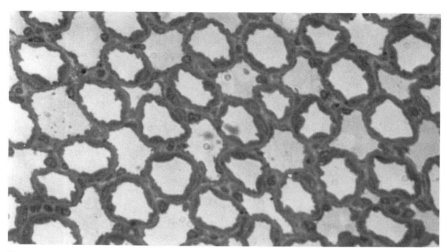

FIGURE 3.1. Light microscopy. Cross-section of the rete mirabile illustrating the large number of capillary vessels, their close apposition, and the absence of any other tissue component. Arterial capillaries alternate with venous capillaries. ×1,000.

an arteriole that provides the swimbladder wall with a diffuse capillary network from which venous blood is drained into a vein. Upon reaching the caudal pole of the rete, the vein divides into thousands of venous capillaries that alternate with the arterial capillaries in a manner that provides maximum contact area (Figure 3.1). At the cephalic pole of the rete, the venous capillaries assemble into a vein that reaches the portal vein.

The electron microscopic features, as studied by transmission and scanning microscopy as well as by freeze–fracture replicas of the capillary walls, are comparable to those observed in the capillaries of mammalian species (Figures 3.2–3.6).

The arterial capillaries of the rete are continuous and have high endothelial cells. The venous capillaries of the rete are fenestrated and the endothelial cells are thin. Both types of endothelium have tight intercellular junctions, a rich population of interconnected cytoplasmic vesicles, and well-defined basement membranes. The capillaries are arranged to gain the greatest possible area of exchange, in a non-staggered pattern that gives rise to a checkerboard. Consequently, the interstitial space is minimal: It includes pericytes, usually seen between leaflets of the arterial endothelial cells' basement membrane, and sparse bundles of collagen fibers (Figure 3.3). The total surface area to volume ratio of the rete is many times higher than that of the human lung and attests to its extraordinary capacity for exchange.

Metabolic Studies

Significant amounts of capillary tissue can be isolated and incubated in vitro (23). Their oxygen consumption is linear at a rate in the order of 100 µl/g wet weight/h,

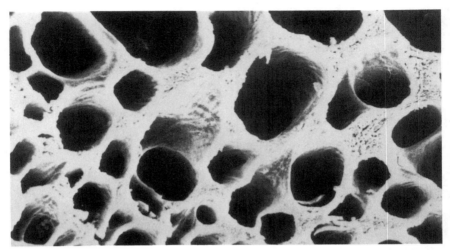

FIGURE 3.2. Scanning electron microscopy. Cross-section view of the rete capillaries. ×4,000.

for up to 3 h, at 37°C (eels can survive in water or air at temperatures ranging from 0 to 40°C). Glucose is the preferential energy substrate of endothelial cells and pericytes. Glucose uptake is 25 μmol/g wet weight/h. More than 95% of the glucose consumed is released as lactic acid in the medium. A small percentage of glucose uptake is oxidized to CO_2 and water. The estimated ATP yield of 66 μmol/g wet weight/h is calculated to be two-thirds from glycolysis and one-third from the Krebs cycle. The energy production of the rete capillary tissue is high: Per unit of wet weight, it is four times that of rabbit aortic tissue and compares with that of rat adipose tissue. Glucose utilization is not modified by the addition to the medium of pyruvate 5 mM, β-hydroxybutyrate 5 mM, or palmitate 1.2 mM. When incubations are carried out at 5°C, glucose conversion to CO_2 is almost abolished while glycolysis is maintained at a lower rate (28). The hexose monophosphate shunt and the sorbitol pathways are operative in the rete capillary tissue. Capillaries incubated for 2 h at 37°C in Krebs–Ringer bicarbonate buffer with 5 mM glucose contain 29.3 ± 4.6 nmol of sorbitol, and release in the medium 725 ± 60 nmol of fructose per gram of wet weight (36). When the mechanisms of glucose entry into the rete microvascular cells are investigated by competition and saturation experiments, no carrier-mediated transport of glucose can be seen. The results of studies of countertransfer and inward or outward fluxes of closely related radiolabeled sugars indicate that glucose passage across the membrane into the cell is compatible with the kinetics of free diffusion (24). At high medium glucose concentrations, glycolysis (23), Krebs cycle oxidation (23), sorbitol pathway (36), and glucose incorporation into basement membranes (4) are stimulated. In contrast, the addition of insulin in vitro (crystalline beef, or

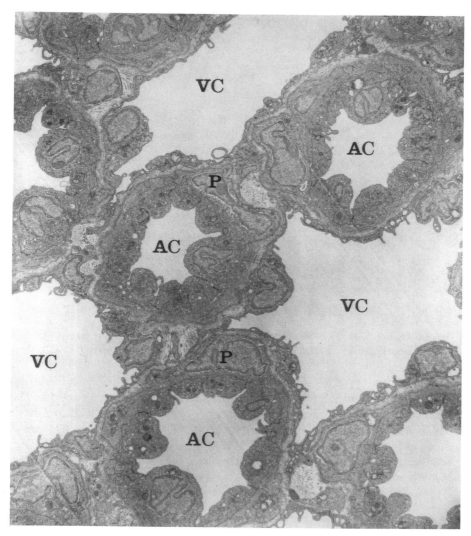

FIGURE 3.3. Transmission electron microscopy. Cross-section of the rete capillaries at low magnification, illustrating the arterial capillaries (AC) with their high endothelial cells and the venous capillaries (VC) with their thin endothelium. The connective tissue, composed of basement membranes, bundles of collagen fibers, and pericytes (P), is situated between arterial and venous capillaries. ×8,000.

amorphous bonito and cod), at a concentration of 0.1 U/ml, has no effect on glucose (or alanine) utilization (23, 24).

Since glycolysis is a major contributor to energy production by the microvascular endothelium, hypoxia or poisons of the cytochrome oxidase complex, such as potassium cyanide, have no effect on the ATP yield. When KCN 1 mM is added to

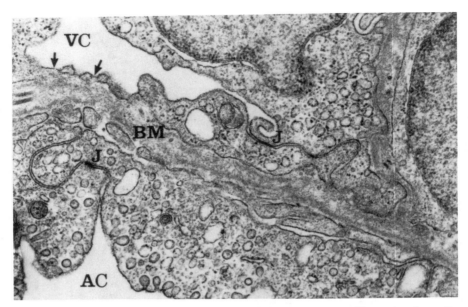

FIGURE 3.4. Transmission electron microscopy. High magnification illustrating the wall between arterial (AC) and venous (VC) capillaries. Both are resting on well defined basement membranes (BM). The arterial endothelial cells are high while the venous are thin and fenestrated (arrows). Endothelial cells are joined by intercellular junctions (J). ×20,000.

the medium, or when the oxygen tension is reduced, glucose uptake and lactate output are stimulated by approximately 30% above control values, while the oxidation of glucose to CO_2 is reduced by an average of 80%; consequently, ATP continues to be generated at a normal rate. In contrast, iodoacetate 1 mM, an inhibitor of glycolysis, effectively reduces both the anaerobic and aerobic catabolism of glucose, and the total yield of ATP drops to 10% to 20% of the control value (27).

Permeability Studies

The rete can be studied in situ or isolated and perfused in vitro, either in concurrent fashion, where flows are in the same direction (41), or in countercurrent fashion (25), to mimic the physiological opposite direction of the flows. The preparative procedures are simple and rapid. Diffusion permeability coefficients are measured when flow is equally maintained in both the arterial and venous systems, under a comparable head pressure, and with a medium of similar osmotic pressure. Various substances, usually radiolabeled, are added to the medium at the arterial input. From measurements of their concentration at the arterial and venous outputs during steady-state conditions, permeability values can be derived without approximations, apart from the use of an estimate of surface area (1 mg of wet weight of capillary tissue is equivalent to a capillary surface area of exchange of

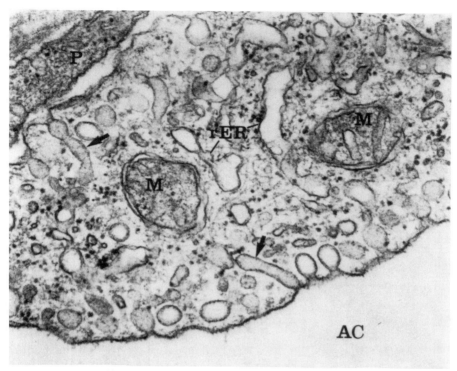

FIGURE 3.5. Transmission electron microscopy. High magnification of an arterial endothelial cell demonstrating the developed vesicular system formed by plasmalemmal vesicles which occasionally constitute endothelial tubular formations (arrows). rER, rough endoplasmic reticulum; M, mitochondria. ×25,000.

1 cm²). This assumption requires that all capillaries of the rete be perfused. Such is the case at high flow (0.5 ml/min) or with a medium containing vasodilators (41).

At infinite permeability in the concurrent flow perfusion, tracer concentrations in the arterial and venous outputs will be equal, whereas in the countercurrent flow perfusion all of the arterial tracer will emerge at the venous output (30). We have used the countercurrent approach, because the larger permeability values are more easily measured.

In this system, the coefficient of permeability $P = F(A_{IN} - A_{OUT})/S(A_{IN} - V_{OUT})$, where F is the unidirectional flow, which is the same in both directions (0.5 ml/min), S is the surface area, A_{IN} is the tracer concentration in the constant arterial infusion, and A_{OUT} and V_{OUT} are the resulting steady-state tracer concentrations at the arterial and venous outputs, respectively. Any loss of indicator substances can be easily detected and the experiment is then discarded. When there is no loss of indicator, $A_{IN} - A_{OUT} = V_{OUT}$ and $A_{IN} - V_{OUT} = A_{OUT}$. Therefore, $P = F \times V_{OUT}/S \times A_{OUT}$ (units are cm/min). The use of this equation enables the measurement of P for highly as well as poorly diffusible substances. All solutes are exposed to the same functional capillary exchange surface area (25).

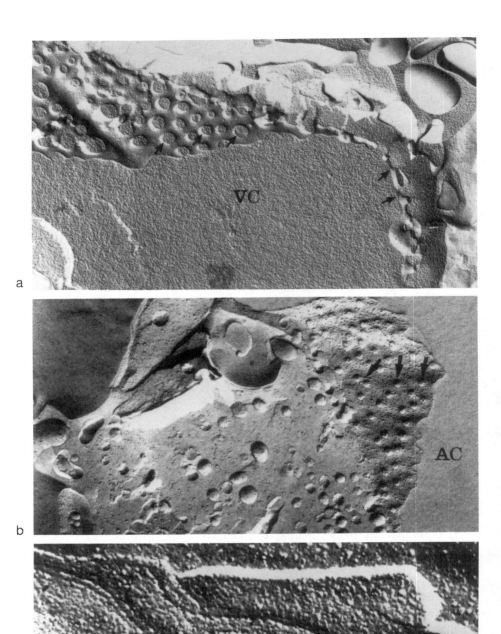

FIGURE 3.6. Freeze–fracture replica of the capillary wall. Figure 3.6a shows the fenestrations (arrows) of the venous endothelial cells either in en-face view or in cross-section. In Figure 3.6b, the luminal plasma membrane of the arterial endothelial cells is presented in en-face view with the opening of the vesicular system (arrows). Figure 3.6c illustrates the intercellular tight junction (arrows) between two endothelial cells. a: ×15,000; b: ×10,000; c: ×22,000.

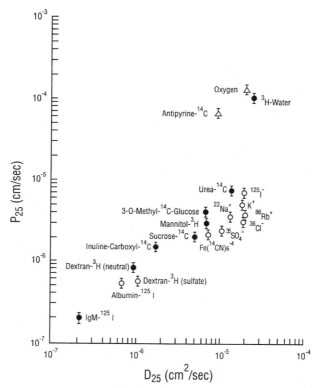

FIGURE 3.7. Relationship between diffusion coefficients (D) and permeability coefficients (P) for various substances in the isolated perfused rete, at 25°C. Mean values ± SEM. ● Neutral water soluble molecules. ○ Charged water-soluble substances. △ Lipid-soluble substances.

The permeability value for tritiated water at 25°C is of the order of 10^{-4} cm/s, which is similar to that reported in the blood–brain barrier of mammals (8, 25). P values for lipid-soluble compounds such as oxygen (34) and ^{14}C-antipyrine (28) are slightly higher and lower, respectively, in regard to the P value of water (Figure 3.7). For this group of highly diffusible substances, the absence of total exchange between the arterial and venous capillaries of the countercurrent perfused rete indicate that the exchange is barrier-limited rather than flow-limited. The P value for oxygen at 25°C is $11.8 \pm 1.9 \times 10^{-5}$ cm/s. The limiting effect of the capillary wall on oxygen transfer is of such a magnitude that it must be taken into account in describing the factors affecting the distribution of oxygen to tissues.

For water-soluble neutral compounds ranging in size from ^{14}C-urea ($r = 2.6 \times 10^{-8}$ cm) to ^{3}H-dextran ($r = 25 \times 10^{-8}$ cm), and comprising 3-O-methyl-^{14}C-glucose, ^{3}H-mannitol, ^{14}C-sucrose, and ^{14}C-carboxylinulin, there is a linear relationship between the diffusion coefficient and the permeability coefficient (Figure 3.7) (25, 29, 35). This indicates that there is no significant restriction imposed upon their passage from the arterial to the venous capillary lumen. For this group of compounds, P values range from $0.739 \pm 0.088 \; 10^{-5}$ cm/s for ^{14}C-urea to $0.085 \pm 0.025 \; 10^{-5}$ cm/s for ^{3}H-dextran.

The rete permeabilities of small inorganic cations and anions are approximately one-half the values expected for neutral solutes with similar diffusion coefficients (Figure 3.7). The decrement is interpreted to reflect the presence of both positive and negative charges along the transendothelial pathway that is accessible to the microions and to nonelectrolytes of comparable size (35).

Current studies in our laboratory aimed at measuring permeability values for various proteins ranging in size from insulin (MW 5,700) to IgM (MW 900,000) indicate no significant additional restriction to their passage. For this group of substances, the relation between P and D values is linear and in a continuum with that of water-soluble, uncharged, nonprotein compounds from ^{3}H neutral-dextran to ^{14}C-urea. Morphological examinations using the protein A–gold postembedding immunocytochemical technique (3) show that insulin, albumin, and macromolecules such as IgG and IgM are localized predominantly in the plasmalemmal vesicles and not at the intercellular junctions (Figure 3.8). These observations are

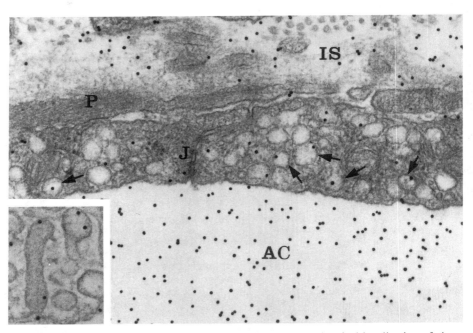

FIGURE 3.8. Transmission electron microscopy. Immunocytochemical localization of circulating albumin. The rete capillaries were perfused for 20 min with a buffer solution containing 4 g/L of bovine serum albumin. The tissues were then fixed by immersion and processed for electron microscopy immunocytochemistry for the demonstration of albumin using the protein A–gold complex. The gold particles revealing albumin antigenic sites are present in the capillary lumen, the interstitial space (IS), and the endothelial cells particularly at the level of the vesicular system (arrows). The intercellular junctions (J) are devoid of any labeling. Inset: presence of labeling in endothelial tubules. ×35,000.

compatible with a predominant mode of transport by transcytosis. Recent data indicate that some proteins, such as albumin and insulin, are transported by different sets of plasmalemmal vesicles, possibly through specific receptor-mediated processes (5).

Is it possible to reconcile diffusion kinetics, as observed in the permeability studies, with transcytosis, in bulk or receptor-mediated, as seen in morpho-cytochemical analyses? Vesicles may be viewed not as migrating structures but as permanent tree-like tubules invaginating from the luminal and abluminal sides of the endothelial cell, with occasional membrane fusion at their nearest surfaces (17). With this type of structure, molecular transit will be mostly extracellular, by diffusion in the aqueous phase through the pores of linked vesicles, rather than intracellular, by an energy-dependent vesicular shuttle.

Various paths of transcapillary passage are recognized on the basis of data provided by functional and morphological studies with various organs and single capillaries. Permeability data for water and lipophilic substances indicate that these diffuse through most of the plasmalemmal membrane of the endothelium (12, 19). The aqueous diffusion pathway (12), in addition to the intercellular junction, would be through the endothelial cell crossing two plasma membranes and the cell cytoplasm. The lipid diffusion pathway (19) would encompass the same modality of transport as water, with the additional potential path of lateral diffusion in the lipid phase of the cell membrane, as well as through the intercellu-lar junction and plasmalemmal vesicular channels. The concept of preferential water and lipid paths of transcapillary diffusion is in keeping with the data of our countercurrent perfusion studies, which show that the permeability values for water (25), antipyrine (28), and oxygen (34) lie substantially above the extrapo-lated line relating permeability versus diffusion coefficients for paracellular tracers such as inulin, sucrose, and sodium (Figure 3.7). The paracellular hydro-philic pathway most likely is represented by the interendothelial junction (19); its length and width vary from tissue to tissue and with consecutive segments of the same capillary. In general, this pathway is too narrow to allow for the passage of macromolecules in physiological conditions. Macromolecules are conveyed across the microvascular endothelium either through leaks in postcapillary ven-ules, fenestrations in fenestrated capillaries, gaps in discontinuous capillaries, or plasmalemmal vesicles in continuous capillaries (22, 37). Vesicles may coalesce into transendothelial open channels or into tubules interconnected by membrane fusion. A matrix of net negative charge extending from the surface of the endo-thelium to the interstitium may coat various paths of transport (13), including those in the kidney glomerulus (9). Other capillary membranes, such as those in the lung (21) and in the intestine (20), seem to carry net positive charges for macromolecular transport, while in the rete there is a staggering of positive and negative charges along the transcapillary pathway of inorganic anions and cations (35).

The identification of the various paths of transcapillary transport with their morphological substratum remains fragmentary and controversial. Also un-resolved is the recognition of possible regulatory agents of these pathways. Many factors have been proposed as modulators of capillary permeability; these may act through a common final path involving structural and functional changes of the cytoskeleton and anchoring proteins in the endothelial cells (39, 11). Thus, contraction or relaxation of the actin and myosin fibers of the cytoskeleton could alter the shape of the endothelial cell, and affect the geometry of the inter-endothelial junction, the surface of the cell membrane, and the disposition and eventual movement of the vesicles. A comprehensive and coherent scheme of the relations between the condition of the cytoskeleton and the permeability change for every pathway involved has yet to emerge. With the rete, we have conducted a series of experiments designed to demonstrate that the various paths of trans-capillary transport can be selectively altered and possibly regulated independently. A summary of the effects observed with various experiments is shown in Table 3.1.

When the rete was perfused with a hyperosmolar solution (by addition to the Krebs–Ringer bicarbonate buffer of 350 mOsm/L sucrose) both at the arterial and venous inputs or in the arterial circuit alone, the capillary diffusion capacity (permeability × surface area) for ^{125}I-albumin and ^{14}C-urea increased two to three times above baseline values, while that for ^{3}H-water was reduced by 30% (26). The intercellular junctions, as examined by electron microscopy, remained intact, but other significant structural changes were observed (Figure 3.9). Morphometric analysis revealed swelling of the extracellular space, membrane vesiculation, a significant decrease in the volume density of the arterial and venous endothelial cells, and a significant increase of the arterial and venous capillary lumina. The volume density of the vesico-tubular system increased, and the increase was

TABLE 3.1. Summary of effect observed with various agents on the permeability and structure of eel rete capillaries.

Agent	Permeability changes			Morphological changes
	^{125}I-Albumin	^{14}C-Sucrose	^{3}H-Water	
Hyperosmolality	↑	↑	↓	Cell shrinkage · ↑ Vesicles
Hyperthermia	↑	↑	↑	None
Hypothermia	↑	↔	↓	None
Second messengers	↑	↑	↓	Vacuolation of cell membrane · Interstitial edema
Hypoxia	↑	↑	↔	Vacuolation of cell membrane · Mitochondria
+ Reperfusion	↑↑	↑↑	↔	swelling · Interstitial edema · Pericytes detached
+ Phalloidin	↔	↔	↔	None
Chronic hyperglycemia				
Early	↑	↔	↔	None
Intermediate	↑	↑	↔	Basement membrane thickening
Late	↑	↑	↑	Basement membrane thickening · Glycogen deposits

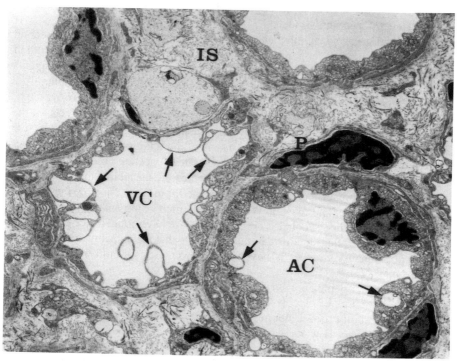

FIGURE 3.9. Transmission electron microscopy. Cross-section of the rete capillaries at the end of a bipolar hyperosmolar perfusion. The volumes of both arterial and venous endothelial cells are diminished. The interstitial space (IS) appears edematous. The endothelial cells, particularly the venous ones, are vesiculated and swollen (arrows) and the nuclear chromatin is condensed. ×6,000.

larger than would be accounted for by shrinkage in nonvesicular cytoplasm, as if new vesicle formation had been initiated. This could explain, at least in part, the increased solute transport found after hyperosmotic exposure. On the contrary, water permeability could have been reduced by increased intracellular viscosity and cell membrane changes induced by dehydration. When acute temperature shifts were applied to the rete perfusate, selective effects on the paths of transport were elicited (28). An abrupt temperature increase, from 25 to 35°C, induced a continuous and irreversible rise of permeability to ^{125}I-albumin, ^{14}C-sucrose, and ^{22}Na, whereas the cellular pathways for oxygen, ^{14}C-antipyrine, and ^{3}H-water were not modified. A gradual increase of temperature was accompanied by a smaller and reversible rise in ^{14}C-sucrose and ^{22}Na permeability and no change in ^{125}I-albumin permeability. When the perfusate temperature was lowered abruptly from 25 to 5°C, the permeability for ^{14}C-sucrose, ^{22}Na, and oxygen did not change; that for ^{3}H-water and ^{14}C-antipyrine decreased to plateau values; and that for ^{125}I-albumin increased progressively. No significant morphological changes were detected as a consequence of the temperature stresses, irrespective of their direction or amplitude.

Investigations using endothelial cells in culture have linked the action of second messengers on monolayer permeability to changes in cytoskeletal structure and in integral membrane proteins (1, 40). In the rete, we have tested various agents aimed at inducing cyclic nucleotide, phosphoinositide, and intracellular Ca^{2+} effects, at concentrations generally used for in vitro and in vivo studies of vascular endothelium (31). In general, we observed characteristic selective effects on the paracellular and transcellular pathways of transport and significant nonselective alterations of capillary structure. For instance, the addition to the perfusate of dibutyryl cAMP (10^{-6} M), phorbol 12-myristate 13-acetate (PMA 10^{-5} M), or calcium ionophore A23187 (5×10^{-7} M) increased the permeabilities to ^{125}I-albumin, ^{14}C-sucrose, and ^{22}Na while not modifying or reducing the permeabilities to ^{3}H-water and oxygen. When present, morphological change consisted of plasma membrane vacuolation (Figure 3.10). We have not detected any widening of the intercellular junctions, but it is possible that the detachment of contacts at the junctional complexes is beyond the power of resolution of the electron microscope.

When the energy metabolism of the rete capillaries is reduced by the combined effects of inhibitors of glycolysis and Krebs-cycle oxidation, a twofold increase of the permeability to ^{125}I-albumin, ^{22}Na, and ^{14}C-sucrose was observed, while the permeability to ^{3}H-water was unchanged (27). For the group of tracers exhibiting the increased permeability, the change was perceptible within 10 min and progressed until plateau values were reached after 30 min. The electron microscopic examination revealed the presence of large vacuoles protruding into the capillary lumen, and swelling of the mitochondria. The intercellular junctions were intact.

The effect of cessation of flow, followed by reperfusion, altered the structure of the venous endothelium but not the permeability of the intercapillary barrier. Stasis with hypoxic medium containing the inhibitors of energy generation, followed by reperfusion with oxygenated control medium, amplified the effects of reduction in energy generation on permeability and morphological changes. Both functional and structural damages were to a large extent reduced by the addition to the medium of 10^{-6} M phalloidin, a stabilizer of filamentous actin (33).

All of the above acute experiments where various physical or chemical agents were used in the rete have evoked diverging effects on the permeability of the various tracers, thereby demonstrating the presence of different paths of transport across the endothelial cells of the rete capillary. Unequivocal evidence that these effects may be mediated by changes of the structure of the cytoskeleton cannot be provided by these in vivo studies. Possible alterations in the number or configuration of the filamentous actin may have opened the tight junctions, reduced the number of vesicle channels or fenestrations, and modified the surface of the plasma membrane, thus accounting for some of the effects observed.

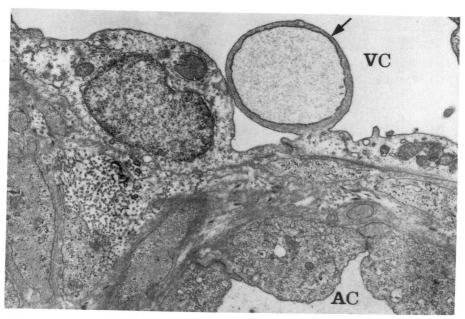

FIGURE 3.10. Transmission electron microscopy. High magnification of the rete capillary wall upon perfusion with phorbol myristate acetate (10^{-5} mol/L). Vacuolation of the venous endothelial cells is apparent with large cytoplasmic vacuoles (arrow) protruding into the capillary lumen. ×10,000.

Rete Capillaries of the Hyperglycemic Eel

Spontaneous diabetes appears in various species of animals. These have been studied extensively, and significant knowledge has been gained in the etiology and pathophysiology of human diabetes. In fishes, sekoke disease has been described in carp raised in fish farms and fed commercial food containing oxidized oils (43). The disease is characterized by hyperglycemia, β-cell atrophy, diffuse glomerulosclerosis, retinal microangiopathy, and muscle degeneration.

Carbohydrate levels are modified in fishes by acclimation to low temperatures (42). Some species, among which is the eel, develop a significant hyperglycemia. In these fishes, the number of α-cells increases in winter, while that of β-cells decreases. Decreased insulin release is typical of cold-acclimated fishes and could be mediated by the action of thermoreceptors on the central nervous system. The pituitary, adrenal cortex, and thyroid glands do not seem to play a clear role in the genesis of hyperglycemia.

When eels are adapted to water at 2°C they become hyperglycemic within 1 month (4). The level of blood glucose varies among eels, but remains stable

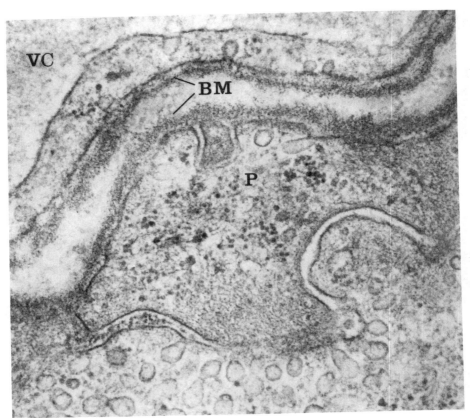

a

FIGURE 3.11. Transmission electron microscopy. Rete capillary wall sampled from a normoglycemic (Figure 3.11a) and an hyperglycemic (Figure 3.11b) eel. Under normoglycemic condition, the basement membranes (BM) appear thin and well defined. In hyperglycemic condition, basement membrane material thickens and fills the entire intercapillary interstitial space (IS). a: ×35,000; b: ×25,000.

within the same eel. With glucose, blood levels of α-amino-nitrogen and free fatty acids also rise. Plasma immunoreactive insulin levels are low and β-cells are degranulated. After a period of 5–6 months of adaptation to water at 2°C, the basement membranes of the rete capillaries were thickened (Figure 3.11), and their composition was altered. Pericytes contained large deposits of glycogen, and in vitro glucose carbon incorporation into basement membrane proteins was enhanced. The morphometric changes were reversible with the progressive rewarming of the water. In keeping with the concept that we have established over the years that the various paths of transcapillary passage can be selectively modified by external agents, current investigations in our laboratory indicate that chronic exposure to elevated ambient glucose concentration also induces specific initial

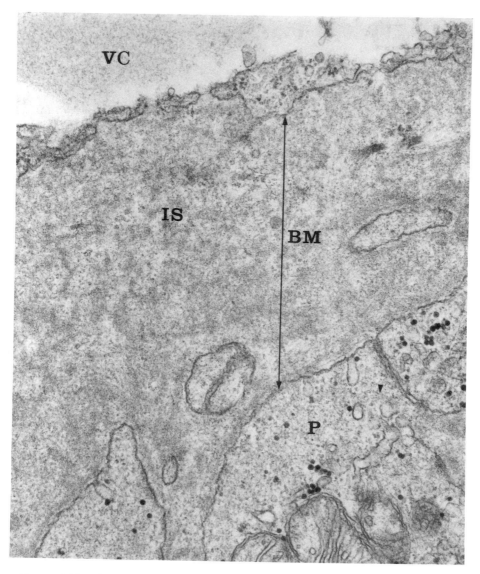

b

FIGURE 3.11. (*Continued*)

changes in permeability that evolve in time to more global alterations (Table 3.1) (32). The rete model in the cold-adapted eel provides an interesting tool for the study of the nature and temporal sequence of structural, biochemical, and functional lesions; their interrelationships; and their reversibility in the microangiopathy associated with hyperglycemia.

References

1. Alexander, J.S. Modulation of endothelial monolayer barrier function and cytoskeleton by second messenger analogs in vitro. Doctoral dissertation. Boston, MA: Boston University, 1989.
2. Bassingthwaighte, J.B., and C.A. Goresky. Modeling in the analysis of solute and water exchange in the microvasculature. In: *Handbook of Physiology,* edited by E.M. Renkin and C.C. Michel. Am. Physiol. Soc., Bethesda, MD, 1994, vol. 4, pp. 549–626.
3. Bendayan, M. Use of protein A-gold technique for the morphological study of vascular permeability. *J. Histochem. Cytochem.* 28:1251–1254, 1980.
4. Bendayan, M., and E.A. Rasio. Hyperglycemia and microangiopathy in the eel. *Diabetes* 30:317–325, 1981.
5. Bendayan, M., and E.A. Rasio. Transport of insulin and albumin by the microvascular endothelium of the rete mirabile. *J. Cell Sci.,* 109:1857–1864, 1996.
6. Bendayan, M., E. Sandborn, and E. Rasio. Studies of the capillary basal lamina. I. Ultrastructure of the red body of the eel swimbladder. *Lab. Invest.* 32:757–767, 1975.
7. Bendayan, M., E. Sandborn, and E. Rasio. Studies of the capillary basal lamina. II. Preparation, chemical composition, and metabolic properties. *Lab. Invest.* 32:768–772, 1975.
8. Bolwig, T.G., and N.A. Lassen. The diffusion permeability to water of the rat blood-brain barrier. *Acta Physiol. Scand.* 93:415–422, 1975.
9. Brenner, B.M., T.H. Hostetter, and H.D. Humes. Molecular basis of proteinuria of glomerular origin. *N. Engl. J. Med.* 298:826–833, 1978.
10. Carey, F.G. Fishes with warm bodies. *Scientific American* 228:36–44, 1973.
11. Crone, C. Modulation of solute permeability in microvascular endothelium. *Fed. Proc.* 45:77–83, 1986.
12. Crone, C., and D.G. Levitt. Capillary permeability to small solutes. In: *Handbook of Physiology,* edited by E.M. Renkin and C.C. Michel. Am. Physiol. Soc., Bethesda, MD, 1984, vol. 4, pp. 411–466.
13. Curry, F.E. Is the transport of hydrophilic substances across the capillary wall determined by a network of fibrous molecules? *Physiologist* 23:90–93, 1980.
14. Kleckner, R.C. Swim bladder volume maintenance related to initial oceanic migratory depth in silver-phase anguilla rostrata. *Science* 208:1481–1482, 1980.
15. Kobayashi, H., B. Pelster, and P. Scheid. CO_2 back-diffusion in the rete aids O_2 secretion in the swimbladder of the eel. *Respiration Physiology* 79:231–242, 1990.
16. Michel, C.C. The investigation of capillary permeability in single vessels. *Acta Physiol. Scand.* 463:67–74, 1979.
17. Michel, C.C. The transport of albumin: a critique of the vesicular system in transendothelial transport. *Am. Rev. Respir. Dis.* 146:S32–S36, 1992.
18. Olesen, S.-P. An electrophysiological study of microvascular permeability and its modulation by chemical mediators. *Acta Physiol. Scand.* 579 (Suppl. 136):1–28, 1989.
19. Pappenheimer, J.R., E.M. Renkin, and M. Borrero. Filtration, diffusion and molecular sieving through capillary membranes: a contribution to the pore theory of capillary permeability. *Am. J. Physiol.* 167:13–46, 1951.
20. Perry, M.A., J.N. Benoit, P.R. Kvietys, and D.N. Granger. Restricted transport of cationic macromolecules across intestinal capillaries. *Am. J. Physiol.* 245:G568–G573, 1983.

21. Pietra, G.G., A.P. Fishman, P.N. Lanken, P. Sampson, and J. Hansen-Flashen. Permeability of pulmonary endothelium to neutral and charged macromolecules. *Ann. N.Y. Acad. Sci.* 401:241–247, 1982.
22. Predescu, D., and G.E. Palade. Plasmalemmal vesicles represent the large pore system of continuous microvascular endothelium. *Am. J. Physiol.* 265 (*Heart Circ. Physiol* 34):H725–H733, 1993.
23. Rasio, E.A. Glucose metabolism in an isolated blood capillary preparation. *Can. J. Biochem.* 51:701–708, 1973.
24. Rasio, E.A. Passage of glucose through the cell membrane of capillary endothelium. *Am. J. Physiol.* 228:1103–1107, 1975.
25. Rasio, E.A., M. Bendayan, and C.A. Goresky. Diffusion permeability of an isolated rete mirabile. *Circ. Res.* 41:791–798, 1977.
26. Rasio, E.A., M. Bendayan, and C.A. Goresky. The effect of hyperosmolality on the permeability and structure of the capillaries of the isolated rete mirabile of the eel. *Circ. Res.* 49:661–676, 1981.
27. Rasio, E.A., M. Bendayan, and C.A. Goresky. Effect of reduced energy metabolism and reperfusion on the permeability and morphology of the capillaries of an isolated rete mirabile. *Circ. Res.* 64:243–254, 1989.
28. Rasio, E.A., M. Bendayan, and C.A. Goresky. Effect of temperature change on the permeability of eel rete capillaries. *Circ. Res.* 70:272–284, 1992.
29. Rasio, E.A., M. Bendayan, and C.A. Goresky. Le rete mirabile de l'anguille : un modèle unique pour l'étude de la perméabilité capillaire. *Médecine Sciences* 9:593–603, 1993.
30. Rasio, E.A., M. Bendayan, and C.A. Goresky. The isolated rete. In: *Biochemistry and molecular biology of fishes,* edited by P.W. Hochachka and T.P. Mommsen. Elsevier: Amsterdam, 1994, vol. 3, 191–203.
31. Rasio, E.A., M. Bendayan, and C.A. Goresky. Effects of second messengers on the permeability and morphology of eel rete capillaries. *Circ. Res.* 76:566–574, 1995.
32. Rasio, E.A., M. Bendayan, and C.A. Goresky. Specific sequential permeability changes during chronic hyperglycemia in the rete capillaries (Abstract). *Int. J. Microcirc. Clin. Exp.* 16 (suppl. 1):218, 1996.
33. Rasio, E.A., M. Bendayan, and C.A. Goresky, J.S. Alexander, and D. Shepro. Effect of phalloidin on structure and permeability of rete capillaries in the normal and hypoxic state. *Circ. Res.* 65:591–599, 1989.
34. Rasio, E.A., and C.A. Goresky. Capillary limitation of oxygen distribution in the isolated rete mirabile of the eel (Anguilla anguilla). *Circ. Res.* 44:498–503, 1979.
35. Rasio, E.A., and C.A. Goresky. Passage of ions and dextran molecules across the rete mirabile of the eel. The effects of charge. *Circ. Res.* 56:74–83, 1985.
36. Rasio, E.A., and A.D. Morrison. Glucose-induced alterations of the metabolism of an isolated capillary preparation. *Diabetes* 27:108–113, 1978.
37. Schnitzer, J.E., P. Oh, E. Pinney, and J. Allard. Filipin-sensitive caveolae-mediated transport in endothelium: reduced transcytosis, scavenger endocytosis, and capillary permeability of select macromolecules. *J. Cell. Biol.* 127:1217–1232, 1994.
38. Scholander, P.F. The wonderful net. *Scientific American* 196:96–107, 1957.
39. Shasby, D.M., S.S. Shasby, J.M. Sullivan, and M.J. Peach. Role of endothelial cell cytoskeleton in control of endothelial permeability. *Circ. Res.* 51:657–661, 1982.
40. Shepro, D., and P.A. D'Amore. Physiology and biochemistry of the vascular wall

endothelium. In: *Handbook of Physiology,* edited by E.M. Renkin, C.C. Michel. Am. Physiol. Soc., Bethesda, 1994, vol. 4, 103–164.

41. Stray-Pedersen, S., and J.B. Steen. The capillary permeability of the rete mirabile of the eel, anguilla vulgaris L. *Acta Physiol. Scand.* 94:401–422, 1975.

42. Umminger, B.L. The role of hormones in the acclimation of fish to low temperatures. *Naturwissenshaften* 65:144–150, 1978.

43. Yokote, M. Sekoke disease, spontaneous diabetes in carp, Cyprinus carpio, found in fish farms. I. Pathological study. *Bull. Freshwater Fisheries Res. Lab.* 20:39–72, 1970.

4

Interactions Between Bovine Adrenal Medulla Endothelial and Chromaffin Cells

Fernando F. Vargas, Soledad Calvo, Raul Vinet, and Eduardo Rojas

Many substances normally present in blood and those released during inflammation or tissue damage can, if they reach threshold concentration, stimulate endothelial cells (ECs) to increase synthesis and secretion of nitric oxide (1) and prostacyclin (2). These products induce smooth muscle cell relaxation and consequently vasodilatation (3). Another EC response evoked by these substances consists on increased transvascular permeability to small solutes and water across intercellular junctions (4, 5).

The cellular processes that cause secretion and permeability increase are extremely complex but have in common a rise in EC cytosolic calcium ($[Ca^{2+}]_i$) induced by many agents (agonists) that interact with specific receptors at the EC surface (6, 7). Agonist binding to receptor-coupled G proteins activates phospholipase C, which splits membrane lipids, yielding diacylglycerol and inositol-*tris*-phosphate (IP3). IP3 diffuses into the cytosol and binds to receptors in the endoplasmic reticulum membrane, opening Ca^{2+}-permeable channels. Agonists also induce the opening of plasmalemma nonselective cationic channels that are an additional pathway for Ca^{2+} influx into the cytosol (6). The combined inflow from the above two pathways leads to a rise in $[Ca^{2+}]_i$. These processes are probably present in all ECs as shown by their generalized response to thrombin, bradykinin, interleukin, histamine, and ATP (8–11).

A parallel pathway for increasing $[Ca^{2+}]_i$ consists of voltage-dependent calcium channels (VDCCs) located in the cell membrane. The opening of VDCCs causes Ca^{2+} influx, since the Ca^{2+} concentration gradient across the plasmalemma is as high as three orders of magnitude. This allows VDCC-containing cells to increase $[Ca^{2+}]_i$ in response to membrane potential depolarization, in addition to the agonist-evoked response. In contrast with the latter, VDCC activation does not cause Ca^{2+} release from intracellular compartments (12).

Voltage-dependent Ca^{2+} channels are present in excitable cells (13–15) as well as in fibroblasts (16) and endocrine cells (17–20), but are absent from most ECs (21–24). It is therefore of great interest that they have been reported to exist in microvascular ECs from bovine adrenal medulla (BAMEC) (25–28) and rat (29, 30), but not bovine brain (31). However, it is not clear which specific functions performed by VDCC-containing ECs would require VDCC participation.

Whereas the agonist-evoked $[Ca^{2+}]_i$, signal has been thoroughly investigated in ECs (6, 7, 32, 33), there is scant information about the relevance of VDCC for ECs.

The extremely high Ca^{2+} concentration gradient between the external medium and the cytosol depolarization-mediated VDCC opening causes the $[Ca^{2+}]_i$ to rise. This response is an essential stage of the secretory process in endocrine cells. Thus, insulin secretion by pancreatic β cells is initiated by a rise in ATP that closes ATP-dependent K^+ channels, which causes depolarization, VDCC gating, and a rise in $[Ca^{2+}]_i$ (17, 18). Likewise, chromaffin cell secretory activity is initiated by depolarization caused by Na^+ influx through acetylcholine-gated cationic receptor channels. The ensuing VDCC gating increases $[Ca^{2+}]_i$ to levels that induce chromaffin cell secretion (20). Likewise, BAMEC $[Ca^{2+}]_i$ rises when the cell is depolarized with high $[K^+]_o$ (12, 28, 34), and reaches levels known to induce secretion of NO and PGI_2 (32, 33). The opposite effect is observed in VDCC-lacking ECs, where depolarization, instead of opening a Ca^{2+} pathway, reduces the driving force for Ca^{2+} influx and diminishes the magnitude of an agonist-evoked $[Ca^+]_i$ increase (35, 36).

Valuable information about BAMEC VDCC functions may be gathered from the well-studied participation of VDCCs in endocrine cell secretion. Such a comparison would require knowing BAMEC general electrophysiology and, more specifically, the VDCC types present in its membrane. For carrying out these studies we used acutely isolated as well as cultured BAMECs. Enzymatic dissociation of the adrenal medulla to obtain BAMEC yields a mixture of BAMEC together with chromaffin and other cells. BAMECs were identified using their well-described morphology (37) and uptake of acetylated LDL (38). In addition, we partially purified BAMEC in the mixture by differential plating, which consists of shaking and washing the culture dish to dislodge other cells, except for BAMEC, which strongly adheres to the dish bottom (39).

Electrophysiology of Adrenal Medulla Endothelial Cells

Electrophysiological studies of ECs were decisively advanced with the introduction of the patch-clamp technique (40). The whole-cell version of this technique allows access to the cytosol through a micropipet that does not penetrate into the cell. The pipet is advanced until it touches the cell surface. A slight negative pressure sucks the membrane patch under the pipet tip and forms a tight seal between its rim and the cell membrane, the so-called giga ohm seal. In the whole-cell version, electrical communication between the pipet lumen and the cytosol is attained by rupturing the membrane patch under the pipet tip, leaving the entire cell membrane as a barrier between the pipet tip and the external solution (Figure 4.1). We measured membrane potential, resistance, capacitance, and currents with the above described version of the patch-clamp technique. More specifically, a pipet filled with a solution mimicking the ionic composition of the cytosol was sealed to the surface of an EC, and electrical communication with the cytosol was

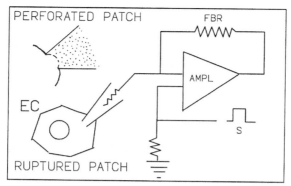

FIGURE 4.1. Diagram of the pipet and electrical circuit for recording current and membrane potential using the whole-cell version of the patch-clamp technique. The upper left pipet contains dots that represent nystatine, and the membrane shows the pores formed by the antibiotic. The lower pipet tip is completely open since the membrane patch was ruptured with suction. FRB = feedback resistance; AMPL = amplifier; S = command pulse; EC = endothelial cell.

obtained by either rupturing the membrane at the pipet tip with negative pressure or by perforating it with the pore former nystatin (Figure 4.1).

BAMEC MP measured with the above technique was −50 mV, which is similar to the values found by Bossu et al. (25), while that of bovine aortic EC has been reported to be higher than −60 mV (41). This membrane potential difference reflects the density and open probability of the voltage-dependent and Ca^{2+}-dependent K^+ channels in each cell.

We recorded the channels present in BAMEC using inside and outside physiological solutions, and found that they contain voltage-dependent K^+ channels of the A-type (K_A^+), which has a low threshold (−60 mV) and a high conductance, as evidenced by a large outward current (Figure 4.2). These characteristics suggest that this channel can prevent sudden membrane depolarization. The K_A^+ channel has been found in fewer than 30% of subcultured pulmonary artery (23) and human umbilical vein (24) ECs, but we found it in 100% of freshly dissociated BAMECs.

BAMEC also contains a K^+ inward rectifier channel, whose current approaches zero in the MP range between −120 and about −10 mV (Figure 4.2). This flat current plateau (Figure 4.2, inset) makes the MP very sensitive to electrogenic transporter activity (42).

Stretch activated channels have been shown to exist in the EC membrane (43). They also may be present in BAMECs, since, in our experiments, pulling the membrane with negative pressure through the patch pipet (after sealing, but before rupturing the membrane) evoked frantic single channel activity.

The above described channels are present in BAMECs and many other ECs (44). What makes BAMEC special is that it also contains voltage-dependent Ca^{2+} channels (25–28).

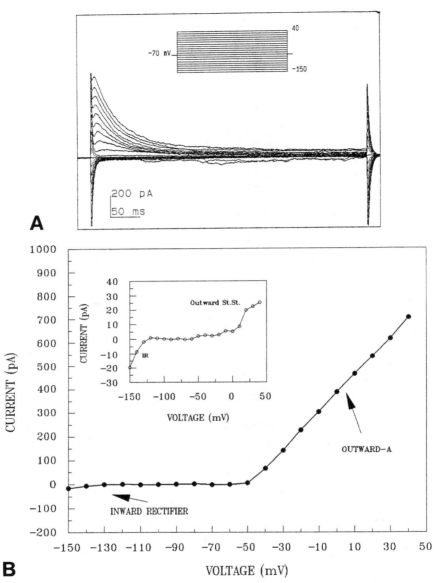

FIGURE 4.2. (A) BAMEC inward (down) and outward (up) currents recorded in physiological external and internal solutions are shown at the top. (B) Lower graph shows
current versus voltage curves (IV curves) for the currents shown above. Inset shows the
same above currents measured after the strong outward component had inactivated in order
to see the K^+ inward rectifier (R).

Voltage-Dependent Ca²⁺ Channels

All of these channels are gated by depolarization and show selectivity for Ca^{2+} as the current carrier in physiological solutions. However, differences in current amplitude, threshold, activation/inactivation kinetics, and response to drugs define specific VDCC types (45). Among these, T- (for transient) and L- (for long) type VDCCs are found together in endocrine cells, where they modulate secretion (17–20). It is therefore interesting to investigate whether ECs would have both channel types, since they also have an intense secretory activity.

A T-type VDCC transient opening and a low threshold are suitable for the generation of MP oscillations and quick cell secretory response (46), while an L-type VDCC can induce sustained cell secretion, since its long-term opening allows prolonged Ca^{2+} influx (46, 47).

VDCC types are also pharmacologically different. Thus, an L-type VDCC shows specific binding for dihydropyridines, which can be antagonists such as nifedipine or agonists such as Bay K 4866 (48, 49). On the other hand, T-type VDCCs are inhibited by amiloride (50). Hence, VDCC pharmacology is a useful complement of electrophysiological measurements for T- and L-channel identification.

Suppression of the relatively large K^+ currents in BAMEC allows detecting a depolarization-activated inward current (25–28) (Figure 4.3). This current was probably carried by Ca^{2+}, since it was sensitive to Ca^{2+} and Ba^{2+} in the external solution and had a reversal potential near +30 mV (27), which is expected from the Ca^{2+} concentration difference between the external medium and the cytosol. Further support for Ca^{2+} as the current carrier came from simultaneous patch-clamp and $[Ca^{2+}]_i$ measurements which showed that a $[Ca^{2+}]_i$ increase was associated with the inward current (12).

The putative Ca^{2+} inward current flowed through T- and L-type VDCCs as shown by the presence of two current components, a transient one with time-dependent inactivation and a noninactivating one (Figures 4.3A and B) (27). The initial kinetics of each current was measured separately by suppressing the T component with a depolarization prepulse. Using this experimental protocol we determined the kinetic parameters of the L-type VDCC, and by subtraction those of the T-type (27). These measurements, in addition to activation and inactivation parameters, strengthen our view that T- and L-type VDCCs present in BAMEC are of the same type as those reported in endocrine cells. Dihydropyridine pharmacology showed that L currents were completely blocked by 10 μM nifedipine, while the transient current was only weakly inhibited. In addition, Bay K 8644 enhanced the steady-state current but did not affect the transient one (27). These pharmacological responses support the conclusion that the observed channels are like the L- and T-type VDCCs in endocrine secretory cells whose characteristics have been described by many research groups (17–20). An opposite view is held by Bossu et al. (34), who claim that a special kind of L-type VDCC is present in BAMEC. Their conclusions are based on the absence of inhibition of the noninactivating current with the dihydropyridine antagonist nicardipine and enhancement with the agonist BAY K8644 (34).

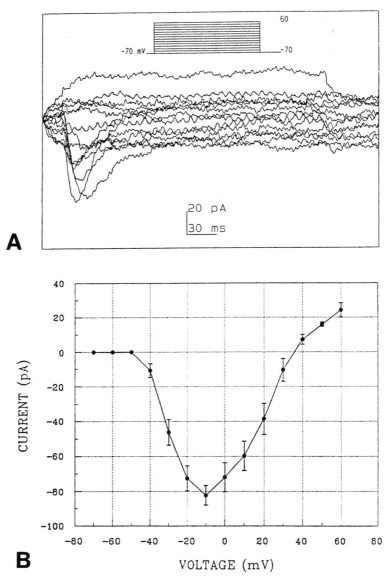

FIGURE 4.3. Inward currents (down) were recorded in K+-free solutions. (A) The currents at different voltage pulses from a holding potential at −70 mV. (B) The current versus voltage curve for the peak current.

Since the T-channel threshold is very low at −50 mV (27), it may be partially inactivated at BAMEC MP, which, as stated before, is also −50 mV. This should diminish the T-current magnitude and thus limit its functional role. However, the MP of an undisturbed BAMEC may be higher than that measured in freshly dissociated BAMECs, since some of the mechanical procedures used during cell dissociation may have lowered its MP. It is known, for example, that treatments such as lifting the cells from the dish and replacing them head down, that is, inverting the cells, causes EC depolarization (51). Therefore, in vivo BAMEC MP may be higher than −50 mV; if this were the case, most T-channels would be in the closed state and thus ready for activation.

The presence of T- and L-type Ca^{2+}-channels and Ca^{2+}-dependent K^+-channels (Vinet and Vargas, unpublished) suggests that BAMECs are capable of generating membrane potential oscillations, like pancreatic β cells (17) and gonadotrophs (52). We have observed oscillatoy behavior in BAMECs stimulated with agonists. However, oscillations also have been observed in ECs that lack VDCCs (53, 54), while depolarization-evoked VDCC openings rarely generated oscillations in our experiments. Hence, further research is needed to ascertain whether VDCCs can modulate agonist-evoked oscillations in BAMECs.

VDCCs can be detected only in acutely isolated BAMECs, and are not functional or cease to be expressed in subcultured BAMECs, as shown in the reduced fraction of BAMEC (<10%) that responded to depolarization after the first passage (12, 28). The VDCC is not the only membrane property that disappears with EC subculture, as shown by the reported loss of muscarinic receptors in arterial ECs (55) and Na^+/Ca^{2+} exchange in coronary ECs (22). VDCC loss caused by subculture is of great interest, but it lies beyond the scope of this chapter. We only propose that it may be associated with the absence of the specific biochemical environment provided by organs or tissues in vivo (56). Also absent is the mechanically evoked Ca^{2+} influx (57), since cell culture is normally done in quiescent fluid. Both biochemical and biomechanical factors may be needed for full expression of cell properties.

Modulation of BAMEC Cytosolic Ca^{2+}

BAMEC $[Ca^{2+}]_i$ measured with either Fura-2 or Indo-1 rose when the cells were depolarized with high $[K^+]_o$ (Figure 4.4). This effect was totally dependent on the external Ca^{2+}, since it was absent in zero external Ca^{2+} and it dropped to near baseline when Ca^{2+} was deleted from the external solution during a trial (12). These results indicate that the $[Ca^{2+}]_i$ rise was caused solely by Ca^{2+} influx through voltage-dependent channels, without an internal release component. Depolarization induced by high $[K^+]_o$ caused about the same rise in $[Ca^{2+}]_i$ as that evoked by acetylcholine, but it was higher than that evoked by ATP. This suggests that VDCC activation can stimulate BAMEC secretion, since the above agonists have been shown to elicit nitric oxide and prostacyclin synthesis in cultured BAMECs (32, 33).

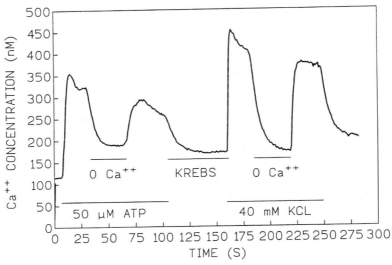

FIGURE 4.4. $[Ca^{2+}]_i$ of acutely isolated BAMEC rose in response to ATP and depolarization with high $[K^+]_o$. Deleting external calcium caused a drop in $[Ca^{2+}]_i$ to an intermediate level under ATP, and to the baseline level under high $[K^+]_o$.

Secretion–Perfusion Coupling

As described above, the circulatory and permeability effects of local pathological changes such as inflammation have received a great deal of attention, while circulatory adjustments associated with endocrine secretion have been largely ignored. They could be especially important for shortening the distribution time of hormones and other secretory products.

A linkage between secretion and organ perfusion was first proposed in relation to ATP-evoked secretion in adrenal medulla ECs (32). This linkage would consist of activation of the gland microcirculatory endothelium by chromaffin cell secretion, and it would be mediated by a BAMEC $[Ca^{2+}]_i$ rise that induced secretion of vasoactive factors and increased microvascular permeability. Higher blood flow through the gland would shorten the distribution time of the secretory products throughout the body. This is particularly important for maintaining an effective plasmatic concentration of catecholamines, since their half-life in blood is only about 3 minutes (58).

We investigated the secretion–perfusion link by testing interactions between chromaffin cells and BAMECs. To this end we measured $[Ca^{2+}]_i$ response to chromaffin cell activation in subcultured BAMECs that, as described above, have lost functional VDCCs. We used human umbilical vein ECs (HUVEC) as controls, since they completely lack VDCCs (24).

Stimulation of Human Umbilical Vein ECs by Chromaffin Cells

We measured the $[Ca^{2+}]_i$ response of a cultured human umbilical vein (HUVEC) to chromaffin cell secretion. HUVECs were grown on a coverslip that was placed at the bottom of a perfusion chamber (Figure 4.5). $[Ca^{2+}]_i$ was measured before and during depolarization with 40 mM $[K^+]_o$, which should activate VDCCs in chromaffin cells but have no effect on HUVEC. Perfusion flow was set to first reach the chromaffin cells and then the HUVEC. High $[K^+]_o$ caused a rise of chromaffin cell $[Ca^{2+}]_i$ but no response from the HUVEC, confirming that the latter lacks VDCCs (24). However, after we put into the dish a coverslip fragment that contained chromaffin cells, the HUVEC showed an increase in $[Ca^{2+}]_i$ when the chamber perfusion was switched from the control solution to one containing 40 mM $[K^+]_o$.

HUVEC $[Ca^{2+}]_i$ rose from a resting level of 82.6 ± 2 (M ± SEM, $N = 3$) to about 250 nM, which is approximately three times the baseline (Figure 4.5).

However, the HUVEC $[Ca^{2+}]_i$ rise was very slow. Thus, the latency time for the effect, from the beginning of the high $[K^+]_o$ perfusion to the initial $[Ca^{2+}]_i$ rise in HUVEC, was 41.7 ± 11.6 (SEM) seconds and the time from the initial $[Ca^{2+}]_i$ rise

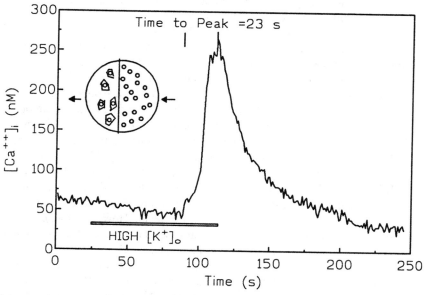

FIGURE 4.5. HUVEC $[Ca^{2+}]_i$ rise evoked by high $[K^+]_o$ when chromaffin cells were present in the same dish. The upper left circle shows one-half of the chamber botton covered with HUEVEC (left), while the other half was covered with chromaffin cells (right). The arrows indicate the direction of perfusion. HUVECs lack VDCCs; therefore the above response was not caused by depolarization but by a messenger released by depolarization-activated chromaffin cells.

to peak was 21.7 ± 6.4 (SEM). These times are much longer than those reported for acutely isolated BAMECs, where depolarization with high $[K^+]_o$ caused the $[Ca^{2+}]_i$ to rise with a latency time of 3.4 ± 0.6 seconds, and a time to peak of 6.6 ± 0.5 seconds (12).

The depolarization-evoked $[Ca^{2+}]_i$ rise in VDCC-lacking HUVEC and the extremely slow kinetics of this response in the present experiments confirm that it was not caused by HUVEC depolarization, but probably by an intermediary molecule released by the activated chromaffin cells. The transmitter between the chromaffin cells and HUVEC (ATP, catecholamine, or another agonist) was most possibly carried by the perfusate flowing from the chromaffin cells to HUVEC, since a reversal of flow direction suppressed the effect.

Stimulation of BAMEC by Chromaffin Cell

In these experiments, the coverslip forming the bottom of the perfusion chamber was covered with subcultured BAMEC. Some BAMECs had chromaffin cells attached so tightly that they were not removed by differential plating. This close relationship between the two cells has been described previously and is thought to be a good model for bidirectional interactions between the two cells in the early developing stages of the adrenal gland (59). In the adult mature gland in vivo, the two cells are separated by a basal membrane (60); thus any interaction between them should involve diffusion of intermediaries.

We measured BAMEC $[Ca^{2+}]_i$ response to high $[K^+]_o$ activation of an attached chromaffin cell. We did not expect the $[Ca^{2+}]_i$ to change in response to depolarization, since BAMECs lose their VDCCs after subculture (12, 28). Therefore, high $[K^+]_o$ should activate the chromaffin cell but not the associated BAMEC. To measure the separate Ca^{2+} response of each attached cell, the excitation UV beam was first focused on the attached chromaffin cell, using a small diaphragm opening, and its response to high $[K^+]_o$ was tested. After washing with fresh control solution we repeated the same experiment, but this time we focused the companion BAMEC to measures its $[Ca^{2+}]_i$ response to high $[K^+]_o$ (Figure 4.6).

Of the six cell couples tested, three showed active normal chromaffin cells that responded to high $[K^+]_o$ with $[Ca^{2+}]_i$ rise. Subcultured BAMECs attached to these responsive chromaffin cells showed a clear increase in $[Ca^{2+}]_i$, which reached a peak at 2.44 ± 0.34 (SEM) times the baseline level. However, the BAMEC response was slow and the $[Ca^{2+}]_i$ rise was staircase-like (Figure 4.6). The latency time was 3.67 ± 2.19 seconds, and the time to peak was 36 ± 6.5 seconds, in contrast with the 6.6-second time to peak of acutely isolated BAMEC (12). The response was not only slower than that of acutely isolated BAMECs to the same stimulus, but also slower than the response to direct ATP stimulation, which had a 2-second latency and 10-second time to peak. These results suggest that BAMEC stimulation was probably mediated by ATP, catecholamines, or another secretion product. The close relationship between the two cells is reflected in the short latency time, which was comparable to that of the $[Ca^{2+}]_i$ response to high $[K^+]_o$ measured in acutely isolated BAMECs (12).

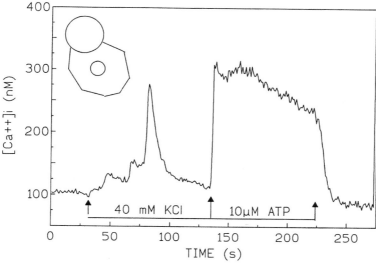

FIGURE 4.6. $[Ca^2]_i$ rise in subcultured BAMEC in response to high $[K^+]_o$. Notice the slow and staircase-like $[Ca^2]_i$ rise, and the sharp response to ATP. The positive response and its kinetics strongly suggest that it was caused by depolarizaton-activated chromaffin cell attached to BAMEC (upper left drawing).

The other three cell couples had chromaffin cells that did not respond to high $[K^+]_o$, nor did any of the attached BAMECs show any $[Ca^{2+}]_i$ change. This confirms that subcultured BAMEC was not directly stimulated by high $[K^+]_o$, but by chromaffin cell activation. Furthermore, the $[Ca^{2+}]_i$ of a single cultured BAMEC without a chromaffin cell attached to it did not rise in response to high $[K^+]_o$, suggesting that, in a medium with few dispersed chromaffin cells, threshold mediator concentration was not reached.

In conclusion, the above experiments showed that in coupled cells an activatable chromaffin cell could induce a $[Ca^{2+}]_i$ increase in the attached BAMEC. On the contrary, no $[Ca^{2+}]_i$ increase was detected in BAMECs attached to nonviable chromaffin cells. Since subcultured BAMECs lack functional VDCCs, the observed response to depolarization in BAMEC with an attached viable chromaffin cell was probably due to an agent released by activated chromaffin cells. In addition to the already mentioned possible intermediaries, we may add K^+, given its high concentration in the secretory granules (61).

Cell–Cell Interaction In Vivo

The ability of chromaffin cells to stimulate endothelial cells in the same dish, as reported in this chapter, suggests that a similar transfer of information may exist in vivo. This interaction would be facilitated by the close proximity between

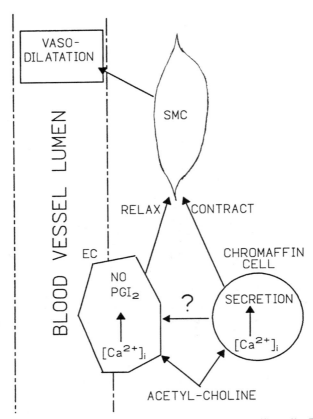

FIGURE 4.7. Diagram of suggested interactions between chromaffin cells, BAMEC, and SMCs. The starting event is ACh activation of chromaffin cells, and possibly the simultaneous stimulation of BAMEC and SMCs. The final event is the vascular response mediated by the balance between contracting and relaxing actions of different factors upon SMCs.

chromaffin cells and the BAMEC in the gland (62). Our results, presented in this chapter, are the first ones to show that chromaffin cells can transmit information to their endothelial counterparts. Cell–cell attachment persists in vitro even after the initial passage.

VDCC activation could occur earlier if the BAMEC were depolarized simultaneously with chromaffin cells by ACh release from nerve terminals, without waiting for the release of an intermediary from the chromaffin cells. This is feasible, since experiments in our laboratory show that BAMEC $[Ca^{2+}]_i$ increases in response to acetylcholine. The different stages that give rise to secretion–perfusion coupling are outlined in Figure 4.7. The first step would consist of simultaneous stimulation of the BAMEC and SMC by products released during chromaffin cell secretion. High $[K^+]_o$ and ATP concentrations may be reached during adrenal gland secretion in the narrow spaces between the chromaffin and endothelial cells. Chromaffin granules contain 80 mM K$^+$ and 150 mM ATP (61).

BAMEC stimulation during secretion would lead to NO release and vasodilatation, raising blood flow through the gland. The counterbalance effect of endothelial cell release of vasoactive factors against smooth muscle cell contraction by an agonist acting simultaneously on both cells has been demonstrated (63). As an example, if the intermediary were ATP, the vessel's abluminal side would be the first one exposed to the agonist. Its binding to smooth muscle cells P_{2x} purinergic receptors would evoke vasoconstriction (64). However, simultaneous stimulation of the BAMEC would induce a $[Ca^{2+}]_i$ rise and endothelial release of vasoactive factors.

Depolarization may be a more direct and faster stimulus for EC secretion than agonists, since stimulation by the latter requires binding to receptors followed by complex biochemical reactions and cytosolic diffusion of intermediaries. This is supported by our data, which show that $[Ca^{2+}]_i$ rise time to peak was 10 seconds under ATP stimulation and 6.6 seconds after depolarization with 40 mM $[K^+]_o$ (12).

ECs from other vascular beds, besides the adrenal medulla and brain, may also contain VDCCs, but this possibility has not been fully explored. Most of the evidence showing lack of VDCCs was collected in large-vessel ECs, and it is therefore necessary to search for VDCCs in ECs from microvascular beds of different organs and tissues. Moreover, VDCCs may have been overlooked in studies performed using cultured ECs after passage 1 or later, since VDCC density is reportedly reduced and becomes zero in cultured BAMECs (12, 28). To our knowledge, the only acutely isolated microvascular ECs measured for VDCC are those from the heart (67) and bovine brain (31), which do not contain VDCCs.

Given the high number of cell processes that are triggered by Ca^{2+}, the search for VDCCs and their functional role in ECs should continue. This research may provide new tools for the control of blood pressure, lipid uptake, inflammatory response, thrombosis (stroke and myocardial infarcts), and so on as shown by the explosive development of research and clinical applications of calcium blockers. Therefore, research in this field has profound implications on many health-related problems.

Acknowledgments. Supported by Grant 1960302 from FONDECYT, Chile, and by the Cystic Fibrosis Foundation.

References

1. Blatter, L. A., Z, Taha, S. Mesaros, P. S. Shacklock, W. G. Wier, and T. Malinsky. Simultaneous measurement of Ca^{2+} and nitric oxide in bradykinin-stimulated vascular endothelial cells. *Circ. Res.* 76:922–24, 1995.
2. Watanabe, K., G. Lam, and E. A. Jaffe. The correlation between rises in intracellular calcium and PGI_2 production in cultured vascular endothelial cells. *Prostaglandins, Leukotrienes and Essential Fatty Acids* 46:211–214, 1992.
3. Jaffe, E. A. Physiological functions of normal endothelial cells. In: *Vascular Medicine*, edited by J. Loscalzo, M. A. Creager, and V. J. Dzau. Boston: Little Brown and Company, pp. 1–19, 1992.

4. He, P., X. Zhang, and F. E. Curry. Ca^{2+} entry through conductive pathway modulated receptor-mediated increase in microvessel permeability. *Am. J. Physiol.* 271:H2377–2387, 1996.

5. Curry, F. E. Modulation of venular microvessel permeability by calcium influx into endothelial cells. *FASEB J.* 6:2456–66, 1992.

6. Himmel, H. M., ARA. Whorton, and H. C. Strauss. Intracellular calcium, currents and stimulus-response coupling in endothelial cells. *Hypertension* 21:112–127, 1993.

7. Johns, A., T. V. Lategan, N. C. Lodge, U. S. Ryan, C. Van Bremen, and J. Adams. Calcium entry through receptor-operated channels in bovine pulmonary artery endothelial cells. *Tissue and Cell* 19(6):733–745, 1987.

8. Colden-Stanfield, M., W. P. Schilling, A. K. Ritchie, S. G. Eskin, L. T. Navarro, and D. L. Kunze. Bradykinin-induced increases in cytosolic calcium and ionic currents in cultured bovine aortic endothelial cells. *Circ. Res.* 61:632–640, 1987.

9. Morgan-Boyd, R., J. M. Stewart, R. J. Vavrek, and A. Hassid. Effects of bradykinin and angiotensin II on intracellular Ca^{2+} dynamics in endothelial cells. *Am. J. Physiol.* 253:C588–C598, 1987.

10. Ryan, U. S., P. V. Avdonin, E. Y. Posin, E. G. Popov, S. M. Danilov, and V. A. Tkachuk. Influence of vasoactive agents on cytoplasmic free calcium in vascular endothelial cells. *J. Appl. Physiol.* 65:2221–2227, 1988.

11. Adams, A. D., R. Lackey, and C. Van Bremen. Ion channels and regulation of intracellular calcium in vascular endothelial cells. *FASEB J.* 3:2390–2400, 1989.

12. Vargas, F. F., S. Calvo, R. Vinet, E. Garde, and E. Rojas. Cytosolic calcium rise evoked by voltage-gated calcium channels activation in adrenal medulla endothelial cells. *Biol. Res.* (in press).

13. Bean, B. P. Classes of calcium channels in vertebrate cells. *Ann. Rev. Physiol.* 51:367–384, 1989.

14. Hess, P. Calcium channels in vertebrate cells. *Annu. Rev. Neurosci.* 13:337–356, 1990.

15. Tsien, R. W., D. Lipscombe, D. V. Madison, K. R. Bley, A. P. Fox. Multiple types of neuronal calcium channels and their selective modulation. *Trends Neurosci.* 11:431–438, 1988.

16. Estacion, M. and L. J. Mordan. Expression of voltage-gated calcium channels correlates with PDGF-stimulated calcium influx and depends upon cell density in C3H 10T1/2 mouse fibroblasts. *Cell Calcium* 14:161–171, 1993.

17. Misler, S., D. W. Barnett, D. M. Pressel, K. D. Gillis, D. W. Scharp, and L. C. Falke. Stimulus-secretion coupling in β-cells of transplantable human islets of Langerhans. *Diabetes* 41:662–670, 1992.

18. Rojas, E., P. Carroll, C. Ricordi, A. Boschero, S. Stojilkovic, and I. Atwater. Control of cytosolic free calcium in cultured human pancreatic β-cells occurs by external calcium-dependent and independent mechanisms. *Endocrinology* 134:771–1781, 1994.

19. Stutzin, A., K. Stojilkovic, J. Catt, and E. G. Rojas. Characteristics of two types of calcium channels in rat pituitary gonadotrophs. *Am. J. Physiol.* 257:C865–C874, 1989.

20. Ceña, V., K. W. Brocklehurst, H. B. Pollard, and E. Rojas. Pertussis toxin stimulation of catecholamine release from adrenal medullary chromaffin cells: Mechanism may be direct activation of L-type and G-type calcium channels. *J. Membr. Biol.* 122:23–31, 1991.

21. Colden-Stanfield, M., W. P. Schilling, L. D. Possani, and D. L. Kunze. Bradykinin-

induced potassium current in cultured bovine aortic endothelial cells. *J. Membr. Biol.* 116:227–230, 1990.

22. Sturek, M., P. Smith, and L. Stehno-Bittel. In vitro models of vascular endothelial cell calcium regulation. In: *Ion Channels of Vascular Smooth Muscle Cells and Endothelial Cells,* edited by N. Sperelakis and H. Kuriyama. New York-Amsterdam-London-Tokyo: Elsevier, pp. 349–365, 1993.

23. Takeda, K., V. Schini, and H. Stoeckel. Voltage activated potassium, but not calcium currents in cultured bovine aortic endothelial cells. *Pflug. Arch.* 410:385–393, 1987.

24. Vargas, F. F., P. Caviedes, and D. O. Grant. Electrophysiological characteristics of cultured human umbilical vein endothelial cells. *Microvasc. Res.* 47:153–165, 1994.

25. Bossu, J., L. A. Feltz, J. L. Rodeau, and F. Tanzi. Voltage dependent calcium transient currents in freshly dissociated capillary endothelial cells. *FEBS Lett.* 255:377–380, 1989.

26. Bossu, J., A. Elhamdani, and L. A. Feltz. Voltage-dependent calcium entry in confluent bovine capillary endothelial cells. *FEBS Lett.* 299:239–242, 1992.

27. Vinet, R. and F. F. Vargas. L- and T-type voltage-gated calcium channels in adrenal medulla microvascular endothelial cells. Submitted to *Am. J. Physiol.* 1997.

28. Vargas, F. F., R. Vinet, and S. Calvo. Voltage-gated Ca^{2+} channels in adrenal medulla endothelial cells and their loss during cell culture. *FASEB. J.* 8:A1061, 1994.

29. Delpiano, M. A. and B. M. Altura. Modulatory effect of extracellular Mg^{2+} ions on K^+ and Ca^{2+} currents on capillary endothelial cells from rat brain. *FEBS Lett.* 394:335–339, 1996.

30. Delpiano, M. A. Ionic currents on endothelial cells of rat brain capillaries. In: *Arterial Chemoreceptors: Cell to system,* edited by R. G. O'Regan, P. Nolan, D. S. McQueen, and D. J. Paterson. New York: Plenum Press, pp. 183–186, 1994.

31. Vargas, F. F., M. E. O'Donnell, and F. E. Curry. Electrophysiology of Brain Micro-vascular Endothelial Cells. *Microcirculation* 4(1):159, 1997.

32. Forsberg, E. J., G. Feuerstein, E. Shohami, and H. B. Pollard. Adenosine triphosphate stimulates inositol phospholipid metabolism and prostacyclin formation in adrenal medullary endothelial cells by means of P_2-purinergic receptors. *Proc. Natl. Acad. Sci. USA* 84:5630–5634, 1987.

33. Gosink, E. C. and E. J. Forsberg. Effect of ATP and bradykinin on endothelial cell Ca^{2+} homeostasis and formation of cGMP and prostacyclin. *Am. J. Physiol.* 265:C1620–C1629, 1993.

34. Bossu, J. L., A. Elhamdani, A. Feltz, F. Tanzi, D. Aunis, and D. Thierse. Voltage-gated Ca entry in isolated bovine capillary endothelial cells: evidence of a new type of BAY K 8644-sensitive channel. *Pflugers Arch.* 420:200–207, 1992.

35. Laskey, R. E., D. J. Adams, A. Johns, G. M. Rubanyi, and C. van Breemen. Membrane potential and $Na^+–K^+$ pump activity modulate resting and bradykinin-stimulated changes in cytosolic free calcium in cultured endothelial cells from bovine atria. *J. Biol. Chem.* 265(5):2613–2619, 1990.

36. Luckhoff, A., and R. Busse. Alcium influx into endothelial cells and formation of endothelium-derived relaxing facror is controlled by the membrane potential. *Pflugers Arch.* 416:305–311, 1990.

37. Furuya, S., C. Edwards, and R. Ornberg. Morphological behavior of cultured bovine adrenal medulla capillary endothelial cells. *Tissue & Cell* 22:615–628, 1990.

38. Voyta, J. C., D. P. Via, C. E. Butterfield, and B. R. Zetter. Identification and isolation of endothelial cells based on their increased uptake of acetylated-low density lipoprotein. *J. Cell Biol.* 99:2034–2040, 1984.

39. Banerjee, D. K., R. L. Ornberg, M. B. H. Youdim, and H. B. Pollard. Endothelial cells from bovine adrenal medulla develop capillary-like growth patterns in culture. *Proc. Natl. Acad. Sci.* USA. 82:4702–4706, 1985.

40. Hamill, O. P., A. Marty, B. Neher, B. Sakman, and F. Sigworth. Improved patch-clamp techniques for high resolution current recording from cells and cell-free membrane patches. *Pfluegers Arch.* 391:85–100, 1981.

41. Olesen, S. P., D. E. Clapham, and P. P. Davies. Haemodynamic shear stress activates a K current in vascular endothelial cells. *Nature* 331:168–170, 1988.

42. Mehrke, G., U. Pohl, and J. Daut. Effects of vasoactive agonists on the membrane potential of cultured bovine aortic and guinea-pig coronary endothelium. *J. Physiol. (London)* 439:277–299, 1991.

43. Lansman, J. B., T. J. Hallam, and T. J. Rink. Single stretch-activated ion channels in vascular endothelial cells as mechano-transducers? *Nature, Lond.* 325:811–813, 1987.

44. Takeda, K. and M. Keppler. Voltage-dependent and agonist-activated ionic currents in vascular endothelial cells. A Review. *Blood Vessels* 27:169–183, 1990.

45. Lori, P., G. Varadi, and A. Schwartz. Molecular insights into regulation of L-type Ca channel function. *NIPS* 6:277–281, 1991.

46. Bertolino, M. and R. R. Llinás. The central role of voltage-activated and receptor-operated calcium channels in neuronal cells. *Annu. Rev. Pharmacol. Toxicol.* 32:399–421, 1992.

47. Stojilkovic, S., A. Torsello, I. Toshihiko, E. Rojas, and K. J. Catt. Calcium signaling and secretory responses in agonist-stimulated pituitary gonadotrophs. *J. Steroid Biochem. Molec. Biol.* 41(3–8):453–457, 1992.

48. Spedding, M. and R. Paoletti. Classification of calcium channels and the sites of action of drugs modifying channel function. *Pharmacol. Rev.* 44:363–376, 1992.

49. Hess, P., B. Lansman, and R. W. Tsien. Different modes of Ca channel gating behaviour favoured by dihydropyridine Ca agonists and antagonists. *Nature* 311:538–544, 1984.

50. Tang, C. M., F. Presser, and M. Morad. Amiloride selectively blocks the low threshold (T) calcium channel. *Science* 240:213–215, 1988.

51. Colden-Stanfield, M., E. B. Cramer, and E. K. Gallin. Comparison of apical and basal surfaces of confluent endothelial cells: Patch-clamp and viral studies. *J. Physiol.* 263:C573–C583, 1992.

52. Stojilkovic, S., M. Kukuljan, M. Tomic, E. Rojas, and J. Catt. Mechanism of agonist-induced [Ca^{2+}]$_i$ oscillations in pituitary gonadotrophs. *J. Biol. Chem.* 268:7713–7720, 1993.

53. Laskey, R. L., D. J. Adams, M. Cannell, and C. van Breemen. Calcium-entry dependent oscillations of cytoplasmic calcium concentration in cultured endothelial cell monolayers. *Proc. Natl. Acad. Sci.* 89:1690–1694, 1992.

54. Neylon, C. B. and R. F. Irvine. Synchronized repetitive spikes in cytoplasmic calcium in confluent monolayers of human umbilical vein endothelial cells. *FEBS Lett.* 275:173–176, 1990.

55. Tracey, W. R. and M. J. Peach. Differential muscarinic receptor mRNA expression by freshly isolated and cultured bovine aortic endothelial cells. *Circ. Res.* 70:234–240, 1992.

56. Stolz, D. B. and B. S. Jacobson. Macro- and microvascular endothelial cells in vitro: Maintenance of biochemical heterogeneity despite loss of ultrastructural characteristics. *In Vitro Cell. Dev. Biol.* 27A:168–182, 1991.

57. Oike, M., G. Droogmans, and B. Nilius. Mechanosensitive Ca^{2+} transients in endo-

thelial cells from human umbilical vein. *Proc. Natl. Acad. Sci. USA.* 91:2940–2944, 1944.

58. Ganong, W. F. *Review of Medical Physiology.* San Francisco, California: Lange Medical Publications, 1985, 295 pp.

59. Mizrachi, Y., P. I. Lelkes, R. L. Ornberg, G. Goping, and H. B. Pollard. Specific adhesion between pheochromocytoma (PC12) cells and adrenal medullary endothelial cells in co-culture. *Cell Tissue Res.* 256:365–372, 1989.

60. Lelkes, P. I. and B. R. Unsworth. Role of heterotypic interactions between endothelial cells and parenchymal cells in organ specific differentiation: A possible trigger for vasculogenesis. In: *Angiogenesis in Health and Disease,* edited by M. E. Maragoudakis, P. Gullino, and P. I. Lelkes. New York: Plenum Press, pp. 27–43, 1992.

61. Ornberg, R. L., G. A. J. Kuijpers, and R. D. Leapman. Electron probe microanalysis of the subcellular compartments of bovine adrenal chromaffin cells. *J. Biol. Chem.* 263(3):1488–1493, 1988.

62. Lelkes, P. I., V. G. Manolopoulos, D. Chick, and B. R. Unsworth. Endothelial cell heterogeneity and organ-specificity. In: *Angiogenesis, Molecular Biology, Clinical Aspects,* edited by M. E. Maragoudaku, P. Gullino, and P. I. Lelkes. New York: Plenum Press, pp. 1–15, 1994.

63. Cohen, R. A., J. T. Shepherd, and P. M. Vanhoutte. Inhibitory role of the endothelium in the response of isolated coronary arteries to platelets. *Science* 221:237–238, 1983.

64. Ralevic, V. and G. Burnstock. Role of P_2-purinoceptors in the cardiovascular system. *Circulation* 84(1):1–14, 1991.

5

Studies of the Glomerular Filtration Barrier: Integration of Physiologic and Cell Biologic Experimental Approaches

Melvin Silverman

Introduction

The human kidney receives approximately 25% of the cardiac output and on a daily basis filters 200 l of plasma at the glomerulus. To maintain volume homeostasis and hemodynamic stability in the face of such large potential outflow "losses" requires that there be rigorous control of the balance between the glomerular filtration rate (i.e., rate of urine formation) and the rate of tubular fluid reabsorption along the nephron. Under normal circumstances, this control is achieved through a combination of hormonal (i.e., renin–angiotensin–aldosterone) and glomerulotubular feedback (macula densa—afferent arteriole resistance) mechanisms. At the level of the glomerular filtration barrier, the classic view of the physiology of urine formation is that it is an entirely passive process (5.1), described by:

$$\text{GFR} = k_f S(\Delta P - \Delta \pi), \tag{5.1}$$

where GFR is the glomerular filtration rate, k_f is the hydraulic conductivity coefficient of the glomerular filtration barrier, S is the surface area available for filtration, ΔP is the hydrostatic pressure gradient across the glomerular filtration barrier, and $\Delta \pi$ is the oncotic pressure gradient across the glomerular filtration barrier.

The 200 l of plasma ultrafiltrate produced each 24 hours also contains ~0.01–0.1 mg/ml albumin. The plasma from which this ultrafiltrate is derived has an albumin content of ~40 mg/ml, which means that the blood/urine albumin concentration ratio is normally maintained at 400–4000/L—a crude but meaningful measure of the solute restrictive properties of the glomerular filtration barrier. Since <100 mg of albumin is eventually excreted in the final urine each day, while during this same period $\sim 200 \times (10{-}100) = 2000{-}20{,}000$ mg of albumin enters Bowman's space, it follows that 1900–19900 mg of albumin undergoes tubular reabsorption every 24 hours.

A relatively small change in the plasma/ultrafiltrate albumin concentration ratio of 40/0.05 to, say, 40/0.1, that is, from 800/L to 400/L, will lead to an increase in albumin filtration of 10 g. Since the tubular reabsorptive capacity for protein is easily saturated, even such minor changes in glomerular capillary permeability as described above can easily result in daily urine albumin losses of >3 g. This degree of albuminuria, if it persists, will eventually lead to a severe, full-blown, clinical condition known as the "nephrotic syndrome," with hypoalbuminuria, edema, and hyperlipidemia in addition to albuminuria. This example serves to dramatize the physiologic importance of maintaining the integrity of the glomerular permeability barrier.

As in the case of solvent (water) flux across the glomerulus, Eq. (5.1), solute flux across the glomerular filtration barrier has also traditionally been described in terms of the standard passive driving forces of diffusion and convection (1), that is,

$$J_s = J_v(1 - \sigma_s) + \omega S \Delta C_s , \qquad (5.2)$$

where J_s is the solute (in our previous example, albumin) flux across the glomerular filtration barrier, J_v is the water flux (effectively equal to the GFR), σ_s is the reflection coefficient of the solute s, ω is the permeability coefficient of the glomerular filtration barrier, S is the surface area available for solute flux across the glomerular filtration barrier, and ΔC_s is the solute concentration gradient across the filtration barrier.

The physical concepts underlying the transglomerular exchange of solvent and solute embodied in Eqs. (5.1) and (5.2) have dictated the traditional view of the biologic behavior of the glomerulus as being that of a passive filter.

Although it is true that the transglomerular exchange of solvent and solute can be accounted for by the phenomenological equations (5.1) and (5.2), nevertheless, over the last decade there has been increasing recognition that glomerular cells contribute directly to these processes by modulating some of the parameters previously assumed to remove constant determinants—such as s surface area available, k_f, hydraulic conductivity, as well as σ_s and ω_s—of solute sieving and permeability.

The major emphasis in this chapter will be to highlight new concepts of glomerular function that give more prominence to the active participation of glomerular cells in determining glomerular filtration and permeability.

Glomerular Ultrastructure

The anatomic organization of the glomerulus is optimized to achieve maximal ultrafiltrative capacity—a beautiful example of the intimate linkage that exists between structure and function in biologic systems. As illustrated schematically in Figure 5.1, blood enters the glomerulus via an afferent arteriole, which immediately branches into an anastomosing capillary network forming a tightly knit vascular tuft of about 20 capillary loops. Blood emerging from the efferent arteriole then enters the peritubular capillaries, which completely envelope adjacent cortical nephrons.

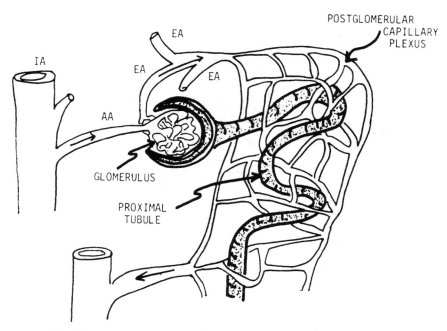

FIGURE 5.1. Schematic representation of glomerular and postglomerular microcirculation. AA, afferent arteriole; EA, efferent arteriole.

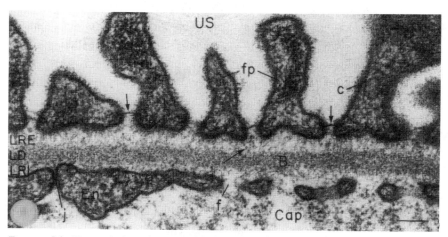

FIGURE 5.2. Transmission electron micrograph of glomerular filtration barrier. B, basement membrane; C, cell coat; CAP, capillary; F, fenestra; FP, foot process; J, junction between two endothelial cells; LD, lamina densa; LI, lamina rara interna; LRE, lamina rara externa; US, urinary space. Arrows point to filtration slit diaphragms. [Reprinted from Fig. 11-4 in Farquhar 1991 (Ref. 13), with permission.]

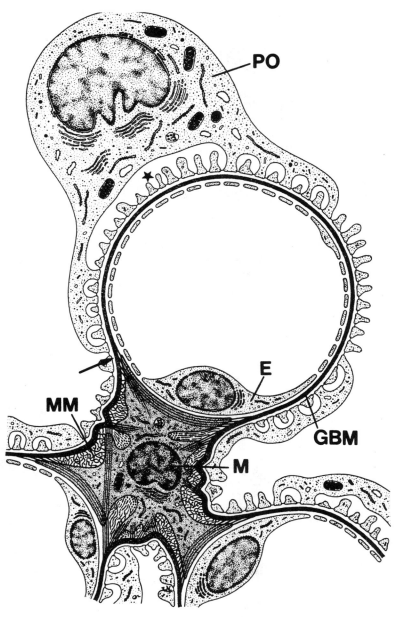

FIGURE 5.3. Schematic representation of the glomerular tuft. E, endothelium; GBM, glomerular basement membrane; M, mesangial cell; MM, mesangial matrix; PO, podocyte. Note the fact that the mesangium is contiguous with a portion of the capillary and that the mesangium effectively forms a "collarlike" structure joining glomerular capillaries. (Reprinted from Benninghoff, *Anatomie*, 15th ed., Urban & Schwarzenberg, München, with permission.)

The glomerular filtration barrier is a heterogeneous structure consisting of three components in series: endothelium, basement membrane, and epithelium. The relationship of these three layers to each other is shown in the transmission electron micrograph in Figure 5.2. Actually, the anatomy of the glomerulus is somewhat more complicated. As depicted schematically in Figure 5.3, interspersed between the capillary loops is an interstitial region composed of mesenchymal-derived mesangial cells and their secreted extracellular matrix products, including collagens (mainly collagen IV), fibronectin, laminin, and various proteoglycans, such as heparan sulfate.

As can be inferred from Figure 5.3, the glomerular cellular constituents are structurally coupled to each other either by direct cell–cell contacts or indirectly through linkages that extend throughout the extracellular fibromatrix. The overall effect is to create a functionally coupled cell network that can be viewed as operating as a single unit in health and disease, and which can modulate the parameters governing transglomerular solute and solvent fluxes.

The glomerular endothelium is fenestrated, with no occluding diaphragm and with an average pore diameter of about 70 nm (see Figure 5.2). This pore size is somewhat larger than is found in other fenestrated capillaries. Routine transmission EM has failed to yield any information about the regions between the fenestra, the so-called fenestral processes. However, high-resolution scanning electron microscopy (HRSEM), an example of which is shown in Figure 5.4, reveals that these fenestral processes have a finite thickness and are nearly circular in cross section. Figure 5.5 is a schematic reconstruction of the fenestral pore based on a morphometric analysis carried out on HRSEM specimens of rat glomerulus (2), similar to the one shown in Figure 5.4. The results illustrated in Figure 5.5 show that the fenestral entrance has a diameter of about 60 nm facing the blood, but the exit diameter facing the basement membrane is about 120 nm.

The detailed three-dimensional structure of the fenestral pore that emerges from these studies has certain possible functional implications. For example, a relatively small decrease of 30 nm at the fenestral pore entrance can reduce the surface area available for filtration by 75%, even though the actual surface area of the pore exposed to the basement membrane is far less affected.

Ultrastructural examination has also demonstrated the presence of cytoskeletal elements in the glomerular endothelium (3), and other studies have confirmed that endothelial cells possess receptors for vasoactive hormones as well as receptor-operated Ca^{2+} channels. Based on this cumulative information, it is becoming increasingly difficult to imagine that the endothelium plays an entirely passive role in the exchange of solutes and water across the glomerular filtration barrier. For example, one possibility is that endothelial fenestra act as dynamic apertures, thereby actively regulating the filtration surface area (2).

Another way in which endothelial cells can influence glomerular function is through release of the signaling molecule nitric oxide (NO). Various ligands, for example, acetylcholine, stimulate production of endothelial NO, which then acts on neighboring mesangial cells causing changes in their contractility (4).

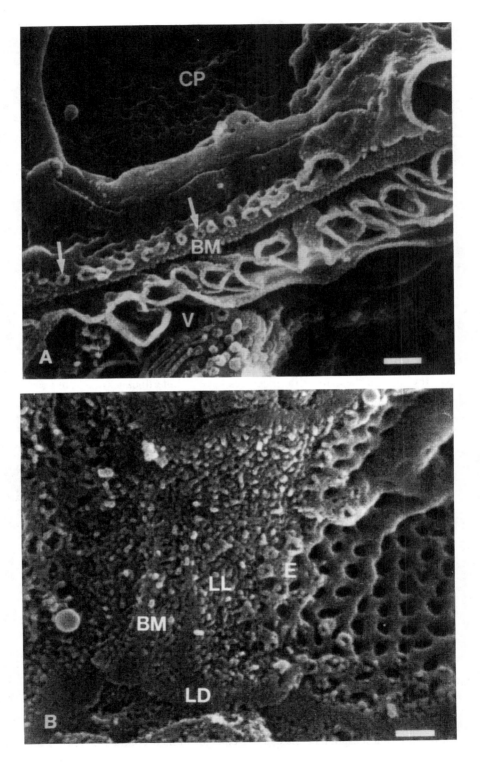

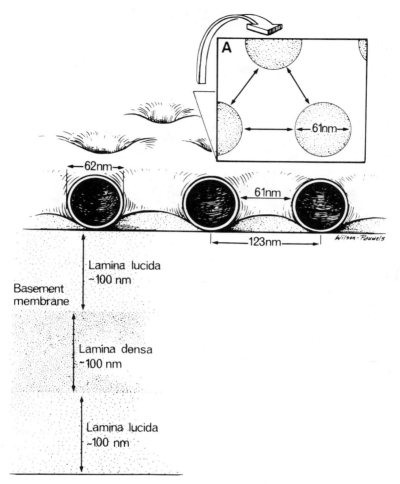

FIGURE 5.5. Structure of a fenestra in the rat glomerular endothelium. A typical cross section, taken from surface view A, shows the ultrastructural relationships of endothelial cell processes to both the fenestra and the glomerular basement membrane. The structures have been drawn to scale in order to reflect more accurately their relative sizes. [Reprinted from Lea et al. 1989 (Ref. 2), with permission.]

FIGURE 5.4. (A) High-resolution scanning electron micrograph of dog glomerular filtration barrier. A transverse fracture shows processes of endothelial cells (CP), basement membrane (BM), and visceral epithelial cells (Podocytes) (V). In cross section, the endothelial cell processes are seen to be nearly circular (arrows). Bar = 0.26 μm. (B) Scanning electron micrograph revealing the granular substructure of the basement membrane (BM). Small, dense granules of the lamina densa (LD) and coarse granules of the lamina lucida (LL) are visible. The fenestrated endothelium (E) is seen partially in cross section, next to the lamina lucida. Bar = 0.22 μm. [Reprinted from Lea et al. 1989 (Ref. 2), with permission.]

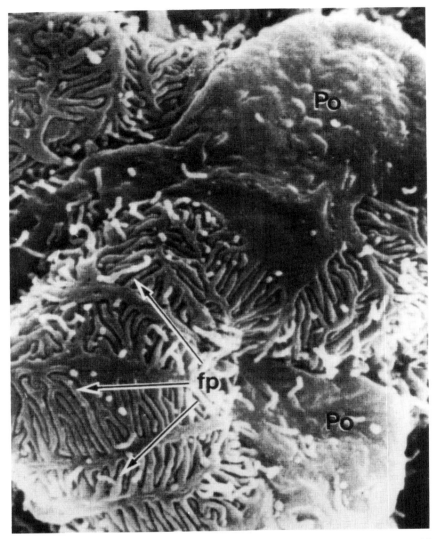

FIGURE 5.6. Scanning electron micrograph of glomerular capillary viewed from outside. Podocytes (Po) send interdigitating foot processes (fp) that cover capillaries from outside. ×5,000. (From Kanwar, Y.S. Biophysiology of glomerular filtration and proteinuria. *Lab. Invest.* 51:7–21, 1984, with permission from Williams & Wilkins.)

The basement membrane is a synthetic product, predominantly of the visceral epithelium, but there is also some contribution from the endothelium. It is made up of collagen IV, fibronectin, laminin, decorin, heparan sulfate, and other proteoglycans. As revealed by HRSEM (Figure 5.2), its ultrastructural organization is predominantly granular and, to a first approximation, it has the appearance of a thin exclusion gel. At the molecular level, there is considerable cross-linking by negatively charged proteoglycan moieties.

The visceral epithelium consists of a cell monolayer that completely envelopes the glomerular capillary (5) (Figure 5.6). These highly differentiated cells are called *podocytes.* Cytoplasmic extensions arise from the main cell body of podocytes and, after several generations of sequential branching, terminate as "foot processes," which are anchored to the basement membrane through integrin mediated attachments. Foot processes from neighboring podocytes interdigitate with each other. The extracellular space between foot processes is a zipperlike structure, called the *slit diaphragm* (Figure 5.7). It functions as a molecular "catcher's mit" and is made up, in part, of ZO1 protein—a well-known constituent of epithelial tight junctions (6). This structure restricts the transglomerular passage of large molecules and also maintains plasma membrane polarization, separating the luminal (urine) podocalyxn domain from the abluminal (basement membrane) integrin domain.

Glomerular mesangial cells (7) play an important role in processing the glomerular filtrate, serving, in effect, as the renal reticuloendothelial system. But of greater interest is the fact that these cells possess surface receptors that are responsive to a variety of circulating vasoactive peptides, hormones, and growth factors as well as to signals from the extracellular matrix. Receptor occupancy at the mesangial cell surface stimulates a variety of intracellular signaling pathways (8) which, in turn, affect the mesangial cytoskeletal contraction, leading to changes in cell shape or actual contraction. As discussed above, there is also evidence of functional coupling between the glomerular endothelial and the mesangial cells mediated by nitric oxide. Consequently, although the mesangial cell cannot be considered an integral part of the filtration barrier, the fact that it is physically contiguous (as shown in Figure 5.3) facilitates functional coupling between the mesangium and the glomerular capillary, ultimately affecting filtration and, perhaps, permeability as well (see below).

Glomerular Permeability

Functional Studies

Functional characterization of glomerular permeability has been carried out by different experimental approaches. The traditional in vivo method has been to determine the fractional clearance of a homologous family of molecular probes such as dextrans, ficoll, or proteins.

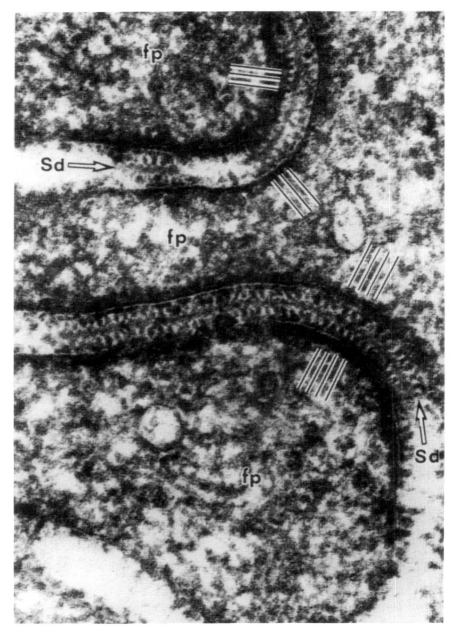

FIGURE 5.7. Electron micrograph of the slit diaphragm (Sd) as viewed en face, revealing highly organized zipperlike substructure. Central filament and cross bridges, indicated by bars, are well resolved. ×150,000. (Reprinted from *The Journal of Cell Biology*, Vol. 60: 423–433, 1974, by copyright permission of The Rockefeller University Press.)

Under steady-state conditions, the clearance of any compound, Cx, over a period τ is given by

$$Cx = \frac{UV}{P} \tag{5.3}$$

where U is the urine concentration of compound x, V is the urine volume over time τ, and P is the plasma concentration of x (assured to be constant over time).

If a compound such as dextran or ficoll is neither secreted nor reabsorbed by the renal tubule, then its clearance calculated from Eq. (5.3) is a measure of its flux across the glomerular filtration barrier. Further, if the clearance of dextran is determined relative to a compound such as inulin, which is freely filtered (i.e., to the same extent as water) at the glomerulus, then the fractional clearance of dextran relative to inulin is a measure of the glomerular filtration selectivity with respect to the test dextran.

Fractional clearance studies in rats using neutral, anionic, and cationic dextrans as molecular weight probes (Figure 5.8) have demonstrated convincingly that size and charge are two important determinants of glomerular permeselectivity. Reduced dextran flux across the glomerular filtration barrier is associated with increasing size and with negative charge. In addition, since proteins or ficoll of equivalent molecular weight are less filterable than dextrans (1), we also can conclude that shape as well as size and charge are also determinants of glomerular permeability—that is to say, the more globular the solute, the greater is the resistance to its movement across the glomerular filtration barrier.

In the early 1980s we undertook a series of investigations using the multiple-indicator dilution technique (MID) to investigate the transcapillary exchange of neutral dextrans in dog kidney in vivo (9). Prior to these studies, it had been controversial as to whether the MID method was capable of distinguishing glomerular from postglomerular extraction of test tracer relative to that of simultaneously injected reference indicators. We were able to establish that this could indeed be achieved experimentally by analyzing simultaneous renal vein and urine fractional recovery outflow curves in response to an intrarenal pulse injection of test tracer. This experimental approach was made quantitative when mathematical models were applied to the renal vein and urine data (10).

Historically, MID had been applied only under conditions when the net capillary ultrafiltration is zero. Our work was intended to measure the "permeability" characteristics of an ultrafiltrative microcirculation. In other words, although the transcapillary flux of solute in the direction perpendicular to blood flow is the sum of convective and diffusive fluxes, the net convective efflux from the capillary to the interstitium is generally assumed to be much smaller than the diffusional flux, and is therefore neglected. But as will become evident below, in an ultrafiltering capillary bed such as exists in the glomerular microculation, the convective flux becomes dominant.

The kinetics of steric interaction (collision) between a permeating solute and the fixed macromolecular structures of the glomerular capillary wall lead to the

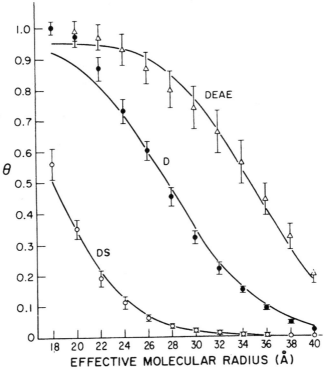

FIGURE 5.8. Filtrate-to-plasma concentration ratio (θ) as function of molecular size for dextran sulfate (DS), neutral dextran (D), and diethylaminoethyl dextran (DEAE). Symbols: mean values ± S.E. measured by Bohrer et al. (*J. Clin. Invest.* 61:72–78, 1978; *Am. J. Physiol.* 233:F13–F21, 1977) in normal hydropenic Munich-Wistar rats; curves are theoretical calculations based on values of Q_A (ΔP), and C_A reported in those studies with $K_f =$ 4.8 nl/(min · mm Hg), r_o = 47 Å, and C_m = 165 mEq/L. (Reprinted from Deen, W.M., B. Satuat, and J.M. Jamieson. Theoretical model for glomerular filtration of charged solutes. *Am. J. Physiol.* 238:F126–F139, 1980, with permission.)

exclusion of a fraction of the molecules attempting to pass through the wall and enter the urine space. Such steric kinetics are characterized by the hydraulic reflection coefficient σ_s. If convection dominates the ultrafiltration process, a fraction equal to $(1 - \sigma_s)$ of the permeating species enters and passes through the capillary wall.

For steady-state ultrafiltration through channels permitting transmembrane movement by diffusion and convection but no chemical reaction,

$$C2/C1 = \frac{1 - \sigma_s}{1 - \sigma_s e^{-k}} \qquad (5.4)$$

where C2/C1 is the relative urine-to-plasma concentration of permeating solute and $k = (1 - \sigma_s)J_V P_S$ with J_V the volume flux and P_S the permeability constant as defined previously.

The left-hand side of the equation is equivalent to the fractional clearance $[(U/P)_D/(U/P)_I]$ or fractional extraction E_D/E_I, measured by MID, where the subscript D refers to the tracer dextran and I refers to the glomerular reference (inulin).

In general, solute flux is the result of both diffusive and convective forces. The contribution of diffusion to the total solute flux can be assessed experimentally by testing for flow dependence of the unidirectional extraction of neutral dextrans from the glomerular microcirculation in anesthetized dogs maintained in auto-regulatory range and under mannitol diuresis. If diffusion represents a significant component of total flux, then, with decreasing flow and consequently decreasing glomerular filtration rate, a disproportionate increase in glomerular extraction of dextrans should occur relative to a simultaneously filtered extracellular reference. However, when we carried out such protocols in anesthetized dogs in contrast to what had been found in the rat, we were unable to observe any change in glomerular extraction of dextran when the renal plasma flow was lowered two- to three-fold.

From inspection of Eq. (5.4), when the contribution from diffusion can be neglected, Eq. (5.4) becomes

$$C2/C1 \equiv (1 - \sigma_s). \tag{5.5}$$

These experimental observations enabled us to exploit the simplified form given by Eq. (5.5) and allowed us to obtain quantitative in vivo estimates of the reflection coefficients for neutral dextrans in vivo.

Results from typical MID experiments are shown in Figure 5.9. The agreement with σ_s values obtained by classical steady-state tracer clearance methods with our MID results is excellent.

Finally, the dextran sieving data shown in Figure 5.10 can be used to calculate an equivalent pore radius to account for the data. It turns out that, to a good first approximation, the glomerular filtration barrier in anesthetized dogs can be modeled as a homogeneous set of pores of ~55 Å (10).

Although such "pore" models (1) account mathematically for the physiologic sieving data, pores of 55 Å, which should be readily identifiable by electron microscopy, cannot be found within the glomerular basement membrane. We shall return to this issue later.

The charge-selective properties of the glomerular capillary wall are maintained by anionic proteins, which are dispersed in the basement membrane and by glomerular polyanions (or GPAs), that restrict the passage of negatively charged molecules more than neutral ones of the same molecular size and facilitate the transport of positively charged molecules. Clearance studies have suggested that the GPAs provide a free energy barrier which prevents excessive ultrafiltration of anionic plasma proteins.

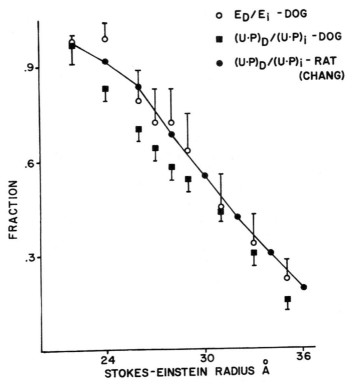

FIGURE 5.9. Neutral dextran glomerular fractional extraction (from single-pass multiple-indicator dilution) and fractional clearance measurements (relative to inulin). Note the very close correspondence of the data obtained in dogs and rats by the two different methods.

However, because of their steady-state nature, clearance methods are unsuited to kinetic studies of the binding events that ligands experience while passing through the glomerular wall. This kinetic information is, however, expressed directly in the dynamics of the MID fractional recovery curves for the urine outflow. Using tracer anionic (11) and cationic dextran probes (12) to elucidate the action of the GPA on these charged solutes, we were able to corroborate that in dogs (as had been found previously in rats) there is a restriction on the transglomerular passage of negatively charged dextrans. We were also able to demonstrate saturable binding of the cationic dextrans to dog glomeruli in vivo.

In summary, the composite picture from physiologic in vivo studies utilizing either clearance or MID methods is that the glomerulus behaves as a more or less isoporous membrane of ~55-Å size imbedded in a polyanionic matrix. The net

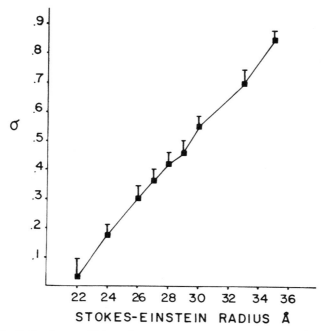

FIGURE 5.10. Reflection coefficients for neutral dextrans as a function of dextran Stokes-Einstein radius calculated from Eq. (5.5), using neutral dextran multiple-indicator dilution data published in reference 9. (Reproduced with permission from Ref. 9.)

result is a barrier that restricts passage on the basis of increasing size, negative charge, and globular shape. Since the major plasma protein, albumin, is a globular anionic protein of Stokes–Einstein radius of ~45 Å, it is not surprising that the normal glomerulus is so impermeant to the passage of albumin.

For more than 25 years there has been a debate about the relative contribution of the glomerular basement membrane and the podocyte slit diaphragm in determining glomerular solute permeselectivity (the endothelium was never considered to be an important factor because of the large fenestra).

Early immunocytochemical studies using relatively large horseradish peroxidase labeled molecular weight markers provided compelling evidence that slit pores were not the limiting barrier. In addition, diffusion gradients of radiolabeled dextrans were demonstrated within the thickness of the basement membrane. In addition, biochemical characterization of basement membrane proteoglycan moieties revealed that the basement membrane macromolecular lattice was liberally "spiked" with anionic components. Taken together, findings such as these led to the suggestion that the basement membrane was the principal barrier to the transglomerular passage of larger molecular weight solutes such as proteins (reviewed in 13).

Studies by Bendayan (14) using a more physiological molecular weight marker, bovine serum albumin coupled dinitrophenol, demonstrated convincingly the restrictive nature of the basement membrane, which seemed to retard molecular entry largely based on charge, whereas the slit pores appear to limit access to the urine space more on the basis of size, as well as charge.

However, some doubt has been cast on the physiological relevance of these ultrastructural investigations based on recent in vitro experiments using isolated rat basement membrane (15). In these latter experiments, both the hydraulic conductivity and the albumin permeability were found to be much larger than the measured values of these same parameters in vivo! The conclusion we can draw is that in vivo glomerular permeability characteristics are probably considerably more dependent on the cellular constituents of the glomerular filtration barrier than has heretofore been recognized.

In particular, there is a growing body of experimental evidence to suggest that both the mesangial cell as well as the visceral epithelial cell (podocyte) each play important roles in maintaining glomerular permeability.

The Glomerular Visceral Epithelium—Its Influence on Glomerular Permeability

One way to illustrate the role of the glomerular epithelium in maintaining the selectivity of the glomerular filtration barrier is to provide an example using an animal model in which the barrier is more or less specifically disrupted. A popular model for this purpose is the so-called *Puromycin Nephrosis Model* in rats. Within 24 hours of parenteral administration of the aminoglycoside puromycin, detectable albuminuria occurs and marked proteinuria ensues by 48 hours (16). There are detectable alterations of the width of the podocyte foot processes within 24 hours, and after 48 hours the loss of foot process structure is marked. Podocyte cytoskeletal disaggregation and detachment of the glomerular epithelial cell from the basement membrane is most severe on day 5 (17). These morphological changes are accompanied by a decrease in proteoglycan and increased collagen I, laminin, and fibronectin mRNA level expression in the glomerular basement membrane within the first 48 hours.

The central observation of the rat puromycin model is that disruption of the visceral epithelium caused by foot process detachment from the GBM is the most important correlate of the development of proteinuria. The evidence from the animal model of puromycin nephrotoxicity combined with the data from ultrastructural studies summarized earlier strongly support the conclusion that the integrity of the visceral epithelium is central to the maintenance of normal glomerular permeselectivity and that the glomerular slit pores are probably an important limiting barrier in the transglomerular exchange of protein solutes.

Mesangial Cell–Extracellular Matrix (ECM) Interactions: Effects on Glomerular Function

What about the effect of mesangial cells on glomerular filtration and permeselectivity? The present view is that mesangial cells exert control on glomerular filtration by a combination of isometric and isotonic contractile behaviors.

Studies of mesangial cells in tissue culture show that isotonic contraction is determined by receptor-mediated changes in intracellular Ca^{2+} signalling, for example, through the interaction of vasoactive agonist hormones such as vasopressin, angiotensin II, and endothelin. Further, in vivo micropuncture studies have shown that vasoactive peptides and hormones influence the GFR through effects that change either or both the capillary surface area available for filtration and the hydraulic conductivity.

Up until recently it was unclear as to what the precise mechanism was at the cellular level by which modulation of the mesangial cell contraction could result in changes of the capillary surface area. However, based on new three-dimensional ultrastructural reconstruction studies (17), approximately 10–15% of the glomerular capillaries are enveloped by mesangial cells. Therefore, stimulation of the isotonic contraction will reduce capillary diameter and effectively exclude blood flow from this region of the glomerular microcirculation. At the macroscopic level such effects would be manifest as a reduced surface area for filtration. The overall impact would not only be reflected in regulation of the GFR, but would also alter the convective component of transglomerular solute flux and would therefore be manifest as a change in glomerular permeability.

On the basis of the foregoing considerations it seems relatively clear how mesangial cell contractility might contribute to regulation of the glomerular function. But what about alterations in the isometric contractility of mesangial cells? Isometric contractility involves alterations in cell shape and is determined through actions of the cytoskeleton. Such cytoskeletal effects dominate mesangial cell interactions with the extracellular matrix, including attachment, spreading, and migration. Given the significant contact area between the mesangium and the capillary, any remodeling of the extracellular matrix in response to mesangial cell attachment, spreading, and migration will cause shortening or lengthening of the mesangial-cell–matrix collar (depicted in Figure 5.3) and thereby exert tension on adjacent capillary walls with the potential to change the intraglomerular flow distribution.

The question then arises as to whether it is feasible to experimentally determine the factors that regulate this type of mesangial cell behavior. It appears that this may indeed be possible by taking advantage of so-called three-dimensional cultures in which mesangial cells are grown *embedded* in collagen as opposed to being grown *attached* to the surface of collagen gels (so-called two-dimensional cultures).

Evidence is accumulating that cells grown in such three-dimensional cultures resemble the in vivo phenotype more closely than cells in two-dimensional culture. One particularly informative experimental cell biological test system involving the use of three-dimensional culture is the *collagen gel contraction assay*. It has been utilized extensively to investigate a variety of physiological and pathological states including wound healing and blood vessel wall repair following angioplasty (19–28). We wished to exploit the gel contraction assay to investigate the regulatory role of the mesangium in determining glomerular function.

The cell biology involved in the gel contraction assay has been studied intensively. When cells are placed in collagen, they attach to it via integrins within 30 minutes of being plated (29), and by 4 hours they begin to spread and develop lammellopodia. Both of these processes are associated with focal adhesion formation (30, 31). As cells spread, the actin cytoskeleton generates force that is transmitted to the extracellular matrix via integrins, resulting in extracellular matrix remodeling (27, 32, 33). Once the extracellular matrix is remodeled, cells migrate and form new cell–substratum attachments. Integrin association and stabilization with the cytoskeleton by focal adhesion formation is essential for cell migration (30, 31, 34). If the collagen gel in which the cells are embedded is not fixed to the culture dish, then the extracellular matrix remodeling that takes place results in gel "contraction."

We have systematically studied the manner in which various biologic agonists stimulate mesangial cells in three-dimensional collagen cultures and initiate gel contraction (35), and found that gel contraction is initiated by agonist-induced activation of various protein tyrosine kinase phosphorylation cascades. But an important unresolved issue is whether the induced gel contraction has any direct relevance to the mechanism by which these same agonists affect glomerular function.

At this stage it would be highly speculative to claim that there is any linkage between the in vitro behavior associated with mesangial cell-induced gel contraction and regulation of glomerular filtration or permeability. Nevertheless, given the intimate physical relationship of the mesangium and the glomerular capillary, it is hard to imagine that changes in the microscopic dimensions of the glomerular interstitium would not impact on the geometry of the capillaries and ultimately effect microcirculatory physiology.

We visualize that forces arising from mesangial–extracellular matrix interactions are transmitted to the microtubules that connect the glomerular basement membrane to the mesangial cell, effectively creating a biomechanical-collar–like unit (represented schematically in Figure 5.3). Under resting conditions, in conjunction with the podocyte, we visualize that the resultant forces maintain the capillaries in an open state. When cells are exposed to "constrictor agents," the actin filaments undergo complex changes that cause a reduction in cell volume (rounding up), which leads to expansion of the mesangial matrix and relaxation of collagen, with a consequent reduction of capillary diameter. Conversely, agonists that interact with mesangial cells in three-dimensional collagen gels causing stimulation of gel contraction will lead to shortening of the mesangial collar and enlargement of the capillary diameter, with a consequent increase in flow. In other words, we are proposing that changes in mesangial cell attachment, spreading, and migration observed in three-dimensional cultures will be translated into changes in the physiology of the glomerular microcirculation.

At this stage there is no direct evidence in support of this conjecture, but there are some intriguing hints that we are on the right track. For example, in recently completed work (35) we have found that low concentrations (~10 ng/ml) of platelet-derived growth factor (PDGF) stimulates gel contraction in mesangial cell three-dimensional collagen cultures through a PI-3 kinase–dependent pathway, but at higher concentrations (~80 ng/ml) gel contraction is inhibited because of disruption of the mesangial cell cytoskeleton. Interestingly, PDGF infusion at rates of 125 ng/kg/min (36) in vivo causes a decrease in the glomerular filtration rate. It is hard to imagine that this effect of the PDGF on glomerular function in vivo derives from its potential to stimulate mesangial cell proliferation. The more plausible explanation is that the PDGF is exerting this action on the glomerular filtration rate by altering ECM–mesangial cell signaling, as reflected in the in vitro gel contraction assay.

It remains to be seen whether the glomerular regulatory action of other agonists such as vasoactive peptides are also mediated at the cellular level through effects on the attachment, spreading, and migration of mesangial cells.

Reactive Oxygen Species (ROS) as Physiologic Signaling Molecules Influencing Mesangial Cell Function

In the final section of this chapter we discuss the role of reactive oxygen species (ROS) as potentially important physiologic signaling molecules regulating glomerular function. Much has been written about nitric oxide as a potent signalling molecule, but so far very little has been written about ROS.

Reactive oxygen species are by-products of oxidative metabolism. Under normal circumstances, the cellular level of ROS is regulated and there is a balance between the production and degradation of ROS by reducing enzymes such as catalase, glutathione reductase, and bcl-2. Over the last few years, there has been mounting evidence implicating ROS as causative agents in glomerular disease. In the glomerulus, the source of ROS may be either the mesangial cell itself or "passenger" leukocytes such as neutrophils, monocytes, or macrophages. ROS have also been implicated in the pathogenesis of puromycin aminonucleoside nephrosis discussed above (37–40).

Constitutive production of ROS by mesangial cells originates from an intrinsic NADPH oxidase (41) that normally functions at a low level but increases in response to inflammatory stimuli such as cytokines (42). Neutrophils, monocytes, and macrophages possess a similar NADPH oxidase, but their production of ROS is considerably greater than that of mesangial cells (37). Glomerular epithelial cell production of ROS is considerably less than that of mesangial cells (35). Whatever the source of glomerular ROS, whether endogenously produced by resident glomerular cells or contributed exogenously by "passenger" inflammatory cells, the question arises as to whether ROS effects on tissue are uniformly hazardous. The answer is no!

Although ROS are highly reactive molecules with the potential to cause cell injury, exposure to ROS also initiates a number of physiologic responses, including tyrosine phosphorylation of growth factor receptors (43) and p42/p44^{MAPK} in vascular smooth muscle cells and neutrophils (44–46). Furthermore, both the proliferative and migratory effects of PDGF on vascular smooth muscle cells are, at least in part, dependent on the production of ROS (47). In cultured mesangial cells, ROS induce tyrosine phosphorylation of the PDGF receptor as well as pp60^{c-src} (48). In addition, high concentrations of ROS have been used as phosphatase inhibitors and, in the presence of sodium vanadate, induce formation of focal adhesions (49). The chemical basis of these effects is unknown but may be due to the direct oxidation of critical protein sulfhydryl groups or through the formation of transition metal complexes (50).

It is not surprising, therefore, that constitutively expressed ROS have been postulated to play a role as physiologic signaling molecules analogous to nitric oxide because they are rapidly diffusible and highly reactive (43, 44, 46). In addition, both intra- and extracellular concentrations of ROS are tightly controlled by several intracellular enzymes, including superoxide dismutase, catalase, and glutathione peroxidase, allowing for the rapid termination of cell signals.

In view of the increasing evidence that ROS are potentially modulating cell–extracellular matrix interactions, we sought to assess the effects of ROS on mesangial cells grown in three-dimensional collagen gels. When mesangial cells in three-dimensional culture are exposed to H_2O_2 at low concentrations, gel contraction is stimulated, reaches a maximum, and then decreases to control levels. Moreover, ROS stimulation of gel contraction is associated with a marked increase of tyrosine phosphorylation of several intracellular proteins, including 125FAK (35).

Therefore, ROS are able to modulate mesangial cell behavior in three-dimensional culture in two different ways: at low concentrations of ROS, a protein tyrosine kinase signaling cascade is initiated with an accompanying gel contraction that is consistent with stimulation of the mesangial cell attachment, spreading and migration. However, as the concentration is raised, we move to the opposite extreme, where now the ROS lose their potential to stimulate gel contraction. Eventually, at higher concentrations, the ROS actually inhibit gel contraction (51), likely through the initiation of apoptosis (52).

Thus, the ROS appear to be capable of dual regulatory behavior—at one extreme stimulation of protein tyrosine kinase signaling, and at the other extreme stimulation of programmed cell death. In a very real sense, therefore, the ROS have the potential to modulate the mesangium as well as the glomerular function.

It is also possible that the ROS produced by mesangial cells can influence the function of neighboring glomerular epithelium in a paracrine fashion.

Returning to the rat puromycin nephrotoxicity model described earlier in this chapter, ROS have been implicated in the pathogenesis of this animal model. Increased amounts of ROS have been measured in whole glomeruli in adriamycin-induced nephrotic syndrome, which may be analogous to PAN (38), and ROS are also produced by cultured kidney slices exposed to puromycin (52). Cultured glomerular epithelial cells exposed to puromycin produce H_2O_2, and toxicity produced by this oxidizing agent can be reversed by the antioxidants catalase and desferri-oxamine (53). In the rat puromycin model, administration of the antioxidants SOD and allopurinol reduced proteinuria and prevented glomerular injury (40). Taken together, there is substantial evidence implicating ROS production in the pathogenesis of puromycin nephrotoxicity with the glomerular epithelial cell as the main target. Similarly, there is also new evidence that ROS are responsible for proteinuria in the rat model of membranous nephritis (Heymann Nephritis). In this model as well, the principal target appears to be the glomerular epithelial cell. Therefore, as stressed earlier, there seems little doubt that the integrity of the visceral epithelial cell attachment to the basement membrane is a critical determinant in maintaining normal glomerular permselectivity.

But how can we link the puromycin-stimulated increase in mesangial cell ROS to the detachment of epithelial foot processes and the production of proteinuria? In an attempt to answer this question we again turned to the mesangial cell three-dimensional culture system to explore the effects of puromycin on mesangial cell-induced gel contraction (51). We were able to show that puromycin, in concentrations similar to those that cause pathological changes to glomerular epithelial cells in culture and in vivo, also produced mesangial cell dysfunction. In particular, puromycin inhibits FBS-induced mesangial cell-collagen gel contraction. Moreover, we found that these effects of puromycin on mesangial cells were caused by a puromycin-mediated induction of increased ROS production by mesangial cells.

These results suggest that the effects of puromycin-stimulated mesangial cell ROS production may not be limited to the mesangial cell itself, but may also impact on adjacent glomerular epithelial cells causing glomerular epithelial cell toxicity and glomerular epithelial cell detachment. Thus, mesangial cells may play a more important role in the pathogenesis of puromycin nephrosis than has heretofore been recognized.

We have identified a new potential mechanism for altering glomerular permselectivity and causing proteinuria—namely, stimulation of mesangial cell ROS production leading to secondary cytotoxic effects on glomerular epithelial cell attachment to basement membranes. It should be noted that there are a number of biologically significant inflammatory activators, such as cytokines, that are capable of stimulating ROS production in all types of glomerular inflammatory states.

In summary, we have demonstrated how glomerular changes in cell–cell (MC–GEC) or cell–ECM (MC–ECM, GEC–GBM) interactions can lead to proteinuria and altered glomerular permselectivity. Moreover, these changes need not be associated with significant cytotoxicity leading to cell death, but likely will lead to cellular dysfunction through changes in intracellular signaling.

The cumulative evidence is therefore compelling that, in order to better understand the regulatory mechanisms controlling glomerular filtration and permselectivity, we need to give greater emphasis to cell biologic rather than pure physicochemical models. It will be especially important that these new models take into consideration the functional linkages between all of the cellular and noncellular constituents of the glomerulus.

Acknowledgments. This work was supported by grants from the Medical Research Council and the Kidney Foundation of Canada.

References

1. Maddox, D.A., W.M. Deen, and B.M. Brenner. Glomerular filtration. In: *Handbook of Physiology, Section 8, Renal Physiology, Volume 1,* edited by E. Windhager. New York: Oxford University Press, pp. 545–638, 1992.
2. Lea, P.J., M. Silverman, R. Hegele, and M.J. Hollenberg. Tridimensional ultrastructure of glomerular capillary endothelium revealed by high-resolution scanning electron microscopy. *Microvasc. Res.* 38:296–308, 1989.
3. Vasmant, D., M. Maurice, and G. Feldman. Cytoskeleton ultrastructure of podocytes and glomerular endothelial cells in man and in the rat. *Anat. Rec.* 210:17–24, 1984.
4. Raij, L. and C. Baylis. Glomerular actions of nitric oxide. *Kidney Int.* 48:20–32, 1995.
5. Tisher, C.C. and Madsen, K.M. Anatomy of the kidney. In: *The Kidney,* edited by B.M. Brenner and F.C. Rector. Philadelphia: W.B. Saunders Co., pp. 3–131, 1991.
6. Schnabel, E., J.M. Anderson, and M.G. Farquhar. The tight junction protein ZO-1 is concentrated along slit diaphragms of the glomerular epithelium. *J. Cell Biol.* 111: 1255–1263, 1990.
7. Latta, H. An approach to the structure and function of the glomerular mesangium. *J. Am. Soc. Nephrol.* 2:S65–S73, 1992.
8. Kreisberg, J.I., M. Venkatachalam, and D. Troyer. Contractile properties of cultured glomerular mesangial cells. *Am. J. Physiol.* 249:F457–F463, 1985.
9. Whiteside, C. and M. Silverman. Determination of glomerular permselectivity to neutral dextrans in the dog. *Am. J. Physiol.* 245:F485–F495, 1983.
10. Lumsden, C.J. and M. Silverman. Multiple indicator dilution and the kidney: Kindetics, permeation, and transport in vivo. *Meth. Enzymol.* 191:34–72, 1990.
11. Whiteside, C. and M. Silverman. Glomerular and postglomerular permselectivity to anionic dextrans in the dog. *Am. J. Physiol.* 247:F965–F974, 1984.
12. Whiteside, C.I. and C.J. Lumsden. Transglomerular cationic macromolecular flux is mediated by a convection-binding mechanism. *Am. J. Physiol.* 256:F882–F893, 1989.
13. Farquhar, M.G. The glomerular basement membrane. A selective macromolecular filter. In: *Cell Biology of Extracellular Matrix,* edited by E.D. Hay. New York: Plenum Press, pp. 365–418, 1991.
14. Ghitescu, L., M. Desjardins, and M. Bendayan. Immunocytochemical study of glomerular permeability to anionic, neutral and cationic albumins. *Kidney Int.* 42:25–32, 1992.

15. Daniels, B., E. Hauser, W. Deen, and T. Hostetter. Glomerular basement membrane: In vitro studies of water and protein permeability. *Am. J. Physiol.* 262:F919–F926, 1992.

16. Whiteside, K., R. Prutis, R. Cameron, and J. Thompson. Glomerular epithelial detachment, not reduced charge density, correlates with proteinuria in adriamycin and puromycin nephrosis. *Lab. Invest.* 61:650–660, 1989.

17. Whiteside, C.I., R. Cameron, S. Munk, and J. Levy. Podocyte cytoskeletal disaggregation and basement-membrane detachment in puromycin aminonucleoside nephrosis. *Am. J. Pathol.* 142:1641–1653, 1993.

18. Inkyo-Hayasaka, K., T. Sakai, N. Kobayashi, I. Shirato, and Y. Tomino. Three-dimensional analysis of the whole mesangium in the rat. *Kidney Int.* 50:673–683, 1996.

19. Emerman, J.T. and D.R. Pitelka. Maintenance and induction of morphological differentiation in dissociated mammary epithelium on floating collagen membranes. *In Vitro Cell. Dev. Biol.* 13:316–328, 1977.

20. Hall, H.G., D.A. Farson, and M.J. Bissell. Lumen formation by epithelial cell lines in response to collagen overlay: A morphogenetic model in culture. *Proc. Natl. Acad. Sci. USA.* 79:4672–4676, 1982.

21. Kitamura, M., N. Maruyama, T. Mitarai, R. Nagasawa, H. Yoshida, and O. Sakai. Extracellular matrix contraction by cultured mesangial cells: Modulation by transforming growth factor b and matrix components. *Exp. Mol. Pathol.* 56:132–142, 1992.

22. Kitamura, M., T. Mitarai, N. Maruyama, R. Nagasawa, H. Yoshida, and T. Sakai. Mesangial cell behaviour in a three-dimensional extracellular matrix. *Kidney Int.* 40:653–661, 1991.

23. Lin, C.Y. and F. Grinnell. Decreased level of PDGF-stimulated receptor autophosphorylation by fibroblasts in mechanically relaxed collagen matrices. *J. Cell Biol.* 122:663–672, 1993.

24. Marx, M., T.O. Daniel, M. Kashgarian, and J.A. Madri. Spatial organization of the extracellular matrix modulates the expression of PDGF-receptor subunits in mesangial cells. *Kidney Int.* 43:1027–1041, 1993.

25. Marx, M., R.A. Perlmutter, and J.A. Madri. Modulation of platelet-derived growth factor receptor expression in microvascular endothelial cells during in vitro angiogenesis. *J. Clin. Invest.* 93:131–139, 1994.

26. Montesano, R. and L. Orci. Transforming growth factor b stimulates collagen-matrix contraction by fibroblasts: Implications for wound healing. *Proc. Nat. Acad. Sci. USA* 85:4894–4897, 1988.

27. Stopak, D. and A.K. Harris. Connective tissue morphogenesis by fibroblast traction. *Develop. Biol* 90:383–398, 1982.

28. Zent, R., M. Ailenberg, T.K. Wadell, G.P. Downey, and M. Silverman. Puromycin aminonucleoside inhibits mesangial cell-induced gel contraction of collagen gels by stimulating production of reactive oxygen species. *Kidney Int.* 47:811–817, 1995.

29. Grinnell, F. and C.R. Lamke. Reorganization of hydrated collagen gels by human skin fibroblasts. *J. Cell. Sci.* 66:51–63, 1984.

30. Burridge, K., C.E. Turner, and L.H. Romer. Tyrosine phosphorylation of paxillin and pp125[FAK] accompanies cell adhesion to extracellular matrix: A role in cytoskeletal assembly. *J. Cell Biol.* 119:893–903, 1992.

31. Romer, L.H., N. McLean, C.E. Turner, and K. Burridge. Tyrosine kinase activity, cytoskeletal organization, and motility in human vascular endothelial cells. *Mol. Biol. Cell* 5:349–361, 1995.

32. Ehrlich, H.P. and D.J. Wyler. Fibroblast contraction of collagen lattices in vitro: Inhibition by chronic inflammatory cell mediators. *J. Cell. Phys.* 116:345–351, 1983.

33. Mochiate, K., P. Pawelek, and F. Grinnell. Stress relaxation of contracted collagen gels: Disruption of actin filament bundles, release of cell surface fibronectin, and downregulation of DNA and protein synthesis. *Exp. Cell Res.* 193:198–207, 1991.

34. Schmidt, C.E., A.F. Horwitz, D.A. Lauffenburger, and M.P. Sheetz. Integrincytoskeletal interactions in migrating fibroblasts are dynamic, assymetric and regulated. *J. Cell Biol.* 123:977–991, 1993.

35. Zent, R. Signaling in mesangial cells grown in three-dimensional culture. Ph.D. Thesis, University of Toronto, 1997.

36. Abboud, H. Role of platelet-derived growth factor in renal injury. *Ann. Rev. Physiol.* 57:297–309, 1995.

37. Shah, S.V. 1995. The role of oxygen metabolites in glomerular disease. *Ann. Rev. Physiol.* 57:245–262.

38. Ueda, N., B. Guidet, and S.V. Shah. Measurement of intracellular generation of hydrogen peroxide by rat glomeruli in vitro. *Kidney Int.* 45:788–793, 1994.

39. Shah, S.V. Role of reactive oxygen metabolites in experimental glomerular disease. *Kidney Int.* 35:1093–1106, 1989.

40. Diamond, J.R., J.F. Bonventre, and M.J. Karnovsky. A role for free oxygen radicals in aminonucleoside nephrosis. *Kidney Int.* 29:478–483, 1986.

41. Radeke, H.H., A.R. Cross, J.T. Hancock, O.T.G. Jones, M. Nakamura, V. Kaever, and K. Resch. Functional expression of NAPDH oxidase components (alpha- and beta-subunits of cytochrome b558 and 45-kDa flavoprotein) by intrinsic human glomerular mesangial cell. *J. Biol. Chem.* 266:21025–21029, 1991.

42. Radeke, H.H., B. Meier, N. Topley, J. Floge, G.G. Habermehl, and K. Resch. Interleukin 1-a and tumor necrosis factor-a induce oxygen radical production in mesangial cells. *Kidney Int.* 37:767–775, 1990.

43. Huang, R., J. Wu, and E.D. Adamson. UV activates growth receptors via reactive oxygen intermediates. *J. Cell Biol.* 133:211–220, 1996.

44. Baas, A.S. and B.C. Berk. Differential activation of mitogen-activated protein kinases by H_2O_2 and O_2^- in vascular smooth muscle cells. *Circ. Res.* 77:29–36, 1995.

45. Fialkow, L., C.K. Chan, S. Grinstein, and G.P. Downey. Regulation of tyrosine phosphorylation in neutrophils by the NADPH oxidase. Role of reactive oxygen intermediates. *J. Biol. Chem.* 268:17131–17137, 1994.

46. Brumell, J.H., A.L. Burkhardt, J.B. Bolen, and S. Grinstein. Endogenous reactive intermediates activate tyrosine kinases in human neutrophils. *J. Biol. Chem.* 271:1455–1461, 1995.

47. Sundaresan, M., Z. Yu, V.J. Ferrans, K. Irani, and T. Finkel. Requirement for generation of H_2O_2 for platelet-derived growth factor signal transduction. *Science* 270:296–299, 1995.

48. Gonzales-Rubio, M., S. Voit, D. Rodriguez-Puyol, M. Weber, and M. Marx. Oxidative stress induces tyrosine phosphorylation of PDGF a- and b-receptors and pp60^{c-src} in mesangial cells. *Kidney Int.* 50:164–173, 1996.

49. Chrzanowska-Wodnicka, M. and K. Burridge. Tyrosine phosphorylation is involved in reorganisation of the actin cytoskeleton in response to serum or LPA stimulation. *J. Cell Sci.* 107:3643–3654, 1994.

50. Hecht, D. and Y. Zick. Selective inhibition of protein tyrosine phosphatase activities by H_2O_2 and vanadate in vitro. *Biochem. Biophys. Res. Comm.* 188:773–779, 1992.

51. Zent, R., M. Ailenberg, T.K. Waddell, G.P. Downey, and M. Silverman. Puromycin aminonucleoside inhibits mesangial cell-induced contraction of collagen gels by stimulating production of reactive oxygen species. *Kidney Int.* 47:811–817, 1995.

52. Ricardo, S.D., J.F. Bertram, G.B. Ryan. Reactive oxygen species in aminonucleoside nephrosis: In vitro studies. *Kidney Int.* 45:1057–1069, 1994.

53. Kawaguchi, M., M. Yamada, H. Wada, and T. Okiaki. Roles of active oxygen species in glomerular epithelial cell injury in vitro caused by puromycin aminonucleoside. *Toxicology* 72:329–340, 1992.

6

Endothelial Barrier Dynamics: Studies in a Cell–Column Model of the Microvasculature

Frederick R. Haselton

Introduction

Endothelial cells form continuous monolayers that restrict the transport of solutes across the endothelial layer (Renkin, 1952; Simionescu, 1983; Albelda et al., 1988; Haselton et al., 1989, 1992a, 1992b). Experimental evidence continues to accumulate in support of the hypothesis that the endothelial cell monolayer lining the intravascular space is an important determinant of normal transcapillary barrier. The majority of transendothelial solute exchange appears to take place paracellularly and is thought to occur at interendothelial junctions (Shasby et al., 1982; Shasby and Shasby, 1986; Haselton et al., 1989, 1992). Many physiological agents modify the endothelial barrier, but maintenance and regulation of solute transport across the endothelium remains poorly understood. In vivo measurements of microvascular transcapillary permeability are difficult to obtain and quantify. One of the principal difficulties is that, as in other microvascular systems, the measurement of microvascular permeability can be affected by changes in capillary recruitment and/or hydrostatic pressure. For these reasons, investigators have sought to study vascular barrier maintenance and regulation in vitro. One of the approaches that has proven useful is the in vitro investigations of the properties of cultured endothelium. A key property of endothelium that makes these studies feasible is that, with the proper substrate conditions, in vitro endothelial cells retain their in vivo growth characteristics and form confluent monolayers. One method of investigation that has proven useful is the study of the dynamic properties of the endothelial in vitro transport barrier through the use of systems that detect and quantify changes in endothelial monolayer permeability. This chapter focuses on 1) a discussion of the in vivo dynamics of interest, 2) a comparison of the design of current in vitro methods, 3) examples of results derived from the cell–column chromatography method, and 4) prospects for future development.

Current thinking views the endothelial cell as a critical element of the homeostatic intravascular transport barrier. Fundamental unanswered questions continue to drive investigation of the endothelial in vitro barrier. There is no longer any question that endothelial monolayers in vitro can modulate their monolayer permeability characteristics in response to treatments. It remains unclear, however, which endogenous signals have a role in the normal regulation of the endothelial barrier, that is: *What are the signaling pathways or, for that matter, what are the important signals that normally modulate the endothelial barrier?* The mechanism of barrier modulation is still unresolved. It remains unclear whether changes in the endothelial barrier involve junction protein redistribution or active cell contraction/relaxation via the cytoskeleton, that is: *How is permeability of the endothelial barrier modified in response to endogenous signals?* The endothelial cell is thought to be a major component of the transport barrier; however, there may be as yet undiscovered factors, cells, or features that need further investigation, that is: *What other vascular components are important for normal barrier maintenance? How do they interact?* Similar and equally important questions also apply to understanding how the endothelial barrier is lost during the progression of diseases that apparently increase transcapillary leak. As described below, each of the in vitro methods has made progress toward providing answers to these fundamental questions.

Specific aspects of the dynamic behavior of the transcapillary transport barrier that might be of interest include changes in the barrier in response to an agent, the time to respond, the magnitude of the response, reversibility of the response, dose response, saturability, and desensitization.

Barrier Response. The most often posed question is simply: Is there a change in the barrier in response to exposure to an agent? As was mentioned in the Introduction, there is still no clear consensus on the major changes that produce a change in transcapillary transport. Direct evidence of endothelial involvement obtained using in vitro systems that contain only endothelial cells is one piece of evidence that the endothelium is involved.

The Time to Respond to Stimuli. The length of time that is required to produce a response in the barrier function is useful in designing interventions and may be important in identifying the cellular mechanism evoked by the stimulus. For example, does the endothelial cell respond by the release of a stored intracellular product, or is de novo protein synthesis required? Observations of changes in in vivo barriers suggest some outside limits to the response time, at least for the special case of breakdown in the transport barrier. A cursory reading of the literature suggests that there is a considerable range in the time required to produce a breakdown in the transvascular transport barrier. At one end of the scale is diabetic retinopathy, which is produced as a consequence of diabetes and is one of the longest, which is on the order of decades. An intermediate scale event is the increased vascular leak associated with IL2 treatment regimens, which is on the order of days. Some shorter scale events might be injury associated with endotoxin, preeclampsia, thrombocytopenia, or disseminated intravascular coagulation, all of which produce a breakdown in transvascular barrier on the order of hours. At the other extreme, in vivo evidence suggests that the breakdown can occur much more rapidly. For example, histamine can produce effects on the order of seconds. Most of these result in a breakdown of barrier, but there may be endogenous signals that also act to maintain the normal in vivo barrier. Unfortunately, whole organism or whole organ in vivo data suggesting a time scale for a decrease in transvascular barrier have been more difficult to obtain.

The Magnitude of Response. One can speculate that there are limits to the change in endothelial barrier which can be produced by endogenous signaling. However, in vivo barrier data from whole organ studies are difficult to interpret. One of the only approaches to obtaining this information is through the use of indicator dilution measurements in animal models. However, the analysis of these data is further complicated by possible concomitant changes in the perfused vascular surface area. The existing measurements suggest that the outer limit to observations from short-term investigations is on the order of ± 20%. The interpretation of the importance of an absolute change is somewhat diminished without additional information about the role of compensatory mechanism(s) that may blunt these effects.

Reversibility of Response. Certainly the in vivo data suggest that the normal barrier function can be restored. For example, local injury associated with histamine clearly has no long-term effects. Although in some of the pathological alterations in vivo the increase in vascular leak can be fatal, many patients survive apparently without any long-term effects. This suggests that mechanisms in these patients can be triggered to reverse the damage or injury.

Dose Response, Saturability, and Desensitization. Again, although data are sketchy in this regard, one can speculate that in some range the response to endogenous signals is dependent on the dose of the agent. Certainly, if the effects of these agents are transduced by receptors, one would expect dose-dependence, saturability, and desensitization all to occur to some extent. In this case, receptor density, second messenger linkage, and down-regulation phenomena all would be important determinants of barrier-modifying phenomena.

The time course, magnitude, reversibility, dose requirements, saturability, and desensitization aspects of in vivo transcapillary exchange are often investigated in studies of the in vitro dynamics of the transport barrier formed by the in vitro endothelium. As other areas of science, an important aspect of in vitro studies of the dynamics of the endothelial barrier is the design of the measurement system. As discussed below, some limitations to dynamic studies are imposed by the measurement system itself.

Review of Available In Vitro Methods of Characterizing Endothelial Barrier Function

There are four basic methods currently used to investigate the in vitro transport barrier formed by endothelial cells. These are a modified Boyden chamber, electrical resistance measurements, dye absorption by microcarrier beads, and a second microcarrier-based technique that we have termed *cell–column chromatography.* Each of these approaches is discussed below.

TABLE 6.1. Comparison of four in vitro permeability methods.

Permeability method	Surface area on cells (cm^2)	Tracers	Minimum response time (sec)	System stability (hours)	System resolution	Magnitude of detectable change
Two-chamber	1	Albumin	1800	48	20%	−80% to +300%
Electrical	0.001	Ions	0.1	5	?	−15% to +15%
Dye absorption	23	Evans blue	120	1	20%	−50% to 100%
Cell–column	100	Small solutes	30	2 to 4	10%	−90% to +1000%

Modified Boyden Chamber Methods

The most common approach is a modification of the Boyden two-chamber method. In this system, transport data are collected by timed sampling of the concentration of a labeled material that diffuses between two chambers across an endothelial monolayer cultured on one side of a porous support. Two-chamber experiments are typically done using small chambers with approximately 1 cm^2 of endothelial surface area separating the two chambers. The most often used tracer is labeled albumin (MW 66,000 D). Most investigations collect data over long intervals, which limits the response time to ~1800 s. However, with this design it is possible to study transport over long intervals, and some have reported results for measurements spanning up to 96 h. Typically, data from this assay technique are reported as percentage of applied tracer that appears in the lower chamber (Garcia et al., 1986), although a number of investigators have estimated permeability from these measurements (Albelda et al., 1988; Cooper et al. 1987; Casnocha et al., 1989). Based on the published reports, the limit of permeability change that one could hope to resolve with this approach appears to be ~ ±20%, and the magnitude of detectable change is estimated to be in the range of −80 to +300%. A review of many of the endothelial findings using this approach is given in Lum and Malik (1994). Table 6.1 summarizes the characteristics of this method and compares it to the other approaches discussed below.

Electrical Resistance Methods

An approach that has been utilized extensively in the investigation of epithelial barriers is the electrical resistance method. The method is simple and straightforward. Unfortunately, its use with endothelial cells is limited by the low electrical resistance of endothelial monolayers (~ 20 Ω/cm^2; Albelda et al., 1988), which is much lower than epithelial monolayers (>500 Ω/cm^2; Gumbiner et al., 1986). However, the method continues to be employed in some studies (Gillies et al., 1995). Recently, Tiruppathi et al. (1992), based on previous studies of cell shape changes (Mitra et al., 1991), reported a modification of this approach for study of endothelial barriers. In this new design, 0.001 cm^2 of endothelial surface area is grown on a gold support membrane. The transport of ions across the cell layer is modeled to predict a permeability estimate. The authors claim a response time on the order of 10^{-1} s and a system stability of up to 5 h. The resolution of this approach is unknown, and the method has a low total system response of ±15%.

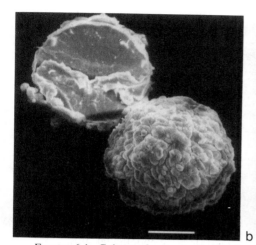
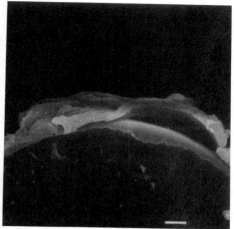

FIGURE 6.1. Culture of retinal endothelial cells on fibronectin-coated microcarrier beads. (a) Whole bead adjacent to a razor sliced bead (bar 50 μm). (b) Magnification of cell layer (bar 5 μm). Photos by Loren Hoffman (Haselton et al., 1996a).

Microcarrier-Based Methods

Two permeability assays have been developed based on measurements of endothelial barriers formed by endothelial cells cultured on microcarrier beads.

Dye Absorption

In the first of these, the permeability of the endothelial monolayer is inferred from either the amount of dye recovered from microcarriers coated with endothelial cells (Killackey et al., 1986) or from the loss of dye in microcarrier suspensions (Bottaro et al., 1986; Alexander et al., 1988). A measurement requires approximately 23 cm^2 of endothelial surface area and depends on dye binding to the interior of the microcarrier bead. The response time in these measurements is approximately 120 s, and it is stable for up to 1 h. Typical data have been reported as a fraction of dye absorbed to microcarriers at a particular time point (Killackey et al., 1986). System resolution is estimated to be ~±20% with a range of −50 to +100% total change.

In Figure 6.1, an SEM of bovine retinal microvascular cells cultured on fibronectin-coated microcarrier beads is shown. Note the confluent nature of the monolayer and the single-layer thickness shown in the higher magnification in Figure 6.1(b).

Chromatographic Cell–Column

A second permeability assay based on the growth of endothelial cells on the surface of microcarrier beads is the cell–column chromatographic approach reported by Haselton et al. (1989) (see Figure 6.2). Other groups have also reported results with this approach (Eaton et al. 1991; Waters et al., 1996). As shown in

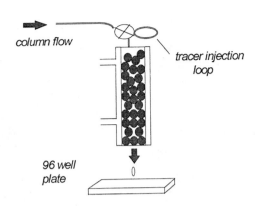

column flow

tracer injection
loop

96 well
plate

FIGURE 6.2. Schematic of cell–column experimental design.

Table 6.1, this method requires a minimum of ~100 cm² of endothelial surface per experiment. It utilizes small solutes with molecular weights less than 5000 D to characterize the permeability and has a minimum response time of approximately 30 s. This system is stable for between 2 and 4 h and has a resolution estimated to be on the order of 10%. Data range from −90 to +1000% in detectable change.

Design Details of Chromatographic Cell–Column Method. Since the examples of dynamic endothelial cell monolayer behavior described below are based on the cell–column chromatographic system, additional details are described in this section.

Microcarrier cultures. Endothelial cells are cultured on microcarrier beads as previously described (Haselton et al., 1989). Cells are seeded on microcarrier beads at a density of 2×10^4 cells/cm². Cell attachment is achieved by intermittent stirring overnight. Microcarrier cultures are maintained at 60 rpm continuously and fed three times weekly. Cultures are used for cell–column assays between 9 and 30 days postseeding.

Endothelial cells from different organs or vessels may require specific culture methods. One major difference is in the cell substrate requirements for endothelial attachment and spreading on microcarrier beads. Almost all of the endothelial cells tested to date grow well on denatured Type I collagen coated onto porous cross-linked dextran microcarrier beads from Pharmacia (Cytodex-3). Approximately 60 μg/cm² of collagen is bound to the surface of these microcarriers, and the molecular weight cutoff is ~100 kD (*Microcarrier cell culture:* Pharmacia Fine Chemicals, Uppsala, Sweden, 1981.) We and others have grown endothelial cells on Cytodex-3 microcarriers from various sources including bovine aortic, bovine fetal aortic, bovine pulmonary microvessel, bovine pulmonary artery, and human umbilical vein (Haselton et al., 1989; Eaton et al., 1991; Killackey et al., 1986; Waters et al., 1996). One of the problems that we have had to address with this approach is that all of the endothelial cells do not attach and grow on Cytodex-3 microcarrier beads. We have recently developed a second microcarrier bead of 100% cross-linked gelatin with a surface coating of fibronectin for use with bovine and human retinal microvessel endothelial cells (Haselton, 1996a). Nonporous beads also have been used for some specialized cell–column studies; these include an uncoated plastic microcarrier (Biosilon) and a gelatin-coated plastic microcarrier (Sigma).

Chromatographic column methods. We currently use a modification of our previously reported assay of endothelial monolayer permeability. The method uses a chromatographic column filled with cell-covered microcarrier beads. The permeability of the endothelial monolayers covering the beads is determined from a comparison of the elution curves of tracers injected into the flow at the top of the column. Most details of this method have been previously described (Haselton et al., 1989). A brief description and modifications are given below.

Chromatographic cell–columns are made from water-jacketed glass columns (0.65-cm diameter, Rainin). Cell-covered beads are poured to a column volume of approximately 0.67 cm³, which provides 130 cm² of endothelial cell culture surface or approximately 1×10^7 cells. The column is washed and equilibrated with Hank's balanced salt solution containing either 10% fetal bovine serum (Montfort Biologicals) or 0.5% bovine serum albumin (Sigma). Constant pressure perfusion is maintained by a Gilson peristaltic pump, at approximately 0.9 ml/min. This flow approximates the flow rate due to gravity. A retrograde perfusion has been adapted by Waters et al. (1996). A tracer bolus described below is applied by a rotary injection valve (Rainin) using a 50-μl loop. The cell–column and all perfusate solutions are held at 37°C. Multiple measurements of permeability are made on each cell–column, but each cell–column is used in only one experimental protocol.

One of the applied tracers (blue dextran, MW 2,000 kD, 10 n*M*) is impermeant and follows the mobile phase, that is, a flow tracer. Two other tracers, sodium fluorescein (MW 376, 0.33 m*M*) and cyanocobalamin (vitamin B_{12}, MW 1355, 1.49 m*M*) are used as permeant tracers. These tracer concentrations were chosen to ensure a linear relationship between the concentration and optical absorbance. The latter two tracers are relatively small, compared to blue dextran, and can diffuse between the cells into the bead matrix beneath the cells. The physical cell–bead interaction properties of the tracer of course influence their pathway across the endothelium. Neither sodium fluorescein nor B_{12} interacts with endothelial cell membranes or with the beads beneath the cells. For each tracer this must be verified by growing cells on nonporous microcarriers and examining the elution patterns under these conditions (see Haselton et al., 1989) and by examining the total recovery of injected material.

Elution profiles were constructed from 66 samples of the column eluant. A fraction collector (Model 203, Gilson, Middleton, WI) equipped with a drop counter collected 2 drops of eluant per well for 66 wells of a 96-well microtiter plate. The sample size and number of samples can be adjusted to balance the elution profile resolution and signal-to-noise ratio. The absorbance of each of the 96 wells was recorded at 620 (blue dextran), 540 (B_{12}), and 492 nm (sodium fluorescein) on a plate reader (Titertek MCC 340, Labsystems, Marlboro, MA) and stored on a computer for analysis. Absorbances were corrected for overlap among tracers. The optical absorbances were used to calculate the fractional recovery per sample of each of the optically absorbing tracers.

Data analysis. Estimating permeability based on the column elution patterns of multiple tracers is an adaptation and extension of techniques used in vivo to assess capillary permeability (Renkin, 1952; Crone, 1963; Sangren and Sheppard, 1953; Harris and Roselli, 1989). To apply these techniques to these experiments, a mathematical model of tracer motion has been developed based on the physical picture of tracer behavior described above. In this previously described model (Haselton et al., 1989), it is assumed that the elution profiles of cyanocobalamin and sodium fluorescein depend on the properties of the mobile phase of the column plus the paracellular permeability properties of the endothelial monolayer and the diffusive motion of the tracer within the microcarrier beads. In contrast to these permeant tracers, the elution of blue dextran depends only on the flow (mobile) phase properties of the column.

Elution profile graphical results and the simple Crone/Renkin extraction estimate of permeability–surface area product (*PS*) are obtained virtually "on-line" (using less than 50 s of initial data), which allows tracking of the experiment's progress. We have found this simple estimate to parallel the more complex modeling calculations based on analysis of the complete column elution patterns using the Sangren–Sheppard model, which were performed "off-line" after all of the data from each experiment were collected (Haselton et al., 1989; Sangren and Sheppard, 1953). A modified Marquardt iteration scheme is used to estimate the monolayer permeability that best approximates the experimental data by varying the two free model parameters, permeability–surface area and tracer distribution volume (Haselton et al., 1989). Best fit is determined by a computational algorithm that minimizes the coefficient of variation between a computer-generated prediction of the permeant tracer's elution profile and the experimental observed elution profile. A baseline permeability value for any one column is computed as the average of two consecutive measurements (spanning 5–7 min). During experimental treatment protocols a single measurement is used.

Statistics of change. A permeability value for any one column and time point is computed as the mean of two consecutive measurements (spanning 5–7 min). Significant differences in endothelial permeability observed at various time points under different cell–column conditions is usually determined using repeated measure ANOVA and post-hoc testing.

Example of elution profile changes. It is relatively easy to discern qualitative changes in consecutive measurements of the elution profiles from a single cell–column. Changes are apparent since, in cell–columns, permeability changes in the endothelial monolayer are accompanied by a shift in the elution pattern of the small molecular weight markers and no shift in the elution of the mobile phase tracer blue dextran (Haselton, 1989). Figure 6.5 was constructed from two separate experiments to illustrate this effect for an agent that we interpret as decreasing permeability in retinal cell–columns (isoproterenol) or increasing permeability

FIGURE 6.3. Example of elution profile changes produced by addition of isoproterenol or cytochalasin D to the perfusate of a cell–column (Haselton et al., 1996a).

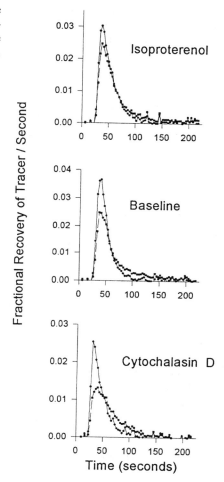

(cytochalasin D). The center panel of Figure 6.3 shows the elution pattern observed for blue dextran (MW 2,000,000) and cyanocobalamin (MW 1355) under baseline conditions. Analysis of these data found an endothelial permeability to cyanocobalamin of 5.55 ($\times$ 10^{-5} cm/s). After perfusing this cell–column with 1 μM isoproterenol for 15 min, the elution pattern in the upper panel of Figure 6.3 was observed. Permeability decreased to 3.04. Note that in this case the elution pattern of cyanocobalamin is shifted up in the peak closer to that of blue dextran than under baseline conditions. Qualitatively, this can be interpreted as indicating a decrease in the rate of transfer across the endothelial monolayer since blue dextran elution represents "zero" permeability behavior and the more the permeating tracer approximates this elution pattern, the lower its permeability. In a separate cell–column with a similar baseline (5.55 $\times$ 10^{-5} cm/s, data not illustrated), treatment with 1 µg/ml of cytochalasin D for 30 min produced a shift opposite to that observed with isoproterenol (lower panel of Figure 6.3). The pattern of cyanocobalamin observed with this treatment is a shift to the right and a lower peak value, indicating an increased rate of transfer across the monolayer. For this treatment, permeability increased to 8.20 (Haselton et al., 1996a). The effects of isoproterenol and cytochalasin D on BAEC are both reversible, and the baseline is restored after removal of the agents. However, the time courses of the reversals are different, with less than 15 min required for isoproterenol and at least 60 min for cytochalasin D (Haselton et al., 1989).

Examples of Dynamic Studies

Most endothelial barrier studies are directed at the question posed in the example above, namely, does a particular agent increase or decrease endothelial monolayer permeability? All of the methods discussed above have been applied to answer this question for a number of agents. However, the reliability of the results depends on the agent producing a monolayer change that can be detected within the resolution time of the particular method. For example, a permeability increase produced by histamine is thought to be short, possibly 30 s or less. If it takes 30 min to measure permeability, then a change must be much larger to appear significant. In this section, some detailed examples of studies of endothelial dynamical behavior are discussed. Each of these is uniquely adapted for the cell–column approach, but an evaluation of alternative methods is described for each. For the sake of discussion, the examples are grouped as extracellular and inter-cellular component studies, and studies utilizing the cell–column as a bioassay model of a microvessel.

Extracellular and Intercellular Components of the Endothelial Barrier

In this first group of studies, we have used cell–columns to study changes in endothelial monolayer permeability in response to the removal of an agent nor-mally present in the cell–column perfusate. Two examples are given. The first of these examines the response to removal of albumin from a cell–column perfusate that is normally maintained at 0.5% (w/v) in the cell–column perfusate. These studies are still ongoing. The second, more complex, experimental design is a study of endothelial cell junction proteins by antibody insertion during conditions of low calcium concentration in the cell–column perfusate.

Albumin Concentration Effects

Rationale. Investigators have examined the effects of albumin concentration on in vivo permeability. The most extensive of these are by Huxley and Curry (1985).

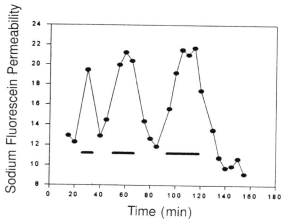

FIGURE 6.4. Alterations in albumin concentration alter permeability (cm/s, × 10^{-5}). Bars indicate periods of perfusate albumin at 0% albumin.

The basic observation is that switching vessel perfusion to a Ringer perfusion increases vessel permeability in individually perfused frog capillaries. Initially, these studies were carried out in gathering data in support of a fiber matrix theory of permeability (Curry, 1986).

Methods. We have observed changes in endothelial monolayer permeability when albumin is removed from a cell–column perfusate similar to those reported in vivo. In this simple example, the permeability of a cell–column is measured under normal perfusion conditions (Hanks balanced salts containing 15 n*M* HEPES and 0.5% bovine serum albumin). The perfusate albumin concentration is lowered to 0% albumin and permeability measurements obtained in the same cell–column.

Results. The response, which is illustrated in Figure 6.4, shows the increase in sodium fluorescein permeability produced by repeated removal and addition of albumin from the cell–column perfusate. At the first time point measured (<5 min of the removal of albumin from the perfusate), permeability is nearly double. After albumin is restored to 0.5%, as in the baseline perfusate, permeability returns to near initial values. This modulation is repeated for 0% albumin perfusion times of 10, 20, and 30 min.

Interpretation. These in vitro dynamics are similar to those reported by Huxley and Currey (1985) for Ringer perfusion in single frog capillaries, although their report also found a hysteresis effect when low albumin concentrations were tested. At albumin concentrations of 10 mg/ml, no hysteresis in single vessels was observed that was similar to these in vitro studies.

Alternative Designs. Because the response to albumin removal and re-addition is on the order of minutes, most alternative methods would be suitable only for following a single-step change and not repeated changes in conditions. The one exception is the newly described electrical resistance measurement system of Tiruppathi et al. (1992).

Junctional Protein Function

A somewhat more complex example that capitalizes on reversible changes in permeability in the same population of cells is a recent study of junction protein involvement in the formation of the endothelial barrier (Alexander et al., 1993). In these studies we sought to identify cadherins in endothelium, and in particular to examine the role of cadherins in the endothelial barrier.

Rationale. To better understand how the barrier function of the endothelium is related to the junctional structure, a more detailed analysis of the proteins constituting the junction is needed (Franke et al., 1988; Heimark, 1991; Albelda, 1991). In this area of cell biology, the proteins forming the epithelial barrier are better characterized and barrier maintenance is thought to depend on the homotypic association of several specialized junction molecules found in "tight" and "adherens" junctions (Gumbiner and Simons, 1986; Gumbiner et al., 1988; Volk and Geiger, 1986a, 1986b). One of the reasons for the lack of evidence for junctional protein involvement is that the functional significance of proteins involved in the barrier has been difficult to demonstrate. A method that has been used previously to demonstrate this property in epithelial cadherins is termed the "calcium switch assay" (Gumbiner and Simons, 1986; Volk and Geiger, 1986a). The approach in epithelia has been to measure the electrical resistance of the monolayer. However, since endothelial junctions are not electrically "tight," like epithelial junctions (Shasby et al., 1986, Albelda et al., 1988), electrical resistance methods do not reliably detect barrier changes.

Methods. This experimental design is based on the observation of Gumbiner et al. (1986) and Alexander (1993) that reduction in extracellular calcium triggers a release of homotypic binding of cadherin junction proteins on adjacent monolayer cells. A similar design is followed in the example shown. In these experiments, we demonstrated the role of cadherins in the maintenance of the endothelial barrier of an anti-cadherin antibody (L-7; Blaschuk et al., 1990), which was tested for its ability to interfere with junctional assembly using the calcium-switch assay described by Gumbiner and Simons (1986) with modifications for cell–column chromatography. Briefly, two simultaneous endothelial cell–columns were prepared and baseline permeability measurements made at 15 min following column setup. The baseline perfusate consisted of Hanks balanced salt solution (HBSS), which contains a "normal" Ca^{2+} concentration of 1.2 mM, 25 mM HEPES (pH 7.35), and 0.5% albumin. Following baseline measurements, "low" calcium (0.12 mM) perfusate (HBSS with 0.12 mM Ca^{2+}, 25 mM HEPES (pH 7.35) + 0.5% albumin) was substituted for the baseline perfusate and permeability measurements were made at 5-min intervals for approximately 20 min. After 20 min of low-calcium treatment, cell–column permeability typically increases to 2 to 3× the baseline values. After approximately 20 min of low calcium, the baseline (normal calcium) perfusate was restored in the presence of either a control (pre-

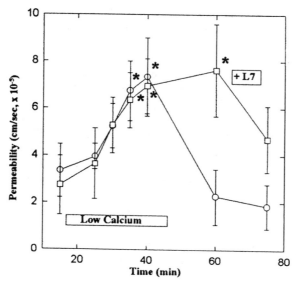

FIGURE 6.5. Cell–column calcium dynamics imply cadherin junctional protein participation in endothelial barrier ($n = 4$, mean ± SEM) (Alexander et al., 1993. © Wiley-Liss, Inc. Reprinted by permission of John Wiley & Sons, Inc.).

immune) or polyclonal post-immune (L-7) cadherin antibody diluted 1:100. Cell–column permeability was measured after 15 and 30 min following addition of either the control or anti-cadherin antiserum to determine the effect of these treatments on barrier recovery.

Results. Figure 6.5 shows the permeability response of endothelial monolayers treated with low (0.12 mM) calcium perfusate for 30 min, followed by reincubation with perfusate containing normal (1.2 mM) calcium containing a 1:100 diluted pre-immune rabbit serum (open circles). In low calcium, permeability to sodium fluorescein increases from 3.35 ± 1.14 to 7.33 ± 1.67 at 20 min. Permeability returns to 2.26 ± 1.17 (67% of baseline) within 20 min ($n = 4$). The other curve (open squares) shows the same calcium switch assay in an additional four cell–columns but the recovery phase contains anti–L-7 antibody (diluted 1:100). Permeability is increased from a baseline of 2.74 ± 1.25 to 6.94 ± 1.17 by 20 min of low calcium. Following a 15-min recovery in the presence of L-7 antibody, permeability is still significantly elevated above baseline (7.63 ± 1.96) and remains elevated, although not significantly (4.7 ± 1.39), following 30 min of recovery.

Interpretation. From these studies, we inferred that at least one endothelial cadherin participates in the maintenance of the endothelial barrier. The cell–column results support this hypothesis since the polyclonal antibody directed against the external sequence FHLRAHAVDINGNQV of mouse N-cadherin

(Blaschuk et al., 1990) interferes with the recovery of the barrier in an endothelial calcium switch assay (Figure 6.5). This peptide contains the sequence HAV, which has been reported to mediate homotypic adhesion in N, P, and E cadherins (Blaschuk et al., 1990; Nose et al., 1990). Therefore, our results suggest that this antibody may interfere with the barrier by blocking N-cadherin pro-adhesive interactions in the endothelium during recovery. Interestingly, although we were able to show blockade, this effect on barrier recovery appears to be temporary rather than permanent, since 30 min after restoring normal calcium, the barrier is still not significantly elevated compared to baseline levels. This result parallels those observed in epithelial calcium switch experiments where electrical resistance recovers over time despite the continued presence of an antibody (Gumbiner and Simons, 1986). Additional studies based on this design with monoclonal antibodies directed against cadherin 5 and 13 have recently been completed (Haselton and Heimark, 1997).

Alternative Methods. Because the timing of the lower calcium concentration exposure depends on feedback from changes in permeability, a detection system with a resolution better than ~7 min is required. Control cell–columns containing endothelial cells from the same microcarrier culture are exposed for the same length of time to low calcium to test for a return to normal barrier. This "overshoot" for the low-calcium response is a major problem with this experimental design, as has been observed by others in epithelial studies (Gumbiner et al., 1986). One promising alternative method might be the new electrical resistance methods described by Tiruppathi et al. (1992). If these measurements could be made on parallel samples, the claimed system time resolution should be able to detect the initial changes produced by a lowered calcium concentration. An attractive advantage of this method is that a much smaller quantity of antibody would be needed.

Microvessel Model—Elements of Barrier Maintenance

A second major use of the cell–column approach is to regard the cell–column as a model of the microvessel and to examine the influence of the addition of nonendothelial microvessel components to the cell–column.

Pericytes

This example capitalizes on the indicator dilution measurement strategy and allows the readdition of a second cellular component without directly affecting the measurement of the endothelial monolayer permeability.

Rationale. One of the vessel components we have begun to investigate is the pericyte, a vessel support cell that surrounds capillary vessels in a number of organs. Of particular interest is the role that this cell plays in the progression of diabetic retinopathy. Evidence suggests that pericyte–endothelial interactions may be critical to maintaining normal retinal microvascular integrity (Sims 1986; Orlidge and D'Amore, 1989). The cooperation between the pericyte and the endothelial cell may involve secreted factors such as, for example, transforming

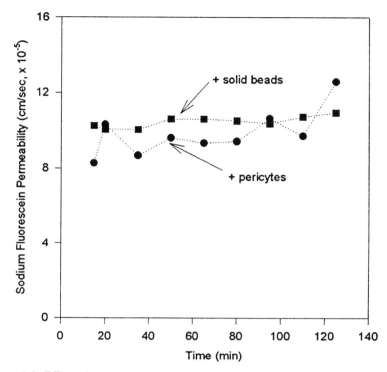

FIGURE 6.6. Effect of pericytes on retinal endothelial monolayer permeability.

growth factor-beta (TGF-B). Active TGF-B is made by co-cultures of endo-thelium and pericyte and inhibits endothelial growth in vitro (Baird and Durkin 1986). Thus, pericytes may regulate capillary structure and permeability through production of TGF-B and suggests that the loss of barrier in diabetes may be related to pericyte dropout through TGF-B.

Methods. In this example, from an ongoing study, we have separated pericytes from endothelial cells using flow cytometry. Separate cultures of endothelial cells and pericytes are maintained. The endothelial cells are cultured on porous micro-carrier beads and pericytes on gelatin-coated solid microcarrier beads. Two simul-taneous cell–columns are set up, each containing a 1:1 mixture of retinal EC and solid beads, either without pericytes or with pericytes cultured on their surface.

Results. As the data in Figure 6.6 illustrate, neither the presence of pericytes nor uncoated gelatin beads under baseline conditions alters the permeability of the endothelial monolayer on the surface of the porous microcarriers. The design of the experiment has taken advantage of the indicator dilution design in that all of the tracers are influenced equally by the presence of solid beads and no additional separation of a large molecular weight compound from the endothelial monolayer permeant tracer is produced.

Interpretation. These studies are still in progress, but they suggest that the mere presence of pericytes is not a contributing factor. Release of a soluble factor that triggers a decrease in permeability in neighboring endothelial cells within the 2 h tested was not observed.

Alternative Methods. One alternative method that has been explored in addressing similar questions is the Boyden two-chamber method. Typically, an endothelial layer is grown on one side of the porous filter and then a second cell type is cultured on the opposite side. This design includes the second cell but depends on the diffusion of a signal across the cell layers and does not include any contact between cell types. An alternative microcarrier method that we are exploring is culturing the pericytes on the surface of microcarrier beads and, once established, adding endothelial cells. This in vitro tissue model would require validating (for example) that sufficient nutrients are delivered to the subluminal pericytes under these conditions.

Platelets

These studies are part of a signal identification investigation and explore rapid changes in monolayer permeability produced by an endogenous platelet signal and also the response of the endothelium to continuous exposure to this signal.

Rationale. We hypothesize that platelets have a beneficial role in the maintenance of vascular endothelial integrity through the release factors, directly affecting endothelial permeability. This is of interest in a number of organs, but of particular interest in the lung, where inflammatory injury is associated with entrapment of platelets and an increase in the flux of protein and fluid across the lung microvascular barrier. Platelets accumulate in response to endothelial damage and are thought to contribute to the increased leak of proteins and fluid through the release of factors that either increase vascular pressure, directly increase endothelial permeability, or attract other damaging cells and/or other aggregating platelets. Recent evidence suggests that platelets may also release factors that are important in the maintenance of the normal transport barrier formed by the endothelium. These factor(s) may also be important as anti-inflammatory factors but currently remain unidentified. The identification of platelet release factors with anti-inflammatory properties may lead, for example, to the development of new treatments to blunt transcapillary leak associated with inflammatory injuries.

We have previously reported that perfusing a cell–column with nonstimulated, nonactivated platelets at a concentration of 6000/µl decreases the permeability of in vitro endothelial monolayers by 35% (Haselton et al., 1993). The presence of platelets is not required to produce this effect, since platelet incubates also produce a similar decrease. In this previous report, we used the cell–column as a bioassay, to partially characterize an unknown platelet release factor with permeability reducing activity.

Methods. In these studies, human platelets are isolated by venipuncture using 3.2% sodium citrate (10% by vol.) as an anticoagulant with 5 µ*M* PPACK (Chemica Alta, Edmonton, AB) added to inhibit thrombin-mediated platelet activation. Blood is centrifuged at $300 \times g$ for 5 min, and the platelet-rich plasma is

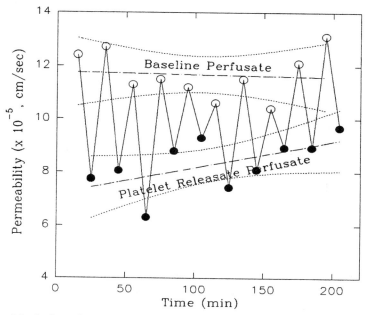

FIGURE 6.7. Sodium fluorescein permeability decreases with many on/off treatments of platelet material. Baseline perfusate conditions (open circles); platelet perfusate conditions (closed circles) (Alexander et al., 1997).

removed. The platelet concentration is measured by Coulter-counting, and the platelet-rich plasma is supplemented with leupeptin (1 µg/ml) and incubated for 2 h at room temperature. The platelets are then removed by centrifugation at 2500 × g for 10 min, and the supernatant is decanted and frozen at −20°C until used. In the experiments described here, platelet material is added to cell–column perfusates in a concentration equivalent to 6000 platelets/µl.

Results. As Figure 6.7 shows, cell–columns still respond to platelet released material after ten cycles of platelet material treatment alternated with baseline treatments. In this experiment, exposure of cultured endothelial cell monolayers to platelet released material produces the expected rapid decrease in cell–column permeability. This effect is completely reversible, and it can be reproduced many times by reexposing the monolayers to platelet released material (Alexander et al., 1997).

A second related example is the endothelial monolayer permeability response to continuous exposure to platelet released material. Cell–columns continually exposed to platelet supernatants showed a rapid initial decrease in endothelial permeability within 10 min (Figure 6.8). Continuous perfusion with platelet supernatant produced a sustained decrease in endothelial solute permeability for the time period tested.

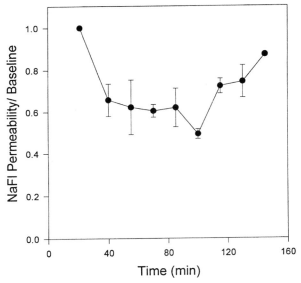

FIGURE 6.8. Continuous exposure to platelet released material produces a sustained decrease in monolayer permeability ($n = 4$, mean ± SEM; Alexander et al., 1997).

Interpretation. In this example, the permeability state alternates between baseline and decreased permeability. The data suggest that the initial response is rapid and reversible and in addition can be maintained by continuous exposure to platelet material. However, we cannot exclude the possibility that the supernatant may contain two or more factors with different time courses of action, e.g., a short-time-course decreasing factor and a long-time-course decreasing factor. This possibility is currently under investigation.

Alternative Methods. Because of the rapid changes in permeability produced by the addition or removal of this factor, and the apparent instability produced by continuous exposure, similar results could not have been obtained by most alternative methods. The one possible exception is the high time resolution described by Tiruppathi et al. (1992).

Microvessel Injury

A third application of the cell–column as a blood vessel model is the characterization of the dynamics of endothelial monolayer breakdown in injury. Two examples of this are discussed below.

Effects of PMN

One example of the use of a cell–column is the endothelial monolayer response to polymorphonuclear leukocyte (PMN) activation (Haselton et al., 1996b).

Rationale. Neutrophil-mediated edema is a life-threatening complication associated with adult respiratory distress syndrome, sepsis, and ischemia/reperfusion

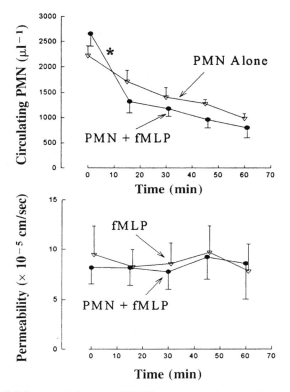

FIGURE 6.9. fMLP increased the rate of PMN entrapment (top panel) but did not change the permeability of endothelial monolayers (Haselton et al., 1996b).

injury. The edema is thought to be produced by the activation of circulating neutrophils (PMN) which then adhere to the microvasculature and break down the solute barrier maintained by the endothelium (Weiss, 1989; Nuijens et al., 1992; Inauen et al., 1990; Zimmerman and Granger, 1990). Most models of PMN-mediated edema have been developed in in vivo systems (Adkins et al., 1993; Seekamp et al., 1992). However, the interpretation of the in vivo response over time to PMN–endothelial adhesion and the loss of the endothelial barrier has often been difficult to quantify and interpret. In vitro methods are an attractive alternative that has been pursued by a number of investigators to more thoroughly examine interactions between PMN and endothelial cells (Killackey and Killackey, 1990; Harlan et al., 1985; Shasby et al., 1983, 1985; Springer, 1994).

Methods. In these experiments, 2×10^6 PMN/ml were added to a 10-ml recirculating cell–column perfusate. Recirculating PMN concentrations and endothelial monolayer permeability characteristics were measured at 15-min intervals for 60 min with and without formyl-methionyl-leucyl phenylalanine (fMLP) treatment.

Results. Figure 6.9(a) (top panel) shows the decrease in circulating PMN in protocols that contained both PMN and fMLP (solid circles). Although there was no significant difference between the initial PMN concentrations, the addition of fMLP (10^{-5} M) to circulating PMN significantly increased the initial rate of PMN entrapment in the cell–column compared to the columns containing PMN alone. In Figure 6.9(a), the slope of the PMN + fMLP concentration curve is significantly steeper between 0 and 15 min than the slope of the PMN concentration curve (open triangles). In the first 15 min, cell–columns perfused with 10^{-5} M fMLP stimulated PMN entrapped $27 \pm 3\%$ of the circulating PMN per 65 cm^2 of surface area (Figure 6.4(a), filled circles, $n = 7$). As shown by the parallel slopes in Figure 6.9, this rate was not different from PMN alone in the other three measurement periods ($-8 \pm 4\%$). Figure 6.9(b) (lower panel) shows the time course of permeability changes in endothelial cell–columns perfused with 2×10^6 PMN/ml in the presence of 10^{-5} M fMLP alone ($n = 6$, open triangles; $n = 5$ at 60 min). Permeability in cell–columns perfused with PMN and 10^{-5} M fMLP ($n = 7$, solid circles; $n = 6$ and 4 at 45 and 60 min) was not significantly elevated at 30 min compared to baseline values ($0.99 \pm 0.19 \times$ baseline).

Interpretation. These results parallel in vivo studies of PMN entrapment. Bureau et al. (1989) reported that 69% of nonstimulated, radiolabeled PMN became entrapped in the lung microvasculature within 10 min, possibly as an artifact of their bolus injection technique, and there was no additional baseline entrapment. Similarly, Doerschuk et al. (1990) reported that approximately 98% of PMN became entrapped in the lungs in a single pass, but after 10 min only 27% remained entrapped. As in both of these studies, we also found a significant initial entrapment; however, we also observed a continuous, although diminished, net increase in PMN entrapment over time.

Alternative Methods. To date, most in vitro studies of PMN mediated edema have been based on nonflowing, two-chambered methods in which PMN are layered onto endothelial monolayers cultured on porous filters separating two compartments. While these studies have demonstrated that activated PMN can increase the flux of permeant tracers across the endothelium (Shasby et al., 1983; Harlan et al., 1985), these systems exclude fluid flow and shear stress, which regulate normal PMN–endothelial interactions in vivo. For example, in vitro PMN detachment studies have shown that even at very low levels of fluid shear, PMN adhesion to endothelial monolayers is greatly reduced compared to static systems, and these studies suggest that static systems may not accurately model in vivo PMN–endothelial interactions (Gallik et al., 1989; Worthen et al., 1987). Previously described in vitro systems that incorporate flow have been mainly designed to study the attachment or detachment of PMN to endothelium (Lawrence et al., 1987) and not to investigate the effects of adhered, activated PMN on endothelial barrier function.

Endotoxin

Rationale. Treatment with endotoxin is widely used in animal models to investigate the mechanism of vessel injury (Brigham, 1994; Yi and Ulich, 1992). The

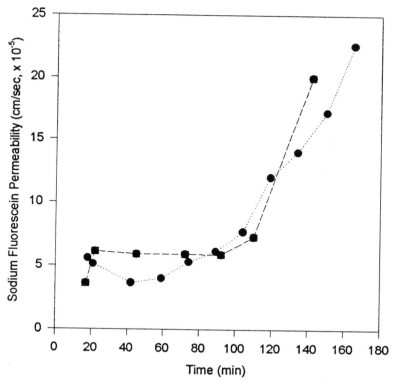

FIGURE 6.10. Time course of endotoxin effects on BAEC monolayer permeability.

basic finding of these animal studies is a biphasic response with an initial increase in vessel leak (<1 h), which resolves, and then a second sustained increase in vessel leak (>4 h). We sought to investigate the direct effects of endotoxin as an injury-producing agent in a pure endothelial system.

Methods. In these studies, endotoxin (3 µg/ml, gift from Richard Parker) was added to cell–column perfusate containing 10% fetal bovine serum (not heat inactivated) and cell–column effuent recirculated back to the cell–column reservoir.

Results. As Figure 6.10 illustrates, there was no early increase in endothelial monolayer permeability but a sustained and large increase after approximately 2 h.

Interpretation. These studies are part of a continuing investigation but suggest that the early response (<1 h) seen in vivo is not a consequence of the direct action of endotoxin alone on endothelial cells.

Alternative Methods. It would be difficult to reproduce these time course findings with an alternative method that is able to track permeability changes over time with a resolution better than 15 min. Only the new electrical method proposed by Tiruppathis et al. (1992) has the potential to achieve this resolution.

Summary and Prospects for Future Developments

The dynamics of the change in endothelial monolayer permeability are an important and useful avenue for continued investigation of the normal maintenance of the endothelial barrier and its response to injury. Despite difficulties in interpreting whole organ results, this approach continues to be pursued as a valuable research tool. Another elegant approach, which has not been discussed in any great detail in this chapter, is the single-capillary model. This has the clear advantage of retaining vessel integrity and dynamics, but none of the disadvantages of whole organ data interpretation. Continued evolution of the in vitro techniques promises to provide an additional source of insight into the identification of the critical elements of barrier formation under normal conditions, how the barrier is maintained, and how these elements are affected by disease.

Acknowledgments. The authors gratefully acknowledge the excellent technical support provided by E. Dworska, S. Bowen, R. Faulks, J. Woodall, and V. Blackwell. This research was supported in part by NIH Grants HL40554 and EY10086, the Whitaker Foundation, and the Juvenile Diabetes Foundation.

References

Adkins, W.K., J.W. Barnard, T.M. Moore, R.C. Allison, V.R. Prasad, and A.E. Taylor. Adenosine prevents PMA-induced lung injury via an A2 receptor mechanism. *J. Appl. Physiol.* 74(3):982–988, 1993.

Albelda, S.M. Endothelial and epithelial cell adhesion molecules. *Am. J. Resp. Cell Mol. Biol.* 4:195–203, 1991.

Albelda, S.M., P.M. Sampson, F.R. Haselton, J.M. McNiff, S.N. Mueller, S.K. Williams, A.P. Fishman, and E.M. Levine. Permeability characteristics of cultured endothelial monolayers. *J. Appl. Physiol.* 64(1):308–322, 1988.

Alexander, J.S. and F.R. Haselton. An N-cadherin-like protein contributes to solute barrier maintenance in cultured endothelium. *J. Cell. Physiol.* 156:610–618, 1993.

Alexander, J.S., H.B. Hechtman, and D. Shepro. Phalloidin improves endothelial barrier function and protects against inflammatory agent mediated permeability in vitro. *Microvasc. Res.* 35:308–315, 1988.

Alexander, J.S., W.F. Patton, B. Christman, L. Larsen, and F.R. Haselton. Platelet-derived lysophosphatidic acid decreases endothelial permeability in vitro. *Am. J. Physiol.* (in press).

Baird, A. and T. Durkin. Inhibition of endothelial cell profileration by type B transforming growth factor: Interaction with acidic and basic FGF. *Biochem. Biophys. Res. Comm.* 138:476–482, 1986.

Blaschuk, O.W., R. Sullivan, and Y. Pouliot. Identification of a cadherin cell adhesion recognition sequence. *Develop. Biol.* 139:227–229, 1990.

Bottaro, D., D. Shepro, and H.B. Hechtman. Heterogeneity of intimal and microvessel endothelial cell barriers *in vitro. Microvasc. Res.* 32:389–398, 1986.

Brigham, K.L. (ed) *Endotoxin and the Lungs.* New York: Marcel Dekker, Inc., 1994.

Bureau, M.F., E. Malanchere, M. Pretolani, M.A. Boukili, and B.B. Vargafig. A new method to evaluate extravascular albumin and blood cell accumulation in the lung. *J. Appl. Physiol.* 67(4):1479–1488, 1989.

Casnocha, S.A., S.G. Eskin, E.R. Hall, and L.V. McIntire. Permeability of human endothelial monolayers: Effect of vasoactive agonists and cAMP. *J. Appl. Physiol.* 67(5):1997–2005, 1989.

Cooper, J.A., P.J. Del Vecchio, F.L. Minnear, K.E. Burhop, W.M. Selig, J.G.N. Garcia, and A.B. Malik. Measurement of albumin permeability across endothlial monolayers *in vitro. J. Appl. Physiol.* 62(3):1076–1083, 1987.

Crone, C. Permeability of capillaries in various organs as determined by use of the indicator diffusion method. *Acta Physiol. Scand.* 48:292–305, 1963.

Croughan, M.S., J.F. Hamel, and D.I.C. Wang. Hydrodynamic effects on animal cells grown in microcarrier cultures. *Biotech. and Bioeng.* 29:130–141, 1987.

Curry, F.E. Determinants of capillary permeability. A review of mechanisms based on single capillary studies in the frog. *Circ. Res.* 59:367–380, 1986.

Doerschuk, C.M., G.P. Downey, D.E. Doherty, D. English, R.P. Gie, M. Ohgami, G.S. Worthen, P.M. Henson, and J.C. Hogg. Leukocyte and platelet margination within microvasculature of rabbit lungs. *J. Appl. Physiol.* 68(5):1956–1961, 1990.

Eaton, B.M., V.J. Toothill, H.A. Davies, J.D. Pearson, and G.E. Mann. Permeability of human venous endothelial cell monolayers perfused in microcarrier cultures: Effect of flow rate, thrombin, and cytochalasin D. *J. Cell. Physiol.* 149:88–99, 1991.

Franke, W.W., P. Cowin, C. Grund, C. Kuhn, and H.P. Kapprell. The endothelial junction. The junction and its components. In: *Endothelial Cell Biology in Health and Disease,* edited by N. Simionescu and M. Simionescu. New York: Plenum Press, 1988.

Gallik, S., S. Usami, K.M. Jan, and S. Chien . Shear-stress induced detachment of human polymorphonuclear leukocytes from endothelial monolayers. *Biorheology* 26:823–834, 1988.

Garcia, J.G.N., A. Siflinger-Birnboim, R. Bizios, P.J. Del Vicchio, J.W. Fenton, and A.B. Malik. Thrombin induced increase in albumin permeability across endothelium. *J. Cell. Physiol.* 128:96–104, 1986.

Gillies, M.C., T. Su, and D. Naidoo. Electrical resistance and macromolecular permeability of retinal capillary endothelial cells in vitro. *Curr. Eye Res.* 14(6):435–442, 1995.

Gumbiner, B. and K. Simons. A functional assay for proteins involved in establishing an epithelial occluding barrier: Identification of a uvomorulin-like polypeptide. *J. Cell Bio.* 102:457–468, 1986.

Gumbiner, B., B. Stevenson, and A. Grimaldi. The role of the cell adhesion molecule uvomorulin in the formation and maintenance of the epithelial junctional complex. *J. Cell Biol.* 107:1575–1587, 1988.

Harlan, J.M., B.R. Schwartz, M.A. Reidy, S.M. Schwartz, H. Ochs, and L.A. Harker. Activated neutrophils disrupt endothelial monolayer integrity by an oxygen radical independent mechanism. *Lab Invest.* 52(2):141–150, 1985.

Harris, T.R. and R.J. Roselli. Exchange of small molecules in the normal and abnormal lung circulatory bed. In: *Respiratory Physiology: An Analytical Approach,* edited by H.K. Chang and M. Paiva, Vol. 40. (Lung Biol. Health Dis. Ser.). New York: Dekker, 1989.

Haselton, F.R. and J.S. Alexander. Platelets and platelet releasate enhance endothelial barrier. *Am. J. Physiol.* 263(*Lung Cell. Molecular Physiol.* 7):L670–L678, 1992.

Haselton, F.R. and R.L. Heimark. Role of cadherins 5 and 13 in the aortic endothelial barrier. *J. Cell Physiol.* 171:243–251, 1997.

Haselton, F.R., S.N. Mueller, R.E. Howell, E.M. Levine, and A.P. Fishman. Chromatographic demonstration of reversible changes in endothelial permeability. *J. Appl. Physiol.* 67:2032–2048, 1989.

Haselton, F.R., J.S. Alexander, S.N. Mueller, and A.P Fishman. Modulators of endothelial permeability: A mechanistic approach. In: *Endothelial Cell Biology in Health and Disease,* edited by N. Simionescu and M. Simionescu. New York: Raven Press, pp. 103–126, 1992a.

Haselton, F.R., J.S. Alexander, S.N. Mueller, and A.P Fishman. Modulation of endothelial paracellular permeability: A mechanistic approach. In: *Endothelial Biology in Health and Disease,* edited by M. Simionescu and N. Simionescu. New York: Plenum Press, 1992b.

Haselton, F.R., J.S. Alexander, and S.N. Mueller. Adenosine decreases permeability of endothelial monolayers. *J. Appl. Physiol.* 74(4):1581–1590, 1993.

Haselton, F.R., J.S. Alexander, E. Dworska, S.S. Evans, and L.H. Hoffman. Modulation of retinal endothelial barrier in an in vitro model of the retinal microvasculature. *Exp. Eye Res.* 63:211–222, 1996a.

Haselton, F.R., J. Woodall, and J.S. Alexander. Neutrophil effects in a cell-column model of the microvasculature: Effects of fMLP. *Microcirc.* 3:329–342, 1996b.

Heimark R.L. Calcium-dependent and calcium-independent cell adhesion molecules in the endothelium. *Ann N. Y. Acad. Sci.* 614:229–239, 1991.

Huxley, V.H. and F.E. Curry. Albumin modulation of capillary permeability: Tests of an adsorption hypothesis. *Am. J. Physiol. (Heart Circ. Physiol. 17)*:H1264–H272, 1985.

Killackey, J.J.F. and B.A. Killackey. Neutrophil-mediated increased permeability of microcarrier-cultured endothelial monolayers: a model for the *in vitro* study of neutrophil-dependent mediators of vasopermeability. *Can. J. Physiol. Pharmacol.* 68:836–844, 1990.

Killackey, J.J., M.G. Johnston, and H.A. Mozat. Increased permeability of microcarrier-cultured endothelial monolayers in response to histamine and thrombin. *Am. J.Pathol.* 122:50–61, 1986.

Lawrence, M.B., L.V. McIntire, and S.G. Eskin. Effect of flow on polymorphonuclear leukocyte/endothelial cell adhesion. *Blood* 70(5):1284–1290, 1987.

Lum, H. and A.B. Malik. Regulation of vascular endothelial barrier function. *Am. J. Physiol.* 267*(Lung Cell. Mol. Physiol. 11)*:L223–L241, 1994.

Mitra P., C.R. Keese, and I. Giaever. Electric measurements can be sued to monitor the attachment and spreading of cells in tissue culture. *Biotechniques* 11(4):504–510, 1991.

Nose, A., K. Tsuji, and M. Takeichi. Localization of specificity determining sites in cadherin cell adhesion molecules. *Cell* 61:147–155, 1990.

Nuijens, J.H., J.J. Abbink, Y.T. Wachtfogel, R.W. Colman, A.J. Berenberg, D. Dors, A.J. Kamp, R.J. Strack van Schijndel, L.G. Thijs, and C.B. Hack. Plasma elastase alpha-1-antitrypsin and lactoferrin in sepsis: Evidence for neutrophils as mediators in fatal sepsis. *J. Lab. Clin. Med.* 119(2):159–168, 1992.

Orlidge, A. and P.A. D'Amore. Inhibition of capillary endothelial cell growth by pericytes and smooth muscle cells. *J. Cell Biol.* 105:1455–1462, 1987.

Renkin, E.M. Capillary permeability to lipid-soluble molecules. *Am. J. Physiol.* 168:538–545, 1952.

Sangren, W.C. and C.W. Sheppard. Mathematical derivation of the exchange of a labelled substance between a liquid flowing in a vessel and an external compartment. *Bull. Math. Biophys.* 15:387–394, 1953.

Seekamp, A., A. Dwenger, M. Weidner, G. Regel, and J.A. Sturm. Effect of recurrent endotoxemia on hemodynamics, lung function and neutrophil activation in sheep. *Eur. Surg. Res.* 24(3):143–154, 1992.

Shasby, D.M., S.S. Shasby, J.M. Sullivan, and M.J. Peach. Role of endothelial cell cytoskeleton in control of endothelial permeability. *Circ. Res.* 51:657–661, 1982.

Shasby, D.M. and S.S. Shasby. Effects of calcium on transendothelial albumin transfer and electrical resistance. *J. Appl. Physiol.* 60(1):71–79, 1986.

Shasby, D.M., S.E. Lind, S.S. Shasby, J.C. Goldsmith, and G.W. Hunninghake. Reversible oxidant-induced increases in albumin transfer across cultured endothelium: Alterations in cell shape and calcium homestasis. *Blood* 65(3):605–614, 1985.

Shasby, D.M., S.S. Shasby, and M.J. Peach. Granulocytes and phorbol myristate acetate increase permeability to albumin of cultured endothelial monolayers and isolated perfused lungs. *Am. Rev. Respir. Dis.* 127:143, 1983.

Simionescu, N. Cellular aspects of transcapillary exchange. *Physiol. Rev.* 63(4):1536–1578, 1983.

Sims, D.E. The pericyte—A review. *Tissue and Cell* 18:153–174, 1986.

Springer, T.A. Traffic signals for lymphocyte recirculation and leukocyte emigration: The multistep paradigm. *Cell* 76:301–304, 1994.

Tiruppathi, C., A.B. Malik, P.J. Del Vecchio, C.R. Keese, and I. Giaver. Electrical method for detection of endothelial cell shape change in real time: Assessment of endothelial barrier function. *Proc. Nat. Acad. Sci.* 89:7919–7923, 1992.

Volk, T. and B. A. Geiger. CAM: A 135-kD receptor of intercellular adherens junctions. I. Immunoelectrom microscopic localization and biochemical studies. *J. Cell Biol.* 103:1441–1450, 1986a.

Volk, T. and B. Geiger. A-CAM: A 135-kD receptor of intercellular adherens junctions. II. Antibody-mediated modulation of junction formation. *J. Cell Biol.* 103(4):1451–1464, 1986b.

Waters, C.M. Flow-induced modulation of the permeability of endothelial cells cultured on microcarrier beads. *J. Cell. Physiol.* 168(2):403–411, 1996.

Weiss, S.J. Tissue destruction by neutrophils. *New Engl. J. Med.* 320(6):365–376, 1989.

Worthen, G.S., L.A. Smedly, M.G. Tonnensen, D. Ellis, N.F. Voelkel, J.T. Reeves, and P.M. Henson. Effects of shear stress on adhesive interaction between neutrophils and cultured endothelial cells. *J. Appl. Physiol.* 63:2031–2041, 1987.

Yi, E.S. and T.R. Ulich. Endotoxin, interleukin-1, and tumor necrosis factor cause neutrophil-dependent microvascular leakage in postcapillary venules. *Am. J. Pathol.* 140(3):659–663, 1992.

Zimmerman, B.J. and D.N. Granger. Reperfusion-induced leukocyte infiltration: Role of elastase. *Am. J. Physiol.* 259(*Heart Circ. Physiol.*):H390–H394, 1990.

3
Metabolism in the Heart and Skeletal Muscle

7

Strategies for Uncovering the Kinetics of Nucleoside Transport and Metabolism in Capillary Endothelial Cells

James B. Bassingthwaighte, Keith Kroll, Lisa M. Schwartz,
Gary M. Raymond, and Richard B. King

Whole Organ Modeling

For the analysis of signals obtained by external detection techniques such as positron tomographic imaging (PET), magnetic resonance imaging (MRI), and X-ray computed tomography (X-ray CT), investigators and diagnosticians usually obtain a sequence of images. For physiological interpretation in terms of the underlying physical and chemical events, it is essential to use models when one wants to learn more than the simplest measures. Among the simplest measures one can often include volume and flow estimates, but not always, for it commonly occurs that the distinctive estimation of these two parameters simultaneously requires using knowledge of the anatomy or of other properties of the tissue. The two most accessible measures of indicator transport are the areas under dilution curves and their mean transit times following a pulse injection. Mass conservation for substances that are not destroyed, such as radioactive tracers, relies on the general expression:

$$q(t) = F\int_0^t C_{in}\, dt - F\int_0^t C_{out}\, dt,$$

where $q(t)$ represents the mass of indicator in the tissue at time t, F is the flow of indicator-containing fluid, and $C_{in}(t)$ and $C_{out}(t)$ represent the concentration–time curves for the indicator at the inflow and outflow. Such expressions represent whole organ behavior exactly when the organ is supplied by a single artery and drained by a single vein. When there are multiple inlets and multiple outlets, then the first term on the right is replaced by a sum of similar terms for each inlet, and likewise the second term on the right is replaced by a similar sum of the outputs (Lassen and Perl, 1979). When the input is a brief pulse input, theoretically infinitely short, then $C_{in}(t)$ is a scaled Dirac delta function, $q_0\delta(t)/F$, where q_0 is the dose of indicator injected and $C_{out}(t)$ is $q_0h(t)/F$, where $h(t)$ is defined as the impulse response or transport function of the organ; see Chapter 1 and the general overview by Bassingthwaighte and Goresky (1984). This expression defines a basis for using the principle of conservation of mass to be applied as a check in the analysis. Up to this point, the approach is not dependent on any model. Such an equation can be used for the "model-free" estimation of flow, for example,

$$F = q_0 \bigg/ \int_0^\infty C_{out} \, dt,$$

which is true even when the input is dispersed, with the proviso that the indicator has passed through the system only once, that is, when there is no recirculation.

Another conservation statement, also model-free, is for the mean transit time volume, V', of the system:

$$V' = F \cdot \bar{t},$$

where the mean transit time $\bar{t}$ is the first moment of the impulse response. However, to model a whole organ and obtain estimates of parameters of the permeabilities and reaction rates, one must make some basic assumptions about the structures or arrangements of the components of the organ. There are several reasons for this: If a substrate undergoes a reaction to form a product and the form of $C_{out}(t)$ for the product differs from that of the substrate, then the two molecules must have undergone different transports or other processes in traveling from the reaction site to the outflow. These might be due to differing membrane permeabilities, different solubilities or volumes of dilution, different degrees of binding, or differences in diffusivity. All of these require interpretation in terms of structure–function relationships.

Network Representation of Organs

Vascular systems branch almost solely by bifurcating (Kassab et al., 1993), but since the daughter branches are unequal in lengths and diameters, the patterns are not those of simple binary trees. While binary tree models have some utility in understanding physiological organ behavior (van Beek et al., 1989; Glenny and Robertson, 1995), much better realism is provided by models that account for the heterogeneity of structural arrangements, since they can account for the patterns of flow distribution (Beard and Bassingthwaighte, 1997). The distributions of flows to small regions of tissue are consequently broad, and while the tissue flows (or perfusion, in ml min^{-1} g^{-1}) are heterogeneous, there is local correlation. How strong the correlation is between near-neighbor regions was found through fractal analysis (Bassingthwaighte, 1988; Bassingthwaighte et al., 1989). It was found that the fractal dimension in the heart is about 1.2, as illustrated in Figure 1.4 of Chapter 1, which is less than the value of 1.5 that would indicate a random variation. This means that the correlation coefficient, r_1, between flows in nearest-neighbor pieces of tissue is about 0.5. This correlation is independent of size since this is a fractal process with self-similarity independent of the actual length scale over a wide range of tissue volume element sizes. The relation between nearest-neighbor correlation r_1 and the fractal D is $r_1 = 2^{3-2D} - 1$ (Bassingthwaighte and Beyer, 1991). The beauty of this is that, if nearest-neighbor capillaries are nearly

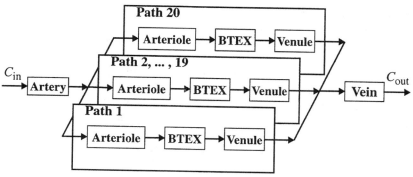

FIGURE 7.1. A 20-path multicapillary model for blood–tissue exchange in a whole organ. The model consists of single common entrance and exit vessels, a large artery and a large vein, and a set of parallel independent units having differing flows, each unit consisting of an arteriole, a capillary–tissue exchange unit, and a venule. All of the components of the system are dispersive, i.e., their impulse responses show a spread of transit times through passage from entrance to exit.

alike with respect to their blood velocities and lengths, then we can make a fundamental and greatly simplifying assumption:

Assumption 1

There is no net exchange of indicator between neighboring capillary–tissue units. This appears to be justified for a wide range of small hydrophilic solutes (Goresky, 1963; Bassingthwaighte, 1970; Kuikka et al., 1986). Consequently, we view the organ as consisting of parallel independent pathways, each differing from the others only with respect to the local flow, as in Figure 7.1. To characterize this model in a computable form, one now has to define several variables. These fall into the following groups:

1. The description of the probability density function of regional flows, $w(f)$, as defined in Chapter 1, and how to represent it in the model: The regression as a finite number of pathways requires care to avoid numerical inaccuracy. King et al. (1996) provide algorithms to select good approximations to $w(f)$ by histograms of $N \leq 20$.
2. The descriptions of transport through the large- and medium-sized vessels: Simple linear operators (King et al., 1993) have proven useful in describing intravascular dispersion and delay.
3. The descriptions of the exchange units: These are described next.

A Basic Blood–Tissue Exchange Unit

Beginning with Bohr's (1909) demonstration that the concentration of a solute escaping from a long tube should diminish exponentially as a function of distance from the entry to the tube, there has been a continuing development of mathematical models of the exchange processes between the blood in a capillary and the fluid in the surrounding tissue. Early models used compartmental analysis with one instantaneously mixed compartment representing the blood (or plasma) space and a second representing all of the extravascular tissue space, e.g., Sapirstein (1958). Through the years, the modeling of blood–tissue exchange has evolved in several ways: Additional anatomical regions (e.g., endothelial and parenchymal cells) are now included, distributed models that can exhibit axial (i.e., arterial to venous) concentration gradients and describe the axial diffusion that dissipates the gradients have been developed, and analytical and numerical techniques have been improved so as to increase both accuracy and computational speed.

Sangren and Sheppard (1953) first gave the analytic solution for a blood–tissue exchange model containing two regions, capillary and extravascular tissue, separated by a permeable membrane. The solution assumed that the axial diffusion in both regions was zero and that the radial diffusion was infinitely fast. The return flux from the extravascular region to the capillary was taken into account; thus, mass was conserved. Goresky et al. (1970) extended this to include uptake from the interstitial fluid (ISF) into the parenchymal cells, without return flux from this third region. Rose et al. (1977) developed a solution for the analogous model that included the parenchymal cell and its membrane, a two-barrier three-region model with return from all regions back to the capillary space. Bassingthwaighte (1974) developed a whole organ model that accounted for dispersion in arteries and veins. This model was extended into a multicapillary model by Kuikka et al. (1986), who also added axial diffusion, and by Lumsden and Silverman (1986), who added terms for the consumption of the solute in all three regions.

We have previously described a three-barrier four-region blood–tissue exchange (BTEX) model (Bassingthwaighte et al., 1989b) that accounts for capillary plasma, endothelial cells, ISF, and parenchymal cells, and axial diffusion and consumption in all regions. The solutions are numerical, using computationally efficient algorithms (Bassingthwaighte et al., 1992) that result in computational speed about 10^7 times faster than an analytical solution and provide good accuracy at long solution times, even with parameter values making computation "stiff" (e.g., $PS/F > 1000$).

In order to provide generality in concept and in application, it has proven useful to work with a basic model form that allows much generality. The model form is also such that it can be reduced to a minimal model, so preserving the strengths and weaknesses of a large variety of model forms. This is MMID4, a model form that does most everything with respect to organ blood flow, diffusion, vascular transport, exchange by permeation, and tracer removal by reaction.

MMID4 is an acronym for multiple-path, multiple-tracer, indicator dilution, 4-region model. It is used to examine the behavior of three types of tracers: one that remains in the vasculature, one that can leave the vasculature but remains extracellular, and a fully permeant tracer that enters cells. The four regions are plasma, endothelial cells, interstitial fluid, and parenchymal cells. The model can be used to study the physiology of the exchange process or as an analytical tool to

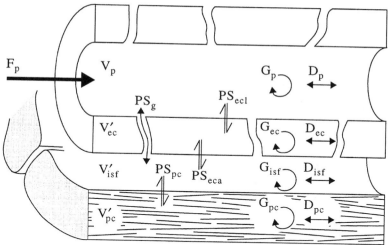

FIGURE 7.2. Basic blood–tissue exchange unit. Parameters affecting exchange of the permeant tracer. F_p and V_p are the same for all tracers. (Reproduced with permission from Bassingthwaighte, Wang, and Chan, 1989. Copyright 1989 American Heart Association.)

help analyze actual experimental data. The exchange process may be examined by viewing the instantaneous outflow concentration, the instantaneous extraction, or the amount of indicator remaining in the exchange unit (i.e., the residue function). The model includes provisions for examining the effects of indicator delay and dispersion between the injection site and the target organ and the effects of flow heterogeneity on exchange within the organ.

The extension of these equations to a multicapillary model with a set of capillaries having different flows is discussed by Bassingthwaighte et al. (1989b). They also present the numerical methods used to obtain solutions to these equations that are analytic within each space and time step.

MMID4 includes within each pathway the model shown in Figure 7.2 and two reduced forms of it. These reduced forms are used for modeling the "vascular" and "extracellular" reference tracers. The "vascular" model permits both an intravascular and an "interstitial" region, which is seldom used but is available to account for leakage or the presence of a receptor on endothelial cells. The extracellular reference solute chosen is normally one with the same permeability through the clefts between endothelial cells and the same V_{isf} as the permeant tracer targeted in the study. The use of these "in-built" controls greatly reduces the number of free parameters for the permeant sought in the optimization to fit data. The plasma flow, F_p, the plasma volume, V_p, the flow heterogeneity, and the intravascular dispersion are the same for all tracers.

Exchange units in the model are arranged in a number of parallel flow pathways. Before entering these pathways, the flow passes through a large vessel component. Each flow pathway consists of a small vessel in series with a blood–

tissue exchange unit that encompasses the parameters shown in Chapter 1. Including the large and small vessel components provides a means of accounting for delay and dispersion of the indicator between the injection site, the organ, and the outflow. The parameters governing large and small vessel behavior are the volume of the vessels and a parameter defining the degree of delay and dispersion that the input function incurs before reaching the exchange unit.

Flow heterogeneity in the target organ is taken into account by making a number of parallel pathways available to the tracers and defining the flow distribution, $w(f)$, through the pathways. MMID4 can be used with up to 20 pathways.

While this BTEX model permitted the consumption of the solute in any region, the solute consumed in a region undergoes no further convection or diffusion. The amount consumed may be considered to either disappear from the system or, since that quantity is calculated by the model, it may be considered to be sequestered in the region in which it was consumed. The former is appropriate for a tracer undergoing radioactive decay and the latter for a tracer such as deoxyglucose that undergoes only the first step of glucose metabolism and then is trapped within the cell. Neither of these is, however, appropriate for a solute that undergoes a metabolic transformation that produces a product that can diffuse, or be transported, back out of the cell or that may be further metabolized to produce a diffusible product.

Consider the case of $[^{14}C]$-labeled adenosine. Wangler et al. (1989) have shown that tracer adenosine injected into the inflow of the isolated guinea pig or rabbit heart is metablized during its single transit of the microvasculature. When the metabolites in the outflow samples are separated by HPLC, the label appears not only on adenosine but also on its metabolites inosine, xanthine, hypoxanthine, and uric acid. The appropriate model for this situation is one in which the solute reacts to form a product that is really the input to a model for the product with its own set of parameters governing its exchange processes. For adenosine and its products, a sequence of such interrelated models is required. This is described next as MSID4.

A Whole Organ Model, MSID4, for a Reactant/Product Sequence

Our purpose here is to present a multisolute, three-barrier, four-region, axially distributed model for blood–tissue exchange that accounts for convection along the capillary and exchanges, across permeable membranes and through the inter-endothelial cell gaps between endothelial cells, ISF, and parenchymal cells. Reactions leading to both the sequestration and production of a metabolite are permitted in all regions within cells or extracellular spaces; the consumption of a tracer-labeled substrate is considered to be a linear process. A fraction goes into forming a metabolite that can exchange and be further metabolized, and the remainder is sequestered permanently in the region in which it is formed. However, in this model no reactions are allowed to result in the return of a tracer to an antecedent species, that is, the reaction sequence is unidirectional. **Note:** In the subsequent discussion, we will use the term "metabolite" for each of the product chemical species produced from the original substrate.

This Multipath, multi*Substrate Indicator Dilution, four*-region model, MSID4, is based on the MMID4 model discussed above. The arrangement of the nonexchanging vascular units has the structure shown in Figure 7.1. However, each of the 1 to 20 flow paths contains a set of exchange units, one for each substrate modeled, rather than a single exchange unit for each path. Therefore, for a 20-path, 6-solute reactant-metabolite model there are 120 BTEX units plus another 20 each for the vascular reference tracer $h_R(t)$ and for the extracellular reference tracer $h_E(t)$. The probability density functions of regional flows and the parameters for intravascular transport in the large and medium size vessels are the same for all substrates. Each solute, substrate, or metabolite has a separate set of parameters for the exchange unit.

A BTEX model for transport and exchange between the plasma (p), endothelial cells (ec), interstitial fluid (isf), and parenchymal cells (pc) in a single tissue–capillary unit is diagrammed in Figure 7.1. Depending on the assumptions made, this general model can be described mathematically in a variety of ways. For the situation in which the only input to the system is by convection into the upstream end of the plasma, an explicit mathematical description is given by assuming steady plug flow in the plasma; the absence of radial (i.e., perpendicular to the direction of flow) concentration gradients in each physical region; and linearity of the coefficients for convection, diffusion, and consumption. The equations of the model account for conductances, PS, across the barriers between the regions, the regional volumes of distribution (V'), the dispersion in the axial direction (D), and first-order consumption (G).

The equations for concentration of the tracer as a function of time, t, and the axial position along the unit, x, are, in the plasma,

$$\frac{\partial C_p}{\partial t} = -\frac{F_p L}{V_p} \cdot \frac{\partial C_p}{\partial x} - \frac{PS_g}{V_p}(C_p - C_{isf}) - \frac{PS_{ecl}}{V_p}(C_p - C_{ec}) - \frac{G_p}{V_p}C_p + D_p\frac{\partial^2 C_p}{\partial x^2};$$

(7.1)

in the endothelial cells,

$$\frac{\partial C_{ec}}{\partial t} = -\frac{PS_{ecl}}{V'_{ec}}(C_{ec} - C_p) - \frac{PS_{eca}}{V'_{ec}}(C_{ec} - C_{isf}) - \frac{G_{ec}}{V'_{ec}}C_{ec} + D_{ec}\frac{\partial^2 C_{ec}}{\partial x^2},$$

(7.2)

in the interstitial fluid,

$$\frac{\partial C_{isf}}{\partial t} =$$
$$-\frac{PS_g}{V'_{isf}}(C_{isf} - C_p) - \frac{PS_{eca}}{V'_{isf}}(C_{isf} - C_{ec}) - \frac{PS_{pc}}{V'_{isf}}(C_{isf} - C_{pc}) - \frac{G_{isf}}{V'_{isf}}C_{isf} + D_{isf}\frac{\partial^2 C_{isf}}{\partial x^2};$$

(7.3)

and in the parenchymal cells,

$$\frac{\partial C_{pc}}{\partial t} = \frac{PS_{pc}}{V'_{pc}} (C_{pc} - C_{isf}) - \frac{G_{pc}}{V'_{pc}} C_{pc} + D_{pc} \frac{\partial^2 C_{pc}}{\partial x^2}. \qquad (7.4)$$

The terms are as follows: C, concentration in moles per milliliter; PS, permeability–surface area product in ml g^{-1} s^{-1}; F, plasma flow in ml g^{-1} s^{-1}; L, capillary length in centimeters; G, regional clearance or gulosity in ml g^{-1} s^{-1}; D axial diffusion or dispersion coefficient in cm^2 s^{-1}; V, volume in ml g^{-1}; and V', volume of distribution in a region in ml g^{-1}. A volume of distribution is the region-to-plasma partition coefficient (dimensionless) times the actual anatomic volume of the region (ml g^{-1} of tissue). Subscripts on concentrations or volumes refer to a region, with the exception of the conductance across the luminal and abluminal surfaces of the endothelial cells, PS_{ecl} and PS_{eca}, respectively, and the conductance through the gaps between the endothelial cells, PS_g. The existence of this direct path for plasma–isf exchange that bypasses the endothelial cells makes this model different than one with a set of concentric volumes.

Parameters Governing Tracer Exchange

The parameters affecting the concentration of the permeant tracer in each region are shown in Figure 7.2. This tracer can move to all regions, and has an apparent volume of distribution (V') in each. In the intravascular region, this is the plasma volume (V_p). For other regions, the equilibrium condition is defined by $V'_r/V_p = C_r(t = \infty)/C_p(t = \infty)$, where the subscript "r" indicates region, and the concentration ratio C_r/C_p shows that V'_r is a virtual volume of plasma–equivalent concentration. The tracer may be metabolized or undergo chemical reactions that result in its being cleared from the region. The degree to which this happens is indicated by the intraregional consumption or "gulosity" (G). Movement of the tracer between regions is governed by the permeability–surface area product (PS) of the various barriers. A molecule of tracer may move, for example, between the capillary and the interstitium. Two routes are available: It can either move through the clefts between the endothelial cells, or it can move through the endothelial cell. In the first case, movement depends on one PS product (PS_g, where the subscript "g" denotes gap on cleft). In the second, movement depends on two PSs, the PS product at the luminal side of the endothelial cell (PS_{ecl}) and the PS product on abluminal side of the cell (PS_{eca}). The plasma flow affecting the permeant tracer exchange is the same flow (F_p) for all of the tracers. In formal notation, the outflow concentration–time curve seen from an exchange unit in response to an impulse input of tracer is denoted as $h(t)$, as in Chapter 1. We use $h(t)$ as a general symbol to denote the normalized (unity area) outflow dilution curve even when the input has forms other than an impulse. A subscript is added to designate specific tracers. Thus, $h_D(t)$ denotes the "permeating" or "diffusible" tracer; $h_E(t)$ signifies the extracellular tracer; and $h_R(t)$ is used to represent the intravascular "reference" tracer.

Substrate Metabolism by Sequential Reactions

The impetus for the development of this model was the need to analyze data from a series of experiments aimed at understanding the utilization of adenosine in the myocardium. These experiments used radiolabeled adenosine at tracer concentration (i.e., a concentration low enough that it could be assumed that the preexisting concentration of unlabeled adenosine was not changed by the addition of the labeled adenosine) and the data were collected over a period of 1–2 min. This led to a design that featured a linear sequence of nonreversible reactions.

A set of six substrates in a linear reaction chain are modeled in MSID4, but this may be extended arbitrarily. The sequence can be diagrammed as:

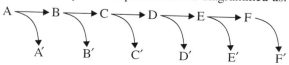

The primed letters represent tracer trapped in some form but not reacted as a part of the sequence. In the description of the four-region BTEX model given above, the consumption term accounts for all processes that cause clearance of the tracer from the region (e.g., sequestration and chemical transformation). In the MSID4 model, however, these processes are separated rather than being lumped together. Focusing on a single reaction in the diagram above, the process can be represented as:

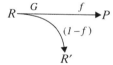

The total clearance of reactant (R) during each time step of the solution is governed by the first-order rate constant G. Of the total amount cleared, some fraction, f, is transformed into the product (P). This product appears in the regions of its own BTEX model where the reaction occurred; it can diffuse across membranes and may undergo further reactions. The remainder of the reactant cleared during the time step is transformed into a sequestered form of the reactant (R') that neither undergoes further chemical transformation nor diffuses across barriers. This reaction sequence modifies Eqs. (7.1)–(7.4) by adding an additional term that accounts for the input from chemical transformation. A general equation for the concentration of a product in any region is

$$\frac{\partial {}_P C_r}{\partial t} = -{}_P X_r + {}_P D_r \frac{\partial^2 {}_P C_r}{\partial x^2} - \frac{{}_P G_r}{{}_P V_r'} {}_P C_r + {}_P f_r \frac{{}_R G_r}{{}_P V_r'} {}_R C_r, \qquad (7.5)$$

where the leading subscript specifies either the reactant or the product, and the trailing subscript, "r," specifies the region. The first term on the right in Eq. (7.5), ${}_P X_r$, is the summation of all permeation terms for the product in that region and, in the case of the plasma region, the convection term as well. The second and third terms are the axial diffusion and consumption terms, respectively, from those equations. The last term is the input from chemical transformation of the reactant that produces this substrate as its product. Thus, the production is the consumption of the preceding reactant substrate in the reaction sequence. Since substrate A is not a reaction product, $_RG_r$ is equal to zero for substrate A.

Using the same nomenclature as in Eq. (7.5), the equation for the sequestered form of the reactant is

$$\frac{\partial_{R'}C_r}{\partial t} = {_{R'}}D_r \frac{\partial_R^2{}_{'}C_r}{\partial x^2} + (1 - {_R}f_r)\frac{_RG_r}{_{R'}V_r'}\,{_R}C_r. \tag{7.6}$$

The first term accounts for axial diffusion within the region and the second is the production term. The latter is the same as in Eq. (7.5), except that the multiplier is $(1 - {_R}f_r)$, thus assuring mass balance. For substrate F, $_Rf_r$ is equal to zero; thus, all cleared tracer is sequestered. The transmembrane diffusion term is absent since the sequestered tracer does not cross barriers. Equation (7.6) must, however, be modified in the case of the plasma region. Even though tracer in the sequestered form does not cross the capillary barriers, it is carried down the capillary and into the outflow by convection. Thus, for the plasma, the equation is

$$\frac{\partial_{R'}C_p}{\partial t} = -\frac{F_pL}{_{R'}V_p} \cdot \frac{\partial_{R'}C_p}{\partial x} + {_{R'}}D_p\frac{\partial_R^2{}_{'}C_p}{\partial x^2} + (1 - {_R}f_p)\frac{_RG_p}{_{R'}V_p'}\,{_R}C_p. \tag{7.7}$$

Several assumptions are implicit in this formulation of the model, and we now state them explicitly. All reactions are unidirectional, and the reaction rates are all first-order. The reaction rates are not subject to control by the concentration of the reactant or the product, or by any other mechanism. The reactions form a linear sequence without feedback, and the pool size for the sequestered form of the substrates is infinite (i.e., there is no return flux). We will come back to some of these limitations in the Discussion.

Basic Modeling Analysis of the MID Experimental Data Set

The modeling analysis can proceed immediately when an input function has been defined. When deconvolution is needed to produce the input function, then the intravascular transport must first be defined, as in the next section. If the input is known, or if it is approximated by an arbitrary input waveform such as a lagged-normal density curve (Bassingthwaighte, 1966) or a gamma variate function (Thompson et al., 1964), then we proceed directly to the modeling of intravascular transport.

Intravascular Transport

As shown in Figure 7.1, there are three components of the blood of the organ: The first is a common large vessel representing large arteries and veins and leading to a set of parallel paths. Each of up to 20 parallel paths consists of an operator representing vascular transport through medium-sized vessels without exchange (component two) and the blood–tissue exchange operator (component three). The total blood volume of the heart is about 0.15 ml/g; of this, the common path volume, V_{LV}, is about 0.03 ml/g, the medium vessels, V_{MV}, 0.07 ml/g, and the capillaries, V_p, about 0.05 ml/g, depending on the driving pressures, flows, and vasodilatory state.

To account for both the dispersion and delay due to transport through medium-sized nonexchanging arteries and veins, an operator is included in series with each capillary tissue unit. The "medium vessel operator" is composed of a pure delay line in series with a dispersive operator, a fourth-order differential operator made up of two second-order operators in series. The parameters of the fourth-order operator are those given by Bassingthwaighte et al. (1970), which result in its relative dispersion (coefficient of variation of the impulse response) being 48%. By setting the delay operator to provide 60% of the total mean transit time of the medium vessel operator, the relative dispersion of the medium vessel operator is set at 0.48 (1.0–0.6) or 19.2% (King et al., 1993). This is approximately the level of dispersion found for arteries by Bassingthwaighte (1966). The operator accounts for both arterial and venous dispersion (convolution being commutative), and if venous dispersion is relatively greater than the arterial, the total dispersion may be a bit smaller than is realistic. The same fractional volume was assumed for high and low flow regions, $V_{MV} = 0.07$ ml/g; the transit times were inversely proportional to the flows in the individual pathways. This medium vessel operator has the advantage of accounting for both intravascular dispersion and for regional variation in transit times. By being dispersive it offers an improvement over the pure delay lines used by Rose and Goresky (1976) while simultaneously accounting for the association between the medium vessel and the capillary transit times that they advocate.

Transport within capillaries is dispersive (Crone et al., 1978). The Lagrangian numerical method matching time and space steps (Bassingthwaighte, 1974; Bassingthwaighte et al., 1989b) is nondispersive, but the dispersion process is applied at each time step using a classical diffusion expression (Bassingthwaighte et al., 1992) where D represents the dispersion due to all operative processes: molecular diffusion, eddy currents at branches, velocity profiles, and disturbed flow.

Finding the Input Function When It Was Not Measured

When the input function is not measured, the model input at the point where the parallel pathways diverge, that is, the exit from the large vessel operator, is determined either by deconvolution from the observed $h_R(t)$ or by optimizing the form of an assumed input function until the observed output vascular $h_R(t)$ is matched by the model solution. An example is shown in Figure 7.3. The approach is to estimate the form of the intravascular transport from inflow to outflow through the set of parallel paths, and then to deconvolute by a smoothing, stabilized method to get the input function. The result is then checked by convolution of the input with the intravascular transport function. That the model solution using the input curve obtained by deconvolution should fit the actual $h_R(t)$ should

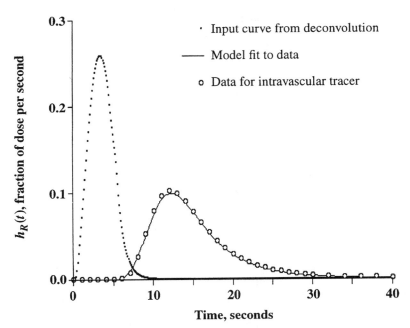

FIGURE 7.3. Deconvolution from the curve of reference data produces a smooth unimodal input. Capillary and small vessel volumes are assumed from anatomic data and the observed mean transit time volume, $F \cdot \bar{t}_R$. Reconvolution, giving a good fit, is merely a test of the numerical methods and does not guarantee that C_{in} is correct. (From Bassingthwaighte, King, and Kroll 1993.)

evoke no surprise, since that simply amounts to reconvolution. The deconvolution process itself is complex. We use a variation of the method of Bronikowski et al. (1980) that constrains the technique to producing wave forms without negative or highly oscillatory components. The use of numerical deconvolution (Maseri et al., 1970) is not usually workable because the result is ordinarily a highly oscillatory function. Deconvolution is a kind of differentiation process, and is generally unreliable unless considerable care is taken. Often it is better to guess the input function than to deconvolute. The safest guess is to assume a narrow pulse input, and then account for the dispersion between the inflow and the outflow as intravascular dispersion. Since it is assumed that all of the tracers undergo the same intravascular dispersion, this introduces no error into the estimation of the values of the parameters for the tracer exchange processes when we are dealing with tracer dilution by a linear stationary system.

The Extracellular Reference Indicator

The diffusion of the extracellular reference tracer through the clefts between the endothelial cells is the event by which this tracer becomes separated from the intravascular reference tracer. The extracellular tracer transport function is labeled

$h_E(t)$. While its area must be identical to that of the intravascular reference tracer, the form of its outflow curve is not: It is lower peaked than the intravascular marker, because of the partial and transient loss from the intravascular region. The tracer that escapes through these clefts is diluted in the volume of the ISF. Its return from the ISF is again retarded by the process of diffusion from the ISF through the clefts back into the capillary space. Its mean transit time is necessarily longer and, by conservation, can be exactly stated:

$$\bar{t}_E = (V_{cap} + V_{isf}) / F_p.$$

There is a usually unstated assumption built into our thinking with respect to the permeating tracers: The exchange is occurring solely between the blood in the capillary and the surrounding region. In fact, it has been recognized repeatedly that the small venules are also permeable to small molecules; this means in reality that, while we estimate "capillary permeability," this is really a composite of capillary and small venular permeability. The next assumption that is ordinarily glossed over is that the intravascular indicator does not escape from the vascular space. This is really a pretty good assumption for most such experiments and can be checked by mass balance. The sum of the tracer concentrations times the volumes of exchange outflow sample should equal the dose of each of the tracers injected. For both the intravascular and the extracellular tracers, this check should be done on every dilution curve as a matter of routine. Acceptable accuracy is within ± 2% of the injected dose. While single-pass extraction of albumin is normally a small fraction of a percent, permeation does actually occur and plasma and ISF albumin concentrations equilibrate after many minutes.

The Permeant or Reactant Substrate

Data on the permeant solute are normalized in exactly the same way, but in this case the total dose may not be recovered in the outflow if there is retention within the organ and the collection period is not extended for a very long time. Of course, if there is intraorgan transformation and total retention of the product of the reaction, then it will never emerge. The fractional recovery then may give some clue as to the fate of the permeating solute. In experiments where external detection of a gamma-emitting permeant is feasible, the total dose is equivalent to the sum of the accumulated outflow fractional doses plus what is retained within the organ:

$$\text{Total dose} = \int_0^T h_D(t)\, dt + R(T),$$

at any time T during the outflow. When beta-emitting tracers are being used, one can estimate $R(t)$ by taking the whole of the tissue at the time of the last outflow sample, that is, at time t, freezing it, and grinding it up to find the value for $R(t)$. When transformation occurs within the tissue one will normally want to refine this idea by separating the various forms of the reaction products to obtain values for the retention of the individual reaction products. These individual values then become a part of the data set for model analysis, and allow a more refined estimation of the rates of the various reactions.

Constraints on Parameter Ranges

Mass Conservation

Finding the dose calibration and the fractional recoveries of each of the tracers is the first step in making sure of mass conservation. The second is that, for the permeant solute which participates in reaction sequences, the sum of the unreacted solute and the amounts of all of the reaction products in the outflow samples and in the tissue should add up to the dose injected, neither more nor less. While this is simply stated, in practice it requires deliberate care to achieve, and when there are several chemically separated reaction products it is difficult to achieve 5% accuracy.

Another aspect of mass balance is in the modeling itself. The model solutions should provide mass balance within a small fraction of 1% when one performs the summations analogous to what one does for the experimental data. A mass balance check, for example, is the flow times the integral of the concentrations in the inflow, and the same for the outflow. Both should be checked. If the integrals are not identical, and if there is no "material" remaining in the model operator itself in one form or another, then there is a computational error and the model is invalid. Since an accumulation of numerical errors can give a result summing up to either too much or too little material, this is an essential check on the model implementation.

Anatomical and Physiological Data

Volumes of distribution should add up appropriately. The sum of the fractional water contents of the several components of the tissue should add up exactly to the total water content of the tissue, as measured by weighing the tissue when wet, drying it at 50 to 60°C for 48 hours, and reweighing. This calculation should take into account the expected water fractions of red blood cells and plasma, the volumes of nonexchanging vessels, of capillaries, of the ISF, and the cells. Red blood cells are the driest cells of the body, about 65% water; plasma is 94% water, and cells and ISF are more variable. The heart water content, if it is not edematous, is remarkably uniform, about 0.78 ml/g water with a standard deviation of about 0.01 ml/g. With edema, usually expansion of the interstitial space, the water fraction rises to a maximum of about 0.84 ml/g, above which the heart ceases to beat. This is actually a formidable dilution of the solid content, from 0.22 down to 0.16 ml/g, so there is little wonder that isolated perfused hearts simply stop functioning when they get this soggy. Other constraints come from data on the volumes of capillaries and on the total blood volume of the organ. With capillarity of 3000 capillaries/mm^2 cross-section and average diameters of 5 μm, the

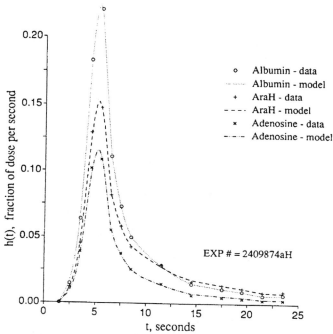

FIGURE 7.4. Graph showing adenosine, arabinofuranosyl hypoxanthine (AraH), and albumin outflow dilution curve [$h(t)$] in an isolated perfused guinea pig heart. $F_p = 5.2$ (ml/g)/min. Model-to-data coefficients of variation were 5.0% for albumin curve, 3.1% for sucrose, and 5.7% for adenosine. Flow heterogeneity was accounted for by using five capillary tissue units in parallel with a relative dispersion of 50%. $V'_{isf} = 0.19$ ml/g for AraH and adenosine. The cleft PS_g was 1.34 (ml/g)/min for AraH and adenosine. Other adenosine parameters were $PS_{ecl} = 2.25$, $PS_{eca} = 20$, $G_{ec} = 30$, $PS_{pc} = 6$, and $G_{pc} = 10$ (ml/g)/min, and $V'_{pc} = 0.62$, $V_p = 0.4$, and $V'_{ec} = 0.02$ ml/g. (From Bassingthwaighte, Wang, and Chan, 1989, their Fig. 12, with permission. Copyright 1989 American Heart Association.)

capillary volume is about 0.055 ml/g (using tissue density of 1.063 g/ml) (Yipintsoi et al., 1972). An important factor in this approach is that the assumption of a value such as $V_p = 0.055$ ml/g is not critical to the subsequent analysis: Wangler et al. (1989), in their Fig. 4, showed that a large error in choosing fixed values of V_p had little influence on estimates of the transport parameters. Since the total blood volume in a heart is about 0.12 to 0.15 ml/g at normal flows but higher when flows are raised in specific situations, for example, in the presence of vasodilating contrast agents (Koiwa et al., 1986), this estimate also provides a constraint to limit the range of the sum of the volumes of the arteries and the veins.

Likewise, data from stereological studies on tissue composition can provide constraints. The different cell types and the different fractions of nuclear and mitochondrial content of the different cell types can provide additional useful values. So far here we have discussed constraints on estimates of volumes, and therefore on the volumes of distribution that can be used in the modeling analysis. If an estimate of a volume of distribution for a solute appears via modeling analysis to be significantly larger than the anatomical volume, then one must ask if this can be explained by intracellular binding or some other physiological process. Sometimes it is difficult to distinguish retention in the tissue by the intracellular binding of a substrate from the long-term retention of its reaction product. On such occasions one must contemplate undertaking the relevant chemical analysis of tissue samples.

Strategy for Parameter Estimation

Typical solutions for an experiment using a set of three tracers (intravascular albumin, extracellular AraH, and a permeant, adenosine, that is metabolized) are shown in Figure 7.4. The AraH permeates the capillary wall via the intercellular clefts; since it has the same molecular weight as adenosine, it serves to define the fraction of adenosine traversing the cleft and being diluted in the interstitial space.

The sensitivity functions for the permeant, shown in Figure 7.5, give an excellent indication of how strong the estimates of the various parameters can be. Their definition is $S_i(t) = \partial h(t)/\partial p_i$, where S_i = sensitivity to the ith parameter, p_i. The generality is that there is greater sensitivity to parameters governing processes that are physically close to the vascular space when the tracer is injected into the inflow and the outflow is sampled. Parameters at more distant sites, for example, PS_{pc}, are estimated with greater confidence from residue detection, as is also the case when they are produced by intracellular transformation.

The general approach to modeling analysis of the results of multiple indicator dilution experiments for multiple metabolites is based on the following three steps.

The first step is to estimate the organ transfer function for an intravascular reference tracer, such as albumin. This requires that measurements be made to constrain estimates of the factors that influence the transfer function, including flow (graduated cylinder and stopwatch), flow heterogeneity (tracer microspheres), capillary volume (anatomy), and large vessel dispersion and volume (indicator dilution and anatomy). Based on these constraints, the model input function can be determined. Ideally, the input function is measured; however, this is not always practical. The input function may alternatively be determined by deconvolution of the model transfer function with the albumin dilution curve. It is also possible to optimize an input function, by assuming a particular waveform function, such as lagged-normal density, Gaussian, random walk, or gamma-variate functions.

The second step is to estimate the organ transfer function for an extracellular reference tracer, such as sucrose, L-glucose, or AraH. The parameters governing extracellular transport, PS_g and V_{isf}, are estimated by optimizing the model fit to the dilution curve of the extravascular reference tracer. This requires that a constraint be placed on the relationship between the regional flow and PS. Tradi-

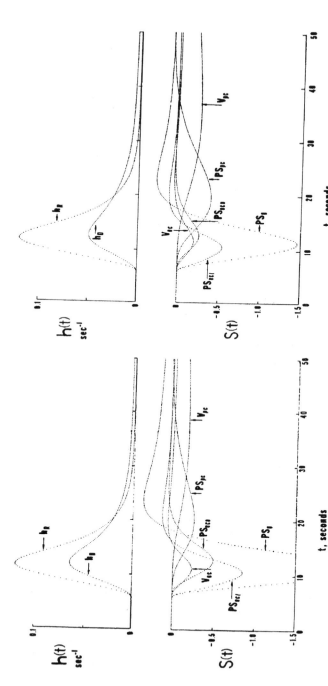

FIGURE 7.5. Graphs showing sensitivity functions for the four-region capillary–endothelial–interstitial cell model. Upper curves show the model responses to a dispersed input function [$h(t)$] for both a vascular reference solute (h_R) and a permeating solute h_D. The lower panel shows sensitivity function [$S(t)$] for several parameters. With higher PS_g, the sensitivities to more distant events, governed by PS_{pc} and V_{pc} are increased (not illustrated). (Figure from Bassingthwaighte, Wang, and Chan, 1989b, with permission. Copyright 1989 American Heart Association.)

tionally, it has been assumed that PS is constant, independent of regional flow. However, recent evidence (Caldwell et al., 1994) suggests that PSs for the heart's primary substrate, fatty acid, are proportional to flow regionally.

The third step is to estimate the organ transfer function for permeant metabolites, based on the above input function and extracellular parameters. For a single permeant species, the strategy for fitting measured dilution curves by optimizing model parameters is outlined in a flow diagram at the top of Figure 7.6. First, the parameters having early peaks in their sensitivity functions are adjusted to fit the peak of the dilution curve. Next, the parameters having late peaks in their sensitivity functions are adjusted to fit the tail of the dilution curve. This process (which may be considered as an optimization module, Φ) is iterated to achieve a final fit. For dilution curves of multiple metabolites, a global optimization strategy was developed to take advantage of the constraints that the dilution curves of daughter metabolites place on parameter estimates of parent metabolites. This strategy (bottom of Figure 7.6) uses the basic optimization module Φ to adjust parameters of parent metabolites to improve model fits to dilution curves of daughter metabolites. Thus, the quantity of metabolite C formed is determined by the PS and consumption parameters of metabolite A (and B).

In addition to the optimization strategy, parameter estimates are also constrained to the extent possible by a priori knowledge (about cellular distribution of enzymes, enzyme kinetics, tracer binding and volume of distribution, physiology, anatomy, etc.).

Using MSID4 to Model Adenosine Reactions

The reactions of adenosine and its metabolic products in the guinea pig heart are modeled in MSID4 for endothelial cells and are shown in Figure 7.7. In myocytes there is no xanthine oxidase, so the reactions stop with IMP formation or hypoxanthine loss from the myocyte. Net uptake of hypoxanthine, followed by IMP formation, serves as a purine salvage pathway. As can be seen, the model ignores several features of the pathways, such as recycling of hypoxanthine to adenosine via IMP and AMP, return flux from the AMP and IMP pools, and contributions from the S-adenosyl-homocysteine and adenine ribose-1-phosphate pathways. These assumptions place limitations on the design of experiments for which the model will be used to analyze data. In the experiments described briefly above, a pulse injection of labeled adenosine into the inflow of an organ, these assumptions cause only small errors. The sizes of the AMP and IMP pools are large enough that return flux from these pools is negligible during the time span of the data collection, one to two minutes. The amount of hypoxanthine recycled to adenosine and the contributions from the unmodeled pathways can be ignored for similar reasons.

Module (Φ) for strategy of parameter optimization for a single metabolite

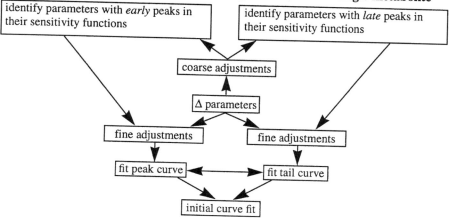

| identify parameters with *early* peaks in their sensitivity functions | identify parameters with *late* peaks in their sensitivity functions |

coarse adjustments

Δ parameters

fine adjustments — fine adjustments

fit peak curve ←→ fit tail curve

initial curve fit

Strategy for parameter optimization for multiple metabolites A, B, and C

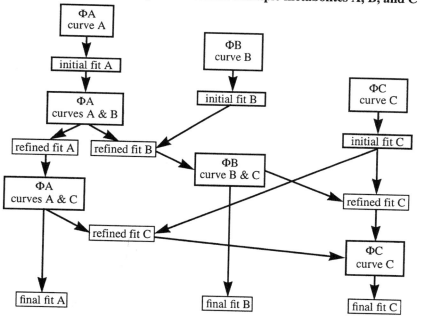

FIGURE 7.6. Flow chart diagram of the strategy employed for parameter estimation based on optimization of the multiple-metabolite model solutions to fit the measured dilution curves. Top panel shows the detail of the strategy for fitting the dilution curve of a single metabolite. This optimization module (Φ) was incorporated into a global optimization strategy (bottom panel) for fitting the dilution curves of the entire sequence of metabolites. A major goal of this approach was to use the results from daughter metabolites to provide constraints on the parameter estimates of the parent metabolites.

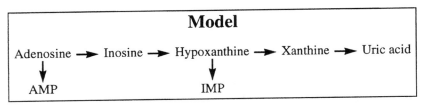

FIGURE 7.7. Reactions of adenosine and its metabolites in endothelial cells modeled in the multisubstrate model, MSID4.

Strategies for the Analysis of Substrate and Product Concentration–Time Curves

Overall Approach

A set of data for analysis is shown in Figure 7.8. In this instance, the HPLC system was not set up to analyze Hx and Xa separately, so these are summed together in the sample analysis and in the modeling analysis. The ^{131}I-albumin and ^{3}H-AraH analyses are not shown; these were done first, giving a fit of $h_R(t)$ to the albumin data and a fit of $h_E(t)$ to the AraH data similar to those in Figure 7.4. This established the adequacy of the descriptions of (1) the multipath system for intravascular transport giving the overall transorgan transport function and including the description of the flow heterogeneity, and (2) the estimates of PS_g and V'_{isf} for AraH. The AraH parameters define several further parameters by inference: (1) the PS_g for adenosine and inosine are taken to be the same as for AraH; (2) the PS_g for Hx and Xa are calculated relative to that for AraH as proportional to their ratios of aqueous diffusion coefficients; and (3) the estimate of V'_{isf} is used identically for all of the solutes.

Manual Optimization: Up and Down the Reaction Sequence

The next step is to seek the parameter estimates for Ado, Ino, Hx, Xa, and UA, given the constraints already provided by the fitting of ^{131}I-albumin and ^{3}H-AraH and the assumptions described above.

For Ado, we assumed that the endothelial cell permeabilities on the two sides, PS_{ecl} and PS_{eca}, were the same. Parameters estimated were therefore PS_{ecl}, PS_{pc}, G_{ec}, and G_{pc}. V'_{pc} was determined from the total water content of 0.8 ml/g, which is $V_p + V'_{isf} + V'_{pc}$. These parameters were adjusted to fit the Ado data curve.

The inosine curve was fitted by using the same PSs as for Ado (since adenosine and inosine use the same transporter and have similar affinities) while adjusting the values for f_{Ado} in the endothelial cell and myocyte, which gives the fraction of Ado metabolized to Ino (while the fraction $1 - f_{Ado}$ is sequestered as nucleotide), and adjusting $G_{ec}(Ino)$. The value for $G_{pc}(Ino)$ was assumed to be identical to that

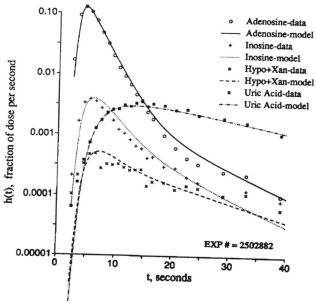

FIGURE 7.8. Solutions for MSID4 fitted to outflow dilution curves for adenosine and its metabolites inosine, hypoxanthine, and xanthine (added together), and uric acid, following injection of [^{14}C]-adenosine, ^{131}I-albumin, and ^{3}H-AraH (an adenosine analog that is not transported and the curve of which is not shown but provides a measure of PS_g and V_{isf}).

of G_{ec}(Ino). For inosine, $f_{Ino} = 1.0$ in both cell types. The Vs are all fixed for Ino, Hx, Xa, and UA. The governing parameters are therefore f_{Ado} and G_{ec}(Ado). The value for G(Ado) · f_{Ado} has to be high enough to provide for the sum of the fractions of dose emerging as Ino + Hx + Xa + UA plus leave something for the hypoxanthine salvage, Hx → IMP, the reaction returning some of the purine base to the heart.

The next steps are to follow down the sequence in similar fashion. For hypoxanthine, f_{Hx} is less than one because of flux into inosine monophosphate (IMP). For xanthine, $f_{Xa} = 1.0$ because there are no side reactions, and for uric acid the Gs are all zero, leaving only the PSs as free parameters.

After this sequence of parameter adjustments, from Ado to UA, any misfitting in the magnitude of the UA signal needs to be compensated for by adjustments in the upstream reactions. These inevitably propagate back up to the adenosine fit, usually being reflected in an adjustment of the Gs and f_{Ado} before a final good fit is obtained for all of the chemical species simultaneously.

Automated Optimization

This is an excellent method, saving much time over manual parameter adjustment by the investigator, but is best done employing constraints on the parameters.

Some of the constraints described above, for example, setting $G_{ec} = G_{pc}$, are artificial and are difficult to refute or justify by physiological measurement. In some sense one is stuck with the problem of failure to separate events in myocytes from those in endothelial cells when one has only a single data set to examine.

The answer is multiple data sets. Instead of [^{14}C]-adenosine, one should in the same heart make injections of [^{14}C]-inosine, then of [^{14}C]-hypoxanthine, and then of [^{14}C]-xanthine. Finally, all of these data sets should be analyzed to determine the common set of parameters that best fits all of the experiments. To be sure, this means making the assumption that nothing in the physiological situation changed during the series of experiments, but this is a minor price to pay for getting an overall descriptive parameter set for a large, comprehensive set of data. The strength is that the primary tracer-labeled substrate (parent of the metabolic products) is influenced first and foremost by the endothelial cells, while the metabolites undergo greater degrees of influence by transport and metabolism in the parenchymal cells. Starting in turn with UA, Hx, Xa, Ino, and Ado is thus a good way to strengthen the distinction between the influences of endothelial cells and myocytes.

Discussion

The multisubstrate model is a useful vehicle for analyzing the data from outflow dilution or residue detection experiments on whole organs or regions of interest from in vivo imaging. While the reaction sequence is a simple linear one with irreversible reactions, it is flexible enough to be usable for a number of substrates of biological interest. The inclusion in the model of flow heterogeneity and operators that model large and medium nonexchanging vessels are important features that make the model more applicable to the analysis of data from both whole organs and regions of interest.

The incorporation of reduced models applicable to substances that are restricted to the vascular and/or extracellular spaces facilitates the analysis of data from multiple-indicator dilution experiments. This also aids in reducing the degrees of freedom when analyzing the data from the substrate that permeates the cell membranes and is metabolized. The model for a metabolized substrate and a series of one or more products with independent transport contains a large number of parameters. In order to obtain meaningful estimates of these parameters, it is essential to obtain high-quality data (i.e., data with good signal-to-noise ratio and with good temporal resolution). It is also important to gain as much information as can be gleaned from the experiment (e.g., using chemical separation to obtain data for each substrate rather than a single, composite signal). Additionally, information from all available sources should be used (e.g., using results from other sources to constrain estimates of volumes and using the analysis of an appropriate extracellular tracer to constrain the PS_g). Finally, the experiment or set of experiments must be designed to facilitate parameter estimation. The use of tracers that start at different points in the reaction sequence can be quite useful in this regard, for example, specific parameters for uric acid (Kroll et al., 1992) to refine the assessment of adenosine metabolism.

While the model is a useful tool for data analysis in its current form, some extensions will make it more useful for other chemical substrates. We have already done some modeling of carrier-mediated membrane transport (Catravas et al., 1988; Bassinghtwaighte et al., 1989b), as have others (Linehan et al., 1987). The second major improvement is the extension of the reaction sequence to include nonlinear kinetics (Achs and Garfinkel, 1977; Goresky and Schwab, 1988) and more complex sequences, especially with feedback inhibition. In this regard, the approach of Palsson seems promising. He analyzes complex reaction networks to identify critical control points and encapsulates a set of intervening reactions in a single operator (Palsson et al., 1987). This may be thought of as a second-order approach to linearization of the problem and is needed as an improvement on the first-order approach taken in MSID4 when there are large numbers of reactions.

Both of these extensions significantly increase the computational complexity of the model and thus the time required to obtain a computer solution. While significant reductions can be obtained with improved hardware, especially the use of parallel processing, further investigation into the algorithms used to obtain numerical solutions may offer hope of much larger computational gains for nonlinear modeling.

Summary

While our earlier blood–tissue exchange models permitted consumption, consumed solute was considered as either sequestered and retained or as having reacted to form a product whose kinetics were not modeled. We have extended the four-region, three-barrier, axially distributed model that accounts for convection in the capillary and diffusion into the endothelial cells, ISF, and parenchymal cells (Bassingthwaighte et al., 1989b) to account for a sequence of reactions in all regions. It uses the same sliding fluid element flow algorithm and analytic method for solving the convection–diffusion equations at each time step. Metabolic reactions are considered to be the sum of two processes: A fraction of the amount consumed goes to a product that can exchange with adjacent regions or may be further metabolized; the remainder is permanently sequestered in the region in which it was formed. The reaction sequence is unidirectional, that is, no reactions are permitted that result in the return of tracer to an antecedent substrate. Each substrate has its own unique parameters for permeabilities, reaction rates, and volumes of distribution. The model is useful for analyzing data from tracer glucose, thymidine, adenosine, and fatty acids. Simulations with the model illustrate that endothelial and parenchymal cell events can be well distinguished and have shown the importance of metabolism in endothelial cells for producing products whose outflow curves have peaks only slightly shifted in time from that of the parent.

Acknowledgements. This work was supported by NIH Grants RR1243, HL38736, and HL19139. The authors are most appreciative of the assistance of Pamela Denchfield in the preparation of the manuscript.

References

Achs, M. J. and D. Garfinkel. Computer simulation of energy metabolism in anoxic per-fused rat heart. *Am. J. Physiol.* 232 (*Regulatory Integrative Comp. Physiol.* 1):R164–R174, 1977.

Bassingthwaighte, J. B. Plasma indicator dispersion in arteries of the human leg. *Circ. Res.* 19:332–346, 1966.

Bassingthwaighte, J. B. Blood flow and diffusion through mammalian organs. *Science* 167:1347–1353, 1970.

Bassingthwaighte, J. B. A concurrent flow model for extraction during transcapillary passage. *Circ. Res.* 35:483–503, 1974.

Bassingthwaighte, J. B. Physiological heterogeneity: Fractals link determinism and ran-domness in structures and functions. *News Physiol. Sci.* 3:5–10, 1988.

Bassingthwaighte, J. B. and R. P. Beyer. Fractal correlation in heterogeneous systems. *Physica D* 53:71–84, 1991.

Bassingthwaighte, J. B. and C. A. Goresky. Modeling in the analysis of solute and water exchange in the microvasculature. In: *Handbook of Physiology. Section 2, The Cardiovascular System. Vol. IV, The Microcirculation,* edited by E. M. Renkin and C. C. Michel. Bethesda, MD: Am. Physiol. Soc., 1984, pp. 539–626.

Bassingthwaighte, J. B., T. J. Knopp, and D. U. Anderson. Flow estimation by indicator dilution (bolus injection): Reduction of errors due to time-averaged sampling during unsteady flow. *Circ. Res.* 27:277–291, 1970.

Bassingthwaighte, J. B., R. B. King, and S. A. Roger. Fractal nature of regional myocardial blood flow heterogeneity. *Circ. Res.* 65:578–590, 1989a.

Bassingthwaighte, J. B., C. Y. Wang, and I. S. Chan. Blood-tissue exchange via transport and transformation by endothelial cells. *Circ. Res.* 65:997–1020, 1989b.

Bassingthwaighte, J. B., I. S. Chan, and C. Y. Wang. Computationally efficient algorithms for capillary convection-permeation-diffusion models for blood-tissue exchange. *Ann. Biomed. Eng.* 20:687–725, 1992.

Beard, D., and J. B. Bassingthwaighte. Fractal nature of myocardial blood flow described by whole organ model of arterial network. *Ann. Biomed. Eng.* 22 (Suppl. 1):20, 1994.

Bohr, C. Über die spezifische Tätigkeit der Lungen bei der respiratorischen Gasaufnahme und ihr Verhalten zu der durch die Alveolarwand stattfindenden Gasdiffusion. *Skand. Arch. Physiol.* 22:221–280, 1909.

Bronikowski, T. A., J. H. Linehan, and C. A. Dawson. A mathematical analysis of the influence of perfusion heterogeneity on indicator extraction. *Math. Biosci.* 52:27–51, 1980.

Caldwell, J. H., G. V. Martin, G. M. Raymond, and J. B. Bassingthwaighte. Regional myocardial flow and capillary permeability-surface area products are nearly propor-tional. *Am. J. Physiol.* 267 (*Heart Circ. Physiol.* 36):H654–H666, 1994.

Catravas, J. D., J. B. Bassingthwaighte, and H. V. Sparks, Jr. Adenosine transport and uptake by cardiac and pulmonary endothelial cells. In: *Endothelial Cells, Vol. I,* edited by U. S. Ryan. Boca Raton, FL: CRC Press, 1988, pp. 65–84.

Crone, C., J. Frøkjaer-Jensen, J. J. Friedman, and O. Christensen. The permeability of single capillaries to potassium ions. *J. Gen. Physiol.* 71:195–220, 1978.

Glenny, R. W. and H. T. Robertson. A computer simulation of pulmonary perfusion in three dimensions. *J. Appl. Physiol.* 79:357–369, 1995.

Goresky, C. A. A linear method for determining liver sinusoidal and extravascular volumes. *Am. J. Physiol.* 204:626–640, 1963.

Goresky, C. A. and A. J. Schwab. Flow, cell entry, and metabolic disposal: Their interactions in hepatic uptake. In: *The Liver: Biology and Pathobiology, 2nd* ed., edited by I. M. Arias. New York: Raven, 1988, pp. 807–832.

Goresky, C. A., W. H. Ziegler, and G. G. Bach. Capillary exchange modeling: Barrier-limited and flow-limited distribution. *Circ. Res.* 27:739–764, 1970.

Kassab, G. S., C. A. Rider, N. J. Tang, and Y. B. Fung. Morphometry of pig coronary arterial trees. *Am. J. Physiol.* 265 *(Heart Circ. Physiol.* 34):H350–H365, 1993.

King, R. B., A. Deussen, G. R. Raymond, and J. B. Bassingthwaighte. A vascular transport operator. *Am. J. Physiol.* 265 *(Heart Circ. Physiol.* 34):H2196–H2208, 1993.

King, R. B., G. M. Raymond, and J. B. Bassingthwaighte. Modeling blood flow heterogeneity. *Ann. Biomed. Eng.* 24:352–372, 1996.

Koiwa, Y., R. C. Bahn, and E. L. Ritman. Regional myocardial volume perfused by the coronary artery branch: Estimation in vivo. *Circulation* 74:157–163, 1986.

Kroll, K., T. R. Bukowski, L. M. Schwartz, D. Knoepfler, and J. B. Bassingthwaighte. Capillary endothelial transport of uric acid in the guinea pig heart. *Am. J. Physiol. (Heart Circ. Physiol. 31)* 262:H420–H431, 1992.

Kuikka, J., M. Levin, and J. B. Bassingthwaighte. Multiple tracer dilution estimates of D- and 2-deoxy-D-glucose uptake by the heart. *Am. J. Physiol.* 250 *(Heart Circ. Physiol.* 19):H29–H42, 1986.

Lassen, N. A. and W. Perl. *Tracer Kinetic Methods in Medical Physiology.* New York: Raven, 1979.

Linehan, J. H., T. A. Bronikowski, and C. A. Dawson. Kinetics of uptake and metabolism by endothelial cell from indicator dilution data. *Ann. Biomed. Eng.* 15:201–215, 1987.

Lumsden, C. J. and M. Silverman. Exchange of multiple indicators across renal-like epithelia: A modeling study of six physiological regimes. *Am. J. Physiol.* 251 *(Renal. Fluid. Elect. Physiol.* 20):F1073–F1089, 1986.

Maseri, A. P., S. Caldini, S. Permutt, and K. L. Zierler. Frequency function of transit times through dog pulmonary circulation. *Circ. Res.* 26:527–543, 1970.

Palsson, B. O., A. Joshi, and S. Ozturk. Reducing complexity in metabolic networks: Making metabolic meshes manageable. *Fed. Proc.* 46:2485–2489, 1987.

Prinzen, T. T., T. Arts, F. W. Prinzen, and R. S. Reneman. Mapping of epicardial deformation using a video processing technique. *J. Biomechanics* 19:263–273, 1986.

Rose, C. P. and C. A. Goresky. Vasomotor control of capillary transit time heterogeneity in the canine coronary circulation. *Circ. Res.* 39:541–554, 1976.

Rose, C. P., C. A. Goresky, and G. G. Bach. The capillary and sarcolemmal barriers in the heart: An exploration of labeled water permeability. *Circ. Res.* 41:515–533, 1977.

Sangren, W. C. and C. W. Sheppard. A mathematical derivation of the exchange of a labeled substance between a liquid flowing in a vessel and an external compartment. *Bull. Math. Biophys.* 15:387–394.

Sapirstein, L. A. Regional blood flow by fractional distribution of indicators. *Am. J. Physiol.* 193:161–168, 1958.

Thompson, H. K., C. F. Starmer, R. E. Whalen, and H. D. McIntosh. Indicator transit time considered as a gamma variate. *Circ. Res.* 14:502–515, 1964.

van Beek, J. H. G. M., S. A. Roger, and J. B. Bassingthwaighte. Regional myocardial flow heterogeneity explained with fractal networks. *Am. J. Physiol.* 257 (*Heart Circ. Physiol.* 26):H1670–H1680, 1989.

Wangler, R. D., M. W. Gorman, C. Y. Wang, D. F. DeWitt, I. S. Chan, J. B. Bassingthwaighte, and H. V. Sparks. Transcapillary adenosine transport and interstitial adenosine concentration in guinea pig hearts. *Am. J. Physiol.* 257 (*Heart Circ. Physiol.* 26):H89–H106, 1989.

Yipintsoi, T., P. D. Scanlon, and J. B. Bassingthwaighte. Density and water content of dog ventricular myocardium. *Proc. Soc. Exp. Biol. Med.* 141:1032–1035, 1972.

8

Norepinephrine Kinetics in Normal and Failing Myocardium: The Importance of Distributed Modeling

Colin P. Rose

Introduction

Tribute to Carl Goresky

As one of the few, fortunate graduate students of Carl Goresky, I would like to show in this tribute to his foresight and leadership that the concepts he originated or promoted have not only illuminated normal blood tissue transport but also led to significant insights into a clinical problem. When, as a medical student and intern, I read the Ziegler and Goresky papers on multiple indicator dilution experiments in the heart of open-chest dogs I was, above all, impressed with the capability of studying nondestructively rapid physiological and biochemical events in the in situ heart. Later, after learning the technique of closed-chest, fluoroscopically guided coronary artery and coronary sinus catheterization, I was able to use the multiple-indicator dilution methodology in humans in the cardiac catheterization laboratory. Rarely does a student have the opportunity to transfer basic physiological concepts and techniques learned in the animal laboratory into the clinical arena with almost no alteration. Carl Goresky's encouragement and advice were invaluable.

The Clinical Problem

To illustrate the application of multiple-indicator dilution (MID) technology and modeling to a significant clinical problem, we chose the kinetics of norepinephrine (NE) in the heart and the aberrations that occur when the myocardium fails. The story of cardiac NE as a significant indicator of cardiac pathology began in the early 1960s with a serendipitous discovery by Eugene Braunwald, then at the National Institutes of Health (1). While investigating the effects of reserpine on cardiac reflexes, he was curious to know exactly how much tissue NE was reduced by the usual clinical doses of reserpine, which were much lower than the doses used in animal studies. He hit upon the idea of giving reserpine to patients before open heart surgery for replacement of a mitral valve, during which the left atrial appendage was amputated. The NE concentration in this piece of tissue could then be compared in controls and patients who had been administered reserpine. Surprisingly, while reserpine reduced NE concentrations to very low levels, as expected, NE concentrations were reduced to the same very low levels in patients who had been in heart failure from mitral regurgitation (volume overload) prior to surgery. Rather than attributing the NE depletion in the failing heart to a localized, reserpine-like effect, Brauwald hypothesized that NE was reduced in all tissues because of prolonged sympathetic activation secondary to heart failure. Since then, most investigators have implicitly adopted this viewpoint. There is a great reluctance to accept the concept of a localized dysfunction of cardiac sympathetic nerves, even though the same phenomenon was later discovered to occur in the pregnant uterus (2); no one seems to be arguing that there is a generalized depletion of NE in pregnancy. Braunwald's discovery remains unexplained, but this author believes that the status of the cardiac sympathetic nerves in heart failure is crucial to the development of a theory of myocardial failure, a conundrum that has puzzled many talented investigators for centuries.

There are a number of problems with the concept of NE depletion secondary to overstimulation of the sympathetic nerve in heart failure, some of which were apparent at the time of the original proposition and some that have been discovered since:

1. In circulatory stress the sympathetic system is activated in the whole body. If NE were depleted in every organ, patients with heart failure would have *low* circulating NE levels and suffer from postural hypertension. However, patients with heart failure generally have *high* circulating NE levels that rise even higher with exercise (3) and do not suffer from postural hypotension secondary to depletion of NE in all peripheral nerves. Thus, it was illogical for Braunwald to propose generalized sympathetic stimulation as the cause of cardiac NE depletion; NE depletion must be localized to the heart. It is still remotely possible that only the cardiac sympathetic nerves are massively stimulated when the myocardium fails, but this has never been demonstrated and is rather unlikely.

2. In later studies of the NE depletion phenomenon, Braunwald's group used a model of heart failure in calves with pulmonary hypertension (4). They used histofluorescence to visualize the distribution of NE-containing nerves in the heart at various stages of heart failure. It is clear in their illustrations (but not commented upon in the text) that NE is preserved around blood vessels and in connective tissue spaces but totally absent among the muscle fibers. Again, if total body sympathetic activation were responsible for NE depletion, one would expect all cardiac nerves to be depleted. It is remotely possible, but unlikely, that only those nerves adjacent to muscle fibers would be selectively activated.

3. NE becomes depleted first in the failing ventricle (4–7). Again, it is remotely possible, but unlikely, that sympathetic nerves could be massively stimulated selectively to either the right or left ventricle.

4. Chronic sympathetic activation is not associated with depletion of NE, but just

the opposite. Cold (8) or immobilization (9) stress, both chronic circulatory stressors like heart failure, lead to an *increase* in tissue NE and its synthesizing enzymes in all organs. NE depletion can occur only with short-term rapid nerve stimulation (10).

The alternative hypothesis is that there is a localized, reserpine-like effect of failing myocardium on sympathetic nerves that first appears around failing myocytes in the affected ventricle. What the mechanism of such an effect might be is unknown, but if cardiac NE depletion is really localized to the heart and not just a secondary side effect of heart failure on cardiovascular reflexes, then the phenomenon might provide some insight into the mechanism of myocyte failure itself. First it was necessary to show that, rather than releasing large amounts of NE, the nerves of the failing heart were dysfunctional.

The Steady-State Assessment of Cardiac Norepinephrine Transport

Rejecting the Braunwald hypothesis also requires rejecting the interpretation of recent data from the laboratory of Murray Esler and collaborators. Esler has perfected a technique that purportedly measures NE release in the in situ heart and other organs (11). His technique uses a steady-state infusion of labeled NE and a single compartmental (point) model of NE transport. Studies in humans with heart failure have been interpreted to show an increased release of NE from the failing heart compared to normals (12–14). If this interpretation of the data were true, we would have to find an explanation for all of the above points refuting Braunwald's hypothesis. What this chapter will emphasize is that the single compartmental model of NE transport is erroneous; it grossly underestimates NE release in the normal heart, so that NE release in the failing heart may appear factitiously elevated.

Functions of Cardiac Sympathetic Nerves

Before considering in detail the modeling of NE transport, the reader should be familiar with some basic concepts in sympathetic nerve physiology. Sympathetic nerves richly innervate the heart, releasing norepinephrine into the interstitial space and stimulating the heart to increase heart rate and contractility at the onset of exercise. Powered by the transmembrane gradient in sodium ion concentration, a neuronal membrane pump for norepinephrine rapidly removes released norepinephrine from the interstitial space (15). Cardiac sympathetic nerves are not necessary for normal heart function; denervation results only in a slower response to the onset of exercise but causes no long-term (over months) myocyte dysfunction. However, both denervated and autologously transplanted hearts eventually become reinnervated to some extent. High concentrations of norepinephrine are contained in vesicles in varicosities scattered along the length of the nerves. (See Figure 8.1.)

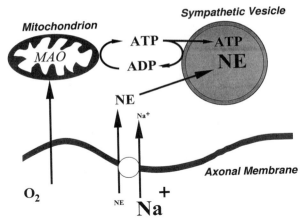

FIGURE 8.1. Energy requiring activities in a sympathetic nerve varicosity. A large amount of ATP must be produced both to create the hydrogen ion gradient that drives the concentrative transport of NE into the vesicle and for storage in the vesicle where it acts to create a complex with NE to increase the concentration without increasing the osmolality. Transport across the axonal membrane is dependent on maintenance of the sodium ion gradient.

When one measures the tissue concentration of NE, one is measuring NE contained in the vesicles. Neuronal cytoplasmic NE is normally only a small fraction of vesicular NE, but it can increase if vesicular uptake is inhibited (15). Braunwald's observation of reduced tissue concentration in heart failure says nothing about either the membrane uptake function of the nerve or the release rate of the vesicles. NE is simultaneously released by vesicular exocytosis and taken up by the membrane uptake pump. In the absence of substantial nerve stimulation, release generally balances uptake and the NE gradient across the heart is zero; it is impossible simply by measuring endogenous arterial and coronary sinus norepinephrine concentrations to detect potentially large changes in release and uptake. One must separate these two nerve functions in order to study the effect of pathological processes on nerve function. Neuronal membrane uptake can be assessed with the use of labeled NE either infused to attain a steady-state arterial activity or as a bolus injection into a coronary artery. There are advantages and disadvantages to both approaches, but this author argues that only the transient bolus method combined with appropriate references in a multiple-indicator dilution analysis yields unambiguous data on cardiac norepinephrine release and uptake in all situations (17).

Advantages of the steady-state infusion technique include not requiring catheterization of a coronary artery (only the coronary sinus has to be selectively catheterized, and an arterial sample can be collected from any peripheral artery) and the use of truly tracer concentrations of labeled NE. Unfortunately, these advantages are completely overshadowed by the lack or interpretability of the resulting data.

Single Compartment

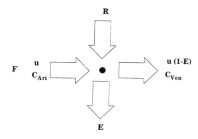

Distributed

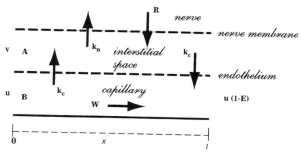

FIGURE 8.2. Comparison of two models for interpretation of NE kinetics in the heart. In the "single compartment" (really a point) model, NE release is calculated by a simple mass balance of material entering and leaving the point. Distributed modeling necessitates the solution of two partial differential equations, and NE release becomes a function of transit time (l/W), the rate constants for exchange across the capillary endothelium, and uptake by the nerve. NE uptake and release occur along the whole length of the capillary so that when uptake is high NE release as calculated from the distributed model will be higher than that calculated from the point model. See text for proof.

Models

Figure 8.2 shows schematically the two models considered. Detailed explanations of the distributed model can be found in the original publications; it would serve little purpose to go into the details of the solutions in this chapter. Elsewhere in this volume, Andreas Schwab outlines the general theory behind the linear modeling of capillary transport. Here we outline the basic assumptions, the differential equations, and their solutions so that the reader may obtain an appreciation of the parameters important to the clinical problem.

Single Compartment

In this model the whole heart (capillaries, interstitial space, nerves) is considered as a single point. Blood flows into and out of, and NE is taken up and released into, this point. If E is the extraction of labeled NE determined by steady-state infusion or integration of bolus outflow relative to a nonextracted reference, C_{Art} and C_{Ven} are the arterial and coronary sinus plasma concentrations of unlabeled NE, ϕ is the plasma flow per unit interstitial space, respectively, and then R_S, the release of NE is given by a simple mass balance of NE flowing into and out of the point:

$$R_s = \phi \left[C_{ven} - C_{Art}(1 - E) \right]. \tag{8.1}$$

Distributed

Single Capillary

Needless to say, the microcirculation in the heart is much more complicated than assumed by the single point. At a very minimum it is essential to account for the length of the capillaries; anatomically, they are 200 to 500 times longer than they are wide. If a substance is highly extracted, as in the case of NE, its concentration will decrease along the length of the capillary in a pattern that depends on the kinetics of its uptake, usually zero- or first-order. Nowhere will its concentration be homogeneous, as assumed by the single-compartment model. Naturally, the microcirculation is composed of millions of capillaries with different lengths and flow rates, but first we must consider events along a single capillary. (See Figure 8.3.) For NE and its appropriate references, four assumptions are adopted:

1. Dispersionless plug flow occurs in the capillary.
2. Longitudinal diffusion can be neglected because the axial dimension is much longer than the radial dimension and flow is faster than diffusion in the capillary.
3. Radial diffusional equilibrium is so rapid that no concentration gradients exist in this direction within any of the spaces.
4. NE sequestration is first-order.

The following symbols are used: A and B are the volumes per unit length for capillary and interstitial spaces, respectively; W is the velocity of blood flow; $u(x,t)$ and $v(x,t)$ are concentrations of tracer at distance x along the capillary–tissue unit and time t in the capillary and interstitial spaces, respectively; k_c is the capillary permeability–surface product per unit interstitial space; k_n is the first-order rate constant for sequestration on NE by the neuron; and R is the rate of release of unlabeled NE from the neuron.

Consideration of the events occurring at each element in space and time leads to two partial differential equations that must be solved simultaneously. The first, for conservation of matter is

$$\frac{\partial u}{\partial t} + W\frac{\partial u}{\partial x} + \gamma\frac{\partial v}{\partial t} = 0, \tag{8.2}$$

where $\gamma = B/A$.

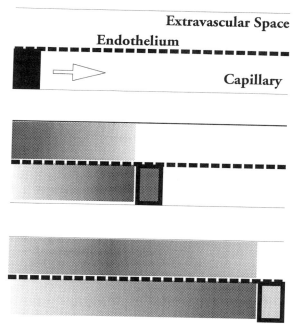

FIGURE 8.3. Schematic diagram of the passage of a bolus of tracer of a substance that permeates the capillary wall, enters the interstitial space, and then returns to the capillary later in time. The bolus is progressively depleted as it proceeds along the capillary. Concentration profiles of the tracer are formed in the vascular and extravascular spaces, depending on the permeability of the capillary and the relative sizes of the spaces. Ordinarily these time- and space-varying profiles are not observable, but the pattern of tracer outflow at the end of the capillary is uniquely determined by the same variables. The outflow will be composed of two components, a depleted bolus emerging at the transit time of the capillary and later material that has left the capillary and returned to it. This component will be of lower magnitude and more prolonged than the first component. If there is no sequestration of material in the extravascular space, all of the material will eventually be recovered after infinite time.

The second, the rate equation for accumulation in the interstitial space, is:

$$\frac{\partial v}{\partial t} = k_c v - k_c u - k_n v. \tag{8.3}$$

Solution of these equations for the impulse injection of an amount q_0 of substance at the origin of a capillary of length L gives an equation of the following form:

$$u(L, t) = \frac{q_0}{F_c} \exp(-k_c \gamma \tau_c) \, \delta(t - \tau_c) + f(\tau_c, k_c, k_n, \gamma) \tag{8.4}$$

where δ symbolizes the Dirac delta (impulse) function. The first term is the throughput or nonexchanging component of the outflow and represents an impulse arriving at the outflow but reduced in magnitude by the factor $\exp(-k_c \gamma \tau_c)$.

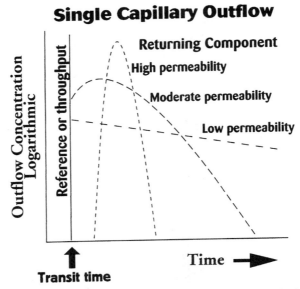

FIGURE 8.4. For a single capillary predicted outflow patterns can be calculated for a given set of exchange parameters. Note the two components of the outflow, one arriving after one capillary transit time simultaneously with the reference tracer but reduced in magnitude and a later component representing material that has left and then reentered the capillary from the extravascular space. It is difficult to graphically represent the magnitude of the through-put component because it has the mathematical form of a delta function, a pulse of infinitesimal duration, but its magnitude will be inversely related to the capillary per-meability. A whole organ model with an array of single capillaries with variable transit times can then be constructed. This model can be optimized—the parameters adjusted—to fit the data. The resulting set of "best-fit" parameters can then be assumed to characterize the exchange state of the organ at that moment.

(See Figure 8.4.) The details of the second term in Eq. (8.4), which describes the outflow of material that has permeated the endothelial wall of the capillary and then returns to it later in time (returning component), are not given because they add nothing to the argument. The interested reader can refer to the original papers.

Whole Organ

The construction of the whole organ model requires three more assumptions:

5. The ratio of interstitial space to capillary volume is constant throughout the heart.
6. There is no interaction between capillaries with different transit times or different directions of flow.
7. There is flow coupling between capillaries and large vessels, that is, short capillary transit times are associated with short nonexchanging vessel transit times.

Multiple-indicator dilution

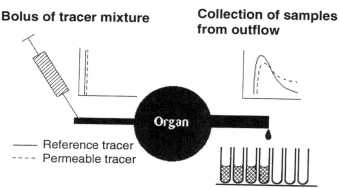

Bolus of tracer mixture

Collection of samples from outflow

——— Reference tracer
- - - - Permeable tracer

FIGURE 8.5. Events underlying the rapid exchange of metabolites between blood and tissue can be studied using the multiple-indicator dilution technique. A mixture of tracers is injected into the inflow to an organ—the coronary artery in the heart, the portal vein in the liver—and samples are collected rapidly from the outflow. By comparing the outflow patterns of exchanging and metabolized tracers with a reference tracer by means of a mathematical model of the capillary, inferences can be made about the rate limiting steps in the exchange and the concentration of metabolites inside the tissue cell.

The whole organ outflow then takes the form

$$u(L,t) = \frac{q_0}{F_c} \exp(-k_c\gamma\tau_c)\, w(\tau_c) + f(w,\tau_c,k_c,k_n,\gamma) \tag{8.5}$$

In general, γ and τ_c cannot be independently estimated and only their product, ϕ, is determinable.

Assumption 6 is problematical. This is not the place to discuss the possibility of diffusional capillary interaction in the heart. Suffice it to say that the present model gives a reasonable first approximation to the data for "barrier-limited" substances, of which sucrose and norepinephrine are examples. Some degree of extracapillary diffusional interaction will not affect the main point of the argument for these substances.

Now, if one also measures endogenous concentrations of NE in the artery, u_0, and the coronary sinus, u_{cs}, one can calculate the release, R_t, of NE from the sympathetic nerve terminals from the following equation:

$$u_{cs} = \sum_{j=1}^{n} w_j \left\{ u_0 \exp\left(-\frac{k_c k_n \gamma \tau_{c(j)}}{k_c + k_n}\right) + \frac{R}{k_n}\left[1 - \exp\left(-\frac{k_c k_n \gamma \tau_{c(j)}}{k_c + k_n}\right)\right] \right\}, \tag{8.6}$$

where w is the fraction of the reference represented at each outflow time.

When $R = 0$, there is no release of unlabeled NE and the steady-state extraction of label NE is given by

$$E = \frac{u_{cs} - u_0}{u_0} = 1 - \exp\left(-\frac{k_c k_n \gamma \tau_c}{k_c + k_a}\right). \tag{8.7}$$

For a single capillary or for a whole organ in which capillary transit time is constant, Eq. (8.6) can be written

$$R_t = \left(u_{cs} - u_0(1 - E)\right)\frac{k_n}{E} \tag{8.8}$$

or, in the notation of Eq. (8.1),

$$R_t = \left(C_{Ven} - C_{Art}(1 - E)\right)\frac{k_n}{E}, \tag{8.8a}$$

where E is again the steady-state extraction of labeled NE. It can be shown that

$$\lim_{E=0, n=0} R_t = \phi\left\{C_{ven} - C_{Art}(1 - E)\right\}. \tag{8.9}$$

It is instructive to compare this formula with Eq. (8.1). When $k_n = 0$, the two definitions of release are equivalent. But for a given observed steady-state extraction of labeled NE, E, and given input and output NE concentrations, the calculated release will be higher in the distributed model compared to the point model in proportion to the neuronal uptake rate constant, k_n. When k_n is very large, NE release can be very large and may not be detectable by simply measuring steady-state extraction.

Elsewhere in this volume, Dr. A. Schwab gives a detailed description of the approach to linear models of capillary transport. Readers interested in the solutions and calculations of these equations should consult his chapter.

Form of the Multiple-Indicator Dilution Data

Figure 8.6 shows the type of data used to investigate NE transport using the multiple-indicator dilution technique (Figure 8.5). In this case, after catheters have been placed in the coronary artery and coronary sinus under fluoroscopy in a patient undergoing cardiac catheterization, three tracers are mixed with blood and injected simultaneously into a coronary artery. At the same time, a sampling rack is started and samples are collected from the coronary sinus at a rate of about one per second. The concentration of each tracer in the outflow is normalized to the amount injected to give the fractional outflow. Labeled albumin, which does not permeate the endothelial barrier to any significant extent during a single transit time, is the intravascular reference, labeled sucrose, which permeates the endothelium through intercellular aqueous pores but does not cross cell membranes, is the reference for the interstitial space, and labeled NE traces the endogenous NE. Because of massive dilution in the larger intraneuronal pools, the tritium label on

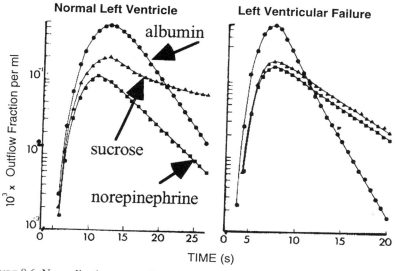

FIGURE 8.6 Normalized coronary sinus outflow curves after injection of mixture of tracers into the left coronary artery in humans, one with a normal left ventricle and one with left ventricular failure (pulmonary edema) secondary to aortic and mitral valve disease. Normally, norepinephrine is sequestered by the sympathetic neurons in the interstitial space via the concentrative membrane uptake pump, part of the mechanism for deactivation of released norepinephrine. At some point in the development of myocardial failure this membrane pump becomes impaired either by destruction of the nerve or because the sodium gradient across the membrane, upon which the pump depends, has disappeared.

the NE can be assumed not to appear in re-released NE or in NE metabolites during a single capillary transit time. Two sets of dilution curves are shown, one from a patient with mitral stenosis and one from a patient with left ventricular (LV) failure secondary to aortic stenosis and aortic regurgitation. A number of inferences can be made by inspection of the data before using any modeling. One can see that, at the peak of the curves, the two sets are identical in the relationship between the three tracers, but on the downslope the patient with the normal left ventricle has an NE curve that diverges progressively from the sucrose curve while the patient with heart failure has an NE curve that is almost superimposed on the sucrose curve. Since we know that sucrose is biologically inert in this system, we can infer that, in the patient with LV failure, NE is not sequestered in the nerve in the same fashion as in the patient with the normal LV.

Comparison of the Models

The parameters in the whole organ model can be optimized as described previously. Briefly, the distribution of total (nonexchanging and exchanging vessels)

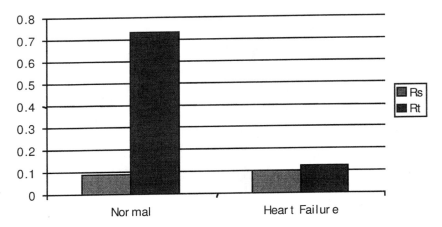

FIGURE 8.7. Estimated release of norepinephrine from cardiac sympathetic nerves in patients with normal left ventricles and patients with heart failure secondary to pressure or volume overload. Note that the single-point model underestimates release from the normal heart by an order of magnitude.

vascular transit times given by the labeled albumin outflow is used as the input to the distributed model, and the sucrose outflow is used to calculate the distribution of capillary (exchanging vessel) transit times. From these two sources of "structural" data an NE outflow can be predicted given a set of transport parameters. These parameters can then be optimized to give as close a fit to the data as possible. The resultant set of parameters is then assumed to represent the system at that moment.

Figure 8.7 shows a comparison of NE release, R calculated using the two models from the data presented previously. As predicted from the above discussion, the single-point model underestimates NE release by an order of magnitude when k_n is high, as it is in the normal heart. When k_n is low, as it is in patients with left ventricular failure secondary to pressure or volume overload, the two models are equivalent. Esler's group has reported that cardiac NE release is actually higher in patients with heart failure than in normals [for review see (18)]. The problem is that their estimate of normal NE release is much too low, and a very slight increase in NE release in heart failure (possibly from nerves around large vessels) can appear to be much larger than normal because of a much reduced uptake.

Figure 8.8 illustrates the problem of trying to estimate uptake from the steady-state extraction. Because of the capillary barrier, extraction must be very low before it is apparent that there is a decrease in neuronal NE uptake.

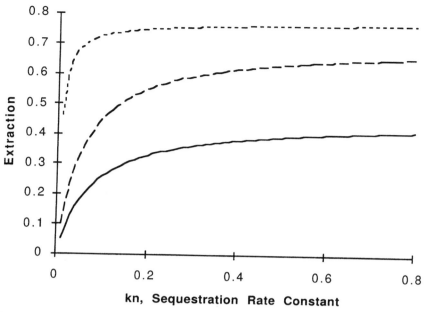

FIGURE 8.8. Single capillary extraction of norepinephrine using representative values for parameters in Eq. (8.7). The three curves from top down represent decreasing capillary permeability ($k_s = 0.3, 0.2, 0.1$, respectively). Note that extraction is relatively unchanged until k_n becomes low. Thus, steady-state extraction of label is very insensitive to events occurring on the abluminal side of the capillary barrier.

Speculation on Reason for Decreased Neuronal Function in Heart Failure

When Braunwald first discovered the absence of cardiac neuronal norepinephrine in volume overload heart failure, knowledge of cardiac microvascular transport was very primitive. He could see no reason (and many investigators still see no reason) why there should be a localized dysfunction of cardiac sympathetic nerves in heart failure secondary to hypertrophy. We now know that even in the normal heart intracellular oxygen levels are an order of magnitude lower than in coronary sinus blood (19–21). Theoretically, an increase in oxygen consumption is associated with a decrease in capillary density, as occurs in cardiac hypertrophy secondary to pressure or volume overload (22). We also now know that the critical oxygen concentration below which cellular metabolism starts to decrease is much higher in whole cells than in isolated mitochondria (23). Thus, it is not inconceivable that both cardiac myocytes and cardiac neurons fail at the same time due to a decrease in extravascular oxygen concentration. Proof of this conjecture must await accurate measurement of tissue oxygen concentration in intact beating hearts. There is some clinical indication that decreased coronary sinus oxygen concentration may be a harbinger of myocardial failure (24).

At any rate, we can see how detailed consideration of microcirculatory events in a distributed fashion leads to a radically different interpretation of data that are potentially of major importance to a classical clinical problem.

References

1. Chidsey, C. A., E. Braunwald, A. G. Morrow, and D. T. Mason. Myocardial nor-epinephrine concentration in man: effects of reserpine and congestive heart failure. *N. Engl. J. Med.* 269:653–658, 1963.
2. Owman, C., P. Alm, E. Rosengren, N.-O Sjoberg, and G. Thorbert. Variations in the level of uterine norepinephrine during pregnancy in the guinea pig. *Am. J. Obstet. Gynecol.* 122:961–964, 1975.
3. Chidsey, C. A., D. C. Harrison, and E. Braunwald. Augmentation of the plasma norepinephrine response to exercise in patients with congestive heart failure. *N. Engl. J. Med.* 267:650–654, 1962.
4. Vogel, J. H. K., D. Jacobowitz, and C. A. Chidsey. Distribution of norepinephrine in the failing bovine heart. *Circ. Res.* 24:71–84, 1969.
5. Krakoff, L. R., R. A. Buccino, J. F. Spann, and J. de Champlain. Cardiac catechol O-methyltransferase and monoamine oxidase activity in congestive heart failure. *Am. J. Physiol.* 215:549–552, 1968.
6. Bristow, M. R., W. Minobe, R. Rasmussen, P. Larrabee, L. Skerl, J. W. Klein, F. L. Anderson, J. Murray, L. Mestroni, S. V. Karwande, M. Fowler, and R. Ginsburg. Beta-adrenergic neuroeffector abnormalities in the failing human heart are produced by local rather than systemic mechanisms. *J. Clinical Invest.* 89:803–815, 1992.
7. Liang, C. S., T. H. M. Fan, J. T. Sullebarger, and S. Sakamoto. Decreased adrenergic neuronal uptake activity in experimental right heart failure—A chamber-specific con-tributor to beta-adrenoceptor downregulation. *J. Clin. Invest.* 84(4):1267–1275, 1989.
8. Fillenz, M., S. C. Stanford, and B. G. Coles. Changes in sympathetic nerve terminals in the heart of cold-exposed rats. *J. Neurochem.* 61(1):132–137, 1993.
9. Lenders, J. W. M., R. Kvetnansky, K. Pacak, D. S. Goldstein, I. J. Kopin, and G. Eisenhofer. Extraneuronal metabolism of endogenous and exogenous norepinephrine and epinephrine in rats. *J. Pharmacol. Exp. Ther.* 266(1):288–293, 1993.
10. Levy, M. N. and B. Blattberg. Progressive reduction in norepinephrine overflow during cardiac sympathetic stimulation in the anaesthetized dog. *Cardiovasc. Res.* 10:549–555, 1976.
11. Esler, M., G. Jennings, P. Korner, P. Blombery, N. Scharias, and P. Leonard. Measure-ment of total and organ-specific norepinephrine kinetics in humans. *Am. J. Physiol.* 247:E21–E28, 1984.
12. Hasking, G. J., M. D. Esler, G. L. Jennings, D. Burton, J. A. Johns, and I. Korner. Norepinephrine spillover to plasma in patients with congestive heart failure: Evidence of increased overall and cardiorenal sympathetic nervous activity. *Circulation* 73:615–621, 1986.
13. Hasking, G. J., M. D. Esler, G. L. Jennings, E. Dewar, and G. Lambert. Norepinephrine spillover to plasma during steady-state supine bicycle exercise. Comparison of patients with congestive heart failure and normal subjects. *Circulation* 78:516–521, 1988.

14. Meredith, I. T., A. Broughton, G. L. Jennings, and M. D. Esler. Evidence of a selective increase in cardiac sympathetic activity in patients with sustained ventricular arrhythmias. *New Eng. J. Med.* 325:618–624, 1991.
15. Fiebig, E. R. and U. Trendelenberg. The neuronal and extraneuronal uptake and metabolism of 3H-noradrenaline in the perfused rat heart. *Naunyn-Schmiedeberg's Arch. Pharmacol.* 303:21–35, 1978.
16. Nakajo, M., K. Shimabukuro, H. Yoshimura, R. Yonekura, Y. Nakabeppu, P. Tanoue, and S. Shinohara. Iodine-131 metaiodobenzylguanidine intra- and extravesicular accumulation in the rat heart. *J. Nucl. Med.* 27:84–89, 1986.
17. Rose, C. P., J. H. Burgess, and D. Cousineau. Tracer norepinephrine kinetics in coronary circulation of patients with heart failure secondary to chronic pressure and volume overload. *J. Clin. Invest.* 76:1740–1747, 1985.
18. Esler, M., G. Jennings, G. Lambert, I. Meredith, M. Horne, and G. Eisenhofer. Overflow of catecholamine neurotransmitters to the circulation—Source, fate, and functions. *Physiol. Rev.* 70(4):963–985, 1990.
19. Whalen, W. J. Intracellular PO2 in heart and skeletal muscle. *Physiologist* 14:69–82, 1971.
20. Schubert, R. W., W. J. Whalen, and P. Nair. Myocardial PO2 distribution: Relationship to coronary autoregulation. *Am. J. Physiol.* 234:H361–H370, 1978.
21. Araki, R., M. Tamura, and I. Yamazaki. The effect of intracellular oxygen concentration on lactate release, pyridine nucleotide reduction, and respiration rate in the rat cardiac tissue. *Circ. Res.* 53:448–445, 1983.
22. Turek, Z., K. Rakusan, J. Olders, L. Hoofd, and F. Kreuzer. Computed myocardial Po2 histograms—Effects of various geometrical and functional conditions. *J. Appl. Physiol.* 70(4):1845–1853, 1991.
23. Rumsey, W. L., C. Schlosser, E. M. Nuutinen, M. Robiolio, and D. F. Wilson. Cellular energetics and the oxygen dependence of respiration in cardiac myocytes isolated from adult rat. *J. Biol. Chem.* 265(26):15392–15399, 1990.
24. White, M., J. L. Rouleau, T. D. Ruddy, T. Demarco, D. Moher, and K. Chatterjee. Decreased coronary sinus oxygen content—A predictor of adverse prognosis in patients with severe congestive heart failure. J. Am. Coll. Cardiol. 18(7):1631–1637, 1991.

9

Metabolic Response Times: A Generalization of Indicator Dilution Theory Applied to Cardiac O_2 Consumption Transients

Johannes H.G.M. van Beek

Introduction

Oxygen consumption, substrate uptake, carbon dioxide and lactate efflux, and other metabolic fluxes in an organ can usually be easily assessed in the steady state. However, the time courses of changing metabolic fluxes are difficult to assess, because the investigator can often only measure outside the organ. Such metabolic transients occur when the organ is in a transition from a metabolic steady state to another level of metabolic steady state. As an example, we will look at the transient response of myocardial oxygen consumption to quick changes in cardiac work load caused by imposing a step in heart rate. Diffusion, membrane permeation, and vascular transport delay the externally measured signal with respect to the internal biochemical events. In this chapter it will be explained how the externally measured signal may be corrected for transport delay to reveal the time course of the internal biochemical event. To this end, a theory was developed that is akin to indicator dilution theory (Van Beek and Westerhof, 1991). We will see that a central part of this theory is played by the mass balance of transport fluxes across the organ's external surfaces, including vascular entrances and outlets on the one hand, and internal metabolic fluxes on the other. Normalizing this mass balance and integrating it over the duration of the transient after the initiation of the stimulus that causes metabolism to change yields the characteristic response time of the changes in metabolite concentration measured in the veins and the response time of the metabolic process. The difference between the internal and external response times is the transport time, which turns out to be directly proportional to the change in amount of metabolite present in the organ. The assessment of the metabolic response time will be demonstrated for the time course of oxygen consumption in isolated hearts. Experimental tests of the validity of the response time of cardiac oxygen consumption calculated with this theory are discussed, for instance by comparing the time course of oxygen consumption with independent measurements of heat generation after cardiac contraction. Examples of physiological and pathophysiological results for the metabolic response time are discussed.

Changing Metabolism

Energy is continuously converted in all organs for them to function and survive. Fuel in the form of carbon substrates, such as carbohydrates and fatty acids, is taken up and metabolized to synthesize ATP. ATP in turn is directly used to power the cell's functions. In muscle cells, ATP is broken down when the muscle contracts. In the heart muscle, the largest part of the resynthesis of ATP is by oxidative phosphorylation and therefore coupled to mitochondrial oxygen consumption. When ATP turnover in the heart does not vary, the steady-state oxygen consumption equals the oxygen uptake, which can be calculated by multiplying the blood flow to the whole organ by the arterial-to-venous oxygen concentration difference. In the steady state, the amount of oxygen present in the heart, either physically dissolved or reversibly bound to myoglobin in the muscle cells or to hemoglobin in the red blood cells, does not change and changes in the tissue oxygen stores do not need to be added to oxygen taken up from the perfusate to account for oxygen consumption. The steady-state uptake of fuel substrates can also be assessed, but not all fuel may be oxidized right away: Some of it may be stored in the form of glycogen and triglycerides to be used later. In such cases, oxygen consumption may not be exactly accounted for by the uptake of substrates.

The work load on the heart often changes quickly: Heart rate, stroke volume, and stimulation via the sympathetic nerves or by epinephrine and other hormones vary through the day. This will lead to changes in ATP hydrolysis in the heart muscle cells. The concentration of ATP would go down when the cardiac work load is increased if extra ATP would not be formed by the enzyme creatine kinase, which transfers the phosphate group of phosphocreatine (PCr) to the ADP that resulted from the breakdown of ATP (Figure 9.1). Because the high-energy phosphate buffer PCr itself is present in limited quantity and is formed only from ATP itself, buffering of ATP by creatine kinase provides only a temporary solution. Oxidative phosphorylation, that is, synthesis of ATP coupled to oxygen consumption in the mitochondria, must adapt: Fast adaptation will enable PCr to be preserved, but slow adaptation will lead to depletion of PCr.

We studied how fast aerobic ATP synthesis adapts after changes in cardiac ATP hydrolysis (Van Beek and Elzinga, 1986) by measuring the transients of mitochondrial oxygen consumption during changes in ATP turnover from one steady-state level to another. The time course of oxygen uptake in the saline-perfused heart as a whole was measured with fast responding oxygen electrodes, which measure physically dissolved oxygen. When the heart is perfused with blood, fiber-optic catheters can be used to measure hemoglobin oxygenation densitometrically. The method to correct the measured external oxygen uptake signal to characterize the true time course at the level of the mitochondria has much in common with indicator dilution theory.

Indicator Dilution Theory for Changing Metabolic Signals

Indicator dilution studies deal with the time course of the venous emergence of an indicator, usually a material foreign to the body, after injection of this indicator

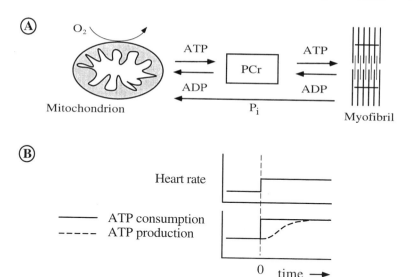

FIGURE 9.1. (A) Scheme of ATP turnover in cardiac muscle. An increase in heart rate and force of contraction leads to immediate changes in rate of ATP hydrolysis to ADP and inorganic phosphate (P_i), both in the myofibrils and by the cellular ion pumps. The changes in ATP are buffered by transfer of the phosphate group from the high-energy phosphate buffer phosphocreatine (PCr) to ATP. However, PCr must be resynthesized from ATP. Therefore, in the long run, the synthesis of ATP must take place mainly in the mitochondria by oxidative phosphorylation, coupled to oxygen consumption. (B) An experiment is schematically depicted where the heart and ATP consumption are changed stepwise. The goal is to follow the time course of ATP production, which is, for practical purposes, instantaneously followed by mitochondrial oxygen consumption.

into an artery supplying the organ under investigation. The goal of such indicator dilution studies is often to determine the time taken by blood to traverse the vascular system in the organ. In this case, an indicator is injected that does not leave the vasculature, for instance by binding a dye molecule to a protein that does not cross the vascular wall to an appreciable extent. The average time taken by the indicator through the organ's vasculature is called the *mean transit time*. The *central volume theorem* is very important in indicator dilution theory (Meier and Zierler, 1954; Lassen and Perl, 1979). It states:

The mean transit time is equal to the volume of distribution of the indicator divided by the flow through the blood vessels.

For an intravascular indicator, the volume of distribution is equal to the blood volume.

Indicator dilution studies are not limited to intravascular indicators, but indicators that permeate the vascular wall and enter the tissue are also used. Some of these extravascular indicators enter only the interstitium, but others also enter the cells, so that the interstitial volume and, if applicable, the intracellular volume contribute to the volume of distribution. To determine the volume of distribution for the extravascular space, the volume of the contributing compartments is multiplied by the ratio of the concentration of indicator in the extravascular compartment to the concentration of indicator in the blood, under the condition that the indicator is in equilibrium between blood and tissue. This ratio is called the *tissue-to-blood partition coefficient*, λ. When the blood flow is known, the volume of distribution can be calculated from the mean transit time of the indicator, using the central volume theorem.

A necessary condition for applying the central volume theorem is that the indicator does not diffuse out of the organ, but leaves the organ in the blood, where the exit of indicator is monitored. A very clear example where this condition is not met is the infusion of cold saline as a heat indicator in the heart (Duijst et al., 1988). Here, the volume of distribution for heat in the myocardium, which is calculated from the mean transit time from coronary artery to the coronary vein, is 22 ± 2 (SD) ml per 100 g wet weight of myocardium. The actual volume of 100 g of myocardium is about 94 ml, the measured temperature is changing everywhere in the myocardium, and the specific heats of blood and heart tissue are almost equal (i.e., the tissue-to-blood partition coefficient is about 1). The explanation for this deviation is that a large portion, about one-quarter, of the infused heat indicator does not leave the heart via the coronary venous blood stream, where it is monitored with a thermistor catheter, but via heat conduction across the endocardium to the left ventricular cavity and across the epicardium to the mediastinum and lungs. A heat "quantum" traveling around in the heart (think of the analogy with the Brownian movement of a diffusing molecule) has a larger chance of being conducted out of the heart the longer it stays in the heart. Because of this, the mean transit time of heat "quanta" leaving the heart via the blood becomes much shorter in the situation with heat conduction compared with the hypothetical situation without heat conduction, and consequently the volume of distribution is markedly underestimated by calculation with the central volume theorem.

A second prerequisite to apply the central volume theorem is that the indicator should be metabolically inert (i.e., it should not be consumed or produced after injection in tissue) and also not irreversibly bound. According to these criteria, oxygen would not seem to be a good indicator, although there are conditions where even oxygen can be used as an extravascular indicator, as will be shown below. In this chapter we discuss how to measure metabolism in the non-steady state. This problem was already analyzed by Zierler in terms of indicator dilution theory (1961). However, the solution proposed in this chapter, with emphasis on the change in amount of the metabolite present in the organ, is very different from Zierler's approach, where the emphasis was on the frequency function of transit times between sites of metabolism and the vascular exit of the organ.

Although classical indicator dilution theory does not apply to an indicator that is metabolized (Lassen and Perl, 1979), its venous outflow curve can be very fruitfully analyzed by applying computer models of the circulation, where the microvascular exchange space is treated as a longitudinally distributed system where reaction, convection, diffusion, and membrane permeation take place

(Bassingthwaighte and Goresky, 1984). The full time course of the venous concentration curve of emerging unchanged injected tracer as well as one or more excreted metabolites of the tracer is often calculated in detail using these models. In contrast, classic indicator dilution theory usually studies only the first statistical moment of the impulse response function, that is, the mean transit time of the concentration–time curve after a short bolus injection of tracer.

When studying the time course of oxygen consumption during transients induced by changes in work load to the heart by measuring the oxygen concentration in the coronary venous outflow, we encounter a whole new facet of tracer kinetic problems. Here metabolism of the "indicator," oxygen, is not in a steady state. However, the response of the coronary venous oxygen concentration after a step in heart rate has a shape that is similar to the venous concentration curve in response to a step change in the arterial concentration of an inert extravascular indicator. The inert indicator enters the heart through the coronary arteries and transits the full length of the capillary system. In contrast, the change in oxygen consumption after a step in cardiac work load is distributed along the length of the capillaries and other microvessels that exchange oxygen. Therefore, the disturbance of local oxygen concentration starts not in an artery to travel the full length of the coronary vascular system, but at many sites along the exchanging vasculature, and travels on average about one-half the length of the coronary vasculature before being detected in the coronary venous effluent. In the next section we present the theory for assessing the response times of metabolic changes and derive the central volume theorem in the same style to highlight the similarities.

The Mean Response Time of Oxygen Consumption

The mass balance of a molecular species for an organ is:

$$\frac{dQ}{dt} = \text{production} - \text{consumption} + \text{inflow} - \text{outflow},$$

where dQ/dt is the rate of change of the amount (Q) of the molecular species inside the organ. The production in this case is taken to mean the metabolic generation of the molecular species plus the gain by diffusion across the surface of the organ. The consumption is the loss by chemical transformation and by diffusion through the surface of the organ. The inflow means transport into the organ by the arterial perfusate, and the outflow means transport out of the organ by the venous effluent.

Because O_2 is not produced and the diffusion of oxygen out of the whole heart is usually negligible relative to vascular transport (Loiselle, 1995; Van Beek et al., 1992), for oxygen this simplifies to:

rate of O_2 consumption = O_2 supply via artery − O_2 removal via veins −

rate of O_2 storage.

TABLE 9.1. List of symbols.

Symbol	Meaning	Unit
C	O$_2$ concentration	mol/L
F	Flow of perfusate	ml/g$_{ww}$/min
$M(t)$	O$_2$ consumption	mol/g$_{ww}$/min
Q	Amount of O$_2$ in organ	mol/g$_{ww}$
Δ	Difference from previous steady state	
λ	Partition coefficient, $\Delta C_{\text{tissue}}/\Delta C_{\text{vessel lumen}}$ in hypothetical equilibrium situation	Dimensionless
w_i	Local M/F divided by overall M/F	Dimensionless
R_i	Factor proportional to local resistance to diffusion	Dimensionless
J	Metabolic flux	mol/s
Subscripts		
a	Arterial	
v	Venous	
ww	Wet weight	

In symbols this becomes:

$$M(t) = F \cdot [C_a(t) - C_v(t)] - \frac{dQ}{dt}. \tag{9.1}$$

The symbols are explained in Table 9.1. This equation shows that if we measure the uptake of oxygen from the perfusate, $F \cdot [C_a(t) - C_v(t)]$, and add the decrease in amount of oxygen present in the organ, the total chemical conversion of oxygen is known.

The oxygen concentration in the arterial and venous perfusates can be measured with oxygen electrodes if the perfusate is a saline solution, or with a fiber-optic catheter connected to a densitometer if hemoglobin is present. It is much more difficult to measure the amount of oxygen present in the organ. One could use a near-infrared tissue oxygen monitor to measure the change in oxygen bound to myoglobin (Van Beek et al., 1996c) and in hemoglobin, but this would not include the physically dissolved oxygen. Using radioactively labeled oxygen with external detection has the disadvantage that the signal from oxygen metabolized to water cannot be discerned from radioactivity in O$_2$. For such reasons it has not been possible so far to measure the change in Q continuously.

If $M(t)$ in Eq. (9.1) is constant, the organ will approach a steady state with $dQ/dt = 0$. During a transition between two steady states, $F \cdot [C_a(t) - C_v(t)]$ lags behind the changes in $M(t)$. This means that $-dQ/dt$ is positive when $M(t)$ increases, because Q is decreasing. The greater the lag of $F \cdot [C_a(t) - C_v(t)]$ with respect to $M(t)$ the greater the accumulated decrease of Q, which can be calculated by integrating dQ/dt during the time taken for the transition between the initial and final steady states of O$_2$ uptake. Next we show that integrating the mass balance of Eq. (9.1) provides the link between the characteristic response times and transit times and the changes in amount of oxygen present in the organ.

Short Derivation of the Central Volume Theorem

First the central volume theorem will be derived by integrating Eq. (9.1). Of course, the central volume theorem has been derived before (Meier and Zierler, 1954), but our purpose is to show how one obtains a brief general derivation by simply integrating the mass balance for the metabolic steady state; the same principle will then be applied when metabolism is not in a steady state (see below). We start with a nonmetabolized indicator, $M(t) = 0$, and with constant flow F. The indicator is not present before $t = 0$, so $C_a(t) = C_v(t) = 0$ at $t < 0$, and the concentration is zero everywhere in the organ for $t < 0$. At $t = 0$ $C_a(t)$ jumps to ΔC_a and stays constant at this value thereafter, yielding a step function. The indicator then fills up the organ until it has reached equilibrium (see Figure 9.2). The net uptake of the indicator becomes zero again, so that $C_v = C_a = \Delta C_a$ in the new steady state (see Figure 9.2). The filling up of the organ with indicator is given by:

$$F \cdot [C_a(t) - C_v(t)] = \frac{dQ}{dt}. \tag{9.2}$$

We integrate Eq. (9.2) from $t = 0$ until equilibrium is reached:

$$F \cdot \int_0^\infty [\Delta Ca - \Delta C_v(t)] \cdot dt = \int_0^\infty \frac{dQ}{dt} \cdot dt. \tag{9.3}$$

Here, $\Delta C_v(t) = C_v(t) - C_v(0)$. In this simple case, $\Delta C_v(t) = C_v(t)$, because the indicator is absent before infusion starts. However, the merit of the general formulation is easily appreciated if one considers the step back when the organ in equilibrated with indicator, and one then stops arterial infusion of indicator.

We divide both sides by $\Delta C_v(\infty) = C_v(\infty) - C_v(0)$, where $t = 0$ indicates the initial steady state and $t = \infty$ the new steady state that is attained when equilibration of the tissue with indicator is complete, in theory very long after the change in venous concentration. $\Delta C_v(\infty)$ equals ΔC_a:

$$F \cdot \int_0^\infty \left[1 - \frac{\Delta C_v(t)}{\Delta C_v(\infty)} \right] \cdot dt = \frac{1}{\Delta C_v(\infty)} \int_0^\infty dQ. \tag{9.4}$$

In this equation, the integral on the left-hand side has the dimension of time and is termed the venous response time, t_v:

$$t_v = \int_0^\infty \left[1 - \frac{\Delta C_v(t)}{\Delta C_v(\infty)} \right] \cdot dt = \overline{t}. \tag{9.5}$$

The t_v after a step in arterial concentration is equivalent to the mean transit time, which is the average time taken by a molecule for transit of the organ between entrance with the arterial blood and exit from the organ via the venous blood. The distribution of such times is given by the frequency function of transit times, $h(t)$, whose shape can be determined by measuring the venous concentration transient after a very brief arterial injection of indicator at $t = 0$ to determine the impulse response $h(t)$ of the system. The mean transit time is therefore

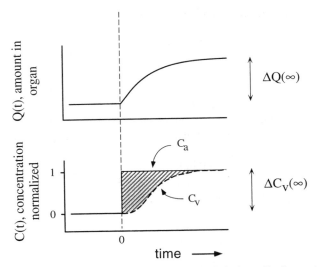

FIGURE 9.2. Hypothetical response to a step to constant infusion of indicator into an organ at $t = 0$. The venous concentration, C_v, increases gradually from zero to a final value of $\Delta C_v(\infty)$. Before infusion of the indicator, the tissue concentration of indicator is zero everywhere in the organ and blood vessels. The hatched area gives the integral which, after multiplication with the flow, equals the increase, $\Delta Q(\infty)$, in amount of indicator in the organ, $Q(t)$, see Eq. (9.3). The hatched area, after normalizing the concentration to zero before the step and to 1 for the final steady state after the step, also equals the mean transit time of the indicator through the organ.

$$\overline{t} = \int_0^\infty t \cdot h(t) \cdot dt. \tag{9.6}$$

For a step in arterial concentration with amplitude $\Delta C_a = \Delta C_v(\infty)$ as input, it can be shown mathematically that integration of the venous concentration response according to Eq. (9.5) yields exactly the mean transit time defined in Eq. (9.6) (Meier and Zierler, 1954). The only necessary assumption is that the transport through the organ behaves linearly and is stationary (Lassen and Perl, 1979), and the venous concentration response to a step in arterial concentration can then be obtained by convolution of $h(t)$ with the Heaviside step function with amplitude $\Delta C_v(\infty)$.

The term on the right-hand side of Eq. (9.4) equals the volume of distribution V_d:

$$V_d = \frac{1}{\Delta C_v(\infty)} \int_0^\infty dQ = \frac{\Delta Q}{\Delta C_v(\infty)}, \tag{9.7}$$

where ΔQ equals the change in amount of indicator present in the organ, which is the change in indicator concentration in the tissue, ΔC, integrated over the volume V of the whole organ:

$$\Delta Q = \int_V \Delta C \cdot dV. \tag{9.8}$$

Thus Eq. (9.4) gives the central volume theorem:

$$\overline{t} = \frac{V_d}{F}, \tag{9.9}$$

which we first mentioned in the Introduction, "Indicator dilution theory."

For this derivation the shape of the step response, $\Delta C_v(t)$, was not specified. Therefore, the details of the system are not important; the central volume theorem is valid irrespective of the geometry of the vascular system and the flow profile in the vessels. Further, the indicator may leave the vessels and be distributed in all of the tissues: the central volume theorem is still valid as long as the total amount of indicator, located both inside and outside the vessel lumina, is taken into account in ΔQ, Eq. (9.8). If the indicator is metabolized, however, the assumptions for the reasoning are no longer valid. The proof given above was published previously (Van Beek and Westerhof, 1991). Justifications of the central volume theorem for diffusible indicators had been given before (Roberts et al., 1973; Lassen et al., 1983), but the derivation given in this chapter has the advantage of being brief and may help insight.

In the next section we will see that the proof may be generalized for the situation in which the "indicator" is metabolized. When metabolism is constant, the central volume theorem can still be valid.

Central Volume Theorem for Metabolized Indicator

Let us now drop the assumption that the indicator is not metabolized, and start from Eq. (9.1) rather than from Eq. (9.2). In Eq. (9.1), C_v is lower than C_a because of substrate uptake. If $M(t)$ is constant, just before the step at $t = 0$ a steady state (ss) will exist with $dQ/dt = 0$, and

$$M(ss) = F \cdot [C_a(ss) - C_v(ss)]. \tag{9.10}$$

This is the well-known equation by Fick (cf. Lassen and Perl, 1979). Now we subtract Eq. (9.10) from Eq. (9.1) and consider changes from the steady-state concentration, $\Delta C(t) = C(t) - C(ss)$, and from the steady-state metabolic rate, $\Delta M(t) = M(t) - M(ss)$:

$$F \cdot [\Delta C_a(t) - \Delta C_v(t)] - \Delta M(t) = \frac{dQ}{dt}. \tag{9.11}$$

For the moment we limit ourselves to the situation where $\Delta M(t)$ is zero (constant metabolism), and consider a change in C_a that starts at $t = 0$ (see Figure 9.3). C_a can change gradually (see Figure 9.3). (This could also have been done in the case of the nonmetabolized indicator above, where a step function was assumed.) Rearranging Eq. (9.11), normalizing to the final step size $\Delta C_v(\infty)$, and integrating over the duration of the response caused by the change in C_a yields

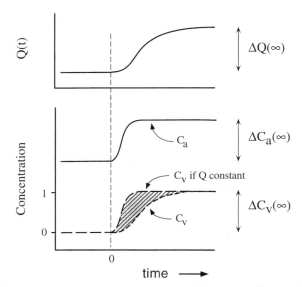

FIGURE 9.3. Hypothetical response of the venous concentration, C_v, to a quick, although not instantaneous, increase in arterial concentration, C_a, of an indicator. Because this indicator is consumed in the tissue in this example, the venous concentration is lower than the arterial concentration. The rates of consumption and flow are constant. Consequently, we can draw the hypothetical time course of C_v assuming that the amount of indicator in the organ, Q, does not change. The hatched area between this hypothetical time course and the real time course of C_v multiplied by the flow is the increase in amount of indicator inside the organ, $\Delta Q(\infty)$, caused by the increased infusion. If the concentration has been normalized to zero before the step and to 1 for the final steady state, the hatched area between the concentration curves also equals the mean transit time of the indicator.

$$\int_0^\infty \left[1 - \frac{\Delta C_v(t)}{\Delta C_v(\infty)} \right] \cdot dt - \int_0^\infty \left[1 - \frac{\Delta C_a(t)}{\Delta C_a(\infty)} \right] \cdot dt = \frac{1}{F} \cdot \int_0^\infty dQ/\Delta C_v(\infty), \qquad (9.12)$$

where $\Delta C_a(\infty)$ was taken equal to $\Delta C_v(\infty)$, as follows from Eq. (9.11) with dQ/dt approaching zero toward the end of the response.

The venous response time, t_v, is equal to the first term. The characteristic time defining the arterial change, t_a, is given by the second term. If $t_v - t_a$ is set equal to the mean transit time, again the central volume theorem emerges:

$$\bar{t} = t_v - t_a = \int_0^\infty \left[1 - \frac{\Delta C_v(t)}{\Delta C_v(\infty)} \right] \cdot dt - \int_0^\infty \left[1 - \frac{\Delta C_a(t)}{\Delta C_a(\infty)} \right] \cdot dt$$

$$= \frac{1}{F} \cdot \frac{\Delta Q}{\Delta C_v(\infty)} = \frac{V_d}{F}. \qquad (9.13)$$

Thus the central volume theorem is also valid for a metabolized indicator on the condition that the rate of metabolism of the indicator is constant and changes in concentration from the steady state rather than the absolute concentrations are considered.

Transients of Metabolism

Now we consider the case when metabolism starts to change after an external factor that affects metabolism changes at $t = 0$. Before $t = 0$, metabolism is in a steady state, and metabolism starts to increase gradually after $t = 0$ to reach a new steady-state level. In this chapter we take the time course of mitochondrial oxygen consumption after an upward step in heart rate as an example; see Figure 9.4. Now the cause of the venous concentration transient is a change in metabolism rather than a change in arterial concentration of the "indicator": $\Delta C_a(t)$ equals zero at all times. First it is important to realize that, when $M(t)$ has reached a new steady state, the change in venous concentration is given by

$$\Delta M(\infty) = F \cdot [-\Delta C_v(\infty)]. \qquad (9.14)$$

We add Eq. (9.14) to Eq. (9.11), and then rearrange and integrate over the duration of the response:

$$\int_0^\infty \left[1 - \frac{\Delta C_v(t)}{\Delta C_v(\infty)} \right] \cdot dt - \int_0^\infty \left[1 - \frac{\Delta M(t)}{\Delta M(\infty)} \right] \cdot dt = \frac{1}{F} \cdot \int_0^\infty dQ/\Delta C_v(\infty). \qquad (9.15)$$

All terms have been divided by $F \cdot [\Delta C_v(\infty)]$ or $-\Delta M(\infty)$, which are equal [Eq. (9.14)]. Note the similarity between Eqs. (9.12) and (9.15); the first concerns the response to a change in arterial concentration, and the second is the response to a change in oxygen consumption. The leftmost term is again the venous response time, t_v, which in this case is to a change in metabolism. Now we define

$$t_{\text{mito}} = \int_0^\infty \left[1 - \frac{\Delta M(t)}{\Delta M(\infty)} \right] \cdot dt. \qquad (9.16)$$

This is the characteristic time of the response of metabolism to the change in the stimulus at $t = 0$. Because the mathematical definition is the same as for the mean *transit* time of an indicator, Eqs. (9.5) and (9.6), this has been called the mean *response* time of metabolism. For the response of cardiac aerobic metabolism to a step in heart rate, where $M(t)$ means oxygen consumption, this is called t_{mito}, the mean response time of mitochondrial oxygen consumption to a step change in heart rate. Its definition is equivalent to the transient times characterizing metabolic transitions defined in a mathematical control theory for biochemical systems (Heinrich and Rapoport, 1975).

The "volume of distribution" appearing in Eq. (9.15) is

$$V_{d,m} = \int_0^\infty \frac{dQ}{\Delta C_v(\infty)} = \frac{\Delta Q}{\Delta C_v(\infty)} = \int_V \frac{\Delta C}{\Delta C_v(\infty)} \cdot dV. \qquad (9.17)$$

$V_{d,m}$ is the change in amount of "indicator" (in our case, oxygen) normalized to the change in venous concentration, $\Delta C_v(\infty)$, both resulting from the change in metab-

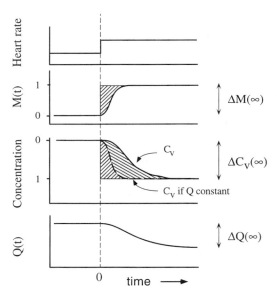

FIGURE 9.4. Hypothetical response of cardiac oxygen consumption, $M(t)$, to a step in heart rate. The venous oxygen concentration, C_v, changes with some delay. Part of the consumed oxygen comes from the decrease in amount of oxygen, Q, in the organ. The hatched area for oxygen consumption is equal to the metabolic response time, t_{mito}, if $M(t)$ has been normalized to 0 before the step and to 1 in the final steady state; see Eq. (9.16). The total hatched area for the venous concentration curve equals the venous response time, t_v, if the concentration has been normalized to 0 before the step and to 1 in the final steady state; see Eq. (9.15). Note that C_v decreases for this upward step in heart rate, but that the normalization according to Eq. (9.15) means that a negative value for $\Delta C_v(t)$ is divided by a negative value for $\Delta C_v(\infty)$, so that the normalized value increases from 0 to 1, while C_v decreases. The hypothetical curve for C_v if Q would not change ($t_{transport} = 0$) is also drawn, and is the mirror image of $M(t)$. The hatched area between these two venous concentration curves equals the transport time, $t_{transport}$, which must be subtracted from t_v to obtain t_{mito}. The similar hatch pattern above $M(t)$ and under the curve for C_v if Q is constant symbolizes that the normalized areas are the same.

olism. This "volume of distribution" is different from V_d, the true, classic volume of distribution, which is determined from a change in arterial concentration. The basic difference between V_d and $V_{d,m}$ is how large vessels and exchange regions are accounted for in terms of volumes of distribution. In the case of $V_{d,m}$, the volume of distribution for metabolism, the volume from where metabolism takes place to the measurement site is taken into account. Therefore, the arterial regions are not included and only one-half the exchange region is included because the mitochondria are distributed throughout the exchange region and oxygen is on average transported through one-half the exchange region. On the other hand, V_d, the total volume of distribution, includes regions from the arterial injection site all the way to the measurement site. To emphasize this difference, the "volume of distribution" for a change in metabolism is indicated with the subscript m.

Equation (9.15) may be written in short form as:

$$t_{mito} = t_v - \frac{V_{d,m}}{F} = t_v - \frac{1}{F} \cdot \frac{\Delta Q}{\Delta C_v(\infty)}. \tag{9.18}$$

This equation means that the response time of metabolism inside the organ to a change in external stimulus can be calculated from the response time of uptake of the metabolite from the blood by determining the change in amount of metabolite in the organ caused by the change in metabolism. The change in amount of metabolite must be divided by the change in metabolism $F \cdot [\Delta C_v(\infty)]$.

Obviously, the last term in Eq. (9.18) represents the delay between the site of metabolism and the venous measurement site, and has the dimension of time. We call this term the *transport time:*

$$t_{transport} = \frac{V_{d,m}}{F} = \frac{1}{F} \cdot \frac{\Delta Q}{-\Delta C_v(\infty)}. \tag{9.19}$$

Thus the problem of deconvolution of the measured venous response time to obtain the characteristic time for changes in metabolism inside the organ reduces to a very simple statement:

$$t_{mito} = t_v - t_{transport} . \tag{9.20}$$

The Transport Time

When metabolism inside the organ changes, the uptake or production of metabolites is measured outside the organ with some delay, characterized by the transport time, $t_{transport}$. For a molecule that is produced inside the organ, $t_{transport}$ is easily interpreted as the mean transit time from the multiple sites of production to the venous site of measurement. In the derivation above we saw that $t_{transport}$ is proportional to the change in amount of metabolite inside the organ, caused by the change in metabolic rate. If, for instance, oxygen uptake follows oxygen consumption with some delay, the difference must be taken from the oxygen stores inside the body. The greater the delay, the greater the change in oxygen stores inside the organ.

Determination of the Change in Oxygen Stores

If the change in oxygen stores, ΔQ, can be measured, the metabolic response time can be determined exactly. The change in oxygen bound to myoglobin can be measured using near-infrared spectroscopy (Van Beek et al., 1996c), but exact quantitation is difficult, although attempts are made (De Groot et al., 1995). It is difficult to measure the change in physically dissolved oxygen inside the tissue. Oxygen electrodes give only a local value that is of limited use given the oxygen gradients inside tissue; although oxygen-sensitive phosphorescence probes can be imaged to yield a spatial distribution, at present they only allow intravascular measurements, albeit with high spatial and temporal resolution (Zheng et al., 1996). Monitoring a metabolite by radioactive labeling has the drawback that the label will also appear in the products of the reaction, and the product and substrate cannot be discerned unless they are chemically or kinetically separated. Indeed, kinetic separation is possible and allows the noninvasive measurement of steady-state oxygen consumption using positron emission tomography (Deussen and Bassingthwaighte, 1996). Although this approach for steady-state oxygen consumption is not directly applicable to measure the time course of the amount of unreacted oxygen during metabolic transitions, which is necessary to assess the full time course of oxygen consumption during metabolic transitions, it is perhaps feasible in the future to measure tissue oxygen content in the steady states before and after the metabolic transition. Using Eq. (9.18), the metabolic response time can then be calculated. The amount of radioactive oxygen label that has been metabolized to water must then be separated from unreacted oxygen using a kinetic model. The sensitivity of nuclear magnetic resonance spectroscopy (MRS) is usually too low to be useful for such experiments, although substrates and products would in general be readily discerned.

When direct measurements of ΔQ are not practical, the change in amount of metabolite can be estimated from a model of the concentration gradients in the tissue. We applied the latter approach in our studies of oxygen consumption transients in heart tissue (Van Beek and Westerhof, 1991). With the same oxygen concentration gradient model, the change in amount of oxygen in tissue was calculated for three interventions that disturb the oxygen supply/oxygen consumption ratio: steps in heart rate at constant flow, steps in arterial oxygen concentration at constant flow, and steps in perfusion flow to the heart at constant arterial oxygen concentration. For the flow step and the arterial concentration step, the change in amount of oxygen in the organ was obtained by measuring the washout of oxygen from tissue following these steps.

Oxygen Concentration Gradient Model

An important characteristic of oxygen transport is the oxygen concentration gradient from arteries to veins that is caused by oxygen uptake in the tissue. In addition, there is a diffusion gradient of oxygen tension from the blood vessels into tissue. Third, the macroscopic heterogeneity of perfusion must be taken into account, which was found both in blood-perfused myocardium (King et al., 1985) and in saline-perfused heart (Gorman et al., 1989).

In our model we assumed a linear concentration gradient in the vessels, from the arterial to the venous side of the exchange region. This is expected to be approximately correct for crystalloid perfusates and for the total oxygen concentration when hemoglobin is present, but the true profile of physically dissolved oxygen will not be linear when hemoglobin is present, because of the S-shaped oxygen–hemoglobin dissociation curve. The oxygen tension, P_{O_2}, is directly proportional to the concentration of physically dissolved oxygen and is the driving force for diffusion into tissue.

Evaluation of this linearized model (Van Beek and Westerhof, 1991) then shows that the volume of distribution for oxygen, determined from washout after arterial concentration reduction, is given by

$$V_{d,O2} = \frac{\Delta Q}{\Delta C_v(\infty)} = \int_V \frac{\Delta C}{\Delta C_v(\infty)} \cdot dV = V_a + V_v + \sum_{i=1}^{N} V_i \cdot \lambda_i, \qquad (9.21)$$

where V_a and V_v denote arterial and venous intravascular volumes, respectively (which may include cannulae and tubing connected to the arteries and veins, respectively). V_i gives the volumes of an arbitrary number N of microvascular exchange regions. The partition coefficient λ_i, which weighs the volumes, is the ratio of the oxygen solubility in the tissue divided by the oxygen solubility in the perfusate. Thus, λ_i is by definition one for V_a and V_v. Equations (9.10)–(9.13) show that $V_{d,O2}$ can be determined by measuring the time course of the venous oxygen concentration during a step change in arterial oxygen concentration, provided that flow and metabolism remain constant.

A "metabolic volume of distribution," $V_{d,m}$, is needed to correct the response time of oxygen uptake measured in the veins to obtain the true response time of oxygen consumption at the level of the mitochondria:

$$V_{d,m} = V_v + \sum_{i=1}^{N} \tfrac{1}{2} V_i \cdot \lambda_i \cdot w_i + \sum_{i=1}^{N} V_i \cdot w_i \cdot R_i. \qquad (9.22)$$

Here, w_i is the local metabolism-to-perfusion ratio, M_i/F_i, normalized by dividing it by the overall metabolism-to-perfusion ratio, M/F. Thus, w_i gives the influence of heterogeneity of metabolism and perfusion on $V_{d,m}$. The factor $\frac{1}{2}$ in the second term on the right-hand side stems from the fact that the average distance between the site of oxygen consumption and the place of measurement includes on average one-half the length of the microvascular exchange region, so that only one-half of the exchange region's volume counts. The average change in oxygen concentration along the microvascular exchange region is only one-half the change in venous oxygen concentration (see Figure 9.5). A rigorous derivation was given by Van Beek and Westerhof (1991). The full venous volume in the organ down to the site of the oxygen measurement is included, represented by V_v in Eq. (9.22). If the oxygen measurement takes place extracorporeally, cannula and tubing must be added to V_v. R_i is proportional to the resistance for oxygen diffusion between microvascular lumen and the mitochondria, and thus the last term of Eq. (9.22) represents the changes in oxygen content in tissue caused by increased diffusion gradients.

Comparing Eqs. (9.21) and (9.22) shows that $V_{d,O2}$ and $V_{d,m}$ differ fundamentally, although they share some components. Therefore, we considered an alternative intervention to disturb the oxygen consumption to supply ratio that would produce an identical arterial-to-venous oxygen concentration gradient. An instantaneous step in perfusion flow would accomplish this, provided that the change in perfusion flow would not cause a change in oxygen consumption. The model's "volume of distribution" for a flow step, $V_{d,f}$, is

$$V_{d,f} = V_v + \sum_{i=1}^{N} \frac{1}{2} V_i \cdot \lambda_i \cdot w_i. \qquad (9.23)$$

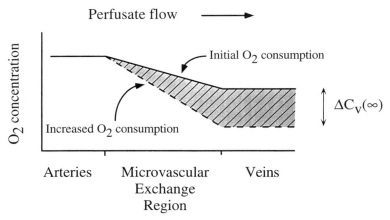

FIGURE 9.5. The intravascular oxygen concentration profile. Oxygen uptake from the perfusate into tissue takes place in the microvascular exchange region: arterioles, capillaries, and venules. The hatched area is proportional to the change in amount of oxygen in tissue, ΔQ, when oxygen consumption increases, assuming that the change in oxygen tension gradient between blood vessel lumen and tissue can be neglected. ΔQ divided by $\Delta C_v(\infty)$, the steady state change in venous oxygen concentration caused by the increase in oxygen consumption, equals the volume of distribution for metabolism.

Thus the "volume of distribution" for a flow step is identical to the "volume of distribution" for a step in oxygen consumption, except for the last term in Eq. (9.22) representing radial diffusion gradients. Using the Krogh model for radial oxygen diffusion from the capillary into tissue, we found that the last term of Eq. (9.22) is usually negligible with respect to the other terms (Van Beek and Westerhof, 1991). Thus, if the washout of oxygen can be determined for a step in perfusion flow, we have not only determined $V_{d,f}$ but effectively also $V_{d,m}$. In the next section it is shown how $V_{d,m}$ may be determined from the venous oxygen transient during a step in perfusion flow.

Step in Perfusion Flow

At $t = 0$, the flow F is changed from $F(0)$ to $F(0) + \Delta F$. We analyze the situation where metabolism does not change. $C_v(0)$ changes to $C_v(0) + \Delta C_v(\infty)$ with

$$\Delta C_v(\infty) = \frac{\Delta F}{F(0) + \Delta F} \cdot (C_a(0) - C_v(0)) = M \cdot \left(\frac{1}{F(0)} - \frac{1}{F(0) + \Delta F} \right). \quad (9.24)$$

Flow F is now considered time-dependent in Eq. (9.1): $(F(0) + \Delta F(t))$. We add Eq. (9.10) (the steady state just before $t = 0$) to Eq. (9.1), divide all terms by $(F(0) + \Delta F(t)) \cdot \Delta C_v(t)$, and then rearrange and integrate over the duration of the response to the step in perfusion flow; arterial concentration and metabolism do not change (i.e., $\Delta C_a(t)$ and $\Delta M(t)$ equal zero):

$$\int_0^\infty \left[1 - \frac{\Delta C_v(t)}{\Delta C_v(\infty)} \right] \cdot dt = \frac{1}{F(0) + \Delta F} \cdot \int_0^\infty dQ/\Delta C_v(\infty). \qquad (9.25)$$

The leftmost term is again the venous response time, t_v, in this case to a step in perfusion flow. The rightmost factor is the "volume of distribution" for a flow step:

$$\int_0^\infty \frac{dQ}{\Delta C_v(\infty)} = \frac{\Delta Q}{\Delta C_v(\infty)} = \int_V \frac{\Delta C}{\Delta C_v(\infty)} \cdot dV = V_{d,f}. \qquad (9.26)$$

The volume of distribution for the flow step is again the change in amount of oxygen normalized to the change in venous concentration, $\Delta C_v(\infty)$, both caused by the change in flow. This volume of distribution is different from $V_{d,O2}$, the "regular" volume of distribution, which is measured from the change in arterial concentration. To emphasize this difference, the volume of distribution for a change in metabolism is indicated with the subscript f, for flow. However, $V_{d,f}$ [Eq. (9.23)] in the oxygen concentration model has the same value as $V_{d,m}$ [Eq. (9.22)] except for the term due to radial diffusional resistance. This latter term was shown to be negligible (Van Beek and Westerhof, 1991). Consequently, $V_{d,f}$ can be calculated from the transient of venous oxygen concentration following a step in perfusion flow, according to Eq. (9.25). This $V_{d,f}$ can then be set equal to $V_{d,m}$ to calculate the transport time for a step in oxygen consumption.

Transport Time from Direct Measurements of Content Changes

The $t_{transport}$ also can be obtained by measuring the change in total amount of oxygen in the organ caused by the increase in oxygen consumption; see Eq. (9.19). So far, only preliminary data have been obtained (Van Beek et al., 1996c; De Groot et al., 1995). In several experiments on isolated perfused rabbit hearts, we measured tissue oxygenation with a near-infrared tissue oxygen monitor using two wavelengths, 760 and 850 nm. We found that for a transition from oxygen tension of about 680 mmHg to zero, about 75% of the signal was due to the deoxygenation of myoglobin, and the remainder was due to the reduction of cytochrome a,a_3, which is part of the enzyme cytochrome oxidase. Given that rabbit heart contains 160 nmol myoglobin per milliliter (Wittenberg, 1970), we estimated from the oxygen concentration gradient model that the change in myoglobin-bound oxygen content was only 20% of the total change in oxygen content, including physically dissolved oxygen. Because the remaining 80% of oxygen could not be measured, we were not able to determine $t_{transport}$ directly from Eq. (9.19) by inserting the change in amount of oxygen. However, the transport times of other metabolites that are less "volatile" than oxygen could conceivably be determined by measuring the changes of metabolite content, for instance nondestructively by NMR spectroscopic methods or by biochemical measurements in frozen tissue samples.

Experimental Determination of the Metabolic Response Time

The theory for the determination of the metabolic response time, explained above, was applied to determine the response time of mitochondrial oxygen consumption to steps in ATP hydrolysis in isolated rabbit hearts. Such steps were brought about by stepwise increases of the electrically paced heart rate, or by increasing the volume of a balloon in the left ventricular lumen, which stretches the heart muscle and thereby increases pressure development and energy turnover. The arterial and coronary venous oxygen tensions were measured with oxygen electrodes, which responded with a time constant of about 1.5 s. The response times of the venous oxygen tension were corrected for the time constant of the oxygen measurement. In Figure 9.6, a recording of the response of the venous oxygen tension to an upward step in heart rate is shown. Oxygen tension decreases from 475 to 447 mmHg, but the response is plotted in Figure 9.6 after normalization according to Eq. (9.5). Consequently, the normalized venous O_2 concentration runs from 0 to 1, and the hatched area in the normalized plot gives t_v. In this example, the measured t_v was 14.7 s, after correction for the response time of the oxygen electrode.

   In the first experiments at 37°C, the metabolic volume of distribution, $V_{d,m}$, was determined by model-based calculation [Eq. (9.22)] from the value obtained for $V_{d,O2}$ from the venous oxygen tension transient after a quick change in arterial oxygen concentration, using Eq. (9.13) (See Figure 9.7). The arterial volume (anatomical value from the literature: 0.038 ml/g) and the aortic cannula volume (0.28 ml) were subtracted from $V_{d,O2}$. One-half the microvascular exchange region part was also subtracted from $V_{d,O2}$ [compare Eqs. (9.21) and (9.22)]. The distribution volume of the microvascular exchange region was found by subtracting the total intravascular volume, measured with an intravascular indicator by application of the central volume theorem, from $V_{d,O2}$ and adding the anatomic capillary volume, 0.035 ml/g, to the remainder. For lack of a method to measure the local aerobic metabolic rate with sufficient spatial resolution, we set the heterogeneity weights w_i equal to 1 in Eq. (9.22), assuming that the metabolism-to-perfusion ratio was constant throughout the tissue. Although the mitochondrial enzyme content is significantly correlated with local blood flow (Bussemaker et al., 1994), direct proof must be obtained by measuring local O_2 consumption in relation to local perfusate flow. Measurement of local O_2 consumption in multiple myocardial samples has only very recently become possible by analysis of [^{13}C]-enrichment of Krebs cycle intermediates and related amino acids (Van Beek et al., 1996a). Initial results suggest that local flow, determined with radioactively labeled microspheres, and O_2 consumption, determined with the new [^{13}C] method, are indeed proportional in isolated rabbit heart. In summary, $V_{d,m}$ could be calculated from the "true" volume of distribution $V_{d,O2}$.

   The second method was also applied: setting $V_{d,m}$ equal to $V_{d,f}$ obtained from the venous O_2 response to a step in perfusion flow [Eqs. (9.23) and (9.25)]. The $V_{d,f}$ calculated for the step in Figure 9.8 was 1.23 ml/g. The right ventricular lumen

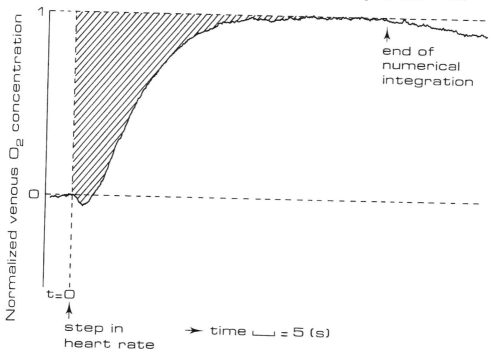

FIGURE 9.6. The response of coronary venous oxygen tension to a step in electrically paced heart rate from 150 to 190 beats/min in a saline-perfused isolated rabbit heart at 37°C. The perfusion flow was kept constant. The oxygen concentration is normalized to 0 for the steady state just before the transient and to 1 for the new steady value. Oxygen tension changes from about 475 (normalized to 0) to about 447 mmHg (normalized to 1). The hatched area gives the integral for t_v in Eq. (9.15). At the end of the trace a secondary change in venous O_2 concentration is seen that is not taken into account in the response times, which therefore only reflect the first phase of the response. (From Van Beek and Westerhof, 1991. By permission of the American Physiological Society.)

and cannula volumes were included in the volume of distribution. The diffusional resistance term of Eq. (9.22) (last term) is 0.01–0.04 times the volume V_i (Van Beek and Westerhof, 1991), which is small relative to the second term on the right-hand side of Eq. (9.22), which is about one-half the volume V_i. Thus, $V_{d,f}$ could be substituted for $V_{d,m}$, and $t_{transport}$ was for example found to be 9.6 s for the experiment of Figure 9.8. Consequently, the venous response time t_v, 16.9 s on average in the experiment of Figure 9.8 for steps between 180 and 220 beats/min, could now be corrected to yield the true response time at the level of the mito-chondria, 7.3 s.

The volumes of distribution $V_{d,02}$ for the arterial oxygen concentration step and $V_{d,f}$ for the perfusion flow step can only be reliably determined if the oxygen consumption remains constant during the steps, because this was assumed for the

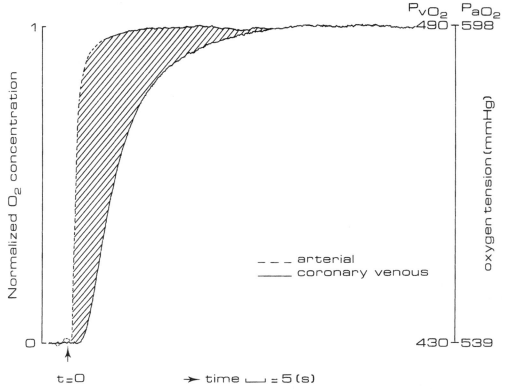

FIGURE 9.7. The response of arterial and venous oxygen tension during a quick change to arterial perfusate containing more oxygen. Experiment on isolated rabbit heart (same heart as in Figure 9.6) at constant perfusion flow. Again the oxygen concentration is normalized to 0 for the steady state just before the transient and to 1 for the new steady value. The oxygen consumption was found not to change despite the change in arterial oxygen concentration. Therefore, the hatched area is the mean transit time of oxygen through the organ; see Eq. (9.13). (From Van Beek and Westerhof, 1991. By permission of the American Physiological Society.)

derivation of Eqs. (9.13) and (9.25). This stability was often a problem in the isolated, crystalloid-solution perfused rabbit heart at 37°C (Van Beek and Westerhof, 1991): 12 of 16 hearts had to be omitted from the study because oxygen consumption changed during these steps. In the remaining four studies, the measured venous response time was 17.6 ± 1.1 s (mean $\pm$ SE). The transport time was 9.9 ± 0.9 s (10.3 ± 1.3 s from three arterial concentration steps and 9.9 ± 2.4 s from one perfusion flow step). Therefore, t_{mito}, the response time of cardiac oxygen consumption to steps in heart rate, corrected for vascular transport and diffusion delay, was 7.7 ± 0.7 s at 37°C. Thus we found the characteristic time of a

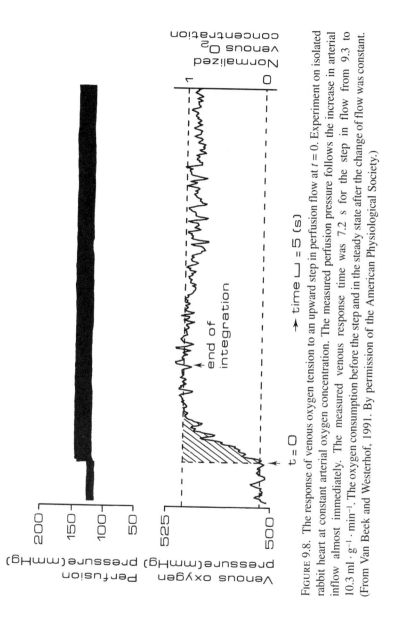

FIGURE 9.8. The response of venous oxygen tension to an upward step in perfusion flow at $t = 0$. Experiment on isolated rabbit heart at constant arterial oxygen concentration. The measured perfusion pressure follows the increase in arterial inflow almost immediately. The measured venous response time was 7.2 s for the step in flow from 9.3 to 10.3 ml · g^{-1} · min^{-1}. The oxygen consumption before the step and in the steady state after the change of flow was constant. (From Van Beek and Westerhof, 1991. By permission of the American Physiological Society.)

metabolic event by applying transport analysis to measurements made exclusively in the venous effluent and arterial inflow of the heart.

The transport times obtained from the arterial oxygen concentration step and from the flow step should give the same result. In the first study at 37°C, none of the 16 hearts was found to have an oxygen consumption that was sufficiently stable during both the step in perfusion flow and the step in arterial O_2 concentration. However, at lower than physiological temperatures the oxygen consumption was usually stable for both steps in the same heart. At 20°C, the transport time obtained from the flow step was 8.5 ± 1.1 s (mean $\pm$ SE, $n = 10$), and from the arterial concentration step it was 7.9 ± 0.4 s; at 28°C, the transport time was 6.2 ± 0.6 s and 6.7 ± 0.3 s ($n = 10$) for flow and concentration steps, respectively (Hak et al., 1992). The transport times obtained from flow steps and concentration steps in the same hearts did not differ significantly, which corroborates the validity of the transport time determination.

Experimental Validation of the Mean Response Time of Oxygen Consumption

The determination of t_{mito} was accomplished using a complicated model. Therefore it seemed appropriate to find an independent measurement of the time course of oxygen consumption to test the procedure critically. Oxygen consumption reflects oxidative phosphorylation because time delays *within* the mitochondria are very small (<100 ms; Chance, 1965). The heat developed during oxidative phosphorylation can be measured with good time resolution in papillary muscles isolated from the rabbit right ventricle (Mast and Elzinga, 1988). Because this preparation is not perfused, oxygen diffusion can only deliver sufficient oxygen at lower than physiological temperatures where metabolic rates are reduced. The time constant of the decay of recovery heat (a monoexponential time course was assumed) after a train of contractions, after correction for heat conduction, was 22–25 s at 20°C. We found that the isolated rabbit heart could be paced at 20°C, and that t_{mito} was 26.9 ± 3.0 s (SE) at this temperature (Hak et al., 1992). The time constant of the heat rate was not significantly different from t_{mito}. We conclude that the method of determining t_{mito} is corroborated by an independent method to measure the time course of oxidative phosphorylation, that is, the measurement of heat rate.

High-Energy Phosphates and the Response Time of Mitochondrial Oxygen Consumption

At 37°C the response time of the increase in oxygen consumption and therefore of the increase in ATP synthesis is about 8 s. ATP hydrolysis is expected to increase instantaneously with heart rate. Aerobic ATP synthesis follows the increase in ATP hydrolysis by a lag that equals t_{mito}. A lag of the venous response of the concentration of an indicator after a step in arterial concentration of the indicator means that the indicator content in the organ is changing (see Eq. 9.13). In analogy, the lag of ATP synthesis after a change in ATP hydrolysis means that the cellular content of high-energy phosphate is changing (see Figure 9.1). The terminal phosphate group of ATP is exchanged very rapidly with creatine to form phosphocreatine (PCr); this reaction, catalyzed by creatine kinase, is poised so that in case of a disbalance of ATP synthesis and ATP hydrolysis one almost exclusively finds a change in PCr. This is described in the following equation:

$$\Delta(\text{PCr}) = \int_0^\infty [\Delta J_{\text{ATP-synthesis}}(t) - \Delta J_{\text{ATP-hydrolysis}}(t)] \cdot dt, \qquad (9.27)$$

where J is the metabolic flux and $\Delta(\text{PCr})$ is the change in PCr content from $t = 0$. For a step change in ATP hydrolysis, Δm, at $t = 0$:

$$\Delta(\text{PCr}) = -\Delta m \cdot \int_0^\infty \left[1 - \frac{\Delta J_{\text{ATP-synthesis}}(t)}{\Delta m} \right] \cdot dt = -\Delta m \cdot t_{\text{mito}} . \qquad (9.28)$$

Here it is assumed that ATP synthesis is linearly related to oxygen consumption, with negligible time delay. The response time for ATP synthesis then equals the response time for oxygen consumption. If ATP synthesis by anaerobic glycolysis does play an appreciable role, the last equality in Eq. (9.28) would not hold.

As an example, for a step in heart rate from 60 to 120 beats/min in isolated rabbit heart at 28°C, t_{mito} was 12.1 s (Eijgelshoven et al., 1993a), and the increase in oxygen consumption 5 μmol/g dw/min. Applying Eq. (9.28) (the mass balance integrated over time of high-energy phosphates) and translating the change in oxygen consumption into the change in ATP synthesis, Δm, using an ATP/O ratio of 2.4, a decrease in PCr content of 5 μmol/g dw is predicted. It should be noted that we followed the response of oxygen consumption to the heart rate step for about 1 min; thus, the calculated change in PCr should also be found at this time. A slow secondary change in PCr partly back in the direction of the original level after the first minute of a heart rate increase has been reported (Elliott et al., 1994). Because the prediction of the change in PCr must be assessed in the first phase of the transient response, we decided to design NMR spectroscopic experiments with about a 2-s resolution, despite the relatively low sensitivity of NMR. The heart rate steps were repeated 64 times and NMR data acquisition was synchronized with the heart rate step. The data were added at fixed time points with respect to the step. A modified oxygen electrode was designed that could be inserted into the NMR magnet without loss of NMR signal (Eijgelshoven et al., 1993b), and the time course of oxygen consumption was determined inside the magnet simultaneously with NMR data acquisition. Both at 28 and 37°C we found a decrease in PCr in the first tens of seconds after the step increase in heart rate (see Figure 9.9), corroborating the prediction based on the response times for oxygen consumption of 5–15 s that we found (Eijgelshoven et al., 1994). However, although the prediction was fulfilled qualitatively, we found a quantitative discrepancy. At 28°C, t_{mito} was 14 s, while the time constant of the change in inorganic phosphate P_i, which mirrored the change in PCr, was 5 s; the change in PCr predicted from t_{mito} was 4.7 μmol/g dw, but for the NMR measurement this was 2.7 μmol/g dw. At 37°C, t_{mito} was 11 s, but the time constant for P_i was about 2.5 s; the decrease

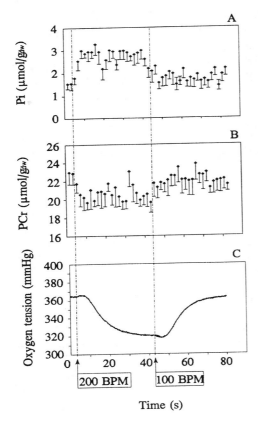

FIGURE 9.9. Time course of contents of inorganic phosphate (P_i), phosphocreatine (PCr), and venous oxygen tension during steps in heart rate from 100 to 200 beats/min and back in isolated rabbit hearts at 37°C. Heart rate is 100 beats/min at $t = 0$. The phosphate metabolites (mean ± SE, seven hearts) were measured with ^{31}P NMR spectroscopy at 1.8-s intervals and quantified using an external methylene diphosphonate standard. Contents are given in per gram dry weight. The heart rate steps were repeated cyclically 64 times in each rabbit heart, and the NMR signal collection was synchronized to the heart rate step and binned. The oxygen tension is an example of a recording of one response in one rabbit heart inside the NMR magnet. (Modified from Eijgelshoven et al., 1994. By permission of the American Heart Association.)

calculated from t_{mito} was 8.5 µmol/g dw, the measured decrease 1.7 µmol/g dw. The discrepancy between the predicted and measured changes in PCr in the first minute after a step in heart rate has been proposed to be explained by appreciable anaerobic glycolytic ATP production during several seconds after the heart rate step, which may transitorily supplement aerobic ATP production (Hak et al., 1993c). It is conceivable that the first changes in phosphate metabolites measured with NMR take place exclusively in or close to the myofibrils, and that such changes reach the mitochondria only with some delay, which would explain the difference between the time constant of the phosphate metabolites and t_{mito} (Van Beek et al., in press).

Physiology and Pathophysiology of the Mitochondrial Response Time

With the method to determine the response time, which we now consider well founded, we explored the behavior of t_{mito} under several circumstances. The t_{mito} increases by 110% per 10°C decrease in temperature (Hak et al., 1992). This might be explained to a large extent by the decrease in aerobic capacity (state 3 respiration) with temperature, found in isolated mitochondria, because this capacity depends similarly on the temperature. When our normal reference substrate in the isolated heart, glucose, is replaced with pyruvate at 28°C no change in t_{mito} at low heart rates is found (Eijgelshoven et al., 1995), but an almost 20% decrease is found at high heart rates (Hak et al., 1993a). At 37°C, t_{mito} is almost halved with pyruvate relative to glucose, but with lactate no such decrease is found, suggesting that the shuttling of reducing equivalents from cytosol into the mitochondrial matrix, which is necessary for glucose and lactate but not for pyruvate, is a major limiting factor for the speed of adaptation of mitochondrial oxygen consumption. When the mitochondrial calcium uptake channels are blocked there is a marked increase in t_{mito}, but when these channels are merely partially blocked t_{mito} is unchanged; thus a minimum amount of calcium is probably needed inside the mitochondria for a fast response, but the mitochondrial calcium uptake channels do not seem to be rate-limiting for the adaptation of oxidative phosphorylation. Indeed, calcium entry into the mitochondria is relatively slow and stimulates NAD^+ reduction, which occurs only about a half-minute after elevation of cardiac work, much too slow to account for our t_{mito} values (Brandes and Bers, 1997). It is likely that the first phase of the response of oxidative phosphorylation to a changed cardiac work load whose speed is characterized by t_{mito} does not depend on Ca^{2+} entry into the mitochondria and is followed by a slower change (Elliott et al., 1994; Brandes and Bers, 1997).

At 28°C, t_{mito} increases if one steps to higher heart rates (Eijgelshoven et al., 1993a), but such an increase is not very clear at 37°C (Eijgelshoven et al., 1995). For the experiments we usually added a maximally vasodilating concentration of adenosine to the perfusate of the isolated rabbit heart. The dependence of t_{mito} on heart rate is decreased when adenosine is omitted, although the average t_{mito} is not changed (Tian et al., 1994). Also, increases in contractility, brought about with β-adrenergic agonists, or changes in left ventricular volume to alter energy turnover in the heart via the Starling mechanism, both do not appear to affect the speed of response of oxygen consumption to metabolic demand.

The t_{mito} increases two- to threefold when the cytosolic pH falls from 7.1 to 6.5, which is a drop in pH that may be found in the ischemic heart (Hak et al., 1993b). This is only partially explained by the approximately 15% inhibition of mitochondrial aerobic capacity that was found in isolated rabbit heart mitochondria for a similar pH drop (Van Wijhe and Van Beek, 1995). At 37°C, an acute reduction of the mitochondrial aerobic capacity by partially blocking mitochondrial ATP synthesis with oligomycin does not appear to lead to an increase of t_{mito} (De Groot, Dijk and Van Beek, unpublished results), although t_{mito} increases with reduction of the mitochondrial capacity at 28°C, at least at low heart rates (Van Beek et al., 1996b). A brief period (15 min at 37°C) of myocardial ischemia or hypoxia did lead to a slower response of mitochondrial oxygen consumption: t_{mito} was increased by ~40% (Zuurbier and Van Beek, 1995). Because the capacity of mitochondria isolated after such brief ischemia or hypoxia is not diminished, and because reduced mitochondrial capacity does not necessarily lead to changes in

t_{mito}, the results of Zuurbier and Van Beek (1995) indicate that the signal from ATP consuming sites to the mitochondria across the cytosol is retarded in post-ischemic or post-hypoxic myocardium. It remains to be investigated whether this retardation of transcytosolic signaling is causally linked with the contractile stunning found in hearts after brief ischemia or hypoxia, or with the reduced contractility during acidosis.

Many physiological and pathophysiological results on the dynamic adaptation of oxidative phosphorylation to demand have already been obtained with the method to determine t_{mito}. Of special interest is the finding that inorganic phosphate and PCr change much more quickly than oxidative phosphorylation, and that signaling from ATP-consuming sites to the mitochondria may take many seconds. This signaling may be retarded after brief ischemia or hypoxia, or during cytosolic acidosis.

Discussion and Summary

The time course of oxygen uptake can be measured at the whole organ level, but is delayed with respect to the time course of oxygen consumption at the mitochondrial level due to diffusion and vascular transport. The characteristic response time of the change in cardiac oxygen uptake during steps in ATP hydrolysis in isolated hearts is measured in the coronary venous effluent and corrected for the transport delay. The theory for the correction is closely akin to indicator dilution theory and is based on the mass balance, taking into account oxygen consumption, oxygen transport into and out of the organ, and the change in amount of oxygen present in the organ. After normalization and integration of this mass balance over the period of the response, a simple relation is obtained between the characteristic response time for the change of oxygen consumption at the mitochondrial level and the characteristic response time measured in the coronary venous effluent. The difference between these two response times is the transport time, which is equal to the decrease in the amount of oxygen within the organ caused by the increase in oxygen consumption divided by the increase in oxygen consumption itself. The analogy with indicator dilution theory is demonstrated here, because $t_{transport}$ is equal to a modified volume of distribution divided by the flow through the organ. The derivations are in line with the integration of the mass balance applied by Meier and Zierler (1954) in their original derivation of "the fundamental relationship, volume = flow multiplied by mean circulation time" (central volume theorem). This derivation has been generalized here to include indicator entering the tissue, indicator that is metabolized, the venous response times to flow steps, and the venous response times to changing metabolism inside the organ (Van Beek and Westerhof, 1991). The venous response times to flow steps or arterial oxygen concentration steps, which are combined with transients in metabolism inside the organ, have also been analyzed (Van Beek, 1995). It appears that the derivation of the central volume theorem for indicator that enters the tissue from the blood vessels may be substantially shorter and easier to comprehend when using the integrated mass balances approach (Van Beek and Westerhof, 1991) than when using the approach of Roberts et al. (1973). The approach explained in this chapter to determine the response time of metabolism may be regarded as a generalization of the indicator dilution theory of Meier and Zierler (1954).

The change in amount of oxygen inside the heart is not easy to measure during metabolic transients, but was estimated by a model calculation based on experimental data on washout of oxygen after downward steps in arterial oxygen concentration or after downward steps in perfusion flow. Of course this method is not only applicable to oxygen, but it is quite general and may also be applied to transients of consumption or production of other metabolites. However, the limitation of the method is that it was derived for a system with linear transport equations, and application to a tissue with nonlinear transport characteristics must be done very carefully. One needs to be especially careful, for instance, if the steps in arterial oxygen concentration or perfusion flow cause substantially bigger or smaller changes in venous oxygen concentration than the step in oxygen consumption. The $t_{transport}$ may not be constant in a nonlinear system, but may depend on the amplitude of the steps. For example, the washout of oxygen for perfusion flow steps of various amplitudes may not be proportional to the change in venous oxygen concentration if there is a substantial contribution of oxygen released from myoglobin; myoglobin has a hyperbolic relation with the cytosolic free oxygen concentration.

Using the method explained above for deconvolution of the venous oxygen measurement, a mean response time ("time constant") of 8 s was found for the change in oxygen consumption after a step in heart rate in isolated rabbit heart at 37°C. Because this means a lag in mitochondrial ATP synthesis after a quick change in ATP hydrolysis, a change in the high-energy phosphate buffer phosphocreatine was predicted, which indeed was demonstrated experimentally by time-resolved ^{31}P nuclear magnetic resonance spectroscopy. At 20°C the mean response time of heat generation during recovery from contraction in isolated heart muscle was the same as the mean response time of oxygen consumption; because both mean response times reflect oxidative phosphorylation and are obtained with different methods, this meant independent confirmation of the method to determine the mean response time. Further confirmation may be obtained from measurements of myoglobin oxygenation in the cardiac tissue, which are now being developed.

It should be investigated whether the full time course of oxygen consumption in the non-steady state may be obtained with computer models for oxygen transport incorporating convection, permeation, diffusion, and metabolism in an axially distributed model (Bassingthwaighte and Goresky, 1984). Such a type of model for oxygen transport has been formulated and applied to analyze the venous time course after tracer injection of oxygen isotopes into coronary arterial blood (Rose and Goresky, 1985). In conclusion, the characteristic response time of oxygen consumption during metabolic transitions has been determined using methods akin to indicator dilution theory applied to the non-steady state of metabolism.

References

Bassingthwaighte, J. B. and C. A. Goresky. Modeling in the analysis of solute and water exchange in the microvasculature. In: *Handbook of Physiology. Sect. 2, The Cardiovascular System, Vol. IV*, edited by E. M. Renkin and C. C. Michel. Bethesda, MD: American Physiological Society, pp. 549–626, 1984.

Bassingthwaighte, J. B., J. H. G. M. Van Beek, and R. B. King. Fractal branchings: The basis of myocardial flow heterogeneities? *Ann. NY. Acad. Sci.* 591:392–401, 1990.

Brandes, R. and D. M. Bers. Intracellular Ca^{2+} increases the mitochondrial NADH concentration during elevated work in intact cardiac muscle. *Circ. Res.* 80:82–87, 1997.

Bussemaker, J., J. H. G. M. Van Beek, A. B. J. Groeneveld, M. Hennekes, T. Teerlink, L. G. Thijs, and N. Westerhof. Local mitochondrial enzyme activity correlates with myocardial blood flow at basal workloads. *J. Moll. Cell. Cardiol.* 26:1017–1028, 1994.

Chance, B. The energy-linked reaction of calcium with mitochondria. *J. Biol. Chem.* 240:2729–2748, 1965.

Deussen, A. and J. B. Bassingthwaighte. Modeling [^{15}O] oxygen tracer data for estimating oxygen consumption. *Am. J. Physiol.* 270:H1115–H1130, 1996.

Duijst, P., J. H. G. M. Van Beek, G. H. M. Ten Velden, G. Elzinga, and N. Westerhof. Shunting of heat in the canine myocardium. In: Cardiac Metabolism and Flow. Thesis, P. Duijst. Free University, Amsterdam, pp. 42–87, 1988.

Eijgelshoven, M. H. J., J. B. Hak, J. H. G. M. Van Beek, and N. Westerhof. Adaptation speed of cardiac mitochondrial oxygen consumption to demand slows with higher heart rate. *Am. J. Physiol.* 265:H1893–H1898, 1993a.

Eijgelshoven, M. H. J., C. Lekkerkerk, J. H. G. M. Van Beek, and C. J. A. Van Echteld. Oxygen measurement inside an NMR magnet with a catheter electrode. *Magn. Res. Med.* 29:559–562, 1993b.

Eijgelshoven, M. H. J., J. H. G. M. Van Beek, I. Mottet, M. G. J. Nederhoff, C. J. A. Van Echteld, and N. Westerhof. Cardiac high-energy phosphates adapt faster than oxygen consumption to changes in heart rate. *Circ. Res.* 75:751–759, 1994.

Eijgelshoven, M. H. J., X. Tian, and J. H. G. M. Van Beek. Exogenous carbon substrate supply affects the time course of adaptation of cardiac mitochondrial oxygen consumption to demand in a temperature dependent fashion. In: Dynamics of Cardiac Energy Metabolism. Thesis, M. H. J. Eijgelshoven. Free University, Amsterdam (ISBN 90-9007857-6), pp. 73–98, 1995.

Elliott, A. C., G. L. Smith, and D. G. Allen. The metabolic consequences of an increase in the frequency of stimulation in isolated ferret hearts. *J. Physiol. (London)* 474:147–159, 1994.

Gorman, M. W., R. D. Wangler, and H. V. Sparks. Distribution of perfusate flow during vasodilation in isolated guinea pig heart. *Am. J. Physiol.* 256:H297–H301, 1989.

De Groot, B., C. J. Zuurbier, and J. H. G. M. Van Beek. Response times of mitochondrial oxygen consumption, tissue oxygenation and venous oxygen concentration during steps in heart rate in isolated rabbit heart. *Pflügers Arch. (Europ. J. Physiol.)* 430 (Suppl.):R93, 1995.

Hak, J. B., J. H. G. M. Van Beek, M. H. van Wijhe, and N. Westerhof. Influence of temperature on the response time of mitochondrial oxygen consumption in isolated rabbit heart. *J. Physiol. (London)* 447:17–31, 1992.

Hak, J. B., J. H. G. M. Van Beek, M. H. J. Eijgelshoven, and N. Westerhof. Mitochondrial dehydrogenase activity affects adaptation of cardiac oxygen consumption to demand. *Am. J. Physiol.* 264:H448–453, 1993a.

Hak, J. B., J. H. G. M. Van Beek, and N. Westerhof. Acidosis slows the response of oxidative phosphorylation to metabolic demand in isolated rabbit heart. *Pflügers Arch. (Europ. J. Physiol.)* 423:324–329, 1993b.

Hak, J. B., J. H. G. M. Van Beek, M. H. van Wijhe, and N. Westerhof. Dynamics of myocardial lactate efflux after a step in heart rate in isolated rabbit heart. *Am. J. Physiol.* 265:H2081–H2085, 1993c.

Heinrich, R. and T. A. Rapoport. Mathematical analysis of multienzyme systems: II. Steady state and transient control. *Biosystems* 7:130–136, 1975.

King, R. B., J. B. Bassingthwaighte, J. R. S. Hales, and L. B. Rowell. Stability of heterogeneity of myocardial blood flow in normal awake baboons. *Circ. Res.* 57:285–295, 1985.

Lassen, N. A. and W. Perl. *Tracer Kinetic Methods in Medical Physiology.* New York: Raven Press, 1979.

Lassen, N. A., O. Henriksen, and P. Sejrsen. Indicator methods for measurement of organ and tissue blood flow. In: *Handbook of Physiology, Section 2: The Cardiovascular System, Vol. III. Peripheral Circulation and Organ Blood Flow, Part 1,* edited by J. T. Shepherd and F. M. Abboud, pp. 21–63, 1983.

Loiselle, D. S., J. H. G. M. Van Beek, D. A. Mawson, and P. J. Hunter. The surface of the heart leaks oxygen. *News Physiol. Sci.* 10:129–133, 1995.

Mast, F. and G. Elzinga. Recovery heat production of isolated rabbit papillary muscle at 20° C. *Pflügers Arch.* 411:600–605, 1988.

Meier, P. and K. L. Zierler. On the theory of the indicator-dilution method for measurement of blood flow and volume. *J. Appl. Physiol.* 6:731–744, 1954.

Roberts, G. W., K. B. Larson, and E. E. Spaeth. The interpretation of mean transit time measurements for multiphase tissue systems. *J. Theor. Biol.* 39:447–475, 1973.

Rose, C. P. and C. A. Goresky. Limitations of tracer oxygen uptake in the canine coronary circulation. *Circ. Res.* 56:57–71, 1985.

Tian, X., J. H. G. M. Van Beek, and M. H. J. Eijgelshoven. Effects of inotropic stimulation and adenosine on the response time of cardiac mitochondrial oxygen consumption to heart rate steps. *J. Physiol. (London)* 479:132P, 1994.

Van Beek, J. H. G. M. and G. Elzinga. Response time of mitochondrial oxygen consumption to heart rate changes in isolated rabbit heart. In: *Proc. Int. Union Physiol. Sci.,* vol. XVI, XXXth Congr., Vancouver, p. 485, 1986.

Van Beek, J. H. G. M. Fractal models of heterogeneity in organ blood flow. In: *Oxygen Transport in Biological Systems: Modelling of Pathways from Environment to Cell,* edited by S. Egginton and H. Ross. (Society for Experimental Biology Seminar Series Vol. 51). Cambridge: Cambridge University Press, pp. 135–163, 1992.

Van Beek, J. H. G. M. Appendix to: Exogenous carbon substrate supply affects the time course of adaptation of cardiac mitochondrial oxygen consumption to demand in a temperature dependent fashion. In: Dynamics of Cardiac Energy Metabolism. Thesis, M. H. J. Eijgelshoven. Vrije Universiteit, Amsterdam (ISBN 90-9007857-6), 1995, pp. 94–96.

Van Beek, J. H. G. M., J. P. F. Barends, and N. Westerhof. The microvascular unit size for fractal flow heterogeneity relevant for oxygen transport. In: *Oxygen Transport to Tissue XV,* edited by P. Vaupel, R. Zander, and D. F. Bruley, pp. 901–908, 1994.

Van Beek, J. H. G. M., D. S. Loiselle, and N. Westerhof. Calculation of oxygen diffusion across the surface of isolated perfused hearts. *Am. J. Physiol.* 263:H1003–H1010, 1992.

Van Beek, J. H. G. M., J. Bussemaker, J. P. F. Barends, and N. Westerhof. A ^{13}C-NMR technique to determine absolute oxygen consumption in quickly frozen small myocardial samples. *FASEB J.* 10:A325, 1996a.

Van Beek, J. H. G. M., J. B. Hak, M. H. van Wijhe, and N. Westerhof. The control exerted by oxidative phosphorylation on respiration in the intact heart. In: *Biothermokinetics of the Living Cell* edited by H. V. Weserhoff, J. L. Snoep, J. E. Wijker, F. E. Sluse, and B. N. Kholodenko. Amsterdam: BioThermoKinetics Press, pp. 119–123, 1996b.

Van Beek, J. H. G. M., M. D. Osbakken, and B. Chance. Measurement of the oxygenation status of the isolated perfused rat heart using near infrared detection. *Adv. Exp. Med. Biol.* 388:147–154, 1996c.

Van Beek, J. H. G. M., S. A. Roger, and J. B. Bassingthwaighte. Regional myocardial flow heterogeneity explained with fractal networks. *Am. J. Physiol.* 257:H1670–H1680, 1989.

Van Beek, J. H. G. M., X. Tian, C. J. Zuurbier, B. de Groot, C. J. A. van Echteld, M. H. J. Eijgelshoven, and J. B. Hak. Analysis of response time of mitochondrial oxygen consumption. Studies of the dynamic regulation of myocardial oxidative phosphorylation (Review). *Mol. Cell. Biochem.* (in press).

Van Beek, J. H. G. M. and N. Westerhof. Response time of mitochondrial oxygen consumption following stepwise changes in cardiac energy demand. In: *Oxygen Transport to Tissue XII*, edited by J. Piiper, T. K. Goldstick, and M. Meyer. New York: Plenum Press, pp. 415–423, 1990.

Van Beek, J. H. G. M. and N. Westerhof. Response time of cardiac mitochondrial oxygen consumption to heart rate steps. *Am. J. Physiol.* 260:H613–H625, 1991.

Wijhe, M. H. and J. H. G. M. van Beek. Effects of respiratory and metabolic acidosis on oxidative phosphorylation in isolated rabbit heart mitochondria. *Pflügers Arch. (Europ. J. Physiol.)* 430:R174, 1995.

Wittenberg, J. B. Myoglobin facilitated oxygen diffusion and the role of myoglobin in oxygen entry into muscle. *Physiol. Rev.* 50:559–636, 1970.

Zheng, L., A. S. Golub, and R. N. Pittman. Determination of PO_2 and its heterogeneity in single capillaries. *Am. J. Physiol.* 271:H365–H372, 1996.

Zierler, K. L. Theory of the use of arteriovenous concentration differences for measuring metabolism in steady and non-steady states. *J. Clin. Invest.* 40:2111–2125, 1961.

Zuurbier, C. J. and J. H. G. M. van Beek. Brief ischemia and hypoxia result in decreased mitochondrial function and contractility in isolated rabbit heart, but not in its right ventricular papillary muscle. *Pflügers Arch. (Europ. J. Physiol.)* 430 (Suppl.):R20, 1995.

10

Quantitative Assessment of Sites of Adenosine Production in the Heart

Andreas Deussen

Introduction

Adenosine, the dephosphorylation product of adenine nucleotide metabolism, has a multitude of physiological and pharmacological actions. Four important functions shall be exemplified for the heart.

1. The strong coronary dilatory effect was described as early as 1931 by Lindner and Rigler. In conjunction with the experimental observation that in hypoxic and ischemic myocardium large amounts of adenosine are produced, this has led to the hypothesis of metabolic control of coronary blood flow via adenosine by Berne (1963) and Gerlach et al. (1963).
2. The negative chronotropic and dromotropic effects of adenosine were described first by Druri and Szentgyörgyi in 1929. These effects have attracted much notice in the clinical arena in recent years for the treatment of supraventricular tachycardia (Di Marco et al., 1983; Clarke et al., 1987; Belardinelli et al., 1989; Olsson and Pearson, 1990).
3. The antiadrenergic effects of adenosine may diminish the inotropic effects (Schrader et al., 1977a; Dobson, 1983) as well as Purkinje fiber automaticity (Rosen et al., 1983) induced by catecholamines in ventricular tissues.
4. Ischemic preconditioning (Murry et al., 1986) may be mediated by adenosine (Downey et al., 1993). This effect, which has been discovered only in recent years, may render cardiac tissue less vulnerable during a period of subsequent myocardial ischemia.

The effects of adenosine are mediated via membrane surface receptors on the respective cell species. Presently, three different surface receptors (A_1, A_2, A_3) are discerned based on the rank order of potency of specific agonists and antagonists (for a recent overview, see Daly and Jacobson, 1995) as well as amino acid sequencing (for a recent overview, see Jacobson, 1995). The classical notion that adenosine receptors mediate their effects exclusively via the adenylate cyclase system has been modified recently. It is now recognized that the A_1-receptor couples to several signal transduction pathways, including the adenylate cyclase system, the guanylate cyclase system, ion-channels (K, Ca), and phospholipase C

235

(Schwabe, 1991). These secondary effects are mediated via G-proteins following receptor activation.

The most potent stimulus of adenosine production is tissue hypoxia or ischemia (Berne, 1980). Enhanced adenosine production during ischemia has been noted for several decades and quantified in detail first by Gerlach and Deuticke in 1963. A major result of their work on the heart is summarized in Figure 10.1. With the onset of myocardial ischemia the total concentration of myocardial nucleotides falls (upper panel of Figure 10.1). This is associated with a rise of the degradation products which in the rat heart include adenosine, inosine, hypoxanthine, xanthine, and uric acid (lower panel of Figure 10.1). The relationship of the tissue concentrations of the different degradation products is not constant over time, but rather exhibits a characteristic kinetic. Adenosine is a major catabolite during early ischemia, but its fraction decreases with the continuation of ischemia. On the other hand, inosine and its degradation products continue to increase over time. Similar results have been reported for different species by others (Berne, 1963; Schrader et al., 1977b; Jennings et al., 1981; Henrichs et al., 1986).

Enhanced tissue adenosine production is the result of a critical myocardial oxygenation. Recently, Smolenski et al. (1991) and Stumpe and Schrader (1994) provided evidence for a critical pO_2 below which the adenosine concentration increases. In isolated cardiomyocytes, this critical pO_2 is in the range of 5 mmHg at the level of the cell membrane. An increased metabolic rate of oxygen per se does not result in an increase of the adenosine concentration. This view is supported by results obtained in isolated cardiomyocytes (Stumpe and Schrader, 1994) and isolated perfused heart preparations (Bardenheuer and Schrader, 1986; Deussen and Schrader, 1991; Headrick et al., 1991) as well as the heart in vivo (Deussen et al., 1988b; 1991; Kroll and Martin, 1994). Improvement of the oxygen supply to oxygen demand relationship blunts the increase of the adenosine concentration despite larger oxygen flux rates.

From such studies it became clear that under conditions of critically limited oxygen supply a pronounced adenosine production ensues. Under conditions of ischemia or hypoxia, the adenosine formed over time can quantitatively originate only from the intracellular pool of adenine nucleotides. What is less well known is to what extent the different sites and biochemical reactions contribute quantitatively to the overall adenosine production under control conditions. For several reasons the quantitative evaluation of the different sites of adenosine production is not a trivial task:

1. Cardiac adenosine metabolism is highly compartmentalized with respect to intra- and extracellular regions (Rubio et al., 1973; Schütz et al., 1981) as well as the various cell species in the heart (Nees et al., 1985; Deussen et al., 1986).
2. Another difficulty arises from the large fraction of adenosine that exists in a protein bound form under physiological conditions and comprises more than 95% of the total tissue adenosine under control conditions (Olsson et al., 1982; Belloni et al., 1984). Hence, changes of the free, biologically active fraction of

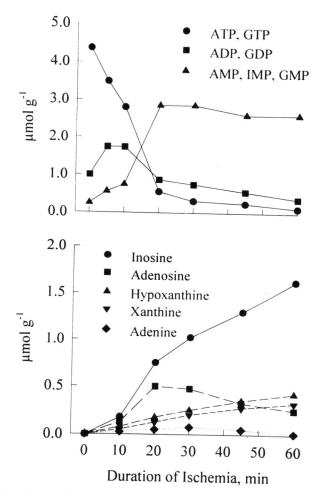

FIGURE 10.1. Myocardial concentrations of adenine nucleotides (upper panel) and degradation products (lower panel). After Gerlach and Deuticke (1963). [With permission of the authors.]

adenosine that may be of physiological importance cannot be detected directly due to the high background signal of protein bound adenosine (Deussen et al., 1988a).

3. Further problems in quantifying cardiac adenosine production rates are brought about by the rapid degradation of adenosine. In human blood plasma, the half-life of tracer adenosine in the presence of a physiological plasma adenosine concentration (50–100 nM) was estimated to be 0.6 s (Möser et al., 1989). Even in the presence of plasma concentrations increased to 1 µM (a plasma concentration that may be found during conditions of tissue ischemia or hypoxia) the half-life was only slightly longer, namely, 1.2 s.

During recent years considerable progress has been made in developing experimental approaches to quantify the different sources of adenosine production. These flux rates may be incorporated into a mass balance model of cardiac adenosine metabolism as it is understood today. The goal of this chapter is to review the experimental strategies that have been developed and apply them to the well-oxygenated heart.

Overview of Adenosine Metabolism

Figure 10.2 summarizes the different biochemical reactions involved in adenosine metabolism. Adenosine is produced in the intracellular region by action of 5′-nucleotidase from 5′-AMP or by action of S-adenosylhomocysteine(SAH)-hydrolase from S-adenosylhomocysteine (SAH). A further source of adenosine is extracellular degradation of 5′-AMP via membrane-bound ecto-5′-nucleotidase. Removal of adenosine in the intracellular region occurs via two pathways: the phosphorylation of adenosine via adenosine kinase to 5′-AMP or the deamination of adenosine to inosine via adenosine deaminase. For some species (e.g., rabbit) an extracellular deamination of adenosine has been suggested (Schrader and West, 1990). However, extracellular deamination of adenosine does not exist or is negligible in the isolated guinea pig heart on which this chapter focusses and therefore is omitted in Figure 10.2. Transmembrane transport of adenosine in the heart is mediated via a nucleoside transporter that permits a facilitated diffusion between the extracellular and intracellular regions along a concentration gradient. Cardiac adenosine metabolism has been reviewed by Schrader (1983), Sparks and Bardenheuer (1986), Belardinelli et al. (1989), and Olsson and Pearson (1990).

For steady-state conditions, the flux rates through the different pathways of adenosine metabolism may be summarized as follows. For the intracellular region,

$$F_c^{nt} + F^{shyd} + F_{in}^{fd} = F^{kin} + F^{ada} + F_{ex}^{fd} + F^{ssyn}, \tag{10.1}$$

and for the extracellular region,

$$F_e^{nt} + F_{ex}^{fd} + F_{in}^{con} = F_{in}^{fd} + F_{out}^{con}. \tag{10.2}$$

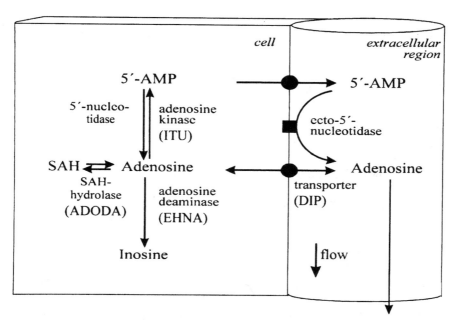

FIGURE 10.2. Overview of biochemical reactions of adenosine metabolism in the guinea pig heart. EHNA = erythro-9-(2-hydroxy-3-nonyl) adenine; ITU = iodotubericidin; ADODA = adenosine dialdehyde; and DIP = dipyridamole. For further details, see text.

The abbreviations are as follows: F_c^{nt} fluxrate through the intracellular 5'-nucleotidase pathway, F_e^{nt} fluxrate through the extracellular 5'-nucleotidase pathway, F^{kin} fluxrate through the adenosine kinase pathway, F^{ada} fluxrate through the adenosine deaminase pathway, F^{ssyn} fluxrate through the SAH-hydrolase pathway in the direction of SAH-synthesis, F^{shyd} fluxrate through the SAH-hydrolase pathway in the direction of SAH-hydrolysis, F_{in}^{fd} facilitated diffusional flux through the cell membrane from extracellular to intracellular, F_{ex}^{fd} facilitated diffusional flux through the cell membrane from intracellular to extracellular, F_{in}^{con} convective inflow via the coronary circulation (coronary arterial supply), and F_{out}^{con} convective outflow via the coronary circulation (coronary venous release rate). The fluxes may be summarized in the following balance equation with $F_{in}^{con} = 0$:

$$0 = F_c^{nt} + F_e^{nt} + F^{shyd} - F^{ssyn} - F^{kin} - F^{ada} - F_{out}^{con}. \qquad (10.3)$$

Figure 10.2 is a simplification of cardiac adenosine metabolism since it considers only two regions, an intracellular and an extracellular region. Although different cell species (e.g., cardiomyocytes, endothelial cells, smooth muscle cells) may contribute to different extents to the overall adenosine production and removal, for reasons of simplicity a reductionist approach using a two-region model is followed for the larger part of this chapter. For the calculation of interstitial adenosine concentrations, a more detailed four region model will be used

(Kroll et al., 1992). Details will be noted in the text. A four-region adenosine model is also considered in Chapter 11 (Kroll and Bassingthwaighte).

Methodical Considerations

The simplest structural organization that contains almost the complete scenario of cardiac adenosine metabolism and simultaneously permits a functional assessment of possible biological effects, for example, on coronary flow, heart rate, or tolerance toward ischemia, is the isolated perfused heart. For the guinea pig heart a rather advanced data set of adenosine metabolism is available. It includes the different apparent K_m- and V_{max}-values of the enzymes involved in adenosine metabolism. Furthermore, measurements of the venous adenosine release from the guinea pig heart have been widely reported in the literature and thus may serve as references. The isolated heart perfusion has been described in detail before (Bünger et al., 1975a; Deussen et al., 1988a) and is summarized here briefly. Hearts are perfused according to the Langendorff technique via the aortic root at a constant pressure of 60–70 cm H_2O (45–53 mmHg), which is the normal systemic arterial blood pressure of guinea pigs (D'Souza and Biggs, 1985). The perfusion medium is usually a modified Krebs buffer (Bünger et al., 1975b) equilibrated with carbogen [O_2:CO_2 = 95:5 (vol:vol)]. A latex balloon in the left ventricular cavity permits the measurement of isovolumic pressure development, dP/dt, and heart rate. Coronary flow is measured with an electromagnetic or Doppler flow probe inserted into the perfusion cannula or by gravimetry after the venous effluent perfusate is collected. A catheter advanced via the pulmonary artery into the right ventricular cavity is used to collect venous effluent perfusate for the measurement of adenosine concentration and, anaerobically, for the measurement of pO_2.

As indices of the cardiac adenosine concentration, adenosine release and the rate of SAH accumulation are determined. Because the arterial inflow of adenosine concentration is zero in the isolated perfused heart, the release rate can be calculated from the venous outflow concentration and the coronary flow ($F_{out}^{con} = F \cdot C_{out}$), where F_{out}^{con} is the venous release rate, F is the average coronary (conductive) flow, and C_{out} is the venous adenosine concentration.

The rate of accumulation of SAH in the presence of homocysteine (200 μM intracoronary concentration) is used as an index of the free intracellular adenosine concentration (Deussen et al., 1988a). The principle of the SAH technique is illustrated in Figure 10.3. Under control conditions, the net flux through this pathway is in the direction of SAH-hydrolysis, because SAH is continuously formed via the transmethylation pathway from S-adenosylmethionine (SAM), and the reaction products adenosine and homocysteine are rapidly metabolized further. In the presence of high homocysteine concentrations, however, the net direction of the SAH-hydrolase reaction is readily reversed toward the production of SAH from adenosine and homocysteine. The free intracellular adenosine con-

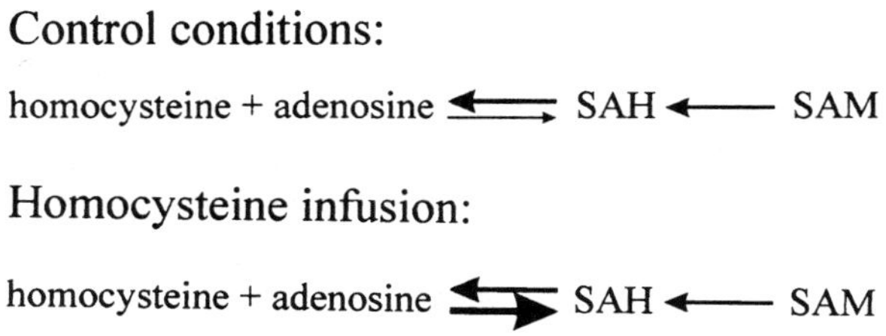

Control conditions:

homocysteine + adenosine $\rightleftharpoons$ SAH $\longleftarrow$ SAM

Homocysteine infusion:

homocysteine + adenosine $\rightleftharpoons$ SAH $\longleftarrow$ SAM

FIGURE 10.3. Schematic representation of the SAH-technique. The upper panel depicts the situation of normal control conditions, i.e., endogenous homocysteine is derived via the transmethylation pathway from S-adenosylmethionine (SAM) via S-adenosylhomocysteine (SAH). The lower panel illustrates the situation when exogenous homocysteine is supplied in excess. Under this condition, the net reaction of the SAH-hydrolase is in the direction of SAH synthesis. The free intracellular adenosine concentration is then a rate-limiting factor for the rate of SAH production. (For further details, see text.)

centration is then a rate limiting factor for the synthesis of SAH. The relationship between the free intracellular adenosine concentration and the rate of SAH accumulation may be described (Kroll et al., 1992) by the equation:

$$V_c \cdot \frac{\partial [S]_c}{\partial t} = + F^{ssyn} - F^{shyd} + \frac{S^S}{1 + [S]/K_i} , \qquad (10.4)$$

where V_c is the intracellular volume of myocardial tissue (0.65 ml/g), $\partial [S]_c/\partial t$ is the change of the SAH concentration over time (mol $\cdot$ ml^{-1} $\cdot$ min^{-1}), F^{ssyn} is the flux rate through the SAH-hydrolase reaction in the direction of SAH synthesis (mol $\cdot$ g^{-1} $\cdot$ min^{-1}), F^{shyd} is the flux through the SAH hydrolase reaction in the direction of adenosine and homocysteine production (mol $\cdot$ g^{-1} $\cdot$ min^{-1}), and S^S is the rate of SAH production from SAM via transmethylation under control conditions. The final term in the equation, $S^S/(1 + [S] / K_i)$, describes transmethylation (Figure 10.2) given in mol $\cdot$ g^{-1} $\cdot$ min^{-1}, which may decrease due to product inhibition by SAH (Borchardt and Wu, 1975).

The SAH-hydrolase reaction may be described by a bi-uni-reactant reaction. This model assumes random order and equilibrium binding of the different substrates (adenosine, homocysteine, SAH). The equations for the unidirectional flux rates of the SAH-hydrolase may be stated as follows:

$$F^{ssyn} = \frac{V^{ssyn} \cdot \dfrac{[H] \cdot [A]}{K^{sH} \cdot K^{sA}}}{1 + \dfrac{[H]}{K^{sH}} + \dfrac{[A]}{K^{sA}} + \dfrac{[H] \cdot [A]}{K^{sH} \cdot K^{sA}} + \dfrac{[S]}{K^{sS}}} \qquad (10.5)$$

and

$$F^{shyd} = \frac{V^{shyd} \cdot \dfrac{[S]}{K^{sS}}}{1 + \dfrac{[H]}{K^{sH}} + \dfrac{[A]}{K^{sA}} + \dfrac{[H] \cdot [A]}{K^{sH} \cdot K^{sA}} + \dfrac{[S]}{K^{sS}}} \qquad (10.6)$$

where F^{ssyn} and F^{shyd} are the unidirectional rates of the SAHH reaction in the synthetic and hydrolytic directions; K^{sA}, K^{sH}, and K^{sS} are the K_m-values (μM) for adenosine, homocysteine, and SAH; and V^{ssyn} and V^{shyd} are the V_{max}-values (mol · g^{-1} · min^{-1}) for the synthetic and hydrolytic reactions, respectively. [A], [H], and [S] denote the substrate concentrations of adenosine, homocysteine, and SAH.

The apparent K_m-values for adenosine, homocysteine, and SAH have been determined (Schrader et al., 1981; Borst et al., 1992). The values used for the model calculations are 2 μM for adenosine, 150 μM for homocysteine, and 3.75 μM for SAH. The V_{max} for synthesis may be taken to be 55 nmol · min^{-1} · g^{-1}. The V_{max} of SAH-hydrolase in hydrolytic direction is calculated from the Haldane equation:

$$\frac{V^{shyd}}{V^{ssyn}} = \frac{K_{eq} \cdot K^{sS}}{K^{sA} \cdot K^{sH}}. \tag{10.7}$$

The K_{eq} has been reported to be 0.8 μM (de la Haba and Cantoni, 1959; Ueland and Saebo, 1979). The ratio of the V_{max} values is then 0.01. This relationship is kept constant.

For the assessment of the different production terms of adenosine, steady-state approaches may be used in which the different pathways of adenosine metabolization and/or the membrane transport are blocked using specific and potent inhibitors. To block adenosine deaminase, erythro-9-(2-hydroxy-3-nonyl)adenine (EHNA) at an intracoronary concentration of 5 μM is used. Adenosine kinase is blocked with iodotubericidin (ITU) at a concentration of 10 μM. SAH-hydrolase may be blocked with adenosine dialdehyde (10 μM). The membrane transport of adenosine is inhibited with dipyridamole (DIP) at a concentration of 2 μM. At these concentrations the inhibitory actions are specific for the respective enzymes with efficacies higher than 95% in all cases (Deussen et al., 1989, 1993; Kroll et al., 1993).

The analytical techniques for adenosine and SAH measurement have been reported in detail before (Deussen et al., 1988a). In brief, the adenosine concentration is measured in concentrated, desalted samples of the effluent perfusate and SAH is measured in acid tissue extracts. Both metabolites are quantitated using sensitive HPLC techniques with a detection limit of approximately 5 pmol per injected sample.

Quantification of Adenosine Production Terms

Total Cardiac Adenosine Production Under Well-Oxygenated Conditions

When isolated hearts are perfused with a buffer lacking the oxygen-carrying capacity of red cells, it is important to define the physiological status of the preparation when quantifying a metabolite as closely related to cardiac energy metabolism as adenosine. At perfusion pressures of 60–70 cm H_2O (arterial pO_2

approx. 600 mmHg), isolated perfused guinea pig hearts exhibit a coronary flow of 5–7 ml · min⁻¹ · g⁻¹, and an oxygen consumption of 50–80 µl · min⁻¹ · g⁻¹ (Deussen et al., 1988a, 1991). Reasonable criteria for a metabolically and functionally stable preparation are: 1) an adenosine release of less than 100 pmol · min⁻¹ · g⁻¹ 30 min after start of the isolated perfusion, and 2) a peak reactive flow response of at least twice the control flow before occlusion following a 20-s coronary flow stop.

Under such (control) conditions, the adenosine release with the effluent coronary perfusate is in the order of 40–80 pmol · min⁻¹ · g⁻¹ (Deussen et al., 1988a, 1991). The concentration of SAH in the absence of homocysteine is 1.1 ± 0.5 (SD) nmol/g and following 20 min of homocysteine infusion (200 µM) it is enhanced to 6.7 ± 1.3 nmol/g. The global tissue adenosine concentration is 2.0 ± 0.5 nmol/g (Deussen et al., 1988a).

In Figure 10.4 the effect of simultaneous inhibition of adenosine deaminase and adenosine kinase on adenosine release and coronary flow is shown for a representative experiment. This intervention results in a profound, more than 15-fold increase of cardiac adenosine release, from 64 to 1081 pmol · min⁻¹ · g⁻¹. Although coronary blood flow doubles, the intervention does not change myocardial oxygen consumption. Typically, there is also a drastic fall in heart rate, which was prevented in this experiment by atrial pacing to avoid secondary effects of increased membrane receptor stimulation. Assuming that the fluxes through the adenosine deaminase and adenosine kinase pathways are essentially zero under these conditions, Eq. (10.3) may be restated:

$$F^{con}_{out} = F^{nt}_c + F^{nt}_e + F^{sahh} - F^{sahs} .$$

(10.8)

If homocysteine is absent as in the experiment shown in Figure 10.4, the net direction of the SAH-hydrolase reaction is toward adenosine production, and thus the venous release equals the sum of the intracellular net production from SAH, cytosolic 5'-AMP as well as extracellular 5'-AMP.

The experiment shown in Figure 10.4 indicates that the total cardiac adenosine production must be at least 1081 pmol · min⁻¹ · g⁻¹ under the given conditions. This is a lower estimate of the total production rate, because it cannot be excluded that a residual flux through the adenosine deaminase or adenosine kinase pathways still occurred. However, given the efficacy of both enzyme inhibitors (Kroll et al., 1993), such a residual flux should be small as long as the adenosine concentration at the catalytic site of the enzyme does not increase drastically. Concerning the preferential site of adenosine production, the effect of dipyridamole on the release of adenosine in the presence of EHNA and ITU is of major importance (Figure 10.4). Since membrane transport block reduces the release, it must be assumed that the larger fraction of adenosine is produced inside the cellular regions.

A comparison of the venous release rate of adenosine under control conditions (64 pmol · min⁻¹ · g⁻¹) with the production rate (1081 pmol · min⁻¹ · g⁻¹) indicates that in this example almost 95% of the produced adenosine are either rephosphorylated to 5'-AMP or deaminated to inosine. A previous study has shown that a

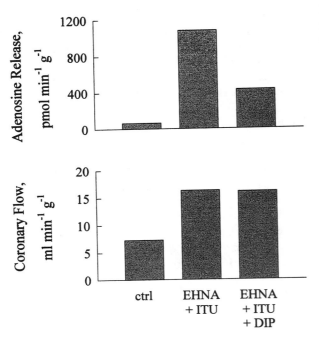

FIGURE 10.4. Effects of inhibition of adenosine deaminase by EHNA (5 μM), adenosine kinase by ITU (10 μM), and additional blockage of membrane transporter by DIP (2 μM) on the coronary venous release of adenosine and coronary flow. Heart paced at 250 bpm. (For abbreviations, see legend of Figure 10.2.)

single administration of EHNA increases the adenosine release from the isolated guinea pig heart rather slightly (Deussen et al., 1989). This suggests that the majority of the adenosine is rephosphorylated to 5'-AMP. This concept is supported by two recent studies in which the blockage of adenosine kinase enhanced cardiac adenosine release markedly (Ely et al., 1992; Kroll et al., 1993). Similar conclusions have been derived from experiments conducted on macrovascular endothelial cells (Deussen et al., 1993).

Extracellular Adenosine Production During Well-Oxygenated Conditions

In Figure 10.5 a typical experiment is shown in which the membrane transport blocker dipyridamole (2 μM) is infused into the arterial inflow of a guinea pig heart. Steady-state release is 42 pmol · min^{-1} · g^{-1} before dipyridamole and increases to 186 pmol · min^{-1} · g^{-1} during blockage of the membrane transporter.

There are several possibilities to explain this effect. The increase of the adenosine release during dipyridamole infusion could be due to an increased cellular production rate of adenosine. An enhanced release rate from the intracellular region under the given conditions would require the intracellular concentration to increase. In order to assess the effect of dipyridamole on the cytosolic adenosine concentration, the accumulation of SAH can be determined in the absence and presence of dipyridamole. Because the SAH concentrations are typically rather

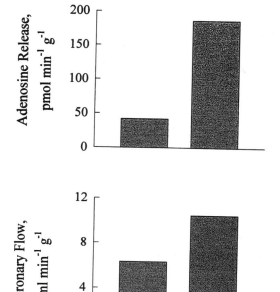

FIGURE 10.5. Effects of the adenosine transport blocker dipyridamole (DIP; 2 μM) on adenosine release and coronary flow. Heart paced at 250 bpm.

similar under these conditions (differences are less than 10%; unpublished data), it is unlikely that an enhanced intracellular adenosine production is a good explanation for the increased adenosine release during dipyridamole infusion.

Another concept previously suggested was that dipyridamole blocks the uptake of adenosine specifically but leaves the cellular adenosine release unaffected (Mustafa, 1979). According to this hypothesis, the enhanced coronary venous release could be interpreted as the result of an asymmetric blockage of the transporter. This possibility, however, seems unlikely because, under the conditions of an enhanced intracellular adenosine concentration, dipyridamole reduced the adenosine release by 60% (Figure 10.4).

Presently, the best concept to explain the enhanced release of adenosine during dipyridamole infusion seems to be the assumption of extracellular production of adenosine that is prevented from reaching its site of intracellular metabolism by the transport blocker. According to uptake studies reported by Clanachan et al. (1987), a dipyridamole concentration of 1 μM should block the adenosine transport in the guinea pig heart almost completely under physiological concentrations of adenosine. Thus, the release rate in the presence of dipyridamole shown in Figure 10.5 may be taken as a minimum figure of the rate of extracellular adenosine production.

Ecto-5'-nucleotidase as a site of extracellular adenosine production in the heart is supported by histochemical studies (Rubio et al., 1973). For endothelial cells as well as for smooth muscle cells, the rapid extracellular degradation of exogenously applied adenine nucleotides to adenosine has been shown (Pearson et al., 1980; Fleetwood et al., 1989). Borst and Schrader (1991) have quantified the adenine nucleotide release of the isolated guinea pig heart to be 430 ± 40 pmol · $min^{-1} \cdot g^{-1}$ under normoxic conditions. This production rate can quantitatively explain the adenosine release rate in the presence of dipyridamole (Figure 10.5). It presently remains unknown how adenine nucleotides are transported across the cell membrane and whether a distinct cell species is responsible for the release. The multidrug resistance (mdr1) gene product, a P-glycoprotein, which has been postulated to mediate nucleotide efflux from cells (Abraham et al., 1993), has so far not been reported for cardiac tissues (Thiebaut et al., 1987; Cordon-Cardo et al., 1989). In this context an early hypothesis also should be called upon. In 1978, Frick and Lowenstein proposed the model of vectorial production of adenosine at the level of the cell membrane. In essence, the concept is that membrane bound 5'-nucleotidase can also act as an adenosine translocase. The concept is based on the experimental observation that, during coronary co-infusion of differently labeled AMP and adenosine, the adenosine derived from AMP enters the cells more readily than preexisting adenosine. It is unclear whether such a mechanism could also facilitate efflux from AMP-derived adenosine from cardiac cells.

Intracellular Adenosine Production During Well-Oxygenated Conditions

In order to support the view of a preferential intracellular adenosine production rate, the measurement of SAH accumulation in the presence of a membrane transport blocker may be helpful. For an experimental condition in which membrane transport, the phosphorylation of adenosine to 5'-AMP, and the deamination to inosine are simultaneously inhibited, Eq. (10.1) takes the form:

$$F_{nt}^c = F_{sahs} - F_{sahh} + d[A]/dt . \tag{10.9}$$

In the presence of enhanced concentrations of homocysteine (200 μM), the net reaction of the SAH-hydrolase is toward SAH synthesis, and the production of adenosine from intracellular 5'-AMP will be largely recovered in the SAH fraction. The last term in Eq. (10.9), $d[A]/dt$, describes the increase of the cellular adenosine concentration that may result under this condition, if the net flux through the SAH-hydrolase pathway toward SAH production is smaller than the flux through the 5'-nucleotidase pathway. In other words, $d[A]/dt$ would indicate an augmentation of the substrate concentration for the SAH-hydrolase reaction. In Figure 10.6 the SAH accumulation during 20 min of homocysteine infusion (Deussen et al., 1988a) is compared with that during homocysteine infusion in the presence of ITU, EHNA, and dipyridamole. While in the presence of homocysteine only the SAH concentration after 20 min of homocysteine was 6.7 ± 1.3 nmol/g, it was 15.1 ± 3.1 nmol/g in the presence of ITU, EHNA, and dipyridamole. The global tissue adenosine concentration was enhanced from 2.0 to

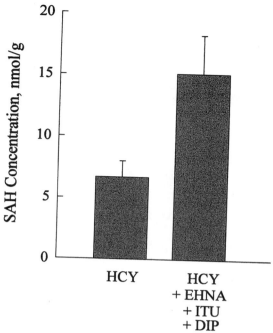

FIGURE 10.6. SAH accumulation during 20-min infusion of homocysteine (HCY; 200 μM intracoronary concentration). In experiments shown on the left, only homocysteine was infused; in those shown on the right, homocysteine was infused in the presence of EHNA (5 μM), ITU (10 μM), and DIP (2 μM). (For abbreviations, see legend of Figure 10.2.)

6.1 nmol/g. Thus, the difference of the SAH concentrations between the two experimental groups plus the difference of the adenosine tissue concentration can serve as a lower estimate of the cytosolic production rate, approximately 625 pmol · min⁻¹ · g⁻¹. This estimate is lower but still in the range of that obtained from the adenosine release experiments (1081–180 = 901 pmol · min⁻¹ · g⁻¹). The lower estimate is probably due to the fact that the adenosine concentration is largely enhanced under this condition and therefore the effectiveness of the enzyme blockage by EHNA, ITU, and dipyridamole diminished. As a result, the rest fluxes through these routes are increased under this condition.

Intracellular Production from SAH

Besides 5'-AMP, SAH may serve as a continuous source of intracellular adenosine (Schrader et al., 1981). The SAH is formed from continuously ongoing cellular transmethylation processes during which S-adenosylmethionine is demethylated to S-adenosylhomocysteine (SAH) (Ueland, 1982). The quantitative contribution of the transmethylation pathway to adenosine production in well-oxygenated conditions has been estimated (Deussen et al., 1989) using the SAH-hydrolase

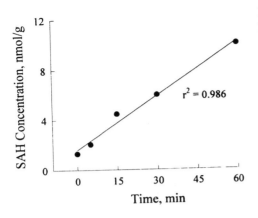

FIGURE 10.7. Effect of the SAH-hydrolase blocker adenosine dialdehyde (10 μM) on the guinea pig heart SAH concentration over 60 min. (Data are from Deussen et al., 1989.)

blocker adenosine dialdehyde (10 μM). In Figure 10.7, the effects of the blocker on the tissue SAH concentration are shown. The time-dependent accumulation of SAH is almost linear over 60 min with an average rate of 160 pmol $\cdot$ min^{-1} $\cdot$ g^{-1}. According to this estimate, the production rate of adenosine from SAH may account for 15% of the global cardiac adenosine production rate.

Contribution of Endothelial Cells to Global Cardiac Adenosine Production

During infusion of radioactive adenosine into the coronary circulation, a considerable fraction is extracted and incorporated into the tissue adenine nucleotide pool (Schrader and Gerlach, 1976). Autoradiographic studies have shown that this uptake of radioactive adenosine is exclusively brought about by the consumption of coronary endothelial cells (Nees et al., 1985). Based on these findings, a prelabeling technique of the coronary endothelial adenine nucleotide pool has been devised to assess the contribution of the coronary endothelium to global cardiac adenosine production (Deussen et al., 1986; Kroll et al., 1987). Isolated hearts from guinea pigs were perfused with tritiated adenosine with a high specific activity (40–60 Ci/mmol) at a low intracoronary concentration (approx. 10 nM) for 35 min. After a washout period the hearts released adenosine at a constant rate and with a measureable specific activity. The total adenosine fraction released under such conditions is the sum of the adenosine released from the endothelial cells that has a certain specific radioactivity and that released from the cardiomyocytes, the specific activity of which is assumed to be zero. Thus, the specific activity measured in the venous effluent perfusate is a weighted average of that released from endothelial cells and cardiomyocytes. In order to obtain a measure of the specific activity of the adenine nucleotide pool of the endothelium, the release of cyclic AMP from the endothelial cell compartment was stimulated using the adenosine A_2-receptor agonist NECA (Kroll et al., 1987). From the relation of the specific radioactivity of cyclic AMP released during NECA stimulation and that of adenosine released under control conditions, the contribution of the endothelial cell region to the global cardiac adenosine release was estimated to be 14% (Kroll et al., 1987). This estimate rests on the assumptions that: 1) the radioactive adenosine and cAMP released into the effluent perfusate have the same precursor pool, and 2) during stimulation the cyclic AMP is released from the endothelial cell region only. Becker and Gerlach (1987) have independently estimated an endothelial contribution to venous adenosine release of 25%, which is in the range of that determined by Kroll et al. (1987). Thus, in comparison to their volume fraction (2.8% of total myocardial water space), endothelial cells contribute excessively to the overall venous adenosine release under such conditions. In absolute terms, however, it must be kept in mind that 75–86% of the adenosine released during well-oxygenated conditions were derived from an unlabeled cell region, most likely the cardiomyocytes.

These indirect estimates are in reasonable agreement with recent measurements of the adenosine production of isolated endothelial cells. Mattig and Deussen (1995) have obtained for macrovascular endothelial cells from pig aorta a global production rate of 9.2 pmol adenosine $\cdot$ min^{-1} $\cdot$ µl^{-1} endothelial cell volume. From this total production 8.8 $\cdot$ min^{-1} $\cdot$ µl^{-1} are further metabolized inside the endothelial cells. Hence, only 0.4 pmol $\cdot$ min^{-1} $\cdot$ µl^{-1} escapes metabolism and is released from the cell column. Since the heart contains about 28 µl of endothelial cell volume per 1 ml of myocardium, the above estimate of adenosine release from isolated endothelial cells would translate to an endothelial adenosine release of approximately 11 pmol $\cdot$ min^{-1} $\cdot$ g^{-1}. This value is in the range of 14–25% of the control venous release of adenosine suggested by the isotope dilution experiments reported above (Becker and Gerlach, 1987; Kroll et al., 1987). Future studies that should ideally be conducted on endothelial cells from guinea pig heart must confirm this view.

Calculation of the Interstitial Adenosine Concentration from the Venous Outflow Concentration of Adenosine

The extracellular adenosine production rate represents about 15% of the total adenosine production rate during well-oxygenated conditions. The question arises whether such a small fraction of the overall adenosine production could be of significance for inducing vasodilation or bradycardia. In order to obtain insight into this, in a first step the interstitial concentration of adenosine is calculated for control conditions as well as for the infusion of dipyridamole. The calculations are based on a five-region, axially distributed model of adenosine transport and metabolism recently published (Kroll et al., 1992; see also Chapter 11). For a given parameter set, which is in agreement with experimental results, this model is used to fit the venous outflow concentrations of adenosine and thereby to predict the respective interstitial concentration.

The five regions of the model are the capillary region, the endothelial cell region, the interstitial fluid region, and the parenchymal cell region arranged in a concentric manner around the capillary region, which is in the center. The fifth region is the transudate region, which is necessary when the adenosine concentration of epicardial transudate (Headrick et al., 1991) is modeled. This region is of no importance for the following considerations and therefore is omitted. Flow in the capillary region is modeled by the Langrangian sliding fluid element approach to solve for the effects of axial convection and radial exchange (Bassingthwaighte et al., 1992). Exchange between regions is described by linear permeability–surface area (PS) products. The production terms within the cellular regions are constant inputs (zero-order terms). The terms that describe the metabolization of adenosine, homocysteine, and SAH are Michaelis–Menten-type kinetics.

The actual coronary flow rates are measured as well as the coronary venous adenosine concentrations. The volumes of the different model regions are chosen in agreement with anatomical and tracer distribution estimates (Bassingthwaighte and Goresky, 1984). The cytosolic production from 5′-AMP is set to 700 and 40 pmol · min^{-1} · g^{-1} for the cardiomyocyte and endothelial cell regions, respectively. These discrete values are chosen because the volumes of distribution of the cardiomyocytes and the endothelial cells are approximately 0.60 and 0.03 ml/g, respectively. Hence, the endothelial contribution is chosen to be 1/20 of the global adenosine production. Similarly, the transmethylation rates in the cardiomyocyte and endothelial cell regions are set to 152 and 8 pmol · min^{-1} · g^{-1} in accordance with experimental results (Figure 10.7). The extracellular production rate of adenosine is assumed to equal the release rate in the presence of dipyridamole and set to 180 pmol · min^{-1} · g^{-1}. Thus, the total adenosine production rate (1080 pmol · min^{-1} · g^{-1}) is constrained to the overall experimental adenosine production rate as taken from the experiment shown in Figure 10.4. For control conditions, the membrane PS-values used for the luminal and abluminal endothelial cell membranes are 4.5 ml · min^{-1} · g^{-1}. The membrane PS-value for the cardiomyocyte membrane is 2.5 ml · min^{-1} · g^{-1}, as is the PS-value for the interendothelial cell clefts. These assumptions are identical to those made earlier (Kroll et al., 1992). The V_{max}-values for the adenosine kinase are adjusted (free parameters) so that the coronary venous concentration of adenosine calculated by the model fits the measured concentration in the absence of dipyridamole within 20%. This is achieved at a V_{max}-value for adenosine kinase in the endothelial cell region of 150 nmol · min^{-1} · g^{-1}. The cardiomyocyte V_{max}-value was set to 60 nmol · min^{-1} · g^{-1}. Smaller V_{max}-values lead to an overestimation of the venous outflow adenosine concentration.

With this parameter set the model calculates a mean capillary adenosine concentration (spatial average along capillary length) of 6.0 nM and an interstitial adenosine concentration of 25.0 nM. The concentration of the endothelial cell region is 2.5 nM, and that of the cardiomyocyte region is 21.0 nM. Thus, the

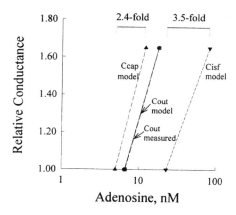

FIGURE 10.8. Modeling results of the capillary and interstitial concentrations of adenosine in the absence and presence of dipyridamole. Adenosine outflow concentration from the heart is measured and the model is used to fit these experimentally measured values. The details of the model assumptions are given in the text. The results indicate that rather small increases of the capillary or interstitial concentration of adenosine are associated with pronounced coronary vasodilation that was measured. This suggests that the dose–response curve for adenosine is steep in the guinea pig heart.

intracellular concentrations are below those of the adjacent extracellular regions; that is, the model describes a concentration gradient of adenosine from extra- to intracellular for endothelial cells as well as cardiomyocytes.

The estimates for the capillary and interstitial concentrations of adenosine in the presence of dipyridamole may be simulated by decreasing the membrane PS-values to zero. Simultaneously, flow is adjusted to that measured during dipyridamole (in the experiment shown in Figure 10.5 an increase from 6.3 to 10.5 ml · min^{-1} · g^{-1}). Hence, the estimates are 11.9 and 81.0 nM for the capillary and interstitial adenosine concentrations, respectively.

The results are illustrated in Figure 10.8 in comparison with the measured increase of the coronary conductance (calculated as the ratio of coronary flow rate/coronary perfusion pressure). It suggests that a 65% increase of coronary conductance is associated with a 2-fold increase of the capillary and a 3.2-fold increase of the interstitial adenosine concentration.

Wangler et al. (1989) have used another approach to simulate the enhanced adenosine release from the isolated guinea pig heart perfused with dipyridamole. They assumed that the membrane transport was inhibited only in the endothelial cell whereas cardiomyocytes continued to release adenosine; this model was originally proposed by Newby (1986). Although this assumption of the mechanism by which the extracellular concentration of adenosine increased during dipyridamole differed from that put forward in this chapter, the modeling results were directionally similar to those shown in Figure 10.8. Their estimate of the capillary and interstitial adenosine concentrations under control conditions were 3.6 and 6.8 nM, while those during dipyridamole infusion were 44 and 191 nM.

Although there is evidence for net uptake of adenosine under resting, well-oxygenated conditions in the isolated perfused guinea pig heart, this does not mean that all of the venous adenosine is derived from an extracellular source. There may be a continuous bidirectional exchange of adenosine between extra-

and intracellular regions. It is emphasized, however, that the sum of the absolute fluxes from extra- to intracellular exceeds those in the opposite direction.

Significance of Extracellular Adenosine Production

The possibility of a causal relationship between the increased adenosine concentration and the enhanced coronary conductivity can be evaluated using intracoronary infusion of adenosine deaminase (Kroll et al., 1993). As evident from Figure 10.9, the flow increase induced by the infusion of dipyridamole can be largely eliminated by the simultaneous infusion of adenosine deaminase (1 U/min). In contrast, when the coronary flow increase is induced by the infusion of bradykinin, which acts in the guinea pig heart via stimulation of nitric oxide production (Kelm and Schrader, 1990), adenosine deaminase does not significantly affect the flow response. Hence, the coronary vasodilation observed during the infusion of dipyridamole is mediated via the enhanced adenosine concentration, which is in agreement with previous studies (Scholtholt et al., 1972; Bünger et al., 1975b). Clemo and Belardinelli (1986) have previously demonstrated that the increased atria-to-*His*-bundle interval prolongation observed in the hypoxic guinea pig heart during application of dipyridamole can effectively be blunted with adenosine deaminase or the adenosine receptor blocker 8-(p-sulfophenyl)theophylline.

The above-calculated dose–response relationship suggests that a 3.5-fold increase of the interstitial and alternatively a 2.4-fold increase of the capillary adenosine concentration are responsible for the observed flow increase. Because adenosine receptors have been localized on endothelial cells as well as smooth muscle cells, it cannot be readily decided whether the enhanced interstitial or the enhanced intracapillary concentration of adenosine better reflects the agonist concentration responsible for the observed vasodilatation. The vasodilation mediated via smooth muscle may be brought about by A_1- or A_2-receptor activation (Olsson and Pearson, 1990; Mustafa et al., 1995). An even more complicating factor is that the dose–response relationships may differ between both receptor subtypes. Alternatively, the coronary dilation may be brought about by an action of adenosine on endothelial adenosine receptors, either from the luminal or the abluminal side. Endothelial cells have adenosine A_2-receptors (Goldman et al., 1983; DesRosiers and Nees, 1987; Schiele and Schwabe, 1994). It is unexplained so far, however, how the endothelial receptor activation is transmitted to the smooth muscle cell. High molecular weight analogues of adenosine (MW ≥ 1000 daltons) cause a strong coronary vasodilatation similar to that induced by adenosine (Olsson et al., 1976). This could indicate that a luminal adenosine receptor is responsible for the vasodilatation seen during adenosine infusion rather than activation of the abluminal (endothelial or smooth muscle) receptors. However, before this conclusion may be drawn it is necessary to obtain a more detailed picture of the microvascular exchange processes of these compounds.

An interesting observation in this context is that the plasma adenosine concentration in humans during intravenous dipyridamole is doubled only when the

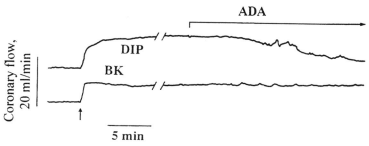

FIGURE 10.9. Effect of intracoronary infusion of adenosine deaminase (ADA; 1 U per min) on coronary flow enhanced by dipyridamole (DIP; 2 μM) or bradykinin (BK; 0.1 μM). Note that adenosine deaminase treatment reduced the flow response during dipyridamole but not the flow response during bradykinin.

flow increase is considerable (Sollevi, 1986; Möser et al., 1989). Besides a rapid metabolism of adenosine in human blood plasma (Möser et al., 1989), the small absolute increase of the adenosine concentration associated with strong vasodilatation may also reflect a steep dose–response curve for adenosine as calculated in this chapter for the isolated perfused guinea pig heart.

The vasodilatation observed during infusion of a membrane transport blocker is close to maximal. The release rate of adenosine determined in the presence of blockage of adenosine deaminase and adenosine kinase exceeds that during blockage of the transporter more than sixfold. This means that, extrapolated to the body circulation, the global adenosine produced under resting control conditions might be far above the rate that would be necessary to provoke severe cardiovascular hypotension. In fact, Figure 10.4 shows that coronary vasodilation in the guinea pig heart fully persists if adenosine release declines following the infusion of dipyridamole in the presence of blocked intracellular metabolization. From this point of view the high capacity and efficacy of the pathways of adenosine metabolization are not only important to salvage adenosine and reduce the energy costs for a possible de novo synthesis of the adenine moiety (Kroll et al., 1993), but they are crucial for the autoregulatory functions of vascular networks within the body circulation.

Conclusions

The above studies show that it is feasible to describe adenosine production rates in the heart in a quantitative manner (Figure 10.10). In the example of the isolated perfused guinea pig heart, the overall cardiac adenosine production rate was deteremined to be at least 1100 pmol · min^{-1} · g^{-1} under well-oxygenated conditions. This global adenosine production can be fractionated into intra- and extracellular contributions. The highest fractional production rate occurs inside the cells. It is estimated to be 0.85 with the global production rate set to 1.0.

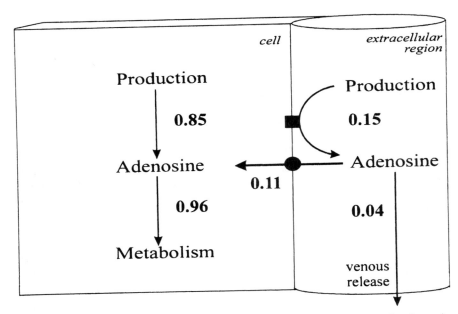

FIGURE 10.10. Flux rates of adenosine metabolism in the well-oxygenated guinea pig heart. Rates are normalized to an overall production rate of 1.0.

Intracellular adenosine may be derived from two biochemical reactions: dephosphorylation of cytosolic 5′-AMP or hydrolysis of SAH. The latter source may account for 160 pmol · min^{-1} · g^{-1} or 0.15 of global production. Thus, the major site of intracellular adenosine production seems to be the cytosolic dephosphorylation of 5′-AMP, which is equivalent to 0.70 of global production. The difference between global and intracellular production is accounted for by the extracellular production from 5′-AMP, which is equivalent to 0.15 of global production.

The continuous formation of adenosine is balanced by an effective metabolization of the nucleoside which in the guinea pig heart occurs exclusively intracellularly. As the intracellular rate of adenosine metabolization exceeds that of intracellular production, and due to the fact that extracellular production occurs in the absence of extracellular metabolization, the intracellular adenosine concentration is normally below that of the interstitial fluid region. Hence, there is a net uptake of adenosine across the cell membrane facilitated by the nucleoside carrier. The net flux rate fraction from extra- to intracellular is estimated to be 0.11 of the global production. Consequently, a fraction of 0.04 is normally released with the coronary effluent perfusate.

Although adenosine production in the extracellular region is small in comparison with the overall cardiac production (15%), extracellular adenosine can evoke significant vasodilatation or bradycardia if access to the intracellular site of metabolism is blocked. This model can explain the well-known clinical observation that dipyridamole induces strong coronary vasodilatation.

In conclusion, quantitative assessment of the cardiac adenosine metabolism provides a powerful approach to evaluating how the different pathways might act synergistically in intact tissue. The adenosine model seems to have predictive value for physiological and pharmacological conditions. The greatest promise for future research, however, may be that after a quantitative model of the adenosine metabolism has been developed it will become feasible to test hypotheses represented by the model in much better detail and specificity.

Acknowledgments. The experimental work underlying this chapter was performed at the Zentrum für Physiologie, Heinrich-Heine-Universität Düsseldorf (chair: Prof. Dr. J. Schrader). The expert technical assistance of Mrs. Margit Becker and Dr. Stefan Schäfer during the isolated heart experiments is greatly appreciated. The work was funded by the Deutsche Forschungsgemeinschaft (Grant De360/6-1) and the Eberhard-Igler-Stiftung. Dr. Andreas Deussen was a Heisenberg fellow of the Deutsche Forschungsgemeinschaft (Grant De360/4-1).

References

Abraham, E. H., A. G. Prat, L. Gerweck, T. Seneveratne, R. J. Arceci, R. Kramer, G. Guidotti, and H. F. Cantiello. The multidrug resistance (mdr1) gene product functions as an ATP channel. *Proc. Natl. Acad. Sci. USA* 90:312–316, 1993.

Bardenheuer, H. and J. Schrader. Supply-to-demand ratio for oxygen determines formation of adenosine by the heart. *Am. J. Physiol.* 250:H173–H180, 1986.

Bassingthwaighte, J. B. and C. A. Gorsky. Modeling in the analysis of solute and water exchange in the microvasculature. In: *Handbook of Physiology, Section 2: The Cardiovascular System, Vol. IV, Microcirculation, Part I* edited by E. M. Renkin, C. C. Michel, and S. R. Geiger. Bethesda, MD: The American Physiological Society, pp. 549–626.

Bassingthwaighte, J. B., I. S. Chan, and C. Y. Wang. Computationally efficient algorithms for capillary convection-permeation-diffusion models for blood tissue exchange. *Ann. Biomed. Eng.* 20:687–725, 1992.

Becker, B. F. and E. Gerlach. Uric acid, the major catabolite of cardiac adenine nucleotides and adenosine, originates in the coronary endothelium. In: *Topics and Perspectives in Adenosine Research*, edited by E. Gerlach and B. F. Becker. Heidelberg: Springer-Verlag, pp. 209–221, 1987.

Belardinelli, L., J. Linden, and R. M. Berne. The cardiac effects of adenosine. *Progr. Cardiovasc. Dis.* 32:73–97, 1989.

Belloni, F. L., R. Rubio, and R. M. Berne. Intracellular adenosine in isolated rat liver cells. *Pflügers Arch* 400:106–108, 1984.

Berne, R. M. Cardiac nucleotides in hypoxia: Possible role in the regulation of coronary blood flow. *Am. J. Physiol.* 204:317–322, 1963.

Berne, R. M. The role of adenosine in the regulation of coronary blood flow. *Circ. Res.* 47:807–813, 1980.

Borchardt, R. T. and Y. S. Wu. Potential inhibitors of S-adenosylmethionine-dependent methyl-transferases: 3. Modification of the sugar portion of S-adenosylhomocysteine. *J. Med. Chem.* 18:300–304, 1975.

Borst, M. M. and J. Schrader. Adenine nucleotide release from isolated perfused guinea pig hearts and extracellular formation of adenosine. *Circ. Res.* 68:797–806, 1991.

Borst, M. M., A. Deussen, and J. Schrader. S-Adenosylhomocysteine hydrolase activity in human myocardium. *Cardiovasc. Res.* 26:143–147, 1992.

Bünger, R., F. J. Haddy, A. Querengässer, and E. Gerlach. An isolated guinea pig heart preparation with in vivo like features. *Pflügers Arch.* 353:317–326, 1975a.

Bünger, R., F. J. Haddy, and E. Gerlach. Coronary responses to dilating substances and competitive inhibition by theophylline in the isolated perfused guinea pig heart. *Pflügers Arch.* 358:213–224, 1975b.

Clanachan, A. S., T. P. Heaton, and F. E. Parkinson. Drug interactions with nucleoside transport systems. In: *Topics and Perspectives in Adenosine Research,* edited by E. Gerlach and B. F. Becker. Heidelberg: Springer-Verlag, pp. 118–129, 1987.

Clarke, B., E. Rowland, P. J. Barnes, J. Till, D. E. Ward, and E. A. Shinebourne. Rapid and safe termination of supraventricular tachycardia in children by adenosine. *Lancet,* February 7:299–301, 1987.

Clemo, H. F. and L. Belardinelli. Effect of adenosine on atrioventricular conduction. II: Modulation of atrioventricular node transmission by adenosine in hypoxic isolated guinea pig hearts. *Circ. Res.* 59:437–446, 1986.

Cordon-Cardo, C., J. P. O'Brien, D. Casals, L. Rittman-Grauer, J. L. Biedler, M. R. Melamed, and J. R. Bertino. Multidrug-resistance gene (P-glycoprotein) is expressed by endothelial cells at blood-brain barrier sites. *Proc. Natl. Acad. Sci. USA* 86:695–698, 1989.

Daly, J. W. and K. A. Jacobson. Adenosine receptors: Selective agonists and antagonists. In: *Adenosine and Adenine Nucleotides: From Molecular Biology to Integrative Physiology,* edited by L. Belardinelli and A. Pelleg. Boston: Kluwer, pp. 157–166, 1995.

De la Haba, G. and G. L. Cantoni. The enzymatic synthesis of S-adenosylhomocysteine from adenosine and homocysteine. *J. Biol. Chem.* 234:603–608, 1959.

Des Rosiers, C. and S. Nees. Functional evidence for the presence of adenosine A_2 receptors in cultured coronary endothelial cells. *Naunyn-Schmiedeberg's Arch. Pharmacol.* 366:94, 1987.

Deussen, A and J. Schrader. Cardiac adenosine production is linked to myocardial pO_2. *J. Mol. Cell Cardiol.* 23:495–504, 1991.

Deussen, A., G. Möser, and J. Schrader. Contribution of coronary endothelial cells to cardiac adenosine production. *Pflügers Arch.* 406:608–614, 1986.

Deussen, A., M. Borst, and J. Schrader. Formation of S-adenosylhomocysteine in the heart. I. An index of free intracellular adenosine. *Circ. Res.* 63:240–249, 1988a.

Deussen, A., M. Borst, K. Kroll, and J. Schrader. Formation of S-adenosylhomocysteine in the heart. II: A sensitive index for regional myocardial underperfusion. *Circ. Res.* 63:251–261, 1988b.

Deussen, A., H. G. E. Lloyd, and J. Schrader. Contribution of S-adenosylhomocysteine to cardiac adenosine formation. *J. Mol. Cell Cardiol.* 21:773–782, 1989.

Deussen, A., Ch. Walter, M. Borst, and J. Schrader. Transmural gradient of adenosine in the canine heart during functional hyperemia. *Am. J. Physiol.* 260:H671–H680, 1991.

Deussen, A., B. Bading, K. Kelm, and J. Schrader. Formation and salvage of adenosine by macrovascular endothelial cells. *Am. J. Physiol.* 264:H692–H700, 1993.

DiMarco, J. P., T. D. Sellers, R. M. Berne, G. A. West, and L. Belardinelli. Adenosine: Electrophysiologic effects and therapeutic use for terminating paroxysmal supraventricular tachycardia. *Circulation* 68:1254–1263, 1983.

Dobson, J. G. Mechanism of adenosine inhibition of catecholamine-induced responses in heart. *Circ. Res.* 52:151–160, 1983.

Downey, J. M., G. S. Liu, and J. D. Thornton. Adenosine and the anti-infarct effects of preconditioning. *Cardiovasc. Res.* 27:3–8, 1993.

Druri, A. M. and A. Szent-Györgyi. The physiological activity of adenine compounds with special reference to their action upon the mammalian heart. *J. Physiol. (London)* 68:213–237, 1929.

D'Souza, S. J. A. and D. F. Biggs. Tartrazine and indomethazin increase firing rates in the carotid sinus nerves of guinea pigs. *Proc. West. Pharmacol.* 28:135–137, 1985.

Ely, S. W., G. P. Matherne, S. D. Coleman, and R. M. Berne. Inhibition of adenosine metabolism increases myocardial interstitial adenosine concentrations and coronary blood flow. *J. Mol. Cell Cardiol.* 24:1321–1332, 1992.

Fleetwood, G., S. B. Coade, J. L. Gordon, and J. D. Pearson. Kinetics of adenine nucleotide catabolism in coronary circulation of rats. *Am. J. Physiol.* 256:H1565–H1572, 1989.

Frick, G. P. and J. M. Lowenstein. Vectorial production of adenosine by 5'-nucleotidase in the perfused rat heart. *J. Biol. Chem.* 253:1240–1244, 1978.

Gerlach, E., F. J. Deuticke, and R. H. Dreisbach. Der Nucleotid-Abbau im Herzmuskel bei Sauerstoffmangel und seine mögliche Bedeutung für die Coronardurchblutung. *Naturwissenschaften* 50:228–229, 1963.

Gerlach, E. and F. J. Deuticke. Biochemische Aspekte der Adenosin-bedingten Koronardilatation. In: *Kreislaufmessungen, 4. Freiburger Colloquium,* edited by A. Fleckenstein München: E. Banaschewski. pp. 126–132, 1964.

Goldman, S., E. S. Dickinson, and L. L. Slakey. Effect of adenosine on synthesis and release of cyclic AMP by cultured vascular cells from swine. *J. Cyclic. Nucl. Prot. Phosphor. Res.* 9:69, 1983.

Headrick, J. P., G. P. Matherne, S. S. Berr, and R. M. Berne. Effects of graded perfusion and isovolumic work on epicardial and venous adenosine and cytosolic metabolism. *J. Mol. Cell Cardiol.* 23:309–324, 1991.

Henrichs, K. J., H. Matsuoka, and W. Schaper. Intracellular trapping of adenosine during myocardial ischemia by L-homocysteine. *Basic Res. Cardiol.* 81:267–275, 1986.

Jacobson, M. A. Molecular biology of adenosine receptors. In: *Adenosine and Adenine nucleotides: from molecular biology to integrative physiology,* edited by L. Belardinelli and A. Pelleg. Boston: Kluwer, pp. 5–13, 1995.

Jennings, R. B., K. A. Reimer, M. L. Hill, and S. E. Mayer. Total ischemia in dog hearts, in vitro. 1. Comparison of high energy phosphate production, utilization, and depletion, and of adenine nucleotide catabolism in total ischemia in vitro vs. severe ischemia in vivo. *Circ. Res.* 49:892–900, 1981.

Kelm, M. and J. Schrader. Control of coronary vascular tone by nitric oxide. *Circ. Res.* 66:1561–1575, 1990.

Kroll, K. and G. V. Martin. Steady-state catecholamine stimulation does not increase cytosolic adenosine in canine hearts. *Am. J. Physiol.* 265:H503–H510, 1994.

Kroll, K., J. Schrader, H. M. Piper, and M. Henrich. Release of adenosine and cyclic AMP from coronary endothelium in isolated guinea pig hearts: Relation to coronary flow. *Circ. Res.* 60:659–665, 1987.

Kroll, K., A. Deussen, and I. R. Sweet. Comprehensive model of transport and metabolism of adenosine and S-adenosylhomocysteine in the guinea pig heart. *Circ. Res.* 71:590–604, 1992.

Kroll, K., U. K. M. Decking, K. Dreikorn, and J. Schrader. Rapid turnover of the AMP-adenosine metabolic cycle in the guinea pig heart. *Circ. Res.* 73:846–856, 1993.

Lindner, F. and R. Rigler. Über die Beeinflussung der Weite der Herzkrankgefäße durch Produkte des Zellkernstoffwechsels. *Pflügers Arch.* 226:697–708, 1931.

Mattig, S. and A. Deussen. Flux through ecto-5′-nucleotidase pathway in macrovascular endothelial cells. Workshop "Extracellular Nucleotides: A novel and universal class of signalling molecules from receptors to clinical function." Magdeburg, 1–2 December, 1995.

Möser, G. H., J. Schrader, and A. Deussen. Turnover of adenosine in plasma of human and dog blood. *Am. J. Physiol.* 256:C799–C806, 1989.

Murry, C. E., R. B. Jennings, and K. A. Reimer. Preconditioning with ischemia: A delay of lethal cell injury in ischemic myocardium. *Circulation* 74:1124–1136, 1986.

Mustafa, S. J. Effects of coronary vasodilator drugs on the uptake and release of adenosine in cardiac cells. *Chem. Pharmacol.* 28:2617–2624, 1979.

Mustafa, S. J., R. Marala, W. Abebe, N. Jeansonne, H. Olanrewaju, and T. Hussain. Coronary adenosine receptors: Subtypes, localization, and function. In: *Adenosine and Adenine Nucleotides: From Molecular Biology to Integrative Physiology,* edited by L. Belardinelli and A. Pelleg. Boston: Kluwer, pp. 229–239, 1995.

Nees, S., V. Herzog, B. F. Becker, M. Böck, C. Des Rosiers, and E. Gerlach. The coronary endothelium: A highly active metabolic barrier for adenosine. *Basic Res. Cardiol.* 80:515–529, 1985.

Newby, A. C. How does dipyridamole elevate extracellular adenosine concentration? Predictions from a three-compartment model of adenosine formation and inactivation. *Biochem. J.* 227:845–851, 1986.

Olsson, R. A., and J. D. Pearson. Cardiovascular purinoceptors. *Physiol. Rev.* 70:761–845, 1990.

Olsson, R. A., C. J. Davis, E. M. Khouri, and R. E. Patterson. Evidence for an adenosine receptor on the surface of dog coronary myocytes. *Circ. Res.* 39:93–98, 1976.

Olsson, R. A., D. Saito, and C. R. Steinhardt. Compartmentalization of the adenosine pool of dog and rat hearts. *Circ. Res.* 50:617–626, 1982.

Pearson, J. D., J. S. Carleton, and J. L. Gordon. Metabolism of adenine nucleotides by ectoenzymes of vascular endothelial and smooth-muscle cells in culture. *Biochem. J.* 190:421–429, 1980.

Rosen, M. R., P. Danilo, and R. M. Weiss. Actions of adenosine on normal and abnormal impulse initiation in canine ventricle. *Am. J. Physiol.* 244:H715–H721, 1983.

Rubio, R., R. M. Berne, and J. G. Dobson, Jr. Sites of adenosine production in cardiac and skeletal muscle. *Am. J. Physiol.* 225:938–953, 1973.

Schiele, J. O. and U. Schwabe. Characterization of the adenosine receptor in microvascular coronary endothelial cells. *Eur. J. Pharmacol.* 269:51–58, 1994.

Scholtholt, J., R. E. Nitz, and E. Schraven. On the mechanism of the antagonistic action of xanthine derivatives against adenosine and coronary vasodilators. *Drug Res.* 22:1255–1259, 1972.

Schrader, J. Metabolism of adenosine and sites of production in the heart. In: *Regulatory Functions of Adenosine,* edited by R. M. Berne, T. W. Rall, and R. Rubio. The Hague: M. Nijhoff, pp. 133–156, 1983.

Schrader, J. and E. Gerlach. Compartmentation of cardiac adenine nucleotides and formation of adenosine. *Pflügers Arch.* 367:129–135, 1976.

Schrader, W. P. and C. A. West. Localization of adenosine deaminase and adenosine deaminase complexing protein in the rabbit heart. Implications for adenosine metabolism. *Circ. Res.* 66:754–762, 1990.

Schrader, J., J. G. Baumann, and E. Gerlach. Adenosine as inhibitor of myocardial effects of catecholamines. *Pflügers Arch.* 372:29–35, 1977a.

Schrader, J., F. J. Haddy, and E. Gerlach. Release of adenosine, inosine and hypoxanthine from the isolated guinea pig heart during hypoxia, flow-autoregulation and reactive hyperemia. *Pflügers Arch.* 369:1–6, 1977b.

Schrader, J., H. Bardenheuer, and E. Gerlach. Role of S-adenosylhomocysteine hydrolase in adenosine metabolism in mammalian heart. *Biochem. J.* 196:65–70, 1981.

Schütz, W., J. Schrader, and E. Gerlach. Different sites of adenosine formation in the heart. *Am. J. Physiol.* 240:H963–H970, 1984.

Schwabe, U. Introduction to adenosine receptors. In: *Role of Adenosine and Adenine Nucleotides in the Biological System,* edited by S. Imai and M. Nakazawa. Amsterdam: Elsevier, pp. 59–69, 1991.

Smolenski, R. T., J. Schrader, H. De Groot, and A. Deussen. Oxygen partial pressure and free intracellular adenosine of isolated cardiomyocytes. *Am. J. Physiol.* 260:C708–C714, 1991.

Sollevi, A. Cardiovascular effects of adenosine in man: Possible clinical implications. *Progr. Neurobiol.* 27:319–349, 1986.

Sparks, H. V. and H. Bardenheuer. Regulation of adenosine formation by the heart. *Circ. Res.* 58:193–201, 1986.

Stumpe, T. and J. Schrader. Adenosine formation and phosphorylation potential in isolated rat cardiomyocytes at different pO_2 and oxygen consumption rates. *Drug Devel. Res.* 31:326 (abstract), 1994.

Thiebaut, F., T. Tsuruo, H. Hamada, M. M. Gottesmann, I. Pastan, and M. C. Willingham. Cellular localization of the multidrug-resistance gene product P-glycoprotein in normal human tissues. *Proc. Natl. Acad. Sci. USA* 84:7735–7738, 1987.

Ueland, P. M. Pharmacological and biochemical aspects of S-adenosylhomocysteine and S-adenosylhomocysteine hydrolase. *Pharmacol. Rev.* 34:223–253, 1982.

Ueland, P. M. and J. Saebo. Sequestration of adenosine in crude extract from mouse liver and other tissues. *Biochem. Biophys. Acta* 587:341–352, 1979.

Wangler, R. D., M. W. Gorman, C. Y. Wang, D. F. Dewitt, I. S. Chan, J. B. Bassingthwaighte, and H. V. Sparks. Transcapillary adenosine transport and interstitial adenosine concentration in guinea pig hearts. *Am. J. Physiol.* 257:H89–H106, 1989.

11

Role of Capillary Endothelial Cells in Transport and Metabolism of Adenosine in the Heart: An Example of the Impact of Endothelial Cells on Measures of Metabolism

Keith Kroll and James Bassingthwaighte

Introduction

Endothelial cells, lying between the blood stream and the parenchymal cells of an organ, are a part of the set of signaling paths for the organ. Sensing blood solute concentrations or sensing intravascular shear can lead to the endothelial production of substances sensed or taken up by other cells. The interactions between endothelium and smooth muscle fall into a special class relating to the regulation of vasomotion. A component of the vasoregulatory system concerns the regulation of interstitial adenosine; understanding of adenosine in endothelial cells and myocytes has come slowly from early beginnings (Berne et al., 1983) and from studies of transport and exchange (Bassingthwaighte et al., 1985a,b; Gorman et al., 1986). In this chapter we provide a further set of ideas on relationships between endothelial cells and cardiac myocytes in vivo, using adenosine as the substrate of interest. These ideas hold for a variety of solutes, substrates, agonists, and pharmacologic agents, which one can choose to contemplate while reading about this local adenosine story.

Adenosine is a local signaling agent in the heart that is produced from the net breakdown of adenine nucleotides due to disturbances in the oxygen supply/demand balance (Bardenheuer and Schrader, 1986), making adenosine a highly sensitive indicator of myocardial ischemia or hypoxia (Deussen et al., 1988). Via membrane receptors on a variety of cell types, adenosine acts to stimulate physiological responses that tend to counteract the disturbance in energy metabolism. Adenosine causes coronary vasodilation during ischemia (Saito et al., 1981), tending to increase blood flow. In addition, adenosine acts to oppose the stimulatory effects of catecholamines (Schrader et al., 1977), and to decrease electrical excitability of the atrioventricular node (Clemo and Belardinelli, 1986). More recently, it was proposed that the hydrolysis of AMP to adenosine improves the free energy of ATP hydrolysis during ischemia via mass action processes by serving as a sink for cytosolic ADP (Kroll et al., 1997).

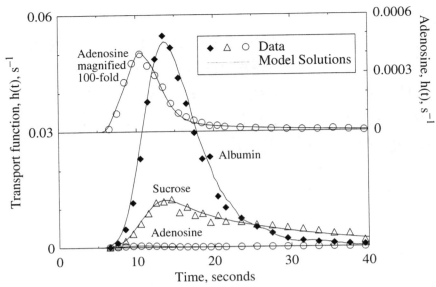

FIGURE 11.1. Multiple-indicator dilution results showing capillary uptake of adenosine in the closed-chest canine coronary circulation. *Main panel:* Albumin, sucrose, and adenosine curves using scale on the left ordinate. *Upper panel:* The adenosine curve is magnified 100-fold using the scale on the right ordinate and the identical time base. The peak of the adenosine curve appears 3 to 4 s earlier than those of the albumin and sucrose curves due to the effects of flow heterogeneity. Symbols show experimental data; curves are model fits (from Kroll and Stepp, 1996).

High Capacity for Adenosine Uptake in Capillary Endothelial Cells

Like many local tissue signaling agents, adenosine in the vicinity of its membrane receptors is regulated by high-capacity systems for membrane uptake and cellular metabolism. What is unique about the adenosine system is the special importance of capillary endothelial cells. Both in skeletal muscle (Gorman et al., 1986) and in the heart (Wangler et al., 1989), there was a striking uptake and retention of tracer adenosine by capillary endothelial cells observed in multiple-indicator dilution experiments. Adenosine uptake is mediated by specific membrane carriers; dipyridamole is a selective inhibitor of the nucleoside transporter that blocks endothelial and myocyte uptake of adenosine in the guinea pig heart (Schwartz et al., 1989) and dog heart (Kroll and Stepp, 1996). Selective adenosine uptake and sequestration in capillary endothelial cells has also been observed using auto-radiography (Stirling, 1983; Nees et al., 1985).

Selective high-capacity membrane transport of adenosine via facilitated diffu-sion has also been observed in many types of cells, including capillary endothelial cells in the lung (Catravas, 1984), cultured vascular endothelial and smooth mus-

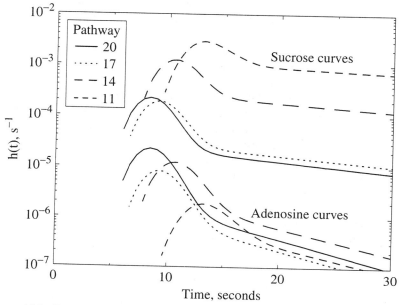

FIGURE 11.2. Weighted outflow curves from four individual flow pathways from a multiple-pathway model fit of sucrose and adenosine dilution curves. Pathway 20 had the highest flow and pathway 11 had flow nearly equal to the mean, while pathways 17 and 14 had intermediate flows, both higher than the mean flow. Outflow curves from low-flow pathways are not shown. The overall model outflow curve such as is shown in Figure 11.1, upper panel, is the sum of all 20 weighted individual pathway curves (from Kroll and Stepp, 1996).

cle cells (Pearson et al., 1978), and cardiomyocytes (Ford and Rovetto, 1987). Indeed, the membrane nucleoside transporter may be ubiquitous, since it has also been observed in erythrocytes (Plagemann et al., 1985) and tumor cells (Plagemann and Wohlhueter, 1985). Interestingly, capillary endothelial cells in rabbit hearts have a relatively low transport capacity for adenosine (Schwartz, et al., 1988).

A particularly striking example of adenosine uptake was obtained in the closed-chest, blood-perfused dog heart, where 99% single-pass extraction of adenosine was observed in multiple-indicator dilution experiments (Figure 11.1) (Kroll and Stepp, 1996). These data, as well as the other indicator dilution results, were analyzed using a multiple-region, multiple-pathway blood–tissue exchange model (Kuikka et al., 1986; Bassingthwaighte et al., 1989). The major reason for such high extraction was that the ratio of capillary permeability–surface area product, *PS*, to blood flow was approximately 7, indicating that, under these in vivo conditions, adenosine uptake was nearly flow-limited. Another reason for the high extraction was that the ratio of endothelial cell consumption to transport, $G_{ec}/(PS_{ecl} + PS_{eca})$, was 2.2, indicating that once tracer adenosine had entered the endothelial cell region it was 2.2 times more likely to be metabolized than to be transported back into the capillary. Adenosine uptake was completely blocked by dipyridamole, an inhibitor of the membrane nucleoside transporter.

The peak of the adenosine dilution curve appeared 3 to 4 s before the peaks of the reference albumin and sucrose curves. The reason for the early peak is the effects of flow heterogeneity on the kinetics of a tracer that exhibits very high peak extraction. This interpretation is based on analyzing the outflow curves from individual flow pathways in the multiple-pathway model (Figure 11.2). High-flow pathways exhibit less extraction of adenosine than low-flow pathways, and high-flow pathways have the earliest arrival times in the venous outflow. Therefore, high-flow pathways make a disproportionate contribution to the earliest phase of the outflow dilution curve of adenosine. Lower flow pathways that arrive at later times exhibit more extraction. The overall outflow curve shown by the model solution in the upper panel of Figure 11.1 is the sum of all 20 weighted individual pathway curves. In contrast to adenosine, for sucrose the high-flow pathways contribute relatively little to the outflow curves, because the fraction of the flow going through pathways 17 to 20, the high-flow paths, is a small fraction of the total flow. The peaks of the adenosine and sucrose curves appear simultaneously in each individual pathway. However, the peak of the overall adenosine curve appears earlier than that of the reference curve sucrose. The reason that the contributions of pathways 20, 17, and 14 are so relatively large for adenosine is that virtually no adenosine escapes from being trapped or consumed in passage through low-flow paths.

Major Pathways of Coronary Endothelial Adenosine Metabolism: Salvage, Degradation, and the SAH Hydrolase Route

Within endothelial cells, there are three major pathways for adenosine metabolism: phosphorylation to AMP via adenosine kinase (salvage), deamination to inosine via adenosine deaminase (degradation), and the reversible S-adenosylhomocysteine (SAH) hydrolase route. Uptake into SAH is important only if homocysteine is supplied in excess. Indicator dilution experiments in isolated guinea pig hearts provide an answer as to whether the primary pathway is via adenosine kinase or adenosine deaminase. The K_m of the kinase is about 1 µM while that for the deaminase is perhaps 80 mM, so at low adenosine concentrations retention by phosphorylation will dominate. When intracellular adenosine concentrations are very high, so that the kinase is saturated, hydrolysis becomes relatively prominent. In the isolated guinea pig heart experiment shown in Figure 11.3, 62% of the injected tracer adenosine was retained in the heart, while 20% of the adenosine emerged unchanged; only 4% was recovered in the venous outflow as labeled inosine, 1% as hypoxanthine + xanthine, and 13% as tritiated water, the final product of the metabolism of tritiated adenosine. In the in vivo dog heart

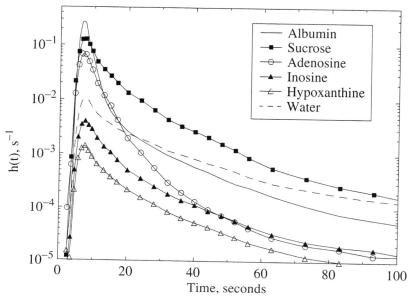

FIGURE 11.3. Capillary endothelial cell metabolism of tracer adenosine in an isolated guinea pig heart. Adenosine plus reference tracers albumin and sucrose were injected into the coronary inflow, and the outflow concentrations of the injected tracers and the metabolic products were measured.

study shown in Figures 11.1 and 11.2, 85% of the injected adenosine was retained in the heart and 14% was converted to membrane-permeable metabolites. The major reason for the difference in retention between dog and guinea pig was the tenfold lower flow in the dog heart. Since most of the labeled adenosine taken up and retained by the heart is incorporated into the adenine nucleotide pool (Schrader and Gerlach, 1976), the results show that the adenosine kinase pathway is normally the major route for adenosine metabolism.

More direct evidence of the importance of the adenosine kinase pathway was the marked reduction in consumption of tracer adenosine due to intracoronary infusion of iodotubericidin (ITC), a selective inhibitor of adenosine kinase. In the in vivo dog heart experiment described above, infusion of 10-μM ITC caused the fraction of injected tracer adenosine recovered in the venous outflow to increase from the control value of 1% (in Figure 11.1) to 35% (in Figure 11.4).

In contrast to the effects of using ITC to block adenosine kinase, there were no significant differences in adenosine kinetics when EHNA, a specific inhibitor, was used to block adenosine deaminase in the guinea pig heart (Figure 11.5, upper panels) (Kroll and Bassingthwaighte, 1995). However, there was adenosine deaminase activity, since EHNA caused a decrease in the venous recovery of the metabolic products of adenosine deamination from a control value of $9 \pm 5\%$ (mean, SD, $n = 6$) to $2 \pm 1\%$ ($n = 4$). A likely explanation is that the endothelial cell

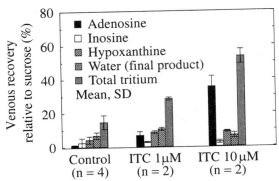

FIGURE 11.4. Inhibition of myocardial uptake of tracer adenosine due to blockade of adenosine kinase via iodotubericidin (ITC) in the dog heart. In the indicator dilution experiments described in Figure 11.1, the venous recovery of injected adenosine and its metabolites was determined relative to the reference sucrose. ITC causes an increase in the recovery of adenosine by blocking the incorporation of adenosine into myocardial adenine nucleotides (from Kroll and Stepp, 1996).

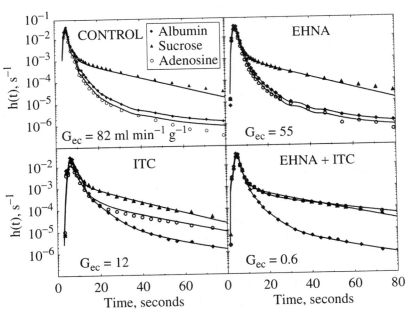

FIGURE 11.5. Effect of metabolic inhibition on capillary adenosine uptake in guinea pig heart. EGNA = adenosine deaminase blocker, ITC = adenosine kinase blocker. Capillary endothelial cell consumption of adenosine (G_{ec}) was estimated by modeling. Together, EHNA + ITC caused nearly complete blockade of adenosine metabolism.

consumption of adenosine is normally so high (G_{ec} = 83 ± 43 ml min^{-1} g^{-1}) compared to the membrane transport (PS_{ecl} = 2.2 ± 0.9 ml min^{-1} g^{-1}) that practically all of the adenosine that enters the endothelial cell region is metabolized, with negligible amounts refluxing back into the venous outflow. Thus, modest reductions in one of the pathways of endothelial consumption (using EHNA) may still result in nearly complete metabolism of the tracer adenosine entering the cells. This explanation is supported by the finding that when ITC and EHNA were infused together, there was a marked reduction of adenosine consumption below what was observed during infusion of ITC alone (Figure 11.5, lower panels). Because the infusion of ITC abolished the metabolic overcapacity, the inhibitory effects of EHNA were revealed. The results demonstrate that the adenosine deaminase pathway is of secondary importance to the adenosine kinase pathway under the normoxic conditions that were studied. This conclusion appears to be valid for the in vivo dog heart as well as the isolated guinea pig heart.

Effects of Endothelial Cell Metabolism of Adenosine

To investigate the physiological role of adenosine in the heart, information must be obtained on the concentration of adenosine at its site of action: membrane adenosine receptors on vascular smooth muscle cells (A_2-subtype) and cardiomyocytes (A_1-subtype), which are bathed in interstitial fluid (Kroll et al., 1987; Kang et al., 1992). Because it is not feasible to measure interstitial concentrations of adenosine directly, the interstitial concentration must be estimated by analyzing measurements of coronary venous adenosine (Wangler et al., 1989; Kroll et al., 1992). Tracer kinetics experiments such as those described above are needed to provide parameter estimates of the endothelial cell processes to account for the concentration gradient between the venous and interstitial regions.

Coronary Flow Regulation by Adenosine

This approach was used to estimate the dose–response relation for coronary flow and endogenous interstitial adenosine concentrations in the perfused guinea pig heart (Kroll et al., 1993). Graded intracoronary infusion of ITC, the adenosine kinase inhibitor, was used to increase the net production of adenosine by blocking the major metabolic pathway while measurements were obtained of coronary flow and venous adenosine concentrations (Figure 11.6). At the highest concentrations of ITC used, endogenous interstitial adenosine concentrations increased to levels producing maximal coronary vasodilation. Interstitial adenosine concentrations causing half-maximal increases in coronary flow were estimated to be approximately 100 nM. The model providing the result shown in the figure is that described by Bassingthwaighte in Chapter 7. (See related results in the chapter by Deussen.) Similar findings were obtained in the in vivo dog heart, although the dose–response curve for adenosine was particularly steep (Stepp et al., 1996).

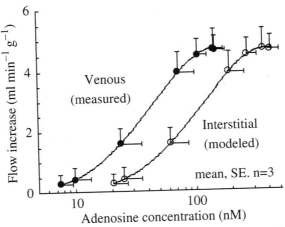

FIGURE 11.6. Estimated dose–response curve for coronary flow and endogenous interstitial adenosine in perfused guinea pig heart. Intracoronary infusion of the adenosine kinase blocker, iodotubericidin (ITC), was used to selectively increase myocardial adenosine production while measuring flow and venous adenosine. Interstitial adenosine was estimated using a standard blood–tissue exchange model with cellular adenosine production. (Reproduced with permission from Kroll et al., 1993. Copyright 1993 American Heart Association.)

Estimating Adenosine Production

Another effect of endothelial cell adenosine uptake is that it greatly complicates the interpretation of a commonly used index of myocardial adenosine production based on the Fick principle, i.e., coronary venous adenosine release: coronary flow × ([adenosine]$_{venous}$ − [adenosine]$_{arterial}$). Investigators of myocardial adenosine have assumed that venous adenosine release indicates *directional* changes in myocardial adenosine production. However, by accounting for the effects of endothelial cell adenosine uptake, it was revealed that this assumption may lead to misleading conclusions, since myocardial adenosine production and venous adenosine release may even change in *opposite* directions if flow is reduced (Figure 11.7). The simulation results were obtained by holding constant the parameters describing adenosine transport and metabolism while flow was reduced (from right to left). Adenosine production in both endothelial and parenchymal cells was either held constant (solid curves) or increased as flow was decreased (dashed and dotted curves) to simulate more realistically the effects of myocardial ischemia. Here, a nonlinear model (multiregion) was used that describes the individual adenosine metabolizing enzymes in parenchymal and endothelial cells (Kroll et al., 1992). In the two cases where the production of adenosine increased as flow decreased (short and long dashes), the curves in the lower panel illustrate that, despite the high production, the amounts released into the venous effluent are reduced at flows less than 2 ml g^{-1} min^{-1}, due simply to the uptake of intracellular adenosine via adenosine kinase. The results show that due to the simultaneous

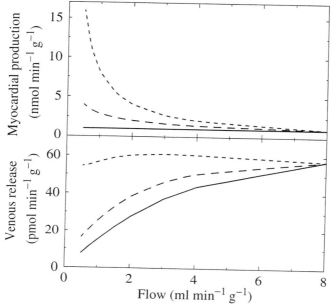

FIGURE 11.7. Simulation of directionally opposite changes in cellular adenosine production (upper panel) and coronary venous adenosine release (lower panel) due to the effects of flow. All model parameters were held constant, except for cellular production and flow. Continuous, dashed, and dotted curves in the two panels represent the same solutions. (After Kroll et al., 1992, with permission. Copyright 1992 American Heart Association.)

production and consumption of adenosine within the same cellular regions, there is no simple relation between the cellular production of adenosine and the coronary venous efflux of adenosine if flow is not constant.

AMP–Adenosine Metabolic Cycle in the Intact Heart

Endothelial cells and myocytes have the same machinery for adenosine metabolism; the events are difficult to distinguish because most data have been acquired in the intact organ. The simultaneous cellular production and consumption of adenosine within the same cells mediated by the enzymes 5'-nucleotidase and adenosine kinase was first described by Arch and Newsholme (1978), and has been termed a "futile cycle" (Bontemps et al., 1983) because the turnover of the cycle requires energy. However, the AMP–adenosine metabolic cycle is not at all futile, but amplifies increases in cytosolic adenosine concentration following increases in the rate of AMP hydrolysis to adenosine (Kroll et al., 1993). The rapid turnover of the intracellular AMP–adenosine cycle was observed directly by measuring a 15-fold increase in coronary venous efflux rate of adenosine ([ade-

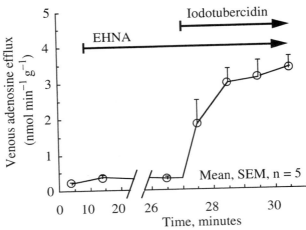

FIGURE 11.8. Direct observation of the high turnover rate of the AMP–adenosine metabolic cycle in the guinea pig heart. Blockade of adenosine kinase via iodotubericidin caused a 15-fold increase in the venous efflux rate of adenosine under normoxic conditions. The results indicate that total myocardial adenosine production is approximately 3 nmol min^{-1} g^{-1}, and that normally more than 90% of the adenosine produced in the heart is salvaged intracellularly via adenosine kinase without escaping into the venous outflow. (Reproduced with permission from Kroll et al., 1993. Copyright 1993 American Heart Association.)

nosine]$_{venous}$ × flow) after blockade of adenosine kinase by ITC (Figure 11.8). Increased cytosolic adenosine concentrations lead to increased interstitial adenosine concentrations because of the transport of adenosine to outside the cell. Cellular adenosine is produced both from AMP hydrolysis in myocytes, which is highly sensitive to changes in the myocardial energy status, and from the hydrolysis of S-adenosylhomocysteine (SAH), which is continually produced via cellular transmethylation reactions and is insensitive to myocardial energetics (Deussen et al., 1989) (Figure 11.9). Because of the AMP–adenosine cycle, the unidirectional flux rate of AMP hydrolysis (via 5′-nucleotidase) is greater than that of SAH, even though there is little net flux through the AMP–adenosine cycle.

To examine the hypothesized function of the AMP–adenosine cycle in more detail, the reactions shown in Figure 11.9 were simulated using a nonlinear adenosine model (Kroll et al., 1992) to observe the changes in cytosolic adenosine concentrations due to a twofold step increase in the rate of AMP hydrolysis (Figure 11.10) (Kroll et al., 1993). All of the other reactions in Figure 11.9 changed in response to the increase in adenosine production, except the rate of SAH hydrolysis, which held constant (see Deussen et al., 1989, for a lower estimate). When the turnover of the cycle was normal (fluxes shown in Figure 11.9), a doubling of AMP hydrolysis caused a 1.8-fold increase in cytosolic adenosine concentration (Figure 11.10). When the turnover of the cycle was decreased to one-tenth the normal value without changing the starting cytosolic

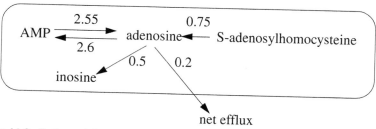

FIGURE 11.9. Estimated flux rates (nmol min^{-1} g^{-1}) for cellular adenosine pathways in the normoxic guinea pig heart. (Reproduced with permission from Kroll et al., 1993. Copyright 1993 American Heart Association.)

adenosine concentration or the rate of SAH hydrolysis, then the effect of doubling AMP hydrolysis was greatly decreased and delayed. Increasing AMP hydrolysis had little effect on cytosolic adenosine concentration because the rate of SAH hydrolysis was much greater. When the turnover of the AMP–adenosine cycle was increased tenfold above normal without changing SAH hydrolysis or the starting cytosolic adenosine concentration, then the effect of doubling the rate of AMP hydrolysis was somewhat increased above normal. Increasing AMP hydrolysis had more effect on cytosolic adenosine, because the effect of SAH hydrolysis on cytosolic adenosine was completely overwhelmed.

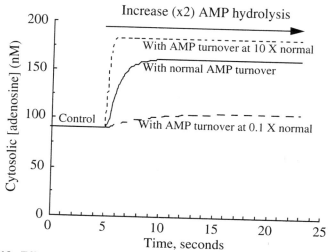

FIGURE 11.10. Effect of the control rate of AMP hydrolysis on the response of [Ado] cell to doubling the rate. When the turnover rate of the AMP–adenosine cycle was decreased to 10% of the normal rate without changing the starting cytosolic adenosine concentration, the effect of doubling AMP hydrolysis was decreased and delayed. When the turnover rate of the cycle was increased tenfold above normal, the effect of doubling AMP hydrolysis was increased. (Reproduced with permission from Kroll et al., 1993. Copyright 1993 American Heart Association.)

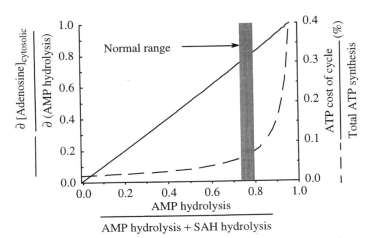

FIGURE 11.11. Cellular [adenosine] rises linearly with AMP hydrolysis but the energy cost is low until high rates of hydrolysis are reached. The abscissa shows the rate of adenosine production via AMP hydrolysis as a fraction of adenosine production via the hydrolysis of AMP and SAH. *Left ordinate and solid line:* [Ado] sensitivity to AMP hydrolysis rate. *Right ordinate and dashed line:* Energy cost.

The results in Figure 11.9 show that when the turnover rate of the AMP–adenosine cycle is high, then the rate of AMP hydrolysis exceeds that of SAH hydrolysis and cytosolic adenosine concentration is controlled by changes in the myocardial energy status. However, the effect of the turnover of the cycle has a limit; even if the turnover rate were increased 100-fold above normal, a doubling of AMP hydrolysis would at most produce a doubling of the cytosolic adenosine concentration. In such a case the energy requirements for the cycle would be 100-fold higher than normal, while the sensitivity for increases in AMP hydrolysis would only be increased by approximately 20%. A comparison of the energy requirement of the AMP–adenosine cycle and the increase in cytosolic adenosine concentration due to an increase in the rate of AMP hydrolysis is shown in Figure 11.11. The left ordinate shows the partial derivative of the cytosolic adenosine concentration with respect to the rate of AMP hydrolysis, expressing both as a percent change from the starting value. This is the sensitivity of cytosolic adenosine to changes in energy status; its maximal value is 1.0. The abscissa is the rate of AMP hydrolysis as a fraction of the total rate of adenosine formation. The sensitivity of cytosolic adenosine to AMP hydrolysis is linear (solid line). The right ordinate shows the energy cost of the cycle in ATP equivalents (dashed curve). Hydrolysis of one AMP costs one ATP equivalent, because a high-energy phosphate bond is lost, which must be ultimately replaced via ATP synthesis. Phosphorylation of one adenosine to AMP also costs one ATP equivalent, since ATP is required for the adenosine kinase reaction. Thus, each turn of the cycle costs two ATP equivalents. When the turnover rate of the cycle is in the normal range (vertical bar), the energy cost is approximately 0.1% of the total myocardial ATP synthesis under normal conditions. The energy cost increases dramatically as the turnover rate of the cycle increases above normal, but the additional increments in sensitivity are small. Thus, the normal heart benefits from most of the potential sensitivity from the AMP–adenosine cycle, while limiting the energy cost. A second conservation feature of the cycle is that purine salvage via adenosine kinase permits the cycle to turn over at a high rate with only a small loss of cellular purines. The result is relatively tight regulation at only moderate cost.

Summary

Endothelial cells and myocytes exchange substrates and metabolites, together serving the functions of the organ. Endothelial cells have dominating influences on the fate of tracer adenosine injected into the inflow to an organ. They also strongly influence the concentrations of nontracer adenosine appearing in the outflow following production inside myocytes or endothelial cells when AMP breakdown exceeds the rate of its formation via adenosine kinase.

References

Arch, J. R. S. and E. A. Newsholme. Activities and some properties of 5′-nucleotidase adenosine kinase and adenosine deaminase in tissues from vertebrates and invertebrates in relation to the control of the concentration and the physiological role of adenosine. *Biochem. J.* 174:965–977, 1978.

Bardenheuer, H. and J. Schrader. Supply-to-demand ratio for oxygen determines formation of adenosine by the heart. *Am. J. Physiol.* 250 (*Heart Circ. Physiol.* 19):H173–H180, 1986.

Bassingthwaighte, J. B., H. V. Sparks, Jr., I. S. Chan, D. F. DeWitt, and M. W. Gorman. Modeling of transendothelial transport. *Fed. Proc.* 44:2623–2626, 1985a.

Bassingthwaighte, J. B., C. Y. Wang, M. Gorman, D. DeWitt, I. S. Chan, and H. V. Sparks. Endothelial regulation of agonist and metabolite concentrations in the interstitium. In: *Carrier-Mediated Transport of Solutes from Blood to Tissue*, edited by D. L. Yudilevich and G. E. Mann. New York: Longman, pp. 191–203, 1985b.

Bassingthwaighte, J. B., C. Y. Wang, and I. S. Chan. Blood-tissue exchange via transport and transformation by endothelial cells. *Circ. Res.* 65:997–1020, 1989.

Berne, R. M., R. M. Knabb, S. W. Ely, and R. Rubio. Adenosine in the local regulation of blood flow: A brief overview. *Fed. Proc.* 42:3136–3142, 1983.

Bontemps, F., G. Van den Berghe, and H. G. Hers. Evidence for a substrate cycle between AMP and adenosine in isolated hepatocytes. *Proc. Natl. Acad. Sci. USA* 80:2829–2833, 1983.

Catravas, J. D. Removal of adenosine from the rabbit pulmonary circulation, in vivo and in vitro. *Circ. Res.* 54:603–611, 1984.

Clemo, H. F. and L. Belardinelli. Effect of adenosine on atrioventricular conduction. I: Site and characterization of adenosine action in the guinea pig atrioventricular node. *Circ. Res.* 59(4):427–436, 1986.

Deussen, A., M. Borst, K. Kroll, and J. Schrader. Formation of S-adenosylhomocysteine in the heart. II. A sensitive index for regional myocardial underperfusion. *Circ. Res.* 63:250–261, 1988.

Deussen, A., H. G. E. Lloyd, and J. Schrader. Contribution of S-adenosylhomocysteine to cardiac adenosine formation. *J. Mol. Cell Cardiol.* 21:773–782, 1989.

Ford, D. A. and M. J. Rovetto. Rat cardiac myocyte adenosine transport and metabolism. *Am. J. Physiol.* 252 (1 Pt. 2):H54–H63, 1987.

Gorman, M. W., J. B. Bassingthwaighte, R. A. Olsson, and H. V. Sparks. Endothelial cell uptake of adenosine in canine skeletal muscle. *Am. J. Physiol.* 250 (*Heart Circ. Physiol.* 19):H482–H489, 1986.

Kang, Y. H., R. T. Mallet, and R. Bunger. Coronary autoregulation and purine release in normoxic heart at various cytoplasmic phosphorylation potentials: Disparate effects of adenosine. *Eur. J. Physiol.* 421:188–199, 1992.

Kroll, K. and J. B. Bassingthwaighte. Capillary endothelial cell adenosine transport in guinea pig heart. *Microcirculation* 2:87, 1995.

Kroll, K. and D. W. Stepp. Adenosine kinetics in the canine coronary circulation. *Am. J. Physiol.* 270 (*Heart Circ. Physiol.* 39):H1469–H1483, 1996.

Kroll, K., J. Schrader, and D. Möllmann. Endothelial activation by adenosine and coronary flow regulation in the guinea pig heart. In: *Topics and Perspectives in Adenosine Research*, edited by E. Gerlach and B. F. Becker. Berlin/Heidelberg: Springer-Verlag, pp. 470–479, 1987.

Kroll, K., A. Deussen, and I. R. Sweet. Comprehensive model of transport and metabolism of adenosine and S-adenosylhomocysteine in the guinea pig heart. *Circ. Res.* 71:590–604. 1992.

Kroll, K., U. Decking, K. Dreikorn, and J. Schrader. Rapid turnover of the AMP-adenosine metabolic cycle in the guinea pig heart. *Circ. Res.* 73:846–856, 1993.

Kroll, K., D. J. Kinzie, and L. A. Gustafson. Open system kinetics of myocardial phosphoenergetics during coronary underperfusion. *Am. J. Physiol.* 272:H2563–H2576, 1997.

Kuikka, J., M. Levin, and J. B. Bassingthwaighte. Multiple tracer dilution estimates of D- and 2-deoxy-D-glucose uptake by the heart. *Am. J. Physiol.* 250 (*Heart Circ. Physiol.* 19):H29–H42, 1986.

Nees, S., V. Herzog, B. F. Becker, M. Böck, C. Des Rosiers, and E. Gerlach. The coronary endothelium: A highly active metabolic barrier for adenosine. *Basic Res. Cardiol.* 80:515–529, 1985.

Pearson, J. D., J. S. Carleton, A. Hutchings, and J. L. Gordon. Uptake and metabolism of adenosine by pig aortic endothelial and smooth-muscle cells in culture. *Biochem. J.* 170:265–271, 1978.

Plagemann, P. G. W. and R. M. Wohlhueter. Effects of nucleoside transport inhibitors on the salvage and toxicity of adenosine and deoxyadenosine in L1210 and P338 mouse leukemia cells. *Cancer Res.* 45:6418–6424, 1985.

Plagemann, P. G. W., R. M. Wohlhueter, and M. Kraupp. Adenosine uptake, transport, and metabolism in human erythrocytes. *J. Cell. Physiol.* 125:330–336, 1985.

Saito, D., C. R. Steinhart, D. G. Nixon, and R. A. Olsson. Intracoronary adenosine deaminase reduces canine myocardial reactive hyperemia. *Circ. Res.* 49:1262–1267, 1981.

Schrader, J. and E. Gerlach. Compartmentation in cardiac adenine nucleotides and formation of adenosine. *Pflügers Arch.* 367:129–135, 1976.

Schrader, J., G. Baumann, and E. Gerlach. Adenosine as inhibitor of myocardial effects of catecholamines. *Eur. J. Physiol.* 372:29–35, 1977.

Schwartz, L. M., T. R. Bukowski, J. H. Revkin, and J. B. Bassingthwaighte. Species differences in capillary transport of inosine and adenosine in rabbit and guinea pig hearts. FASEB J. 2:A1524, 1988.

Schwartz, L. M., T. R. Bukowski, and J. B. Bassingthwaighte. Indicator dilution estimates of adenosine transport kinetics in cardiac capillary endothelial cells of guinea pigs. *FASEB J.* 3:A269, 1989.

Schwartz, L. M., T. R. Bukowski, and J. B. Bassingthwaighte. Indicator dilution estimates of adenosine capillary transport kinetics in guinea pig hearts. *Am. J. Physiol.* 270 (*Heart Circ. Physiol.* 39), in review.

Stepp, D. W., R. van Bibber, K. Kroll, and E. O. Feigl. Quantitative relation between interstitial adenosine concentration and coronary blood flow. *Circ. Res.* 79:601–610, 1996.

Stirling, C. E. Autoradiographic localization of ^{3}H-adenosine. In: *Regulatory Function of Adenosine*, edited by R. M. Berne, T. W. Rall, and R. Rubio. Boston: Martinus Nijhoff, 1983, p. 542.

Wangler, R. D., M. W. Gorman, C. Y. Wang, D. F. DeWitt, I. S. Chan, J. B. Bassingthwaighte, and H. V. Sparks. Transcapillary adenosine transport and interstitial adenosine concentration in guinea pig hearts. *Am. J. Physiol.* 257 (*Heart Circ. Physiol.* 26):H89–H106, 1989.

12
Distribution of Intravascular and Extravascular Resistances to Oxygen Transport

Aleksander S. Popel, Tuhin K. Roy, and Abhijit Dutta

Introduction

One of the major functions of the circulation in mammals is to deliver oxygen molecules from the lung to the mitochondria, where oxygen participates in the production of ATP—the common source of energy in living cells. It has been shown that in many organs, including myocardium and skeletal muscle, the structure of the circulatory system is closely correlated with the function of adequate oxygen delivery (Weibel, 1984; Weibel et al., 1991). Oxygen is carried by the blood primarily in the form of oxyhemoglobin inside red blood cells, and the exchange with the tissue occurs in the small vessels in the microcirculation. The classical view of capillaries as the exchange vessels for oxygen has recently been replaced by a more complex picture of capillaries, arterioles, and, to some extent, venules as being the sites of oxygen exchange (Ellsworth et al., 1994). When an oxygen molecule is released from hemoglobin, it has to traverse the red cell, blood plasma, vascular wall, interstitium, parenchymal cell plasma membrane, and the intracellular space to reach the mitochondria.

The mechanism of transport along this pathway is believed to be passive free or facilitated diffusion; there is no evidence of active transport. Oxygen diffusion can be facilitated by hemoglobin inside the red blood cell and by myoglobin inside some muscle cells. Each of the segments along the pathway can be ascribed a value of transport resistance, with its associated value of the oxygen tension difference necessary for maintaining a given flux of oxygen. The determinants of these resistances are the vascular and tissue geometry, the diffusion and solubility coefficients, the rates of chemical reactions, and the oxygen carrier concentrations. The goal of this chapter is to analyze the distribution of transport resistances in different muscle tissues based on available experimental and theoretical studies. A recent review is devoted to intravascular resistance to oxygen transport in the microcirculation (Hellums et al., 1996). General reviews on oxygen transport in the microcirculation are also available (Popel, 1989; Shiga, 1994; Pittman, 1995; Intaglietta et al., 1996).

Experimental knowledge of O_2 transport parameters is derived from measurements at the whole organ, microcirculatory, and cellular levels. In whole organ experiments, typical measurements include blood flow per unit mass of tissue, Q, systemic hematocrit, H_s, and arterial and mixed venous hemoglobin saturations, $S(P_a)$ and $S(P_v)$, respectively, where P_a and P_v are the corresponding oxygen tensions. Neglecting the amount of free oxygen dissolved in the blood plasma, these variables can be related to the rate of O_2 consumption per unit mass of tissue, $\dot{V}O_2$, by the O_2 mass balance relationship (the First Fick's Law):

$$QH_sC_{Hb}[S(P_a) - S(P_v)] = \dot{V}O_2 , \qquad (12.1)$$

where C_{Hb} is the oxygen-carrying capacity of the hemoglobin solution inside red blood cells at 100% saturation. The product $QO_2 = QH_sC_{Hb}S(P_a)$ is called *oxygen delivery*. Equation (12.1) is used for determination of $\dot{V}O_2$ from the experimentally determined variables appearing in the left side. In addition, the flux of oxygen from the exchange vessels is related to the driving force for oxygen transport, which is assumed to be proportional to the difference between the mean capillary oxygen tension, P_c, and the minimum tissue oxygen tension, P_t,

$$J = \dot{V}O_2 = DO_2(P_c - P_t). \qquad (12.2)$$

Equation (12.2) has the form of the Second Fick's Law of Diffusion applied to the whole organ; in fact, Eq. (12.2) is the definition of DO_2 that is referred to as *tissue diffusing capacity*. DO_2 can be considered to be tissue O_2 transport conductivity or the inverse of the transport resistance. This resistance characterizes the entire pathway from the red blood cell to the mitochondrion. The diffusing capacity has been evaluated experimentally for a number of tissues (Wagner, 1995). Evaluation of this parameter is particularly convenient in muscle that is working aerobically at its maximum capacity, since under these conditions the minimum tissue PO_2 is believed to be close to zero (Severinghaus, 1994). Under these conditions, the flux of oxygen is simply proportional to the mean capillary PO_2. If, in addition, one assumes that the mean capillary PO_2 is proportional to the mixed venous PO_2, then one can determine the maximum oxygen consumption rate, $\dot{V}O_{2max}$, from coupled equations (12.1) and (12.2). Graphically, $\dot{V}O_{2max}$ is determined from the intersection of two curves representing Eqs. (12.1) and (12.2) in a plot of $\dot{V}O_2$ versus P_v (Wagner, 1995). In a series of experiments with isolated canine gastrocnemius muscle, Hogan et al. (1991) demonstrated that DO_2 is not affected by blood flow or arterial PO_2, but is a function of hematocrit. The diffusing capacity increases when hematocrit is raised and decreases when hematocrit is reduced. The hematocrit dependence of DO_2 suggests that part of the total transport resistance from the red blood cell to the mitochondrion is intravascular, most likely intracapillary, since the hematocrit can affect only this part of the pathway. Hogan et al. (1991) estimated that, at a normal hematocrit level, approximately 40% of the total resistance is hematocrit-dependent.

The above experiments do not provide information about the heterogeneity of oxygen delivery and oxygen exchange. An indicator dilution technique has been used in the last four decades that can provide information on perfusion heterogeneity (Rose and Goresky, 1976). In addition, the use of tracer oxygen ($^{15}O_2$ or $^{18}O_2$) in whole organ experiments can help identify oxygen transport barriers along the pathway of O_2 molecules (Rose and Goresky, 1985; Goresky, 1985).

Surface O_2 electrodes sample PO_2 within regions that are typically several millimeters of linear dimension and can provide valuable information on the spatial distribution of O_2. They have been used to obtain PO_2 distributions in skeletal muscle, myocardium, brain, and liver (Messmer et al., 1973). Quantitative interpretation of these measurements is not straightforward because the spatial location of the electrodes with respect to the microvasculature is not known.

Developing new methods to measure microvascular parameters has opened new ways of quantifying tissue oxygen delivery (see Pittman, 1995). Among the most important of these are optical measurements of red blood cell velocity using temporal and spatial correlation techniques (Wayland and Johnson, 1967; Pries, 1988), measurements of microvessel hematocrit based on absorption of light transmitted through the red blood cells (Lipowsky et al., 1982; Pries et al., 1983), the microspectrophotometric method of determining hemoglobin saturation in arterioles and venules (Pittman and Duling, 1975) and capillaries (Ellsworth et al., 1987), and polarographic measurements of PO_2 using recessed microelectrodes (Whalen et al., 1967). A recently developed technique based on phosphorescence quenching of excited phosphors allows optical measurements of PO_2 in plasma in microvessels down to the capillary level (Torres Filho and Intaglietta, 1994; Zheng et al., 1996). The phosphorescence technique has also been used to map PO_2 distribution on the surface of a microcirculatory preparation (Itoh et al., 1994), and to measure PO_2 in tissue (Torres Filho et al., 1994). These techniques are used under conditions of intravital microscopy, which provides the spatial resolution of the corresponding measurements. Cryomicrospectrophotometry is based on the same principles as microspectrophotometry for in vivo measurements, but it uses frozen samples of tissue. This technique has been applied to measurements of myoglobin saturation in samples of skeletal muscle (Honig and Gayeski, 1993) and myocardium (Gayeski and Honig, 1991). According to a recent report, however, the spatial resolution of these measurements is lower than previously assumed, and reinterpretation of the results may be necessary (Voter and Gayeski, 1995).

Despite these experimental developments in the last three decades, microcirculatory data necessary for evaluating oxygen transport resistances, particularly capillary transport resistances, are scarce or nonexistent. Such data would have to include measurements of intracapillary PO_2 or hemoglobin saturation in the red blood cells, measurements of pericapillary PO_2, and measurements of tissue PO_2 away from the capillary.

In this chapter we will present a mathematical model of O_2 transport from the capillaries and apply it to calculating segmental transport resistances along the pathway of O_2 molecules. Numerical values of the transport resistances will be given for several muscles with different levels of metabolism. We start with definitions of variables necessary for a quantitative description of O_2 transport.

General Transport Equations

Free Diffusion

We consider the transport of solute (oxygen) in a solvent (e.g., hemoglobin solution, blood plasma, cell membrane, interstitium, cytoplasm). In what follows we use the molar concentration of oxygen, $[O_2]$, measured in mol/ml. For simplicity, first consider a plane layer of solvent through which oxygen molecules diffuse; we assume that the x axis is normal to the plane and that the concentration is only a function of x (one-dimensional diffusion) and time, t. According to the Second Fick's Law of Diffusion, based on experiments with dilute systems, the mass of O_2 crossing the unit area of the plane per unit time, j, referred to as oxygen flux, is proportional to the concentration gradient

$$j = -D\frac{\partial[O_2]}{\partial x},$$ (12.3)

where the coefficient of proportionality, D, is called the diffusion coefficient. The units of D are cm^2/s. For physiological levels of O_2 concentration, D is independent of concentration but is a function of temperature. From the balance of O_2 mass in a layer of thickness dx, we obtain

$$\frac{\partial[O_2]}{\partial t} = -\frac{\partial j}{\partial x} - M,$$ (12.4)

where M is the O_2 consumption measured in mol/(ml·s). Substituting Eq. (12.3) into Eq. (12.4), we obtain the fundamental one-dimensional diffusion equation

$$\frac{\partial[O_2]}{\partial t} = D\frac{\partial^2[O_2]}{\partial x^2} - M.$$ (12.5)

Generalizing Eq. (12.5) to three dimensions and allowing for convective transport of O_2 by a macroscopic velocity field, $\mathbf{v}(x, y, z)$, and assuming that diffusion is isotropic (i.e., the diffusion coefficient is a scalar), we obtain the general convective diffusion equation

$$\frac{\partial[O_2]}{\partial t} + (\mathbf{v} \cdot \nabla)[O_2] = D\left(\frac{\partial^2[O_2]}{\partial x^2} + \frac{\partial^2[O_2]}{\partial y^2} + \frac{\partial^2[O_2]}{\partial z^2}\right) - M.$$ (12.6)

It is common to characterize O_2 concentration in terms of its partial pressure or O_2 tension, P, according to Henry's law:

$$[O_2] = \alpha P,$$ (12.7)

where α is the Bunsen solubility coefficient. Oxygen tension is commonly measured in mm Hg; thus the units of solubility coefficient are mol/(ml·mm Hg).

As an alternative to molar concentration, the mass concentration of oxygen, c, can be introduced and expressed in physical units of ml O_2/ml; in the conversion from $[O_2]$ to c we use the gas molar equivalent $22.4\cdot10^3$ ml O_2/mol at STP (Standard Temperature and Pressure). If the concentration c is used in Eq. (12.7), then the units of the solubility coefficient become ml $O_2/(ml \cdot mm\ Hg)$ and of the consumption rate ml $O_2/(ml \cdot s)$. The values of the biophysical parameters for different tissue components will be discussed below.

Facilitated or Carrier-Mediated Diffusion

Facilitated or carrier-mediated diffusion is common for many cellular transport processes. Here we consider two carrier molecules that play an important role in O_2 transport: myoglobin (Mb) and hemoglobin (Hb).

Myoglobin is present in cardiac myocytes and in many striated muscle cells. A protein with a molecular weight of approximately 17.5 kD (a single polypeptide chain), Mb contains one heme group capable of binding O_2 noncooperatively. First, we consider the storage function of myoglobin and then its possible function in facilitating O_2 diffusion.

The rate of Mb–O_2 chemical reaction can be expressed as

$$R = k'[O_2][Mb] - k[MbO_2], \qquad (12.8)$$

where k' and k are the association and dissociation reaction constants, respectively. Under conditions of chemical equilibrium, the net rate of chemical reaction equals zero, $R = 0$; thus, from Eq. (12.8) we obtain a hyperbolic relationship between oxymyoglobin saturation, $S_{Mb} = [MbO_2]/([Mb] + [MbO_2])$ and O_2 concentration

$$S_{Mb} = \frac{K[O_2]}{1 + K[O_2]}, \qquad (12.9)$$

where $K = k'/k$ is the equilibrium constant. $S_{Mb} = 0$ when all Mb molecules are free, and $S = 1$ when all Mb molecules are bound. Using Henry's law, Eq. (12.7), we can recast Eq. (12.9) in the form

$$S_{Mb} = \frac{P}{P_{50} + P}, \qquad (12.10)$$

where $P_{50} = (K\alpha)^{-1}$ is the PO_2 at which Mb is 50% saturated. Equation (12.10) is referred to as the *oxymyoglobin dissociation curve*. Thus, at a tissue oxygen tension P, the concentration of Mb-bound O_2 is $[Mb]_T \cdot S_{Mb} \cdot 22.4 \cdot 10^3$ (ml O_2/ml), where $[Mb]_T$ is the total concentration of Mb (i.e., both free and bound), whereas the concentration of free oxygen is αP.

In addition to its storage function, Mb can carry bound O_2 molecules in the presence of an oxymyoglobin concentration gradient. The total flux of O_2 can be expressed as

$$j = -D\nabla[O_2] - D_{Mb}\nabla[MbO_2], \qquad (12.11)$$

where we assume that the diffusion coefficients of Mb and MbO_2 are approximately the same, D_{Mb}, because the molecular weights of the two species are very close. If the Mb–O_2 chemical reaction is in equilibrium, we can substitute $[MbO_2] = [Mb]_T S_{Mb}(P)$ into Eq. (12.11) to obtain

$$j = -D\left(1 + \frac{D_{Mb}[Mb]_T}{\alpha D}\frac{dS_{Mb}}{dP}\right)\nabla[O_2].$$

(12.12)

The dimensionless expression in parentheses is the facilitation factor: It represents the enhancement of O_2 diffusion due to the myoglobin facilitation. We will discuss the numerical values of these variables below.

Hemoglobin, a protein with a molecular weight of approximately 67 kD, is contained predominantly within red blood cells. This globular molecule consists of four polypeptide subunits, each with a heme group capable of cooperative O_2 binding. The storage and facilitation functions of hemoglobin, Hb, can be described in a similar fashion to those of Mb, except the reaction rate similar to Eq. (12.8) should be written for several intermediate compounds with different association–dissociation reaction constants, and the saturation as a function of PO_2 is nonhyperbolic because of the cooperativity of the chemical kinetics. Two expressions for the oxyhemoglobin dissociation curve that are commonly used are the Hill equation,

$$S = \frac{P^n}{P_{50}^n + P^n},$$

(12.13)

and the Adair equation,

$$S = \frac{a_1 P + 2a_2 P^2 + 3a_3 P^3 + 4a_4 P^4}{4(1 + a_1 P + a_2 P^2 + a_3 P^3 + a_4 P^4)},$$

(12.14)

where the coefficients P_{50}, n, a_1, . . . , a_4 are functions of pH, PCO_2, 2,3-diphosphoglycerate concentration [DPG], and temperature (Popel, 1989).

We will now apply the general equations of O_2 transport to the particular problem of transport from capillaries to muscle fibers. In this analysis we will neglect all tissue heterogeneities and assume that all capillaries have identical characteristics.

Capillary–Tissue Oxygen Transport

If heterogeneities in capillary–fiber geometry are not taken into account, two geometrical models of a capillary–fiber complex can be considered: the classical Kroghian geometry of a central capillary surrounded by a cylindrical tissue region and a cylindrical fiber surrounded by several capillaries. The two models are not equivalent: In the first model, the maximal distance from any point to the nearest capillary decreases with increasing capillary density (number of capillaries per unit area perpendicular to muscle fibers), but in the second model this distance can remain fixed if the fiber diameter is constant while the capillary density increases. In muscles, the average number of capillaries per fiber varies between 3 and 7 (Mathieu-Costello, 1993). Groebe (1995) compared two geometrical models for capillary–fiber ratios 3 and 6, and showed that the difference in predicted O_2 distributions was small for the same values of capillary density. Conley and Jones

(1996), among others, argue that the difference between the two models is significant. In the following analysis, we will use the Kroghian geometry since the mathematical analysis can be greatly simplified. However, this issue of which geometrical model is more accurate might have to be revisited in the future.

Intracapillary Transport Resistance

Inside a capillary, oxygen molecules dissociate from Hb, diffuse through the concentrated Hb solution inside the red blood cell (RBC), cross the RBC membrane, diffuse through either the thin sleeve of plasma between the RBC and the endothelium and other structures of the capillary wall or through the plasma gap between adjacent RBCs, and permeate the endothelial cells (Figure 12.1). We can define transport resistances associated with these pathways by relating the O_2 flux and the PO_2 difference across the corresponding region, $J = k\Delta P$, where k is the transport conductance or mass transfer coefficient and k^{-1} is the transport resistance. For simplicity, consider cylindrical RBCs of length L_{rbc} separated by plasma gaps of length L_p; the intracapillary transport resistances corresponding to this geometry are shown in Figure 12.2. Let J_{rbc} (ml O_2/s) be the total flux out of a single RBC. For the intracapillary region, the intracapillary mass transfer coefficient can be calculated from

$$J_{rbc} = k_{cap}(P_c - P_p), \tag{12.15}$$

where P_c is the PO_2 at the RBC core and P_p is the average plasma PO_2 at the inner capillary wall. We can further subdivide the intracapillary resistance into two parallel resistances of the pathways through the lateral and the basal surfaces:

$$J_l = k_1(P_c - P_l) = k_2(P_l - P_p)$$
$$J_b = k_3(P_c - P_b) = k_4(P_b - P_p), \tag{12.16}$$

where k_1 and k_3 are intraerythrocyte mass transfer coefficients, and k_2 and k_4 are the plasma sleeve and plasma gap mass transfer coefficients, respectively. The resultant intracapillary mass transfer coefficient k_{cap} can be expressed as

$$k_{cap} = (k_1^{-1} + k_2^{-1})^{-1} + (k_3^{-1} + k_4^{-1})^{-1}. \tag{12.17}$$

The mass transfer coefficients account for the kinetics of an oxyhemoglobin dissociation reaction taking place primarily in a thin layer inside the RBC adjacent to the RBC membrane. Expressions for J_l and J_b are adapted from the kinetic boundary layer analysis of Clark et al. (1985). In the physiological range of parameters, these expressions can be linearized with respect to PO_2, resulting in the following approximate relationship for k_1:

$$k_1 = \frac{2\pi r_{rbc} L_{rbc}}{P_c} \left[\frac{1 + (P_c/P_{50})^n}{D_{rbc}\alpha_{rbc}k_- P_{50}[Hb]_T \, n(P_c/P_{50})^{n+1}} \right]^{-1/2} \tag{12.18}$$

and a similar relationship for k_3. Here, P_{50} is the PO_2 at which hemoglobin is 50% saturated, r_{rbc} is the RBC radius, D_{rbc} and α_{rbc} are the O_2 diffusion and solubility

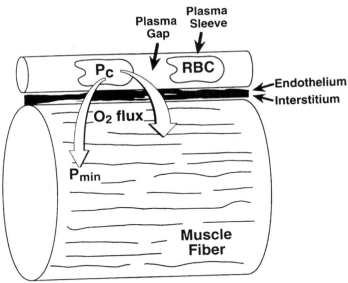

FIGURE 12.1. A schematic diagram showing pathways from the red blood cell to the muscle fiber.

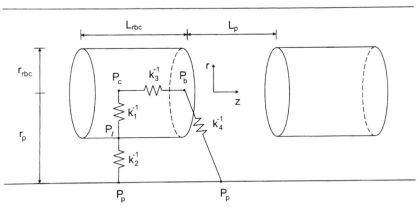

FIGURE 12.2. Intracapillary geometry. Capillary lumen contains plasma and equally spaced red blood cells modeled as cylinders. RBC has a core PO_2 denoted P_c and values of lateral and basal surface PO_2 denoted P_l and P_b. Plasma PO_2 at inner capillary wall is denoted P_p. Value of P_c calculated at venular end of capillary is reported as the critical PO_2 required to sustain $\dot{V}O_{2max}$.

coefficients inside the cell, k_- is the HbO_2 dissociation rate constant, $[Hb]_T$ is the Hb concentration inside the RBC, and n is the Hill parameter in Eq. (12.13) [if necessary, the Hill equation can be replaced by a more accurate expression for the oxygen dissociation curve, e.g., the Adair equation, Eq. (12.14)].

An expression for k_2 can be obtained by solving the axisymmetric one-dimensional diffusion equation, assuming that O_2 diffusion in the plasma sleeve is purely radial:

$$k_2 = \frac{2\pi L_{rbc} D_p \alpha_p}{\ln(r_p/r_{rbc})} \, ,$$
(12.19)

where D_p and α_p are the diffusion and solubility coefficients in the plasma, and r_p is the distance from the capillary axis to the endothelium. The mass transfer coefficient k_4 for the plasma gap has to be calculated numerically.

Extracapillary Transport Resistance

Expressions similar to Eq. (12.19) can be derived for the mass transfer coefficients of the capillary wall, k_w, and interstitium, k_i:

$$k_w = 2\pi(L_{rbc} + L_p)\frac{D_w \alpha_w}{\ln(r_w/r_p)} \, , \qquad k_i = 2\pi(L_{rbc} + L_p)\frac{D_i \alpha_i}{\ln(r_i/r_w)} \, .$$
(12.20)

The diffusion equation for the muscle fiber region is obtained by substituting Eqs. (12.10) and (12.12) into Eq. (12.4); the resulting nonlinear equation describes both free and facilitated diffusion of O_2. This equation can be solved with the boundary conditions $P = P_i$ at $r = r_i$ and $\partial P/\partial r = 0$ at $r = r_t$. After somewhat lengthy transformations, we can express the tissue mass transfer coefficient in the form

$$k_t = \pi(r_t^2 - r_i^2)(L_{rbc} + L_p)c(K,0)^{-1}M$$

$$\left[P_i + 0.5(P_{50}^{Mb} + P_F - q) - 0.5\sqrt{(P_{50}^{Mb} + P_F - q)^2 + 4P_{50}^{Mb}q} \right]^{-1}$$
(12.21)

where

$$P_F = \frac{D_{Mb}[Mb]_T}{D_t \alpha_t} \, , \quad q = P_i\left(1 + \frac{P_F}{P_i + P_{50}^{Mb}}\right) - \frac{M}{4D_t\alpha_t}\left(2r_t^2 \ln\frac{r_t}{r_i} - (r_t^2 - r_i^2)\right). \quad (12.22)$$

Here, D_t and α_t are the oxygen diffusion and solubility coefficients in the muscle fiber, $c(K,O)$ is the stereological notation for capillary tortuosity, r_t is the radius of the tissue cylinder surrounding the capillary, r_i is the distance from the capillary axis to the sarcolemma, $P_F = D_{Mb}[Mb]_T/D_t\alpha_t$ is a myoglobin facilitation factor (it has the units of pressure), D_{Mb} is the diffusion coefficient of myoglobin, P_i is the PO_2 at the sarcolemma, and P_{50}^{Mb} is the PO_2 at which Mb saturation is 50%.

In the absence of myoglobin facilitation, Eq. (12.21) takes the form

$$k_t^0 = 4\pi(r_t^2 - r_i^2)(L_{rbc} + L_p)c(K,0)^{-1}D_t\alpha_t\left(2r_t^2\ln\frac{r_t}{r_i} - (r_t^2 - r_i^2)\right)^{-1}. \quad (12.23)$$

Because the transport equations inside the red blood cell and in the muscle fiber region are nonlinear, determination of the transport resistances requires the numerical solution of a nonlinear system of algebraic equations with respect to P_l, P_b, P_p, and P_i provided that P_c is specified.

In presenting numerical examples of resistance distributions below, we will express the intracapillary resistance fraction as

$$\rho_{cap} = \frac{k_{cap}^{-1}}{k_{cap}^{-1} + k_w^{-1} + k_i^{-1} + k_t^{-1}}, \quad (12.24)$$

where $k_{cap}^{-1} + k_w^{-1} + k_i^{-1} + k_t^{-1}$ is the total resistance along the pathway, the resistance fraction of the combined capillary wall and interstitium as

$$\rho_{w,i} = \frac{k_w^{-1} + k_i^{-1}}{k_{cap}^{-1} + k_w^{-1} + k_i^{-1} + k_t^{-1}}, \quad (12.25)$$

and the tissue resistance fraction as

$$\rho_t = \frac{k_t^{-1}}{k_{cap}^{-1} + k_w^{-1} + k_i^{-1} + k_t^{-1}}. \quad (12.26)$$

Parameter Values

The transport resistances have been expressed explicitly in terms of the parameters that can in principle be measured experimentally. The contribution of each factor can therefore be determined and understood. We emphasize that the apparent complexity of the above equations and the number of physical parameters involved reflect the complexity of the oxygen transport phenomena; the relative importance of different parameters can be assessed by a theoretical sensitivity analysis, and experimental measurements can then be aimed at determining the most important among them.

There are no muscles for which all of the parameters have been measured; thus, extrapolations from other muscles are necessary. Roy and Popel (1996) compiled a complete set of parameters necessary for calculating the transport resistances for four muscles (m. semitendinosus, vastus medialis, diaphragm, and myocardium) at maximal aerobic work in the dog, goat, horse, cow, pony, and calf. Muscle-specific oxygen consumption rates were calculated using the mitochondrial volume fraction $V_v(mt, f)$ for each muscle and values of mitochondrial oxygen consumption $\dot{V}O_2^{mt}$ at $\dot{V}O_{2max}$. $\dot{V}O_2^{mt}$ is approximately the same in all tissues [~4.5 ml O_2/(ml mitochondria · min)]; therefore, the muscle consumption rate is approximately proportional to the mitochondrial volume fraction, with its values increasing from skeletal muscle to diaphragm to myocardium. Sensitivity analysis performed for most of the parameters showed the effects of variation of the parameters on the calculated resistances and therefore on PO_2 distribution from the RBC to the mitochondrion. Here we will discuss experimental data on several critical parameters.

Tissue oxygen diffusion and solubility coefficients (D_t and α_t) have a significant impact on the tissue transport resistance; Eq. (12.21). Practically all of the reported measurements of muscle diffusion coefficients were made in unperfused tissues at room temperature due to the difficulty of keeping the muscle oxygenated at body temperature because of higher O_2 consumption. Measurements of D_t and α_t are reviewed in Dutta and Popel (1995). Bentley et al. (1993) made measurements at 37°C in the hamster retractor muscle and found the diffusion coefficient, $D_t = 2.41 \cdot 10^{-5}$ cm²/s, to be 73% higher than the value calculated by extrapolating the measurement at room temperature to body temperature using the temperature coefficient for aqueous solutions. This value is used in the results presented below. For the solubility coefficient, we use the value $\alpha_t = 3.89 \cdot 10^{-5}$ ml O_2/(ml·mm Hg) obtained for frog sartorius muscle (Mahler et al., 1985).

In previous theoretical studies, the values of the myoglobin diffusion coefficient were taken from experiments on myoglobin diffusion in 18 g/dl aqueous protein solution, the protein concentration of which corresponds to muscle protein content; the typical measured value was $D_{Mb} = 10^{-6}$ cm²/s (Riveros-Moreno and Wittenberg, 1972). Recent measurements in muscle give much smaller values, for example, $D_{Mb} = 1.73 \cdot 10^{-7}$ cm²/s for rat diaphragm at 37°C (Jürgens et al., 1994) and $D_{Mb} = 2.2 \cdot 10^{-7}$ cm²/s for rat soleus muscle (Papadopoulos et al., 1995). The value $D_{Mb} = 1.73 \cdot 10^{-7}$ cm²/s is used in the calculations unless noted otherwise. It should be noted that these values of the diffusion coefficients characterize diffusion along muscle fibers, whereas a more relevant parameter would be the diffusion coefficient across fibers. Homer et al. (1984) have shown that O_2 diffusion is anisotropic with the rate of diffusion along muscle fibers being approximately twice that of the rate across fibers. If similar results hold for Mb, however, then the diffusion coefficients across muscle fibers would be even smaller than those quoted above.

Myoglobin content, $[Mb]_T$, in striated muscle and myocardium varies within a wide range between different muscles and mammals: from $7 \cdot 10^{-10}$ mol/ml in pig longissimus dorsi (O'Brien et al., 1992) to $8.8 \cdot 10^{-7}$ mol/ml in horse vastus intermedius (Armstrong et al., 1992; Conley and Jones, 1996). In diving mammals, the myoglobin content can reach much higher values, for example, $4.4 \cdot 10^{-6}$ mol/ml in Ringed seal longissimus dorsi (O'Brien et al., 1992). Figure 12.3 demonstrates the effect of D_{Mb} and $[Mb]_T$ on the tissue mass transfer coefficient, k_t [Eq. (12.21)] normalized by the mass transfer coefficient, k_t^0 [Eq. (12.23)]. The values are shown as a function of sarcolemmal oxygen tension, P_i; for each set of parameters there exists a critical value of P_i at which the minimum tissue PO_2 equals zero. The mass transfer coefficient is not calculated for P_i below these critical levels when anoxic regions develop. Interestingly, k_t may change significantly with myoglobin content, whereas the critical PO_2 values are not very

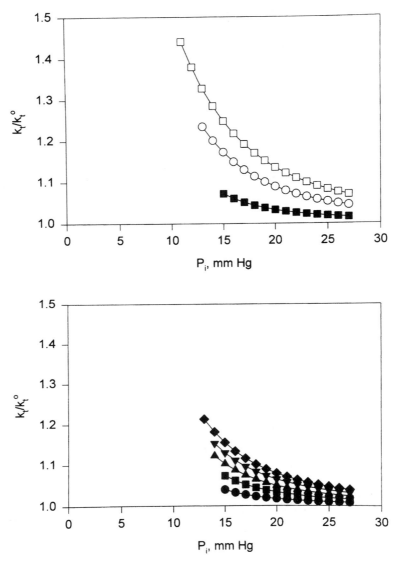

FIGURE 12.3. The ratio of the tissue mass transfer coefficient, k_t, determined by Eq. (12.21), and the tissue mass transfer coefficient in the absence of myoglobin, k_t^0, determined by Eq. (12.23), as a function of PO_2 at the sarcolemma, P_i. Upper panel: $[Mb]_T$ = 4 · 10^{-7} mol/ml. ■ D_{Mb} = 1.73 · 10^{-7} cm²/s; ○ 5 · 10^{-7} cm²/s; □ 8 · 10^{-7} cm²/s. *Lower panel*: D_{Mb} = 1.73 · 10^{-7} cm²/s. ● $[Mb]_T$ = 2 · 10^{-7} mol/ml; ■ 4 · 10^{-7} mol/ml; ▲ 6 · 10^{-7} mol/ml; ▼ 8 · 10^{-7} mol/ml; ◆ 10^{-6} mol/ml.

sensitive to these changes, as has also been demonstrated by Roy and Popel (1996).

Another important parameter is capillary hematocrit or its close correlate, RBC lineal density, which determines the intracapillary geometry shown in Figure 12.1. In adenosine-dilated hamster cremaster muscle, the ratio of capillary and systemic hematocrit has been estimated as $H_{cap}/H_s = 0.75$ (Keller et al., 1994). We assume that this value applies to the animals exercising at maximal aerobic capacity.

Calculations of Transport Resistances

The main goal of this chapter is to discuss the values of the transport resistances defined by Eqs. (12.17)–(12.21). Table 12.1 shows the summary of these calculations for 24 muscles (m. semitendinosus, m. vastus medialis, diaphragm, and myocardium in dog, goat, horse, cow, pony, and calf).

The calculations show that, on average, approximately 50% of the total transport resistance to O_2 resides outside the muscle fiber (in the capillary, capillary wall, and interstitium), in agreement with earlier studies (Federspiel and Popel, 1986). An end-capillary PO_2 in RBCs of approximately 25 mm Hg is necessary to supply the muscle fibers and to prevent them from being hypoxic. Further, approximately 66% of the flux of O_2 is supplied through the lateral surface of the cylindrical RBC, and the rest through its basal surfaces.

The upper panel of Figure 12.4 shows the extracellular (outside muscle fiber) resistance fraction, $1 - \rho_t = \rho_{cap} + \rho_w + \rho_i$, for each of the 24 muscles plotted versus $\dot{V}O_{2max}$. These results show that the extracellular resistance is a substantial fraction of the total resistance, ranging from 35% in horse semitendinosus muscle to 72% in goat myocardium, and these values do not correlate with metabolic rate.

The lower panel of Figure 12.4 shows critical end-capillary PO_2 for each of the 24 muscles considered in this study plotted versus their estimated maximal oxygen consumption, $\dot{V}O_{2max}$. There is no systematic variation of critical end-capillary PO_2 values between similar muscles in athletic animals (horse, pony, dog) and nonathletic animals (cow, calf, goat), or between the group of skeletal muscles and diaphragms and the group of myocardia whose consumption rates are significantly higher.

It should be noted that the calculations presented do not address the issue of PO_2 variation along the capillary; blood flow is not a parameter in the present calculations.

It is also of interest to look in more detail at the sensitivity of the calculated transport resistances to those parameters whose experimental values have been disputed. These calculations are performed for a particular muscle, dog vastus medialis. Under control conditions, the extracellular resistance fraction in the tissue is 0.47. When the tissue diffusion coefficient $D_t = 2.41 \cdot 10^{-5}$ cm²/s from Bentley et al. (1993) is replaced by a ten-times smaller value obtained in isolated cardiac myocytes at 30°C and extrapolated to body temperature, $D_t = 2.41 \cdot 10^{-6}$

TABLE 12.1. Transport resistance calculations.

ρ_{cap}	$\rho_{w;i}$	ρ_t	Critical PO_2 (mm Hg)	Critical SO_2	J_t/J_{rbc}	J_1/J_{rbc}
0.37 ± 0.10	0.16 ± 0.03	0.47 ± 0.10	25.3 ± 6.7	0.20 ± 0.09	0.34 ± 0.06	0.66 ± 0.06

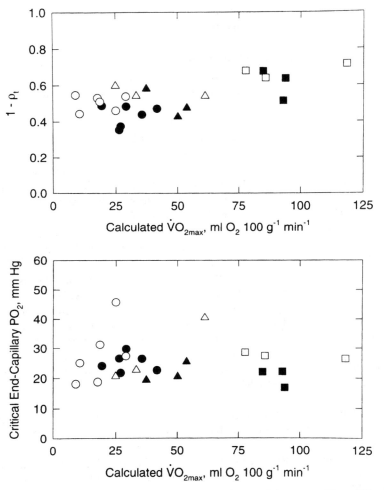

FIGURE 12.4. Extracapillary resistance fraction, $1 - \rho_t$, and end-capillary PO_2 versus $\dot{V}O_{2max}$. ○ skeletal muscle; △ diaphragm; □ myocardium. Closed symbols represent athletic animals (horse, pony, dog) and open symbols represent nonathletic animals (cow, calf, goat). Note that $\dot{V}O_{2max}$ is highest in myocardium, followed by diaphragm and skeletal muscle.

cm²/s (Jones and Kennedy, 1986), the extracellular resistance fraction becomes 0.08. However, the critical end-capillary PO_2 necessary for preventing hypoxia becomes 127 mm Hg, which is physiologically unrealistic. Thus, there appears to be a problem with the experimental values of diffusion coefficients in isolated cells (see the discussion in Dutta and Popel, 1995). The standard calculations assume a thickness of the interstitial fluid layer between the outer capillary wall and the sarcolemma of 0.35 μm; if this thickness is increased to 1.75 μm (keeping the capillary density constant), the extracellular resistance fraction changes from 0.47 to 0.63, but the critical end-capillary PO_2 increases only slightly from its standard value of 22.5 to 23.8 mm Hg. Variation of other parameters within ranges that are consistent with experimentally reported values in different tissues has only minor effects on the calculated values of the transport resistances.

Summary

We have presented expressions for oxygen transport resistances from red blood cells to the core of muscle fibers. Model calculations based on physiologically realistic parameter values show that about 50% of the transport resistance is located outside the muscle fibers, and most of that resistance is intracapillary. The calculated critical end-capillary PO_2 necessary for preventing tissue hypoxia under $\dot{V}O_{2max}$ conditions appears to be independent of muscle type (skeletal muscle, diaphragm, myocardium); the mean value of end-capillary PO_2 is approximately 25 mm Hg. Direct experimental measurements of the transport resistances are presently not available, although the results of whole organ studies are consistent with the above theoretical predictions.

Acknowledgments. This work was supported in part by Grant HL 18292 from the National Heart, Lung, and Blood Institute and a Research Fellowship award (to AD) from the American Heart Association, Maryland Affiliate, Inc.

References

Armstrong, R. B., B. Essen-Gustavsson, H. Hoppeler, J. H. Jones, S. R. Kayar, M. H. Laughlin, A. Lindholm, K. E. Longworth, C. R. Taylor, and E. R. Weibel. O₂ delivery at $\dot{V}O_{2max}$ and oxidative capacity in muscles of Standardbred horses. *J. Appl. Physiol.* 73:2274–2282, 1992.

Bentley, T. B., H. Meng, and R. N. Pittman. Temperature dependence of oxygen diffusion and consumption in mammalian striated muscle. *Am. J. Physiol.* 264:H1825–H1830, 1993.

Clark, Jr., A., W. J. Federspiel, P. A. A. Clark, and G. R. Cokelet. Oxygen delivery from red cells. *Biophys. J.* 47:171–181, 1985.

Conley, K. E. and C. Jones. Myoglobin content and oxygen diffusion: Model analysis of horse and steer muscle. *Am. J. Physiol.* 271:C2027–C2036, 1996.

Dutta, A. and A. S. Popel. A theoretical analysis of intracellular oxygen transport. *J. Theor. Biol.* 176:433–445, 1995.

Ellsworth, M. L., R. N. Pittman, and C. G. Ellis. Measurement of hemoglobin oxygen saturation in capillaries. *Am. J. Physiol.* 251:H1031–H1040, 1987.

Ellsworth, M. L., C. G. Ellis, A. S. Popel, and R. N. Pittman. Role of microvessels in oxygen supply to tissue. *News Physiol. Sci.* 9:119–123, 1994.

Federspiel, W. J. and A. S. Popel. A theoretical analysis of the effect of the particulate nature of blood on oxygen release in capillaries. *Microvasc. Res.* 32:164–189, 1986.

Gayeski, T. E. J. and C. R. Honig. Intracellular PO_2 in individual cardiac myocytes in dogs, cats, rabbits, ferrets and rats. *Am. J. Physiol.* 260:H522–H531, 1991.

Goresky, C. A. Biological barriers: Their effects on cellular entry and metabolism in vivo. *Microvasc. Res.* 29:1–17, 1985.

Groebe, K. An easy-to-use model for O_2 supply to red muscle. Validity of assumptions, sensitivity to errors in data. *Biophys. J.* 68:1246–1269, 1995.

Hellums, J. D., P. K. Nair, N. S. Huang, and N. Ohshima. Simulation of intraluminal gas transport processes in the microcirculation. *Ann. Biomed. Eng.* 24:1–24, 1996.

Hogan, M. C., D. E. Bebout, and P. D. Wagner. Effect of hemoglobin concentration on maximal O_2 uptake in canine gastrocnemius muscle in situ. *J. Appl. Physiol.* 70:1105–1112, 1991.

Homer, L. D., J. B. Shelton, C. H. Dorsey, and T. J. Williams. Anisotropic diffusion of oxygen in slices of rat muscle. *Am. J. Physiol.* 246 (*Regul. Integrat. Comp. Physiol. 15*):R107–R113, 1984.

Honig, C. R. and T. E. J. Gayeski. Resistance to O_2 diffusion in anemic red muscle: Roles of flux density and cell PO_2. *Am. J. Physiol.* 265:H868–H875, 1993.

Intaglietta, M., P. C. Johnson, and R. M. Winslow. Microvascular and tissue oxygen distribution. *Cardiovasc. Res.* 32:632–643, 1996.

Itoh, T., K. Yaegashi, T. Kosaka, T. Kinoshita, and T. Morimoto. In vivo visualization of oxygen transport in microvascular network. *Am. J. Physiol.* 267:H2068–H2078, 1994.

Jones, D. P. and F. G. Kennedy. Analysis of intracellular oxygenation of isolated adult cardiac myocytes. *Am. J. Physiol.* 250:C384–C390, 1986.

Jürgens, K. D., T. Peters, and G. Gros. Diffusivity of myoglobin in intact skeletal muscle cells. *Proc. Natl. Acad. Sci. USA* 91:3829–3833, 1994.

Keller, M. W., D. N. Damon, and B. R. Duling. Determination of capillary tube hematocrit during arteriolar microperfusion. *Am. J. Physiol.* 266:H2229–H2238, 1994.

Lipowsky, H. H., S. Usami, S. Chien, and R. N. Pittman. Hematocrit determination in small bore tubes by differential spectrophotometry. *Microvasc. Res.* 24:42–55, 1982.

Mahler, M., C. Louy, E. Homsher, and A. Peskoff. Reappraisal of diffusion, solubility, and consumption of oxygen in frog skeletal muscle, with applications to muscle energy balance. *J. Gen. Physiol.* 86:105–134, 1985.

Mathieu-Costello, O. Comparative aspects of muscle capillary supply. *Annu. Rev. Physiol.* 55:503–525, 1993.

Messmer, K. L., L. Sunder-Plassmann, F. Jesch, L. Cornandt, E. Sinagowitz, and M. Kessler. Oxygen supply to the tissues during limited normovolemic hemodilution. *Res. Exp. Med.* 159:152–166, 1973.

O'Brien, P. J., H. Shen, L. J. McCutcheon, M. O'Grady, P. J. Byrne, H. W. Ferguson, M. S. Mirsalimi, R. J. Julian, J. M. Sargeant, R. R. M. Tremblay, and T. E. Blackwell. Rapid, simple and sensitive microassay for skeletal and cardiac muscle myoglobin and hemoglobin: Use of various animals indicates functional role of myohemoproteins. *Mol. Cell. Biochem.* 112:45–52, 1992.

Papadopoulos, S., K. D. Jürgens, and G. Gros. Diffusion of myoglobin in skeletal muscle cells—Dependence on fibre type, contraction and temperature. *Pflugers Arch.* 430:519–525, 1995.

Pittman, R. N. and B. R. Duling. Measurement of percent oxyhemoglobin in the microvasculature. *J. Appl. Physiol.* 38:321–327, 1975.

Pittman, R. N. Influence of microvascular architecture on oxygen exchange in skeletal muscle. *Microcirculation* 2:1–18, 1995.

Popel, A. S. Theory of oxygen transport to tissue. *Crit. Rev. Biomed. Eng.* 17:257–321, 1989.

Pries, A. R., G. Kanzow, and P. Gaehtgens. Microphotometric determination of hematocrit in small vessels. *Am. J. Physiol.* 245:H167–H177, 1983.

Pries, A. R. A versatile video image analysis system for microcirculatory research. *Int. J. Microcirc. Clin. Exp.* 7:327–345, 1988.

Riveros-Moreno, V. and J. B. Wittenberg. The self-diffusion coefficients of myoglobin and hemoglobin in concentrated solutions. *J. Biol. Chem.* 247:895–901, 1972.

Rose, C. P. and C. A. Goresky. Vasomotor control of capillary transit time heterogeneity in the canine coronary circulation. *Circ. Res.* 39:541–554, 1979.

Rose, C. P. and C. A. Goresky. Limitations of tracer oxygen uptake in the canine coronary circulation. *Circ. Res.* 56:57–71, 1985.

Roy, T. K. and A. S. Popel. Theoretical predictions of end-capillary PO_2 in muscles of athletic and non-athletic animals at $\dot{V}O_{2max}$. *Am. J. Physiol.* 271:H721–H737, 1996.

Severinghaus, J. W. Exercise O_2 transport model assuming zero cytochrome PO_2 and $\dot{V}O_{2max}$. *J. Appl. Physiol.* 77:671–678, 1994.

Shiga, T. Oxygen transport in microcirculation. *Jap. J. Physiol.* 44:19–34, 1994.

Torres Filho, I. P. and M. Intaglietta. Microvessel PO_2 measurements by phosphorescence decay method. *Am. J. Physiol.* 256:H1434–H1438, 1993.

Torres Filho, I. P., M. Leunig, F. Yuan, M. Intaglietta, and R. K. Jain. Noninvasive measurement of microvascular and interstitial oxygen profiles in a human tumor in SCID mice. *Proc. Natl. Acad. Sci. USA* 91:2081–2085, 1994.

Voter, W. A. and T. E. J. Gayeski. Determination of myoglobin saturation of frozen specimens using a reflecting cryospectrophotometer. *Am. J. Physiol.* 269:H1328–H1341, 1995.

Wagner, P. D. Muscle O_2 transport and O_2 dependent control of metabolism. *Med. Sci. Sports Exerc.* 27:47–53, 1995.

Wayland, H. and P. C. Johnson. Erythrocyte velocity measurement in microvessels by two-slit photometric method. *J. Appl. Physiol.* 22:333–337, 1967.

Whalen, W. J., J. Riley, and P. Nair. A microelectrode for measuring intracellular PO_2. *J. Appl. Physiol.* 23:798–801, 1967.

Weibel, E. R. *The Pathway for Oxygen: Structure and Function of the Mammalian Respiratory System.* Cambridge: Harvard University Press, 1984, 425 pp.

Weibel, E. R., C. R. Taylor, and H. Hoppeler. The concept of symmorphosis: A testable hypothesis of structure-function relationship. *Proc. Natl. Acad. Sci. USA* 88:10357–10361, 1991.

Zheng, Ł., A. S. Golub, and R. N. Pittman. Determination of PO_2 and its heterogeneity in single capillaries. *Am. J. Physiol.* 271:H365–H372, 1996.

4
Metabolism in the Liver

13

Liver Cell Entry In Vivo and Enzymic Conversion

Carl A. Goresky, Glen G. Bach, Andreas J. Schwab, and K. Sandy Pang

Introduction

The liver is a unique organ from the kinetic point of view. The endothelium lining the hepatic sinusoids is so highly permeable that the organ presents only a simple barrier to parenchymal cell entry of substrates. The situation is not simple, however, because interposed between the vascular lumen and cell membrane there is an interstitial space that modifies access of molecules to the liver cell surface. The barrier itself is generally not passive. The cell membrane structure exhibits all of the usual characteristics: enzymes facilitate the utilization of substrate or the conversion of materials to products within the liver cell. We will review here how tracer methodology has provided insight into these processes.

The experimental approach used in these studies is a tracer transient approach within a steady state. In vivo, in the anesthetized dog, a catheter is placed in the portal vein for injection and another in the hepatic vein to provide for serial samples (Goresky, 1963). Tracer introduced into the circulation is used to gain insight into the events underlying its distribution into tissue, using the multiple indicator dilution technique (Chinard et al., 1955). This consists of the simultaneous rapid injection of both a reference substance and the substance under study into the blood flowing into the organ. The reference substance is ordinarily one that does not leave the vascular space; it is assumed to behave in the fashion the study substance would have, in the absence of passage out of the vasculature. The study substance is usually one that does leave the microvasculature. Venous effluent is sampled in serial fashion, beginning at the time of injection, and concentrations of the introduced tracers are measured in the effluent blood. To normalize the curves, the concentration of each substance is expressed as a fractional proportion of the total tracer injected per milliliter of blood, and outflow dilution curves are constructed by plotting fractional recoveries per milliliter versus time. With this normalization, the curves for substances behaving identically will superimpose. Displacements of the normalized dilution pattern of the study substance from that of the reference substance are then assumed to result from passage across the microvasculature, as well as from any consumption processes.

Flow-Limited Distribution

A group of substances expected to behave passively, on the basis of their behavior elsewhere, was examined first (Goresky, 1963). This included labeled red cells, which are confined to the vascular space; labeled albumin, which is predominantly confined to the vascular space during its passage through the microvasculature of organs with a continuous capillary lining; labeled inulin, sucrose, and sodium, which have been used as classical extracellular space labels and have been so viewed in the liver; and labeled water, which is expected to penetrate cells.

A representative set of normalized hepatic venous outflow dilution curves for labeled red cells, albumin, sucrose, and water is displayed in the upper and middle panels of Figure 13.2 (Goresky and Rose, 1977). The labeled red cells emerge first, their outflow dilution curve reaches the highest and earliest peak, and the curve decays the most quickly until recirculation occurs. Since we wish to deal with first passage phenomena for both this curve and the others, recirculation is excluded by linear extrapolation of the downslope of the dilution curves on a semilogarithmic plot (Hamilton et al., 1928). In comparison with the labeled red cells, the other substances show a set of systematically differing outflow profiles. From labeled albumin to labeled water there is, in comparison to the curve for labeled red cells, a progressive delay in the outflow appearance, a diminution in and delay in reaching the peak, and a progressively retarded downslope decay. Labeled albumin and the extracellular space labels (inulin, sucrose, and sodium) are confined to the plasma phase of blood; they either ordinarily do not enter red cells or will not have done so, under the conditions of these experiments. Labeled water, in contrast, rapidly enters the red cells. Despite this, it extends the continuum exhibited by the extracellular space labels. Both the labeled red cells and each of the other substances are completely recovered at the outflow. The areas under their normalized curves are identical.

Bolus flow will be expected within the sinusoids (Prothero and Burton, 1961). The red cells will be centered during their flow, and will fill the sinusoids and be deformed by them. The mixing motion between the cells and the small diameter of the sinusoids (corresponding to the diameter of a red cell) assure rapid cross-sectional equilibration within the sinusoid. Beyond the vascular lumen, the dimensions of the extracellular space and the half-width of the liver cell plate lying between the sinusoids are such that, for most small molecular weight probes, rapid lateral equilibration will be expected for each substance within its respective space.

The modeling of the exchange of albumin and the extracellular substances in the liver was developed before the structure of the sinusoidal lining cells of the liver had been well defined. The development of scanning electron microscopy, in particular, allowed for a much clearer definition of the structure of these cells (Layden et al., 1975) and this, in turn, demonstrated how exchange between plasma and interstitium must take place. Except in the region of the nucleus, the sinusoidal lining cells are occupied by sieve plates, which are groups of open fenestræ that allow free diffusional communication between the sinusoidal plasma and the interstitial space of the liver, the space lying between the sinusoidal lining cells and the surface of the liver cell, the space of Disse. Thus there is no effective barrier between the plasma of the sinusoids and the Disse space. The surface of the underlying liver cells is greatly expanded by the presence of numerous microvilli, an adaptation that will enhance the membrane carrier transport of materials.

Now let us examine what happens in the single sinusoid (Goresky, 1963; Goresky and Groom, 1984). We assume that flow is confined to the vascular compartment, diffusional equilibration takes place so rapidly in the lateral direction in the dimensionally small extravascular space that the concentration at each point along the length is equal to the concentration in the sinusoidal plasma space, and the mechanism transporting material in the longitudinal direction is flow (diffusional transport in the longitudinal direction is negligible compared with the rate at which material is transported in that direction by flow). Suppose that an amount of label q_0* is rapidly introduced at the origin of a sinusoid and that this is then carried along by the sinusoidal flow F_s with the velocity v_F. If we solve the partial differential equation describing this, we find that the expression for the outflow tracer concentration for the vascular reference, a labeled red cell, at the outflow from a sinusoid of length L, at time t, is

$$u*(L,t) = \frac{q_0*}{F_s} \delta\left(t - \frac{L}{v_F}\right),$$ (13.1)

where δ stands for a Dirac delta or impulse function. The red cell propagates along the length as a traveling wave and emerges at the time L/v_F. For a diffusible substance equilibrating within its extravascular space at each point along the length, the behavior changes. If, for such a substance, the ratio of its extravascular to its vascular spaces of distribution along the sinusoid is γ, the expected outflow profile will be

$$u*(L,t) = \frac{q_0*}{F_s} \delta\left(t - [1 + \gamma]\frac{L}{v_F}\right).$$ (13.2)

The impulse function for the diffusible substance travels along the sinusoid as if it were flowing in a larger space, $(1 + \gamma)$ times as large as that of the reference substance. This behavior is best described by the phase-delayed wave flow–limited distribution. The equilibrating tracer flows along the sinusoid with the velocity $v_F/(1 + \gamma)$, which is slower than that of the vascular reference, and consequently emerges at the outflow delayed with respect to it.

The relation between the outflow curves from the whole liver for labeled red cells and for each of the diffusible interstitial substances will depend on the relation between the transit times of the large vessels and those of the sinusoids.

There are two theoretical asymptotic extremes in the relation between these distributions: that in which the sinusoidal transit times are uniform and that in which large vessel transit times are uniform (Goresky et al., 1970). The forms of the curves in Figure 13.1 indicated that the latter might describe the case for the liver. If this is so, the relation between the outflow curves for each of the interstitial substances and the vascular reference, labeled red cells, is described (Goresky and Groom, 1984) by the equation,

$$C_{int}(t - t_0) = \frac{1}{1 + \gamma} C_{RBC}\left(\frac{t - t_0}{1 + \gamma}\right), \tag{13.3}$$

where t_0 is the common large vessel transit time. From Eq. (13.3) the expectation is that, after a common large vessel transit time, each point along the curve for the diffusible interstitial substance (compared with the reference curve) will have its transit time increased by the factor $(1 + \gamma)$ and, because the areas under the normalized curves are all the same, its magnitude decreased by the factor $1/(1 + \gamma)$. When each diffusible curve is transformed by use of its characteristic γ, so as to reverse these effects, it should superimpose on the labeled red cell curve. The curves for the diffusible substances have been appropriately transformed, each with its characteristic γ or space ratio, in the bottom panel of Figure 13.1. Superimposition occurs for all of the labeled diffusible substances in a virtually ideal fashion. Using this superimposition as a criterion, we can infer that there is flow-limited distribution of these substances in the liver. Within the data set, labeled water is found to behave in the same fashion as the substances entering the Disse space (Goresky, 1963). With the long transit times through the liver and the large surface area for exchange (which has been increased further by the formation of microvilli), a retarding effect of the cell membrane on the exchange of tracer water is not perceptible during a single passage with the kind of perfusion that is encountered in vivo.

The volume of distribution of each of the tracers can be calculated as the product of the flow of the medium in which the tracer is distributed and the mean transit time, corrected for the mean transit times of the catheters. The blood volume of the liver is relatively large, of the order of 0.25 ml/g wet weight (Goresky, 1963). The larger proportion of this is sinusoidal, so that it would be expected that the heterogeneity of transit times in the sinusoids would be predominant. Values obtained for sinusoidal blood volume, with the t_0-value obtained after correction of the dilution curves for catheter distortion (Goresky and Silverman, 1964), correspond well with those obtained by morphometry (Hess et al., 1973). For labeled water, the element used in the volume calculation is the flow of blood water. The water content of the liver and its contained blood, which is estimated from the labeled water dilution curve, corresponds to the gravimetric water content of the organ (Goresky, 1963). The interstitial or extracellular tracers are confined to the plasma phase of blood. The element used in volume calculation for these is plasma flow. If it is assumed that the red cells mark out the vascular space, the interstitial spaces of distribution for these substances can be calculated as the product of plasma flow and the difference between the mean transit times of these

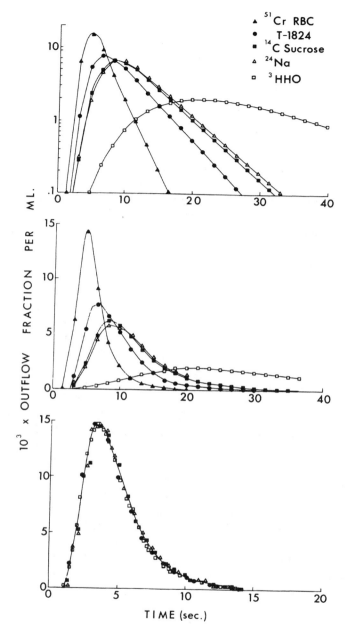

FIGURE 13.1. Normalized tracer hepatic venous outflow curves. T-1824 refers to albumin labeled with Evans Blue dye. The ordinate scale in the upper panel is logarithmic and that in the middle and lower panels, linear. The curves are corrected for recirculation by linear extrapolation on the semilogarithmic plot. The superimposition of the adjusted diffusible label curves upon the labeled red cell curve is illustrated in the bottom panel. (From Goresky and Rose, 1977, with permission.)

TABLE 13.1. Relative volumes accessible in the extracellular space of the liver.

Substance	Percent liver weight	Accessible proportion of sodium space
Sodium	8.9	1.00
Sucrose	8.8	0.99
Inulin	7.7	0.87
Albumin	5.7	0.64

labels and that for labeled red cells. The liver is an organ with a comparatively very large extra plasma space for albumin, above that available to red cells. Estimates of this from the dilution curves and from tissue analysis coincide (Goresky, 1963). Low molecular weight estimates of the interstitial or Disse space from the dilution curves correspond to those obtained by morphometry (Hess et al., 1973). Thus transient, steady-state, and morphometric values correspond.

For the extracellular or interstitial space labels, average spaces of distribution, expressed as a proportion of organ weight, are presented in Table 13.1. The implicit assumption is that albumin as well as the extracellular probes enter the interstitial space of the liver. For albumin, since its space of distribution is proportionately much larger than in organs with continuous capillaries, the data appear consonant with the assumption. Despite the assumption that albumin and the extracellular probes enter a common space of Disse, it is evident that the calculated size of the accessible space varies; it decreases as the size of the probing molecule increases. There is an apparent molecular exclusion phenomenon that increases with molecular weight. An analogous exclusion phenomenon has been demonstrated in vitro for two polymeric substances that are characteristic of the extracellular space, hyaluronic acid gel (Ogston and Phelps, 1961) and collagen (Wiederhelm and Black, 1976). It therefore appears that the probing interstitial molecules are encountering an exclusion effect imposed by collagen and other components of the extracellular matrix present in the Disse space.

The findings indicate that the interstitial space of the liver is immediately accessible to substances dissolved in the plasma. Hence, in studying cell entry and uptake by the liver cells, it will be necessary to use, in addition to labeled red cells, a second substance that behaves in the same fashion as the uptake substance would if it did not enter liver cells, or a group of substances from which such a reference can be derived. For low molecular weight substances not bound to protein, labeled sucrose or sodium is the appropriate second reference; and for substances highly bound to albumin, labeled albumin will be more appropriate.

Liver Cell Entry with Mass Conservation

The first major biological barrier to the distribution of materials in the liver is the hepatocyte cell membrane. The anatomical structure of the liver provides immediate access to the hepatic cell surface through the Disse space, with no intermediate resistance at the level of the liver sinusoidal lining cells. The regime in which the participation of the liver cell membrane in the cell entry process will be most easily characterized is that in which the substance will be neither consumed by a metabolic process nor secreted in bile. Under these circumstances there will be a steady concentration along the length of the hepatic sinusoids from input to output. Concentrations in the sinusoid/Disse space complex and in the liver cells will be the same everywhere along the length. Analysis of the tracer profiles will then provide information concerning the barrier at the liver cell membrane and the intracellular space of distribution.

For important substrates, passage across the cell membrane has evolved in such a fashion that a component of the cell membrane (a carrier protein) facilitates their transfer across the cell membrane. The specificity of these varies with the kind of substrate involved. At a kinetic level, the mechanics of the transfer exhibit a set of common characteristics (Wilbrandt and Rosenberg, 1961). The first is the presence of saturation kinetics. Initial rates of uptake of material are found not to increase proportionately with increasing concentration, but to saturate, with a maximum rate of uptake being achieved at high external concentration. The second is the phenomenon of competitive inhibition. An interacting and penetrating substance that competes for the transfer process will inhibit the cellular entry of the material with which it is competing. The third is the induction of countertransport. Suppose that a first substrate is equilibrated on the two sides of a cell membrane, and that a second substrate is added to the external medium, which can also combine with the transport molecule. The equilibrium for the first substance will be interrupted, and a net outward movement of the initial substrate will occur, an uphill countertransport. If this is labeled, a flush of label from cell to medium will be observed. The transfer process will, for many substances, reach an equilibrium in which cellular concentrations are equal to those outside the cell. Cell entry and exit processes are then kinetically symmetrical. For other substances, the process will be concentrative, with cellular concentrations reaching levels many times those outside the cell. In this case, as the external concentration is raised, the equilibrium concentration ratio (cell/outside medium) characteristically diminishes. For this case, the usual behavior is one in which the entry process is found to saturate, whereas the exit process tends to be linear. As a result, the ratio of the unidirectional permeability surface area products, influx/efflux, decreases as the underlying steady-state concentration is raised.

The finding of analytical descriptions for the behavior of the interstitial reference, to which uptake is referred, and of a substance undergoing flow-limited distribution into the liver cells led to the expectation that it would be possible to model both the upslopes and the downslopes of dilution curves for cases of more limited and more rapid cell entry. The expectation was in contradistinction to the Crone approach, in which return of tracer to the vasculature was neglected, so that only upslopes of dilution curves could be modeled, and then only for the case of limited permeability (Crone, 1963).

A Concentrative Entry Process

One of the major processes at the liver cell membrane is that which steadily maintains the potassium ion concentration in the liver cells at a level many times that in the surrounding extracellular fluid. Rubidium, with a diffusion coefficient almost identical to that of potassium, is concentrated in liver cells in a manner similar to potassium (Love et al., 1954) and has an isotopic species, ^{86}Rb, with a half-life that is convenient for use, unlike [42]K. A set of outflow curves from a labeled rubidium experiment is shown in Figure 13.2 (Goresky et al., 1973a). The injection mixture includes labeled red cells (the vascular reference), labeled sodium (the appropriate low molecular weight second reference, which enters the Disse space freely, but does not enter liver cells to any measurable degree during the time of a single passage), and labeled rubidium. The first part of the labeled rubidium curve is contained within the labeled sodium curve. On the semilogarithmic plot, the initial values for labeled rubidium are slightly lower than those for labeled sodium, and thereafter diverge as the labeled rubidium values become a progressively smaller fraction of the labeled sodium curve with time. The peak of the labeled rubidium curve is therefore lower and slightly earlier than that for the labeled sodium, and the early downslope decays more rapidly than that for the labeled sodium. The final part of the labeled rubidium curve, the late return of rubidium activity (which conservation dictates must occur, but which will be much delayed and decreased in magnitude by the concentrative effect at the liver cell membrane) is obscured by recirculation.

Kinetic analysis of the case in which there is exchange across the liver cell membrane results in the following expression, $C_{exch}(t)$, for the exchanging material. With the time shift $t' = t - t_0$, this becomes (Goresky et al., 1973a, 1992)

$$C_{exch}(t) = e^{-[k_1\theta/(1+\gamma_{ref})]t'} \cdot C_{ref}(t)$$

$$+ e^{-k_2 t'} \int_{t_0}^{t} C_{ref}(\tau') e^{-[k_1\theta/(1+\gamma_{ref}) - k_2](\tau' - t_0)}$$

$$\cdot \sum_{n=1}^{\infty} \frac{\left(\left[\dfrac{k_1\theta k_2}{1+\gamma_{ref}}\right][\tau' - t_0]\right)^n (t - \tau')^{n-1}}{n!(n-1)!} \, d\tau' , \qquad (13.4)$$

where $C_{ref}(t)$ is the outflow curve for the appropriate second reference (labeled sodium, in this case), with an interstitial space ratio γ_{ref} corresponding to that of the substance under study, and a cell to sinusoidal plasma water space ratio of θ. The parameters k_1 and k_2 are the permeability surface area products per cellular water space for cellular influx and efflux, respectively, and the expression $k_1\theta/(1 + \gamma_{ref})$ is a mass transfer coefficient that corresponds to the permeability surface area product for influx, divided by the size of the space from which the flux comes (the sum of the sinusoidal plasma and interstitial spaces). The outflow response is related to that of the second reference. It consists of two parts: a throughput component, a tracer that sweeps past the liver cell surface without entering the liver cells to emerge without any additional delay, which is related to the second reference (the labeled sodium curve) in the labeled rubidium experiments by a falling exponential of time, $\exp[-k_1\theta/(1 + \gamma_{ref})](t - t_0)$; and a returning compo-

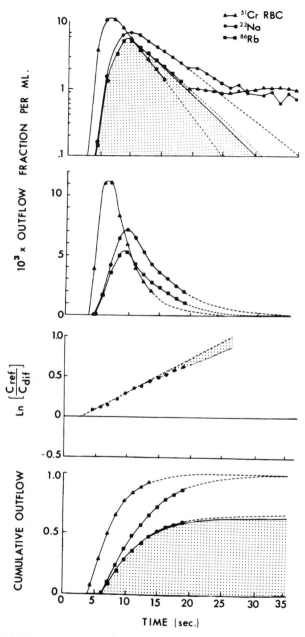

FIGURE 13.2. Rubidium experiment. *Upper two panels:* Dilution curves (outflow fractions/ml) vs. midinterval collection times, with logarithmic and linear ordinates. *Third panel:* ln(^{22}Na/^{86}Rb) vs. time. *Bottom panel:* Cumulative outflow of tracers vs. end-interval collection times. Upper and lower panels early shaded areas, calculated throughput components of rubidium curves; third panel shaded area: deviation of ln ratio plot from linear corresponding to ln(^{22}Na/[^{86}Rb throughput component]). (From Goresky and Rose, 1977.)

nent, consisting of a tracer that has entered the liver cells and returned to the vascular space later in time, a component that in the rubidium case is low in magnitude and prolonged in time. Because of the concentrative effect at the liver cell membrane, the returning component shows a prolonged tailing that is low in magnitude and not easily perceived at the experimental level, such that the throughput component of the curve becomes isolated and laid bare. It becomes the preponderant part of the outflow profile during the interval from injection to recirculation.

In Figure 13.2, the throughput component has been resolved from the outflow curve and has been shaded in the upper and lower panels. The small difference between the throughput and the total for labeled rubidium represents the returning component early in time. The expected relation between the second reference and a response composed only of the throughput component is

$$\ln\left[\frac{C_{\text{ref}}(t)}{C_{\text{exch}}(t)}\right] = \left(\frac{k_1\theta}{1 + \gamma_{\text{ref}}}\right)t. \tag{13.5}$$

For a substance that is highly concentrated in cells, such as rubidium, the plot of the natural logarithm of the ratio between outflow concentrations of the second reference (sodium) and the substance taken up (rubidium) is expected to be a straight line with the slope $k_1\theta/(1 + \gamma_{\text{ref}})$, deviating only as returning material becomes a measurable part of the outflow profile. For rubidium, the log ratio plot, the third panel of Figure 13.1, demonstrates a pattern corresponding to these expectations.

The emergence of the throughput component as a visibly separate entity within an outflow dilution curve for a substance entering and leaving liver cells will not ordinarily be expected when the process is not highly concentrative (for rubidium, the k_1/k_2 ratio is very large; it is of the order of 20 or greater). On the other hand, even for labeled rubidium, which does show this phenomenon, all of the tracer will ultimately be expected to arrive at the outflow (except for the small proportion entering bile); a later part of the curve must be present. Hence the curve for cumulative total activity, in the bottom panel of Figure 13.2, would be expected to rise slowly to a final asymptote near unity.

Nonconcentrative Cell Entry

The intracellular concentration of free glucose in the liver has been demonstrated to be equal to the concomitant plasma concentration (Cahill et al., 1958). The cellular entry of D-glucose is the prototype of a nonconcentrative cellular entry process in the liver. To characterize this, tracer glucose experiments were carried out when the underlying bulk glucose concentrations had been set by infusion and were in a steady state (Goresky and Nadeau, 1974). Characteristic outflow curves are illustrated in Figure 13.3. Labeled red cells are the vascular reference, and labeled sucrose, a substance that is confined to the plasma and interstitial space, is the second reference. The labeled glucose is added to the injection mixture just before its introduction. Since labeled glucose enters dog red cells very slowly, the label may be considered confined to the plasma phase of the blood at the time of the experiment. In all three panels, the sucrose curve is related to the labeled red

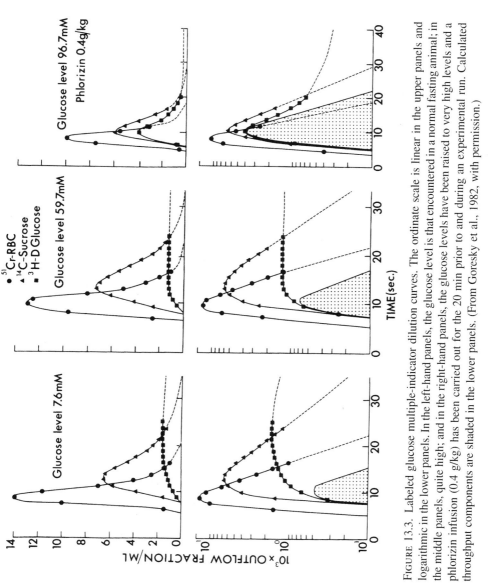

FIGURE 13.3. Labeled glucose multiple-indicator dilution curves. The ordinate scale is linear in the upper panels and logarithmic in the lower panels. In the left-hand panels, the glucose level is that encountered in a normal fasting animal; in the middle panels, quite high; and in the right-hand panels, the glucose levels have been raised to very high levels and a phlorizin infusion (0.4 g/kg) has been carried out for the 20 min prior to and during an experimental run. Calculated throughput components are shaded in the lower panels. (From Goresky et al., 1982, with permission.)

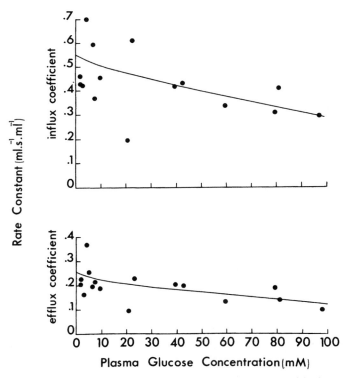

FIGURE 13.4. Changes in transfer coefficients for tracer D-glucose with change in underlying plasma glucose concentration. (Reproduced with permission from Goresky, 1983. Copyright 1983 American Heart Association.)

cell curve in the manner expected for flow-limited distribution. In the panels with the normal glucose level, the labeled glucose curve begins to rise at the same time as the labeled sucrose curve but increases to a peak that is later and substantially lower than for sucrose; it then decreases slowly, crossing the sucrose curve. When the glucose levels are high (middle panels), the glucose curve has a more squared-off initial slope on the semilogarithmic plot; it rises quickly to a low, prolonged, and flattened peak, from which it decays slowly. Model fitting with Eq. (13.4) shows that the change in shape is chiefly the result of an increase in magnitude of the throughput component. At both the lower and higher glucose levels, the predominant part of the tracer glucose curve is the exchanging component. Without model analysis, it would not have been possible to resolve the components of the outflow profiles for tracer glucose. As the plasma glucose level is increased, influx $[k_1\theta/(1 + \gamma_{ref})]$ and efflux (k_2) coefficients decrease (see Figure 13.4). If k is either an influx or efflux coefficient and C is the plasma bulk concentration, we would expect for a cell membrane carrier mediated mechanism

$$kC = V_{max}\frac{C}{K_m + C} \tag{13.6}$$

or

$$k = \frac{V_{max}}{K_m + C}.$$ (13.7)

Least squares fitting of the profile gave K_m values for influx and efflux of 121 mM, and a V_{max} value of 0.028 mmol·s^{-1}·ml^{-1} intracellular liver space, a value about three times the transport maximum for human erythrocytes (Wilbrandt and Rosenberg, 1961), for which the glucose transport rate is exceedingly rapid.

The model analysis would, intuitively, be trusted to a greater extent if the experimental conditions could be manipulated to make the throughput component emerge from the dilution curve. The right-hand panels of Figure 13.4 show such an effect. When the bulk glucose level is increased to very high levels and the competitive inhibitor phlorizin is infused, an early hump emerges from the tracer glucose curve, consisting mainly of a throughput component. In addition to saturation and competitive inhibition, which are demonstrated above, the transfer of tracer glucose has been shown to exhibit stereospecificity (L-glucose enters liver cells very slowly) and isotope countertransport (induced by a bulk glucose bolus) (Goresky and Nadeau, 1974). These in vivo, in situ studies show that the hepatic cellular entry of D-glucose exhibits the major operational characteristics of a membrane carrier transport system.

Liver Cell Entry with Intracellular Sequestration

The process of intracellular sequestration, that is, removal of substrate from the intracellular pool, introduces a new effect. Concentrations in the sinusoidal vascular space diminish from input to output and, concomitantly, so do those within the liver cells. The task of describing the underlying bulk profiles and of relating tracer uptake to these is presently evolving. The ultimate aim of the evolution of this kind of tracer methodology will be the development of the technology to a level where it will be possible to access intracellular metabolism and secretion in situ in an intact organ.

Hepatic Cell Entry and Sequestration with Resistance at the Cell Membrane

D-Galactose, when it accumulates chronically to substantial levels within the circulation, causes central nervous system problems. In normal animals, a system for hepatic removal has developed. D-Galactose is removed by a hepatic intracellular sequestration mechanism, which involves the phosphorylation of galactose to galactose 1-phosphate by galactokinase, its further conversion to glucose 1-phosphate, and its subsequent incorporation into glycogen or release as free intracellular glucose. Net removal of galactose exhibits saturation (Tygstrup and Winkler, 1954). At very low plasma levels, removal is almost complete; whereas at very high levels, the removal rate becomes constant, so that the arterial-hepatic vein difference varies inversely with flow (Tygstrup and Winkler, 1958).

The outflow profiles from a set of multiple-indicator dilution experiments with tracer D-galactose (Goresky et al., 1973b) are shown in Figure 13.5 for low (0.3 mM), intermediate (12.5 mM), and high (17.5 mM) steady-state galactose concentrations; glucose was in the normal range. A linear tracer model was developed to describe this, in which the rate of sequestration was assumed proportional to the concentration (Goresky et al., 1973b). The resultant kinetic analysis, fitted to the tracer galactose outflow curve, $C_{seq}(t)$, was in relation to the labeled sucrose second reference curve, $C_{ref}(t)$,

$$C_{seq}(t) = e^{-[k_1\theta/(1 + \gamma_{ref})]t'} \cdot C_{ref}(t)$$

$$+ e^{-(k_2 + k_{seq})t'} \int_{t_0}^{t} C_{ref}(\tau') e^{-[k_1\theta/(1 + \gamma_{ref}) - (k_2 + k_{seq})](\tau' - t_0)}$$

$$\cdot \sum_{n=1}^{\infty} \frac{\left(\left[\dfrac{k_1\theta k_2}{1 + \gamma_{ref}}\right][\tau' - t_0]\right)^n (t - \tau')^{n-1}}{n!(n-1)!} \, d\tau', \qquad (13.8)$$

where k_{seq} is the rate constant for intracellular sequestration, with dimensions ml · s^{-1} · ml^{-1} intracellular space and $t' = t - t_0$. This differs from Eq. (13.4) in that the intracellular sequestration process reduces the magnitude of the returning component. At the lowest galactose level, the modeling analysis indicates that the tracer galactose curve consists of an early-in-time, low-in-magnitude shaded peak (the throughput component) followed by an abbreviated tailing. At the intermediate galactose level, the early shaded throughput component increases in magnitude and grades over into a much increased tailing, which crosses the downslope of the labeled sucrose curve; a substantially larger proportion of the tracer galactose emerges at the outflow. At the highest galactose concentration, the throughput component becomes substantially larger, producing a perceptibly earlier peak on the labeled galactose curve, and the trailing component also increases in magnitude, so that the outflow recovery is even larger; net uptake of the label becomes much diminished.

In the modeling analysis, setting the sequestration rate constant to zero gave the form the fitted tracer galactose curve would have had if none of the tracer had been sequestered. The forms of the no-sequestration curves correspond more or less to those observed in the tracer glucose experiments. In the illustration, the expected sequestration effect has also been shaded. It is the area between the $k_{seq} = 0$ curve and the experimental data. The sequestration effect is most marked late in time, as expected.

The enzymatic process underlying the sequestration of galactose is characterized by Michaelis–Menten kinetics (Cuatrecasas and Segal, 1965), and thus the sequestration coefficient is expected to diminish with an increase in concentration in the same way that the influx and efflux coefficients do, with saturation of the

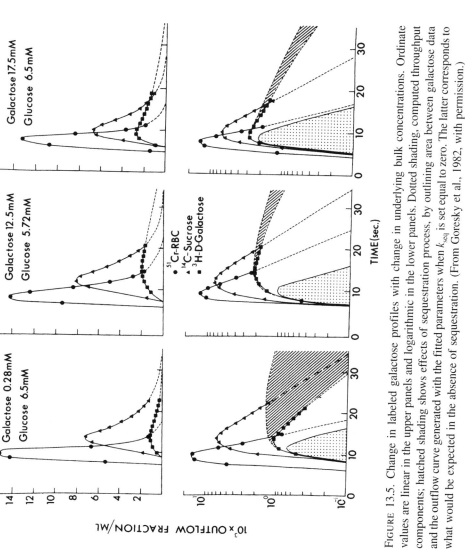

FIGURE 13.5. Change in labeled galactose profiles with change in underlying bulk concentrations. Ordinate values are linear in the upper panels and logarithmic in the lower panels. Dotted shading, computed throughput components; hatched shading shows effects of sequestration process, by outlining area between galactose data and the outflow curve generated with the fitted parameters when k_{seq} is set equal to zero. The latter corresponds to what would be expected in the absence of sequestration. (From Goresky et al., 1982, with permission.)

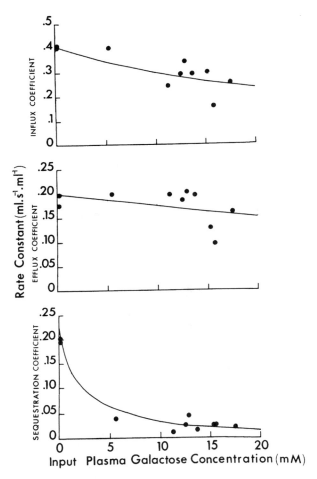

FIGURE 13.6. Manner in which transfer coefficients for tracer galactose changes with corresponding input bulk galactose concentrations. (Reproduced with permission from Goresky, 1983. Copyright 1983 American Heart Association.)

membrane uptake process. The change in all three of these coefficients with change in the input galactose concentration is shown in Figure 13.6. The sequestration rate constant saturates over a lower concentration range, and the influx and efflux rate constants saturate over a much higher concentration range. For the linear case, the intracellular concentration is predicted to be $k_1/(k_2 + k_{seq})$ times the adjacent plasma concentration. We will use this approximation even though the results demonstrate that both membrane transfer and intracellular sequestration are nonlinear. When the influx coefficient is related to the average sinusoidal concentration across the organ and its nonlinear behavior with respect to this is taken into account, its K_m is estimated to be 22.5 mM; and when the sequestration coefficient is related to the estimated intracellular concentration, the K_m was estimated to be 0.4 mM. The latter is of the same order as the K_m for galactokinase from pig liver (Cuatrecasas and Segal, 1965).

A set of steady-state concentration profiles, decreasing from input to output, is expected in both plasma and liver cells, but will be lower in liver cells. In the presence of resistance at the liver cell membrane, the intracellular sequestration of galactose is predicted to create a drop in concentration across the liver cell membrane, so that a lower but parallel gradient in the liver cells will be expected. The presence of this has been documented. On autoradiography, an intravenous injection of labeled galactose has been found to leave an acinar concentration gradient in radioactivity, decreasing from portal vein to hepatic vein, in the glycogen granules in the liver cells, reflecting the underlying cellular concentration gradient of free galactose (Goresky et al., 1973b).

Enzymic Sequestration and Product Formation

In the previous example, the details of the mechanism leading to substrate sequestration were neglected. The process was regarded as linear and, if product had been released at the outflow, it would have been expected to emerge delayed with respect to the precursor (Goresky et al., 1993a). On the other hand, the details of the nonlinear behavior of the enzymic mechanism can be important and can lead to kinetic events that would be completely unexpected if the mechanism itself were not explored and characterized (Goresky et al., 1993a).

Principles Underlying the Enzymic Mechanism

The simplest paradigm for exploration of the enzymic removal mechanism is that in which there is flow-limited presentation of substrate to the hepatic removal site. We will examine this in detail. The general experimental situation is one in which the tracer disposition is explored within a steady state. The problem which presents is that of developing expressions for steady state for a bulk substrate (time-dependent terms are set to zero) and for tracer transient disposition, and of solving these simultaneously.

We will consider Michaelis–Menten type kinetics. Consider that substrate U combines with free enzyme E to form an enzyme–substrate complex ES with the rate constant k_a and that the complex either dissociates to form E and S with the rate constant k_d or proceeds to form product P in an irreversible manner with the rate constant k_s, viz.

$$[E] + [S] \underset{k_d}{\overset{k_a}{\rightleftharpoons}} [ES] \overset{k_s}{\longrightarrow} [E] + [P]. \tag{13.9}$$

In this case, $K_m = (k_d + k_s)/k_a$ and, if $[E_T] = [E] + [ES]$ is the total enzyme in the system, $V_{max} = k_s [E_T]$.

To utilize these relations in the transient case, one must develop the usual equations of motion together with appropriate rate equations for each of the steps underlying the enzymic mechanism, for both bulk and tracer, rather than dealing with the enzymic V_{max} and K_m (Goresky et al., 1993b). In correspondence to the steady state, bulk equations need to be solved for a steady input concentration with the development of appropriate expressions for corresponding lengthwise concentration profiles. When the enzyme is uniformly distributed, in the flow-limited case, for very low input concentrations, the common vascular/tissue profile is exponential; whereas at high concentrations with respect to K_m, it grades over into a linearly decreasing profile that corresponds to a maximal rate of removal. Where a barrier is present at the cell membrane and the membrane mechanism is symmetric, intracellular removal creates a stepdown in concentration at the cell membrane, so that there are parallel profiles in blood and tissue. When the barrier is linear, the profiles are again exponential at very low concentration, and again grade over into a linear decrease at very high concentration (Goresky et al., 1983a, 1993b).

The general tracer transient solution for the flow-limited case resembles Eq. (13.8), with the microscopic rate constants k_a, k_d, and k_s of the enzymic mechanism serving as the rate constants in the equation (Goresky et al., 1993b). The kinetics of the enzymic mechanism effectively introduce into the system an enzymic space that is equivalent to a new compartment behind a barrier, one that is saturable with increasing bulk concentration. When there is a linear barrier at the cell membrane, the form of the solution corresponds to that found for the two-barrier case (Rose et al., 1977); the enzymic space, when large enough to be perceptible, results, in this instance, in a late, slowly decaying final component on the downslope of the dilution curve (Goresky et al., 1993b).

The importance of the details of the enzymic mechanism for understanding tracer kinetics was brought to our attention in an analysis of tracer alcohol data (Goresky et al., 1983a). In that case, a particular asymptote of the enzymic analysis describes the data. It is that in which the microscopic enzymic constants k_a and k_d are so large that they may be regarded as infinite. Removal again is found to saturate with increasing bulk concentration. The enzymic space appears, in this case, as an immediately accessible space additional to the cellular space, which saturates with increase in bulk concentration, rather than a separable compartment with entry and exit resistances (Goresky et al., 1983a, 1993a, 1993b).

The Behavior of Alcohols in the Liver: An Enzymic Space Effect

Ethanol removal by the liver was studied by carrying out tracer indicator dilution curves with labeled ethanol in the situation in which steady-state bulk ethanol levels were set previously by steady infusion (Goresky et al., 1983b). Figure 13.7 demonstrates effluent dilution curves for low, intermediate, and high steady-state ethanol levels in both rectilinear (upper panels) and semilogarithmic (middle panels) formats. The labeled ethanol curve is related to the labeled water curve in each case. At the lowest concentration, on the semilogarithmic plot, the upslope of the labeled ethanol curve is initially close to the labeled water curve, but with time

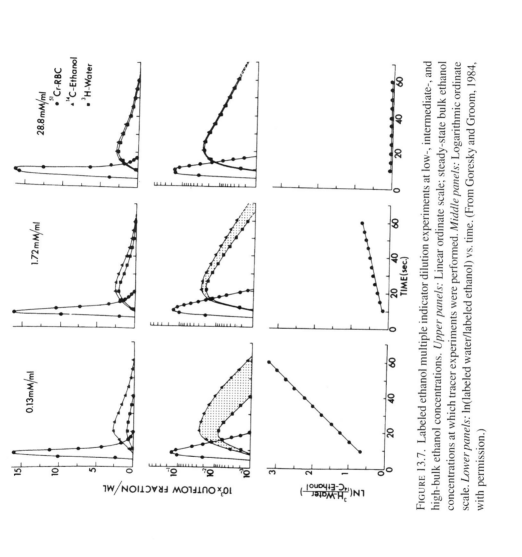

FIGURE 13.7. Labeled ethanol multiple indicator dilution experiments at low-, intermediate-, and high-bulk ethanol concentrations. *Upper panels:* Linear ordinate scale; steady-state bulk ethanol concentrations at which tracer experiments were performed. *Middle panels:* Logarithmic ordinate scale. *Lower panels:* ln(labeled water/labeled ethanol) *vs.* time. (From Goresky and Groom, 1984, with permission.)

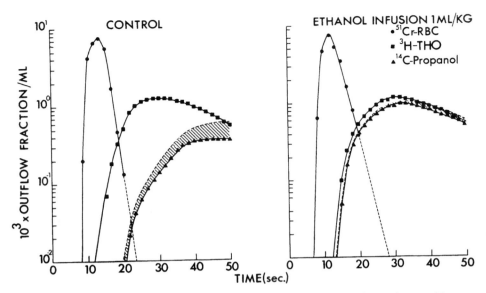

FIGURE 13.8. Outflow tracer dilution curves from a labeled propanol experiment without (control) and with preceding and concurrent saturating bulk ethanol infusions. The calculated appropriate reference is illustrated as a dashed line on each panel. The uptake effect (the area between the appropriate reference curve and the labeled propanol curve) is shaded. (From Goresky et al., 1983a, with permission.)

it progressively diverges from it so that the peak of the labeled ethanol curve is lower and earlier than that of the labeled water curve, and the downslope decays more rapidly than that for labeled water. The whole of the labeled ethanol curve is contained within the labeled water curve and appears related to it by a function that causes progressive divergence of the two curves with time. With increase in the concentration of ethanol the divergence between the two curves diminishes; and when the underlying ethanol concentration is very high, the labeled ethanol curve clearly converges on the labeled water curve.

At the same time, the behavior of the one-, three-, four-, and five-carbon straight chain monohydric alcohols was surveyed (Goresky et al., 1983b). The curves for labeled methanol were close to those for labeled water; little removal was occurring. For propanol, butanol, and pentanol, a common and completely different pattern of behavior was found. Their bulk infusion provokes hemolysis; hence their behavior was explored without and with saturating levels of bulk ethanol. A characteristic set of profiles for labeled propanol is displayed in Figure 13.8. With no preceding ethanol infusion, the labeled propanol emerges substantially delayed with respect to the labeled water. Its appropriate reference clearly has a distribution volume that is substantially larger than that for tracer water. After infusion of bulk ethanol, the labeled propanol emerges much earlier, just after the labeled water, rises to a peak just below and slightly later than that for labeled water, and the two curves are close on the downslope. The appropriate reference for the labeled propanol curve has shifted much closer to the labeled water curve. Similar findings were observed with labeled butanol and pentanol.

The expression describing the behavior of substrate in this data set, which corresponds to the k_a, $k_d \to \infty$ case, is (Goresky et al., 1983a, 1983b, 1993b)

$$C_S(t') = \frac{1}{1 + \gamma_{rel}} C_{THO}\left(\frac{t'}{1 + \gamma_{rel}}\right) e^{-k'_{seq}t'}, \tag{13.10}$$

where $t' = t - t_0$,

$$1 + \gamma_{rel} = \frac{1 + \beta' + \gamma + \theta'}{1 + \beta + \gamma + \theta}, \tag{13.10a}$$

$$\frac{\theta'}{\theta} = 1 + \frac{[E_T]}{[\hat{S}] + K_m} + \frac{\Delta}{\theta}, \tag{13.10b}$$

$$k'_{seq} = \frac{k_{seq}\theta}{1 + \beta' + \gamma + \theta'}, \tag{13.10c}$$

and

$$k_{seq} = \frac{V_{max}}{[\hat{S}] + K_m}. \tag{13.10d}$$

β is the ratio of the red cell water space to the plasma water space in the sinusoids, β' is the red cell ratio for the test substance, θ is the ratio of cell to plasma space for water, θ' is the ratio of liver cell space to plasma space for the substrate, and Δ is the ratio of the nonsaturable increment in cellular space in excess of that available to tracer water (probably a consequence of the solubility of higher alcohols in cellular lipids) to the plasma water space. In the alcohol cases, we assume that $\beta' = \beta$. The expression $\hat{C}$ is the logarithmic average bulk concentration,

$$\hat{C} = \frac{C_{in} - C_{out}}{\ln C_{in} - \ln C_{out}}, \tag{13.10e}$$

where C_{in} is the input bulk substrate concentration and C_{out} is the output bulk substrate concentration. Fitting the data involves defining both the appropriate reference curve, with the ratio $(1 + \gamma_{ref})$, and the rate constant k'_{seq} for the uptake process.

For the labeled ethanol data, except for the very lowest bulk concentrations, the calculated space of distribution corresponded to that for labeled water. Hence a plot of the logarithmic ratio, $\ln [C_{THO}(t)/C_S(t)]$, versus time, illustrated in the lower panel of Figure 13.8, yields a set of straight lines, the slope of which is k'_{seq}. The average value for $\hat{C}$, summed over all of the flow paths in a given experiment, was found to be equal to the mixed venous average $\hat{C}$; hence, the latter was used in the further analysis of the data. The sequestration rate constant k'_{seq} has the form $V_{tot}/(K_m + \hat{C})$. The fit of this expression to the ethanol data is illustrated in

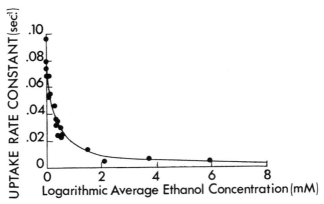

FIGURE 13.9. Change in uptake rate constant for tracer ethanol with change in the underlying steady-state concentration of bulk ethanol, expressed as the logarithmic average ($\hat{C}$) across the organ. (Reproduced with permission from Goresky et al., 1983a. Copyright 1983 American Heart Association.)

Figure 13.9. Best fit values for V_{tot} and K_m are 0.025 μmol $\cdot$ s^{-1} $\cdot$ ml^{-1} and 0.33 mM.

The fit of Eq. (13.10) to a set of representative propanol data without and with saturating ethanol infusion is illustrated in Figure 13.8. On average, the additional enzymic space was, for propanol and butanol, 3.23 and 2.10 times the cellular water space, and the respective nonsaturable lipid spaces were 0.07 and 0.20 times the cellular water spaces. From the comparison of the no-ethanol infusion data for propanol and butanol with that for tracer ethanol, the K_m values were, respectively, 0.0083 and 0.014 mM. The affinity of the enzyme alcohol dehydrogenase for these monohydric alcohol species is much larger than that for ethanol.

Product Formation

Product formation will be expected to occur in a distributed in-space fashion within liver cells, its evolution at each site being governed by local enzymic activity. When the enzymic activity is uniformly distributed, the characteristic lengthwise concentration profiles for precursor substrate evolve, exponential at low concentration, grading over into linearly declining profiles with saturation at high concentration. If product emerges exclusively via the perfusate, the lengthwise profiles for the product are complementary to those for precursor, rising along the length of the sinusoid (Goresky et al., 1993b). The nature of the profile for product in liver cells will depend on the permeability of the liver cell membranes to product. When the membrane is highly permeable, the concentration profile in liver cells and plasma is the same; and when there is a resistance at the membrane, there will be, in the membrane symmetric case, a converse stepdown in product concentration from tissue to blood (Goresky et al., 1993b).

In an indicator dilution experiment in which tracer product formation occurs, it has been found that the following set of guidelines will computationally simplify

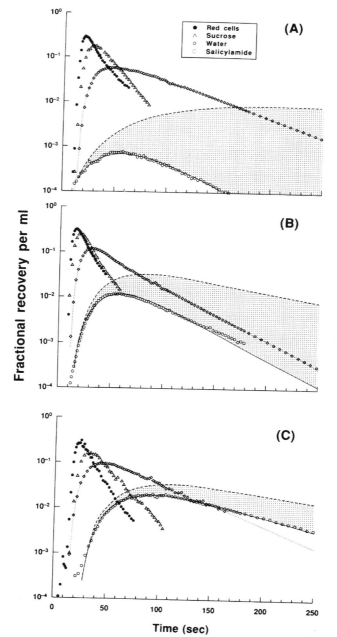

FIGURE 13.10. Outflow tracer profiles at various bulk salicylamide concentrations, displayed in semilogarithmic format. The input concentrations were 9 μM in A, 200 μM in B, and 870 μM in C. The appropriate reference curve for the salicylamide tracer is displayed as a dashed line, and the tracer sequestration by shading between this line and the salicylamide data. (From Pang et al., 1994, with permission.)

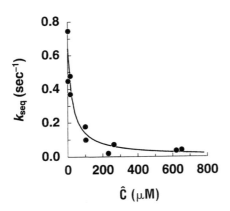

FIGURE 13.11. Plot of k_{seq} against the logarithmic average unbound concentration of salicylamide. (From Pang et al., 1994, with permission.)

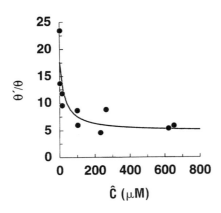

FIGURE 13.12. Plot of θ'/θ (ratio of cellular distribution space for salicylamide/cellular water space) versus logarithmic average unbound concentration of salicylamide. (From Pang et al., 1994, with permission.)

the task of fitting the experimental tracer product data. One should include in the injection mixture tracer product labeled with a second tracer that is distinguishable from that used as a precursor label, and simultaneously inject both tracer precursor and tracer product, and resolve at the outflow not only the two labels but also the precursor label in both the chemical precursor and the chemical product. The resulting description of the handling of preformed product can then be used to simplify the task of defining the behavior of locally generated product. If this procedure cannot be followed, parameter values from parallel experiments in which tracer product is separately injected will prove useful.

Within the enzymic mechanism, although the reversible association of precursor with enzyme results in an enzymic space effect, product does not reassociate with the enzyme and so can potentially leave the system earlier than the precursor. The conversion of salicylamide to salicylamide sulfate by the isolated perfused liver will be illustrated to provide a further example (Pang et al., 1994). This again corresponds to the k_a, $k_d \rightarrow \infty$ case. A set of outflow dilution curves for labeled red cells, sucrose, water, and salicylamide at various bulk concentrations of sali-

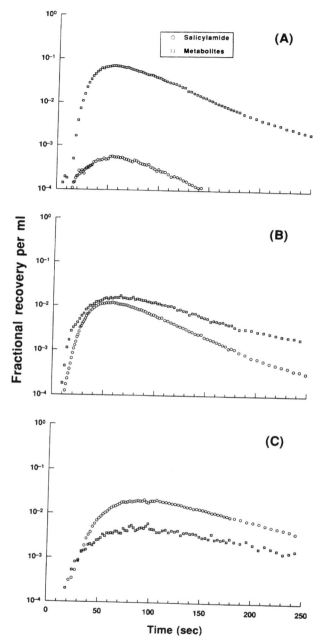

FIGURE 13.13. Outflow curves for tracer salicylamide and total tracer salicylamide metabolites. (From Pang et al., 1994, with permission.)

cylamide are displayed in Figure 13.10. The space of distribution for salicylamide, as described by the curve for its appropriate reference, diminishes with increase in bulk salicylamide concentration. Tracer (and bulk) extraction varies with increase in concentration from almost complete to 0.40.

The variation in k_{seq} with the logarithmic average salicylamide concentration $\hat{C}_u$ is illustrated in Figure 13.11. The best fit derived values for V_{max} and K_m were 17.5 mmol $\cdot$ s^{-1} $\cdot$ ml^{-1} cellular water space and 27 μM. The variation in the ratio (cellular space accessible to salicylamide/cellular water space), θ'/θ, with $\hat{C}$ is illustrated in Figure 13.12. The solid line is the best fit line, with the assumption that the binding site is provided by the enzyme (that the K_m is therefore that of the enzyme). Since the K_m is known, two parameters are provided by the fit: a value for [E$_T$] of 352 μM and a value of 3.8 for the ratio of the nonsaturable component to the cellular water space. The compound has a high lipid partition. The profile for metabolite products, shown in Figure 13.13, was not modeled because analytical separation of metabolites was not performed, due to small sample volumes. Nevertheless the data show, qualitatively, the effects expected for product as a result of the enzymic mechanism for the k_a, $k_d \rightarrow \infty$ case: an early emergence in time with respect to the precursor and a saturation of the conversion process as input concentration was increased. With increasing salicylamide concentration, the downslopes of the metabolite curves become progressively more prolonged.

The foregoing shows the promise of the present approach. It will become possible to look from the circulation through the interstitium and cell membrane to see metabolic events within the cells, and particularly to kinetically characterize the processes by which substrates are converted to products, with their subsequent release to the circulation. The kinetic tools being developed are general. They will apply as well to positron emission tomographic serial imaging as to indicator dilution experiments.

Acknowledgments. The authors wish to thank Kay Lumsden and Bruce Ritchie for their technical assistance, and Mary Ann Adjemian for typing this manuscript.

This work was supported by grants from the Medical Research Council of Canada, the National Institutes of Health, the Quebec Heart Foundation, and the Fast Foundation.

References

Cahill, G. F. Jr., J. Ashmore, A. S. Earle, and S. Zottu. Glucose penetration into liver. *Am. J. Physiol.* 192:491–496, 1958.

Chinard, F. P., G. J. Vosburgh, and T. Enns. Transcapillary exchange of water and other substances in certain organs of the dog. *Am. J. Physiol.* 183:221–234, 1955.

Crone, C. The permeability of capillaries in various organs as determined by use of the "indicator diffusion" method. *Acta Physiol. Scand.* 58:292–305, 1963.

Cuatrecasas, P. and S. Segal. Mammalian galactokinase. *J. Biol. Chem.* 240:2382–2388, 1965.

Goresky, C. A. A linear method for determining liver sinusoidal and extravascular volumes. *Am. J. Physiol.* 204:626–640, 1963.

Goresky, C. A. Kinetic interpretation of hepatic multiple-indicator dilution studies. *Am. J. Physiol.* 245(*Gastrointest. Liver Physiol 8*):G1–G12, 1983.

Goresky, C. A. and A. C. Groom. Microcirculatory events in the liver and spleen. In: *Handbook of Physiology, Section 2, The Cardiovascular System, Vol. IV, Microcirculation*, edited by E. M. Renkin and C. C. Michel. Bethesda, MD: American Physiological Society, 1984, pp. 689–780.

Goresky, C. A. and B. E. Nadeau. Uptake of materials by the intact liver: The exchange of glucose across the cell membranes. *J. Clin. Invest.* 53:634–646, 1974.

Goresky, C. A. and C. P. Rose. Blood-tissue exchange in liver and heart: The influence of the heterogeneity of capillary transit times. *Fed. Proc.* 36:2629–2634, 1977.

Goresky, C. A. and M. Silverman. Effect of correction of catheter distortion on calculated liver sinusoidal volumes. *Am. J. Physiol.* 207:883–892, 1964.

Goresky, C. A., W. H. Zeigler, and G. G. Bach. Capillary exchange modeling: Barrier limited and flow limited distribution. *Circ. Res.* 27:739–764, 1970.

Goresky, C. A., G. G. Bach and B. E. Nadeau. On the uptake of materials by the intact liver: the concentrative transport of rubidium[86]. *J. Clin Invest.* 53:975–990, 1973a.

Goresky, C. A., G. G. Bach, and B. E. Nadeau. On the uptake of materials by the intact liver: the transport and net removal of galactose. *J. Clin. Invest.* 52:991–1009, 1973b.

Goresky, C. A., P.-M. Huet, and J. P. Villeneuve. Blood tissue exchange and blood flow in liver. In: *Hepatology*, edited by D. Zakim and T. Boyer. Philadelphia, PA: Saunders, 1982, pp. 32–63.

Goresky, C. A., G. G. Bach, and C. P. Rose. Effects of saturating metabolic uptake on space profiles and tracer kinetics. *Am. J. Physiol.* 244 (*Gastrointest. Liver Physiol. 7*):G215–G232, 1983a.

Goresky, C. A., K. S. Pang, A. J. Schwab, F. Barker III, W. F. Cherry, and G. G. Bach. Uptake of a protein-bound polar compound, acetaminophen sulfate, by perfused rat liver. *Hepatology* 16:173–190, 1992.

Goresky, C. A., E. R. Gordon, and G. G. Bach. Uptake of monohydric alcohols by liver: Demonstration of a shared enzymic space. *Am. J. Physiol.* 244 (*Gastrointest Liver Physiol. 7*):G198–G214, 1983a.

Goresky, C. A., G. G. Bach, and A. J. Schwab. Distributed-in-space product formation in vivo: linear kinetics. *Am. J. Physiol.* 264 (*Heart Circ. Physiol 33*):H2007–H2028, 1993.

Goresky, C. A., G. G. Bach, and A. J. Schwab. Distributed-in-space product formation in vivo: Enzymic kinetics. *Am. J. Physiol.* 264 (*Heart Circ. Physiol 33*):H2029–H2050, 1993a.

Hamilton, W. F., J. W. Moore, J. M. Kinsman, and R. G. Spurling. Simultaneous determination of the pulmonary and systemic circulation times in man and of a figure related to the cardiac output. *Am. J. Physiol.* 84:338–344, 1928.

Hess, F. A., E. R. Weibel, and R. Preisig. Morphometry of the dog liver: Normal baseline data. *Virchows Arch. B.* 12:303–317, 1973.

Layden, T. J., J. Schwarz, and J. L. Boyer. Scanning electron microscopy of rat liver: Studies of the effects of taurocholate and other models of cholestasis. *Gastroenterology* 69:724–738, 1975.

Love, W. D., R. B. Ronney, and G. E. Burch. A comparison of the distribution of potassium and exchangeable rubidium in the organs of the dog, using rubidium[86]. *Circ. Res.* 2:112–122, 1954.

Ogston, A. G. and C. Phelps. The partition of solutes between buffer solutions and solutions containing hyaluronic acid. *Biochem. J.* 78:827–833, 1961.

Pang, K. S., F. Barker III, A. J. Schwab, and C. A. Goresky. Demonstration of rapid entry and a cellular binding space for salicylamide in perfused rat liver: A multiple indicator dilution study. *J. Pharm. Exp. Therap.* 270:285–295, 1994.

Prothero, J. and A. C. Burton. The physics of flow in capillaries. I. The nature of the motion. *Biophys. J.* 1:565–579, 1961.

Rose, C. P., C. A. Goresky, and G. G. Bach. The capillary and sarcolemmal barriers in the heart: an exploration of labeled water permeability. *Circ. Res.* 41:515–533, 1977.

Tygstrup, N. and K. Winkler. Kinetics of galactose elimination. *Acta. Physiol. Scand.* 32:354–357, 1954.

Tygstrup, N. and K. Winkler. Galactose blood clearance as a function of hepatic blood flow. *Clin. Sci.* 17:1–9, 1958.

Wiederhelm, C. A. and L. L. Black. Osmotic interaction of plasma proteins with interstitial molecules. *Am. J. Physiol.* 231:638–641, 1976.

Wilbrandt, W. and T. Rosenberg. The concept of carrier transport and its corollaries in pharmacology. *Pharmacol. Rev.* 13:109–184, 1961.

14

Probing the Structure and Function of the Liver with the Multiple-Indicator Dilution Technique

K. Sandy Pang, Carl A. Goresky, Andreas J. Schwab, and Wanping Geng

Introduction

The liver is the most important drug eliminating organ that is capable of both metabolism and excretion. The efficiency with which the liver removes a drug is often expressed as the hepatic drug clearance, or the volume of perfusing fluid cleared of its contained drug per unit time. Clearance is affected by the hepatic blood flow rate, the binding to vascular proteins, transport across the sinusoidal membrane, and the K_m and V_{max} of the metabolic and excretory pathways (for reviews, see Gillette and Pang, 1977; Pang and Xu, 1988; Pang et al., 1991a; Goresky et al., 1993a, 1993b). The liver is a highly specialized and complex organ (Rappaport, 1958, 1980; Novikoff, 1959; Miller et al., 1979; de Leeue and Knook, 1984; Gooding et al., 1978; Jungerman and Katz 1982), and its attendant heterogeneities—capillary transit times (Goresky, 1963; Sherman et al., 1990; Almond and Wheatley, 1992; Pang et al., 1994a), transport (Burger et al., 1989; McFarlane et al., 1990), enzyme zonation (Baron et al., 1982; Ullrich et al., 1984; Knapp et al., 1988; Thurman et al., 1987; Pang et al., 1983; Pang and Terrell, 1981a), acinar biliary excretion (Gumucio et al., 1978; Boyer et al., 1979), cosubstrate abundance (Smith et al., 1979; Murray et al., 1986; Chiba and Pang, 1995), and intracellular binding (Braakman et al., 1989; Bass et al., 1989)—must be viewed in an integrative fashion in order to accurately relate the intrahepatic events involved in drug and metabolite processing (Pang and Stillwell, 1983; Goresky and Groom, 1984; Goresky et al., 1994; Pang and Xu, 1988). The overall removal process is highly dependent on these variables as well as the concentration of drug entering the liver.

Within the intact liver, drug disappearance within the sinusoid is a result of uptake by the organ and occurs as a distributed-in-space phenomenon, which is directed by flow along the sinusoid and described according to its position along the sinusoidal flow path (Goresky et al., 1973; Winkler et al., 1973; Pang and Stillwell, 1983). The loss of substrate leads to the observed concentration gradient between the inlet and outlet of the liver (Jones et al., 1980; Gumucio et al., 1981; Weisiger et al., 1986). Drug uptake is governed by time-related events including drug transport into and out of hepatocytes, binding to red cells, plasma, and tissue proteins, and metabolism by enzymes or excretion into bile. The events of substrate influx and efflux and the removal by metabolic/excretory activities are linked to the delivery by flow, and are modulated by substrate binding to red blood cells (Goresky et al., 1975, 1988; Pang et al., 1995) and to plasma (Gillette, 1973; Wolkoff et al., 1979; Weisiger, 1985; Xu et al., 1993; Chiba and Pang, 1993) and tissue (Fleischner et al., 1975; Goresky et al., 1978, 1983; Gärtner et al., 1982; Theilmann et al., 1985; Pang et al., 1994b) proteins. The generally accepted view is that the unbound substrate is the species that is transported across membranes and becomes eliminated (Levy and Yacobi, 1974; Grausz and Schmid, 1971; Barnhart and Clarenburg, 1973; Sorrentino et al., 1989a). The rates with which metabolism and excretion proceed, however, are dependent on the amounts and localization of enzymatic activity for metabolite formation, and the transport activities at the canalicular membrane. Recent investigation adds futile cycling as another variable that influences organ and metabolite clearances (Ratna et al., 1993). The net removal is highly dependent on these variables, and the rate-limiting step can vary at any point x along the sinusoidal flow path.

Metabolites are linked kinetically to their precursors. Furthermore, metabolite processing is also a distributed-in-space phenomenon (Pang and Stillwell, 1983; Pang et al., 1987; Goresky et al., 1993a, 1993b). Generated tracer product is expected to behave differently from a corresponding already formed or preformed tracer metabolite introduced as an input into the system even though events underlying the elimination of metabolite generated from a precursor and preformed metabolite are interrelated because of involvement of the same (enzyme/excretory) system(s) for metabolite elimination. Differences are expected to occur because of differing points of entry of the metabolites in the organ. For the generated metabolite, it exists at each site of formation and is highly dependent on drug uptake and biotransformation, whereas for a preformed metabolite, which already exists at the origin of the sinusoid, access to hepatocytes is dependent only on cell permeability and space of distribution, and is independent of the primary metabolic transformation of the drug. Additionally, if the cell membrane is less permeable to metabolites, this will lead to increased intraorgan metabolite elimination compared with that for preformed metabolite; otherwise, the converse will be true (de Lannoy and Pang, 1986, 1987; Goresky et al., 1992). Acinar metabolic activity has been shown to confer different metabolic fates on preformed and generated metabolites (Pang and Terrell, 1981a, Pang et al., 1984, 1988a; Xu et al., 1990a).

In this chapter, we describe applications of the multiple-indicator dilution (MID) technique in a series of liver perfusion studies to illustrate the importance of zonation of enzymes, vascular drug binding, and the presence of a transmembrane barrier in drug metabolism. The MID technique (Goresky, 1963; Goresky et al., 1973) is an experimental tool that provides information on the physiology and structure of the organ. The physiologic distribution spaces may be assessed from the outflow profiles obtained from the simultaneous injection of a set of noneliminated indicators ^{51}Cr-labeled RBC, a vascular reference, ^{125}I-labeled albumin, a large molecular weight interstitial space reference,

[^{14}C]sucrose or [^{58}Co]EDTA (Pang et al., 1990), a small molecular weight interstitial space reference, and ^3H_2O or D_2O (Pang et al., 1991b), a cellular reference] by
multiplication of the transit time and the appropriate flow rate. The principle of
the MID technique, when extended to include a labeled substrate and/or its metabolite and a full set of noneliminated reference indicators, allows inferences to be
drawn concerning the behavior of the labeled substrate/metabolite under investigation. By comparison of a labeled substance to the hypothetical reference based
on the binding characteristics of the labeled substrate and metabolite, modeling of
the data (see Goresky et al.'s chapter) will furnish the processes underlying influx,
efflux, and sequestration (removal). In some instances, the rate-controlling step in
the overall removal may become evident.

Liver Perfusion and Multiple-Indicator Dilution Technique

The perfused rat liver is an ideal system for investigation since good viability and
stability can be maintained (for details of perfusion, see St-Pierre et al., 1989;
Pang et al., 1994a) when livers are perfused with red blood cells (20%), albumin
(1%), and 3% Dextran and electrolytes (Krebs Henseleit bicarbonate buffer). This
perfused organ has been shown to retain its vasoactivity toward vasopressive
agents and exhibits the hepatic arterial buffer response (Sherman et al., 1996). The
perfusion condition—composition of perfusate, input concentration, and flow
rate—could be readily manipulated within the same liver preparation, allowing
small but finite differences to become detectable. Normally, the portal and hepatic
veins are cannulated for the inflowing and outflowing conduits, respectively.
Typically, recirculation is prevented by single-pass or once-through perfusion,
and bulk (unlabeled) substrate is added in varying quantities to the inflowing
perfusate for the study of concentration-related phenomena. The MID injection,
containing a set of noneliminated reference indicators (^{51}Cr-labeled RBC, ^{125}I-
labeled albumin, [^{14}C]- or [^{3}H]sucrose, and ^3H_2O or D_2O) and labeled tracer
substrate, is introduced during the steady state. The behavior of the tracer will thus
reflect that within the steady state. The pulse tracer experiments (MID) of this
kind reveal not only the same, but more information than comparable steady-state
studies. They potentially provide information on the accessible spaces of distribution; additionally, temporal changes due to vascular and tissue binding, and saturability of influx and removal processes become evident. Much of the mathematical details for drug/metabolite entry and sequestration has been covered in the
accompanying chapters by Goresky et al. and Schwab, this volume.

For the presentation of data in this system, the outflow concentrations are
normalized to the injection dose and are expressed in relation to the outflow
recovery. The labeled red blood cells emerge first and reach the highest and
earliest peak; their outflow curve has the steepest upslope and the downslope
decays most rapidly. The ^{125}I-albumin curve rises slightly less quickly and decays
with a slightly reduced slope; it shows a lower and later peak. The [^{14}C]sucrose

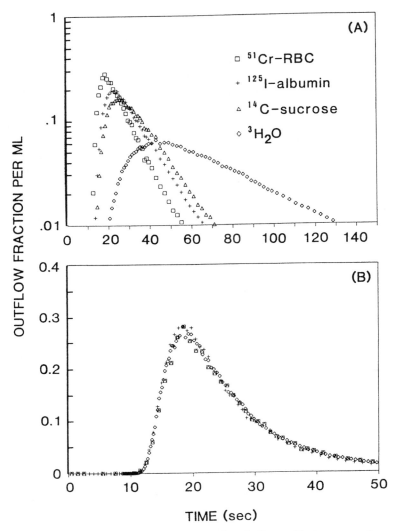

FIGURE 14.1. A representative set of hepatic outflow indicator dilution curves from once-through rat liver perfusions ($10 \text{ ml} \cdot \text{min}^{-1}$). (A) Rat livers were perfused with blood perfusate for 25 min prior to a bolus injection of noneliminated radiolabeled references: ^{51}Cr-labeled RBC, ^{125}I-labeled albumin, [^{14}C]sucrose, and [^{3}H]water into the portal vein. Values for y axis are outflow concentrations that have been normalized by the injected dose, and values on the x axis represent time lapsed after the injected dose was allowed to flow into the liver (the time is uncorrected for catheter and large vessel transit times). (B) Superposition of the adjusted diffusible label (albumin, sucrose, and water) curves on the labeled RBC curve. The outflow indicator dilution curves for the diffusible tracers are now presented with the concentration and time beyond a common t_0 value multiplied by the factor $(1 + \gamma_{Alb})$, $(1 + \gamma_{Suc})$, or $(1 + \gamma_W)$ and $1/(1 + \gamma_{Alb})$, $1/(1 + \gamma_{Suc})$, or $1/(1 + \gamma_W)$, respectively, for albumin and sucrose, and water; γ_{Alb} and γ_{Suc} are the volume ratios of albumin and sucrose Disse space to the sinusoidal plasma space; γ_W is the volume ratio of the sum of Disse water space and intracellular water space to the volume of water in the sinusoid.

curve shows, in comparison to the labeled albumin curve, a slightly more delayed upslope, a slightly lower and later peak, and a more prolonged downslope. The greatest dispersion is seen with ^3H_2O, whose upslope and downslope are very delayed and whose peak occurs much later with a much lower magnitude due to its permeation of the cellular as well as vascular and interstitial spaces (Figure 14.1A). The curve patterns conform to the rank order of the distribution spaces. After correction for the differences in distribution (reduction of magnitude and prolongation of time scale, in relation to those for labeled red cells, due to increased distribution), the labeled albumin, sucrose, and water curves superpose onto the labeled red cell curve (Figure 14.1B). The delayed wave forms indicate that flow limitation is the common mechanism underlying the dispersion of the noneliminated reference indicators (Goresky, 1963). The mean transit time of each noneliminated reference in the system is given by moment analysis: the time integrals of the product of fractional recovery and time divided by the time integrals of the fractions (Zierler, 1965).

Metabolic Zonation

Prograde–Retrograde Perfusion

Enzymatic zonation has been described with direct and indirect approaches (deBaun et al., 1971; Pang and Terrell, 1981a; Baron et al., 1982; Redick et al., 1982; Knapp et al., 1988). Among these is the technique of prograde (portal vein to hepatic vein) and retrograde (hepatic vein to portal vein) livers perfused at ml · min^{-1} (Pang and Terrell, 1981a). Retrograde flow is expected to bring the substrate to enzymatic sites in an order that is the reverse of prograde flow. For proper data interpretation to be made, equal access to hepatocytes (recruitment) in a reversed direction needs to be assured with retrograde flow. Evidence for such is provided when the extraction ratios (removal rate/input rate, at steady state) of compounds with intermediate or low extraction values remain unchanged during prograde and retrograde flows; by contrast, compounds with high extraction ratios must be excluded since the enzymatic system for removal is so large in this instance that not all hepatocytic activity is required. The former was indeed observed for the extraction ratios for acetaminophen sulfation (Pang and Terrell, 1981a) and ethanol oxidation (St-Pierre et al., 1989), which remained at about 0.7 and 0.44, respectively, with prograde and retrograde flows.

Another way of substantiating equivalent access by prograde and retrograde perfusions is provided by estimates of the intracellular water space accessed by ^3H_2O. With MID injections made into the portal vein, the transit time for cell water, $\bar{t}_{cell,eff}$, is the cellular water space V_{cell} divided by the water flow rate (product of total flow Q_{tot} and the fraction of blood that is water f_b), and is related to the catheter-corrected transit times for water ($\bar{t}_w$), sucrose ($\bar{t}_{suc}$), and red blood cells ($\bar{t}_{rbc}$):

$$\bar{t}_{cell,eff} = \frac{V_{cell}}{Q_{tot}f_b} = \bar{t}_w - \frac{f_r}{f_b}Hct\,\bar{t}_{rbc} - \frac{f_p}{f_b}(1-Hct)\bar{t}_{Suc}, \qquad (14.1)$$

where f_r and f_p are the water contents of red cells and plasma (in ml · ml^{-1}); these are taken to be 0.7 and 0.98, respectively, since only 1% albumin was present in perfusate, and f_b was estimated from the relation

$$f_b = f_r \text{Hct} + f_p(1 - \text{Hct}). \qquad (14.2)$$

The cellular water space was found to be identical during prograde and retrograde perfusion in MID studies (St-Pierre et al., 1989), validating the assumption that tracer water accesses equal proportions of hepatocytes, both in forward and reverse fashion (Figure 14.2); however, distention of the vascular and Disse spaces was observed during retrograde perfusion. The technique of prograde and retrograde perfusion was first utilized to explain the lower extent of (upstream) sulfation of acetaminophen when generated from phenacetin by a downstream process for O-deethylation (Pang and Gillette, 1978) that was totally reversed on retrograde perfusion, which then rendered the formation process upstream and the sulfation downstream (Pang and Terrell, 1981a). The technique has been useful in delineating periportal sulfation and an even or perivenously distributed glucuronidation system (Table 14.1). Through this effort in mapping of enzymatic activities, a better understanding of the nature of competing pathways is obtained and of the effects of changes in concentration or flow in the face of varying enzyme stacking and enzymic parameters (the K_m and V_{max}) (Pang et al., 1983; Morris et al., 1988a, 1988b; Morris and Pang, 1987; Xu and Pang, 1989; Xu et al., 1990a; Pang and Mulder, 1990). The technique is now routinely used to define not only relative zonal localization of enzymes, but also uptake mechanisms for macrolides such as proteins and antibodies (McFarlane et al., 1990).

HAPV–HAHV Perfusion

Another technique that probes the localization of enzymatic activities utilizes perfusion of the hepatic artery for steady-state delivery of drug and injection of the MID dose, while blank perfusate medium enters the portal vein (HAPV) or hepatic vein (HAHV) (Pang et al., 1988a). The method is applicable to the rat where there is drainage of branches of the hepatic artery directly into the portal vein and sinusoid, without prior perfusion of the peribiliary plexus (Yamamoto et al., 1985; Motta et al., 1978) and also sequential perfusion of the peribiliary plexus (interstitial and peribiliary cellular spaces) and sinusoid (extracellular and cellular water spaces); the proportion of the cellular space reached by tracer introduced via HA will vary depending on whether PV or HV perfusion is used (Figure 14.3). The expectation is that with HAPV perfusion, arterially delivered elements will be brought across the entire sinusoidal bed, whereas with HAHV, arterially delivered elements will be confined mostly to the periportal region. The confinement of HA-borne elements to the periportal region was shown by the exclusive staining of acridine orange in zone 1 after hepatic arterial introduction during HAHV perfusion (Watanabe et al., 1994); the cellular water space during HAHV is about one-third that for HAPV (Pang et al., 1988a).

The outflow profiles resulting from HA injection of the MID dose for HAPV are essentially similar to those resulting from PV injection, excepting the initial

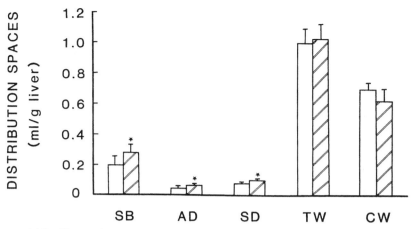

FIGURE 14.2. Changes in sinusoidal blood volume (SB), albumin Disse space (AD), sucrose Disse space (SD), total water (TW), and accessible cellular water space (CW) during prograde (open bars) and retrograde (hatched bars) perfusion of the rat liver at 10 ml · min⁻¹. (Data taken from St-Pierre et al., 1989.)

TABLE 14.1. Zonation of drug metabolizing activities found by prograde and retrograde (PR) and HAPV–HAHV perfusion of the rat liver.

Localization of metabolic activities				
Anterior[a]	Even[b]	Posterior[c]	Drug examples	Reference
				Pang and Terrell, 1981a
Sulfation			Acetaminophen	Pang et al., 1988a
Sulfation	Glucuronidation		Gentisamide	Morris et al., 1988a, 1988b
Sulfation	Glucuronidation		Harmol	Pang et al., 1983
Sulfation	Glucuronidation	Hydroxylation	Salicylamide	Xu et al., 1990a Xu and Pang, 1989
Sulfation	Glucuronidation		7-Hydroxycoumarin	Conway et al., 1987
Sulfation		Glucuronidation	7-Hydroxycoumarin	Conway et al., 1984, 1985
		O-deethylation	Phenacetin	Pang et al., 1988a
		Carboxyester hydrolysis	Enalapril	Pang et al., 1991c
		Glutathione conjugation	Bromosulfophthalein	Zhao et al., 1993
		Glycine conjugation	Benzoic acid	Chiba et al., 1994

[a]Abundance in periportal region or zone 1.
[b]Equal abundance among zonal regions or zones 1, 2, and 3.
[c]Abundance in perivenous region zone 3.

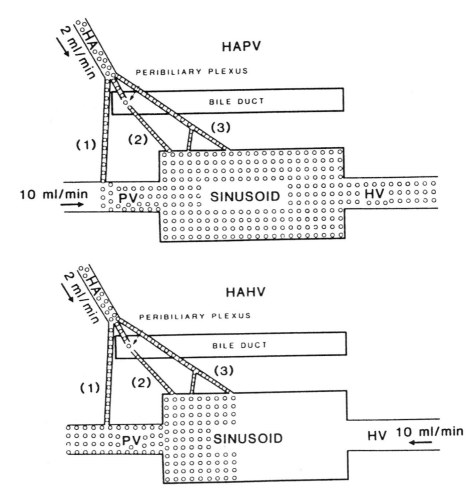

FIGURE 14.3. Schematic presentation of flow patterns in the HAPV and HAHV perfusion systems. HA, PV, and HV denote hepatic artery, portal vein, and hepatic vein, respectively. Pathways 1, 2, and 3 are possible drainages of hepatic artery into liver: Pathways 1 and 2 are indirect ones: Pathway 1 drains into the portal vein whereas Pathway 2 first perfuses the peribiliary capillary plexus before drainage into the sinusoids; Pathway 3 provides direct access of hepatic arterial flow to sinusoids downstream from the portal triads. Drug-free perfusate enters into either the PV or HV (10 ml · min^{-1}), respectively, for HAPV and HAHV perfusion. Drug and MID injection mixtures are introduced into the HA (2 ml · min^{-1}). During HAPV, drug (circles) is swept across the liver and reaches most of the sinusoidal perfusion surfaces (upper); during HAHV, drug (circles) is confined mostly in the periportal region (lower). (Data taken from Pang et al., 1988a.)

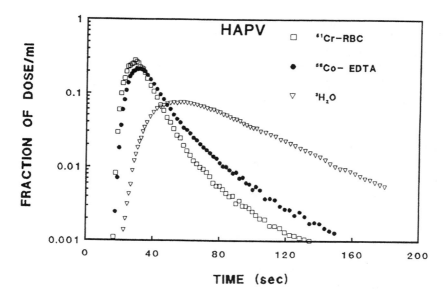

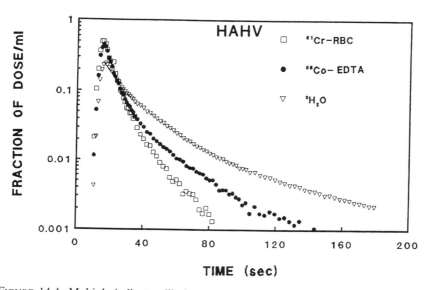

FIGURE 14.4. Multiple-indicator dilution curves of ^{51}Cr-labeled red blood cells, [^{58}Co]-EDTA (a kinetic equivalent of labeled sucrose), and $^{3}H_2O$ during HAPV (upper panel) and HAHV (lower panel) flows after their rapid injection as a single bolus into the hepatic artery flowing at 2 ml · min^{-1} in a once-through perfused rat liver in situ preparation. (Data taken from Chiba et al., 1994, with permission.)

delay in their appearance due to a prolonged traverse across the catheter/tubing because of the lower HA flow rate. The outflow profiles for HAHV injection, however, differ dramatically, showing virtually identical peaks of different magnitudes for labeled red cells, sucrose, and water but with substantial shunting (Figure 14.4). The experimentally determined transit time ($\overline{t}_{comb}$) of the cellular water space accessed by the hepatic artery is now the combination of nonparenchymal (the portion associated with the peribiliary capillary plexus of transit time, $\overline{t}_{ns}$) and parenchymal (sinusoidal cellular transit time, $\overline{t}_{cell,eff}$) water spaces. The contribution of the nonparenchymal portion has been shown to be about 5% of the accessible parenchymal cell water space (Pang et al., 1994a). Nonetheless, in the case of HAHV, where only a portion of the sinusoids (namely periportal) is perfused, the part of the acinus accessed by substrate entering via the hepatic artery is greatly reduced, whereas that of the peribiliary plexus remains the same. This kind of expectation must be taken into account in appraising apparent enzyme activities calculated in this manner. In applying these expressions, however, we utilize $\overline{t}_{cell,eff}$ in place of $\overline{t}_{comb}$ since their difference is negligible.

When the heterogeneity of transit times is neglected and intracellular uptake is assumed to be occurring in a flow limited system, the single-channel expression

$$E_{tot} = 1 - e^{-k_{seq}\overline{t}_{cell,eff}} \tag{14.3}$$

can be used to approximate what is happening in the organ, where E_{tot} is the extraction ratio of the substrate and k_{seq} is a sequestration rate constant with the dimensions ml $\cdot$ s^{-1} $\cdot$ ml^{-1} intracellular water. When the drug concentration is small with respect to the K_m or the Michaelis–Menten constant, the (total) rate constant for sequestration, k_{seq}, is the ratio of the maximal enzyme velocity, V_{max} (expressed in terms of moles substrate converted per unit intracellular volume per unit time), to K_m; it is approximated (Goresky et al., 1983) by the following expression, utilizing the total extraction ratio of substrate, E_{tot}, and the transit time for cell water, $\overline{t}_{cell,eff}$,

$$k_{seq} = \frac{V_{max}}{K_m} = \frac{CL_{int}}{V_{cell}} = -\frac{\ln(1 - E_{tot})}{\overline{t}_{cell,eff}} \tag{14.4}$$

and is an averaged value for the total intrinsic clearance (CL_{int}) per unit intracellular volume (V_{cell}). The extent of biliary excretion of E_{bile} (or ratio of the excretion rate to input rate, at steady state) may be estimated in an analogous fashion. The biliary excretion rate constant (k_b) can be expressed as the biliary intrinsic clearance ($CL_{int,b}$) per unit of accessible water space (V_{cell}):

$$k_b = \frac{CL_{int,b}}{V_{cell}} = -\frac{\ln(1 - E_{bile})}{\overline{t}_{cell,eff}} . \tag{14.5}$$

When metabolites are directly quantified and the fractional removal due to metabolism (E_{met}) is known, the metabolic sequestration rate constant, k_{met}, may be obtained in a fashion analogous to Eq. (14.5). Alternately, the metabolic intrinsic clearance ($CL_{int,m}$) per unit of accessible water space may be estimated from the difference,

$$k_{met} = \frac{CL_{int,m}}{V_{cell}} = \frac{\ln(1 - E_{met})}{\bar{t}_{cell,eff}}$$

$$= \frac{CL_{int} - CL_{int,b}}{V_{cell}} = k_{seq} - k_b . \tag{14.6}$$

As commented previously, these rate constants are underestimates of the true values of V_{max}/K_m corresponding to the averaged enzyme concentration within the accessible intracellular space, due to a lack of consideration of the heterogeneity of capillary transit times (Gonzalez-Fernandez and Atta, 1973; Bass and Robinson, 1981; Bass, 1983, 1985). The underestimation also occurs, especially during HAHV, because the approximated sojourn time for cellular water also includes an overestimate of the small accessible nonparenchymal cellular volume associated with the peribiliary capillary plexus. If, on comparison of the respective total (k_{seq}), metabolic (k_{met}), or biliary (k_b) rate constants (activity per unit cell water space) during HAHV to HAPV, one finds ratios that are greater than unity, this strongly suggests a periportal abundance of the activity, whereas ratios of less than unity infer a perivenous abundance of the activity. Zonation of uni-enzyme activities has been mapped in this way with the HAHV or HAPV technique. Examples are periportal sulfation (Pang et al., 1988a) and perivenous localization of phenacetin O-deethylation (Pang et al., 1988a), glutathione conjugation of bromosulfophthalein (Zhao et al., 1993), glycine conjugation of benzoic acid (Chiba et al., 1994), and esterolysis of enalapril (Pang et al., 1991c). Thus, in addition to providing physiological volumes and rate constants, the MID technique, coupled with drug-metabolite data during HAPV–HAHV perfusions, provides a new approach for the examination of zonal metabolic heterogeneity (Table 14.1).

Microcirculation

The manner in which the hepatic vasculature responds to changes in flow has been thoroughly characterized in the perfused liver system by the combination of a steady delivery of drug probes and the MID technique for examination of the changes in the vasculature with flow. When hepatic arterial perfusion is present additionally, the pattern of intermixing of the hepatic arterial (HA) and portal venous (PV) flows may be studied under constant flow (12 ml · min^{-1}) with varying HA:PV flow ratios and constant delivery of drug probes into the HA and PV.

Flow and Derecruitment

Microcirculatory hemodynamics have been studied with the portal vein- (or prograde flow) and hepatic vein- (or retrograde flow) only perfusions (Pang et al., 1988b; Xu et al., 1990b). Distention of the vasculature is observed with increasing flow to the liver (Figure 14.5). The extravascular spaces—the large vessel volume, sinusoidal blood volume (estimated blood volume minus large vessel volume), and the apparent plasma albumin and sucrose Disse space—are all enlarged with increased prograde or retrograde flow, although retrograde values are consis-

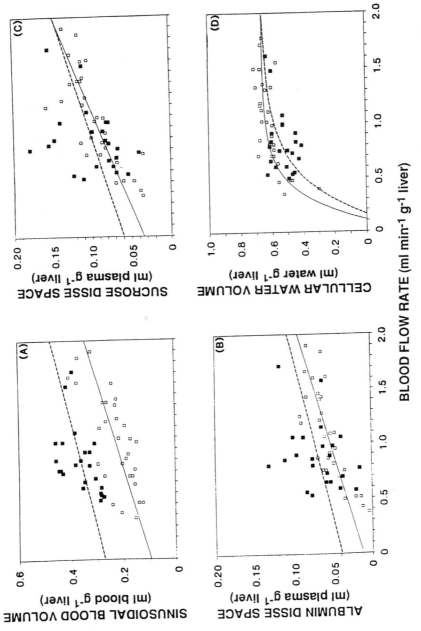

FIGURE 14.5. Changes in the estimated sinusoidal blood volume (A), albumin Disse space (B), sucrose Disse space (C), and the accessible cellular water spaces (D) with changes in prograde (open symbols) and retrograde (closed symbols) blood flow rate (0 to 2 ml · min^{-1} · g^{-1} liver). (Data taken from Pang et al., 1988b, and Xu et al., 1990b.)

TABLE 14.2. Sequestration rate constants (k_{seq}) with changes in prograde and retrograde flows in rat livers perfused at 8 and 12 ml·min^{-1} with drug substrates when an MID dose is injected during the steady state.

	Sequestration rate constants (ml · s^{-1} · ml^{-1})[a] with prograde flow			Sequestration rate constants (ml · s^{-1} · ml^{-1})[a] with retrograde flow		
	12 (ml · min^{-1})	8 (ml · min^{-1})	% change[b]	12 (ml · min^{-1})	8 (ml · min^{-1})	% change
Acetaminophen[c]	0.024 ± 0.005	0.021 ± 0.004[b]	−11%	0.035 ± 0.004	0.021 ± 0.003	−39%
Phenacetin[d]	0.044 ± 0.013	0.028 ± 0.007[b]	−36%	0.115 ± 0.014	0.091 ± 0.019	−21%
Meperidine[d]	0.085 ± 0.008	0.055 ± 0.008	−36%			

[a]Estimated with Eq. (14.4), within the same rat liver preparation; flow rate regime was randomized (data of Pang et al., 1988b and Xu et al., 1990b).
[b]Reduction in metabolic activity upon reduction of flow from 12 to 8 ml · min^{-1}.
[c]Acetaminophen is mainly sulfated upstream during prograde perfusion, and downstream during retrograde perfusion.
[d]Phenacetin and meperidine are metabolized to acetaminophen and normeperidine, respectively, by the cytochrome P-450s downstream during prograde perfusion, and upstream during retrograde perfusion.

tently higher than those at comparable prograde flows. The regressed slopes are slightly less for retrograde than for prograde for the sinusoidal blood volume, large vessel volume, total albumin and sucrose spaces, and for albumin and sucrose Disse spaces whereas the intercepts for the plots with retrograde flow are, however, substantially larger. By contrast, the cellular water spaces attain a rather constant level above a water flow rate of 0.75 ml · min^{-1} · g^{-1} liver. Below that, a derecruitment is evident, accompanied by a reduction in hepatocytic metabolic activities of drug probes given simultaneously (Table 14.2). The derecruitment of hepatocytes at reduced flows explains the paradox of reduced extraction ratios of drugs with reduced flow rates (Pang et al., 1988b). The derecruitment is found to be more severe downstream when probed with drugs that are metabolized in the either periportal region or zone 1 region (acetaminophen) or the perivenous region or zone 3 (phenacetin and meperidine) regions (Table 14.2). The pattern holds regardless of the flow (prograde or retrograde) direction.

HA and PV Mixing

A similar strategy may be used to study the hepatic microcirculation. The spaces accessed by noneliminated reference indicators introduced as a bolus into HA or PV at different HA:PV flow regimes, 0:12, 2:10, and 4:8, have been recently investigated (Pang et al., 1994a). [^{14}C]Phenacetin and [^{3}H]acetaminophen, two drug probes that are metabolized primarily in perivenous and periportal regions of the rat liver, respectively, were delivered into both HA and PV simultaneously, each at the same concentration, in a single-pass fashion. Small but significant decreases in E_{tot} of [^{14}C]phenacetin (from 0.989 to 0.980) and [^{3}H]acetaminophen (from 0.631 to 0.563) were observed with increments of hepatic arterial flow within the same liver preparation. The blood volume, total albumin space, albumin Disse space, total water, and parenchymal cellular water spaces were found to be unchanged following PV injection of an MID dose for all HA:PV flow ratios, suggesting that the presence of arterial flow is ineffective in perturbing average sinusoidal flow dynamics. However, slightly larger total water spaces were associated with HA injection. This excess water space can be accounted for solely by the "nonsinusoidal" extravascular space associated with the peribiliary capillary plexus; it averaged 0.03 ml · g^{-1} and was independent of flow. Adjunct data on sinusoidal velocities of the fluorescently labeled FITC-erythrocytes infused into the HA and PV (at 4 and 8 ml · min^{-1}, respectively) were 327 ± 78 and 301 ± 63 μm · s^{-1}, respectively, while that for the solely PV-perfused liver (12 ml · min^{-1}) was 347 ± 74 μm · s^{-1}. The resultant HA:PV flow-weighted sinusoidal velocity (310 μm · s^{-1}) was highly correlated to the sinusoidal volume for the dually perfused rat liver. Hence the anomaly: a reduced flow-weighted sinusoidal velocity for the dually perfused liver, an unchanged diameter of the terminal hepatic venule (32 μm) among the HA:PV flow ratios, and the reduction in the extraction ratio of the drug probes and oxygen consumption rates suggest that some of the arterial flow must have entered the sinusoids somewhat downstream.

Drug Behavior

Theoretical

Binding to Plasma and Tissue Proteins

The first and necessary step in exploration of substrate entry with the MID technique is the vascular and tissue binding characteristic of the substrate. Information on the distribution of the substrate into red blood cells or the plasma binding to albumin is necessary to define the distribution of the hypothetical reference in the vasculature. This hypothetical reference describes fully the intravascular behavior of the tracer labeled substrate, bound and free. The labeled albumin, labeled red cells, and labeled sucrose curves alone serve as references for the albumin-bound, erythrocyte-bound, and free species, respectively. For substrates that are partially bound, the use of a hypothetical reference (kinetic equivalent of tracer substrate) curve that encompasses both bound and free forms of the tracer is appropriate.

For a substrate that binds only to vascular proteins and not red blood cells (RBCs), the hypothetical reference curve is based on the level of binding in plasma, with a very rapid exchange between the bound and free forms, as described for acetaminophen sulfate (Goresky et al., 1992) and enalaprilat (Schwab et al., 1990). The parameter γ_{ref}, for the appropriate reference, is related to the unbound fraction of tracer in plasma, f_u, and γ_{Suc} and γ_{Alb}, the space distribution ratios (interstitial to sinusoidal plasma) of sucrose and albumin, respectively,

$$\gamma_{\mathrm{ref}} = f_u \gamma_{\mathrm{Suc}} + (1 - f_u)\gamma_{\mathrm{Alb}} \tag{14.7}$$

and is the apparent space distribution ratio (interstitial to sinusoidal plasma) of the substrate. The calculated parameter provides a value for the interstitial space ratio of the reference, which behaves in a manner identical to that for the substrate in the vasculature. The value, however, is expected to change as binding to plasma protein (f_u) changes.

When the tracer equilibrates with red blood cells in a rather rapid fashion such that entry and exit from red blood cells are rapid and not limiting, then γ_{ref}, as shown for salicylamide sulfate conjugate (Xu et al., 1994) and salicylamide (Pang et al., 1994b), is

$$\gamma_{ref} = \frac{f_u \gamma_{Suc} + (1 - f_u)\gamma_{Alb}}{1 + f_u \lambda_R \beta}, \qquad (14.8)$$

where λ_R, the partition ratio relating total to unbound tracer in RBCs, is used to define the total concentration in RBCs, c_{rbc}, by means of the relation $c_{rbc} = \lambda_R c_{p,u}$, Hct is the hematocrit, and β is the ratio of red cell water space to sinusoidal plasma water space:

$$\beta = \frac{f_r \, Hct}{f_p(1 - Hct)}. \qquad (14.9)$$

Vascular protein binding will inadvertently affect the uptake of substrates. The effect of tight binding is expected to decrease uptake and removal, especially with adoption of the conventional view that the unbound form is the species involved in the processes. An example of this, pertaining to tight red cell binding and slow efflux of acetaminophen, will be given later in the chapter. The unbound acetaminophen species in the plasma compartment equilibrates between the vasculature and tissue. On the contrary, there exists much controversy on disequilibria between bound and free species within the vasculature, and the view that the bound substrate could contribute to transmembrane transfer and removal (Weisiger et al., 1989, 1991; Schwab and Goresky 1996). Details on this topic may be found in Weisiger's chapter.

The estimate of the average plasma concentration in liver hinges on assumptions made for the liver with regard to its flow and mixing behaviors (Perl and Chinard, 1968; Levenspiel, 1972; Winkler et al., 1973; Pang and Rowland, 1977; Froment and Bischoff, 1976; Roberts and Rowland, 1986; St-Pierre et al., 1992), as shown in comparative studies of hepatic clearance models. The topic is, however, beyond the present coverage. Due to the reported concentration gradient observed within the liver acinus (Jones et al., 1980; Gumucio et al., 1981, 1984), it is not unreasonable to adopt the logarithmic average of the unbound input ($C_{In,u}$) and unbound output ($C_{Out,u}$) concentrations, at steady state, or $\hat{C}_u$, which is viewed as an appropriate description of the average (unbound) substrate concentration in liver (Xu et al., 1993; Geng et al., 1995a, 1995b):

$$\hat{C}_u = \frac{C_{In,u} - C_{Out,u}}{\ln(C_{In,u} / C_{Out,u})}. \qquad (14.10)$$

Tissue concentrations (C_t) are provided upon homogenization of the liver. When the unbound fraction in tissue, f_t, is known, the unbound tissue concentration in liver, $C_{t,u}$ may be calculated ($f_t C_t$). Quantitation of the former is easier said than done, since higher artifactual levels for C_t may result upon tissue homogenization due to inclusion of bile ductular fragments containing micelles (Scharschmidt and Schmid, 1978; Geng et al., 1995a). Additionally, tissue metabolism will further confound estimates of both C_t and f_t. As will soon become apparent, the ratio of $C_{t,u}/\hat{C}_u$ allow us to discern the concentrative ability of liver cell membrane.

Hepatocytic Drug Entry: Barrier-Limited or Flow-Limited

The plasma membrane of hepatocytes poses a substantial barrier for the entry of substrates. Transmembrane transport is a process that can constitute the rate-determining factor in the net loss of substrate across the organ. For lipophilic substrates, the mode of entry via passive diffusion is rapid, and the flux across the membrane is sufficiently high such that the membrane is seldom rate-limiting (Goresky and Groom, 1984). Ionic substrates are expected to encounter the membrane as a barrier unless specific transport carriers exist (for reviews, see Meijer, 1987; Nathanson and Boyer, 1991; Meier and Stieger, 1993). In the absence of facilitated transport, larger hydrophilic compounds will be barred from entry (Pang et al., 1984; deLannoy and Pang, 1987; Goresky et al., 1992; Ratna et al., 1993) and their uptake is often retarded. Conceptual frameworks for a membrane barrier for polar compounds, which may be applied to preformed polar metabolites, have been presented (Gillette and Pang, 1977; Sato et al., 1986; Miyauchi et al., 1988; Schwab et al., 1990). Instantaneous equilibration at any point x would not occur along the sinusoidal flow path in this instance, that is, the unbound blood concentration at point x would not reflect that in the liver tissue at the same locale.

The following section highlights the effect of such a barrier on the difference in the distribution of substrates (tissue partition coefficient) and methods for estimation of kinetic constants (K_m and V_{max}). An assumption often taken is that the unbound drug concentration in the Disse space ($C_{D,u}$) equals that in sinusoidal plasma ($C_{p,u}$). The unbound species exchanges with that in the cell ($C_{t,u}$), and the rate and extent are governed by the transfer rate constants, k_1 and k_{-1} (Figure 14.6); inside the cell, removal ensues with k_{seq}, the sequestration rate constant. With flow-limited distribution, $C_{t,u}$ equals $C_{D,u}$ and $C_{p,u}$ at any point x when elimination is negligible. For this reason, the equation governing the loss of substrate is substantially simplified. With carrier-mediated transport, with the influx and efflux being saturable processes of distinct capacity (V_{max}) and affinity (K_m), or with diffusion-limitation when the influx and efflux constants represent values much lower than the flow rate (F), and in the presence of sequestration or removal, $C_{t,u}$ will no longer reflect $C_{p,u}$ at x.

Tissue Equilibrium Ratio ($C_{t,u}/\hat{C}_u$) *or Apparent Partition Coefficient.* The mass transfer at steady state shows that the following condition holds:

$$\hat{C}_u k_1 V_{cell} = C_{t,u}(k_{-1} V_{cell} + k_{seq} V_{cell}).$$ (14.11)

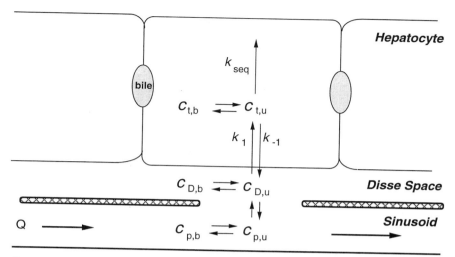

FIGURE 14.6. Schematic representation of drug uptake at the level of a single sinusoid. Equilibria are assumed between bound and unbound forms of substrate in plasma and tissue. The rate constants for influx, efflux, and sequestration (k_1, k_{-1}, and k_{seq}) have been defined with reference to V_{cell}, the accessible cell water volume. $c_{p,u}$, $c_{p,b}$, $c_{D,u}$, $c_{D,b}$, $c_{t,u}$, and $c_{t,b}$ are the unbound and bound concentrations of substrate in sinusoidal plasma, the Disse space, and the cell, respectively. The hatched areas denote the endothelial cells.

Upon rearrangement, the tissue to plasma equilibrium ratio is

$$\frac{C_{t,u}}{\hat{C}_u} = \frac{k_1}{(k_{-1} + k_{seq})}. \tag{14.12}$$

The ratio is also viewed as the apparent partition coefficient with the presence of removal ($k_{seq} > 0$). In the absence of sequestration ($k_{seq} = 0$), the distribution ratio becomes the partition coefficient (k_1/k_{-1}).

The apparent partition coefficients ($C_{t,u}/\hat{C}_u$) of flow-limited, carrier-mediated, and diffusion-limited transport have been compared (Geng et al., 1995b). For carrier-mediated entry, the ratio revealed a concave downward profile with increasing concentration (Figure 14.7A), whereas for flow-limited substrates (influx clearance ≥ 5 times the flow rate), the values will initially be lower than unity and these will gradually plateau at unity when $C_{t,u} = \hat{C}_u$ with increasing concentration when removal no longer constitutes a measurable loss (Figure 14.7B). For diffusion-limited substrates (see curves associated with influx clearances values that are < 2 times the flow rate; Figure 14.7B), low $C_{t,u}/\hat{C}_u$ values that crept higher with increasing concentrations are observed; the $C_{t,u}/\hat{C}_u$ ratio, however, will be much lower than unity. Exceptions to these general patterns do exist. These include systems in which the influx rate greatly exceeds the sum of the efflux and

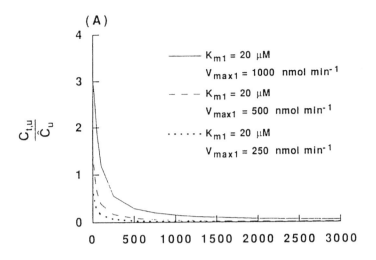

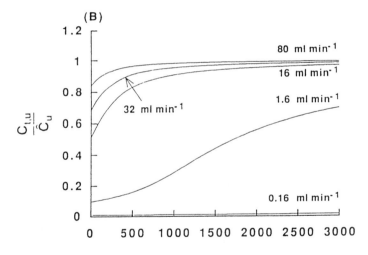

Input Unbound Concentration (μM)

FIGURE 14.7. The relationship between the apparent tissue distribution equilibrium ratio estimated with $C_{t,u}/\hat{C}_u$), and input plasma concentration for hypothetic substrates whose entry is via carrier-mediated transport (A), or flow-limited and diffusion-limited (B). The different influx parameters [expressed as the K_m's and V_{max}'s for carrier-mediated entry (A) or influx clearances for flow-limited (80 ml · min^{-1} or 5× flow rate) and diffusion-limited (0.16, 1.6, 16, and 32 ml · min^{-1} or 0.01, 0.1, 1, and 2× flow rate) entry (B)] were shown. Drug binding is assumed to be negligible, and the flow rate was 16 ml · min^{-1} in this example. (Data taken from Geng et al., 1995b.)

sequestration rates; under this circumstance, there is a tendency for $C_{t,u}$ to increase dramatically and steady state is not attainable in the system.

Estimation of Kinetic Constants, V_{max} and K_m. The next question that remains to be answered is whether the overall removal rate would cast insight into the rate-determining step of the reaction, that is, whether it is solely due to sequestration. The answer is, of course, dependent on the modes of entry and efflux of the substrate. The rate of removal (v) is represented by the Michaelis–Menten equation. The following relationship applies for flow-limited transport, since $C_{t,u} = \hat{C}_u$:

$$v = \frac{V_{max}\hat{C}_u}{K_m + \hat{C}_u} \; ; \tag{14.13}$$

otherwise, for carrier-mediated and diffusion-limited (membrane- or barrier-limited) cases, the tissue unbound concentration should be used:

$$v = \frac{V_{max}C_{t,u}}{K_m + C_{t,u}} \; . \tag{14.14}$$

General Equations. The general solutions for flow-limited and barrier-limited transport of an injected tracer have been developed (Goresky and Groom, 1984). The premise is the comparison of the substrate curve onto the appropriate reference curve. The comparison will fully account for the dispersion in the system due to heterogeneity in capillary transit times. For flow-limited substrates,

$$C(t) = \frac{1}{1 + \gamma_{rel}} \, C_{ref}\left(\frac{t - t_0}{1 + \gamma_{ref}} + t_0\right) e^{-k_{seq}t} \tag{14.15}$$

and for membrane-limited transport,

$$C(t) = C_{ref}(t)e^{-k_1 f_u \theta'(t - t_0)} + e^{-(k'_{-1} + k'_{seq})(t - t_0)} \times \int_{t_0}^{t} C_{ref}(\tau')e^{(-f_u k_1 \theta' + k'_{-1} + k'_{seq})(\tau' - t_0)}$$

$$\times \sum_{n=1}^{\infty} \frac{([f_u k_1 \theta' k'_{-1}][\tau' - t_0])^n (t - \tau')^{n-1}}{n!(n-1)!} \, d\tau' \; , \tag{14.16}$$

where $C(t)$ is the fractional recovered dose (concentration/dose) of tracer related to the behavior of the reference, $C_{ref}(t)$, its unbound fraction to plasma (f_u) and tissue (f_t) proteins, and t_0 the large vessel transit time; $\tau' = (1 + \gamma_{ref})\tau + t_0$, t is the distance x divided by the velocity v_F, and $\theta' = \theta/(1 + \gamma_{ref})$, as shown in the previous chapter of Goresky et al. (1997). It is noteworthy that k_1 is defined with respect to the cellular water space, V_{cell}, such that the permeability surface area product $(P_{in}S)$ is $k_1 V_{cell}$. The first term of Eq. (14.16) represents the throughput component, and the second term is the exchanging or returning component. With fitting of the data, the influx $(f_u k_1 \theta')$, efflux $(k'_{-1}$ or $f_t k_{-1})$, and sequestration $(k'_{seq}$ or $f_t k_{seq})$ coefficients are obtained. The respective rate constants $(k_{-1}$ and $k_{seq})$

may be obtained when f_t is known. After setting $k_{seq} = 0$ in Eq. (14.16) the difference between this and the outflow curve equals the sequestered component.

With the above as background, we may now proceed to examine the temporal behavior of several substrates. The study design is optimized when single-pass perfusion is carried out with a steady delivery of unlabeled substrate and the MID dose containing radiolabeled tracer substrate injected during the steady state, with the expectation that the steady-state extraction ratio $[(C_{In} - C_{Out})/C_{In}]$ of the bulk will equal [1 − integral of tracer outflow recovery, normalized to dose].

Examples

Barrier-Limitation

Diffusion-Limited. Polar substrates that do not enter cells readily are represented by enalaprilat (Schwab et al., 1990) and acetaminophen sulfate (Goresky et al., 1992). For both enalaprilat and acetaminophen sulfate, little excretion is noted. The outflow profiles are mostly similar to that of labeled sucrose, but precesses the labeled sucrose curve due to its slight albumin binding. The throughput component, the portion that has not entered liver cells, is very large in comparison to the returning component, that portion which enters the cells and returns later to the vasculature (see Figure 14.8). For enalaprilat, the partition coefficient remained less than unity (0.2); for acetaminophen sulfate, which also enters the cell poorly, concentration-independence in influx was demonstrated.

Carrier-Limited. The glutathione conjugate of bromosulfophthalein (BSPGSH), similar to its precursor, bromosulfophthalein (BSP), is an anionic dye that is known to exhibit carrier-mediated entry and ATP-dependent transport at the sinusoidal and canalicular membranes, respectively (Sorrentino et al., 1989b; Kitamura et al., 1990). BSPGSH is normally excreted unchanged into bile; little metabolism occurs in the perfused rat liver preparation (Zhao et al., 1993).

When exploring carrier-mediated transport, the background steady-state BSPGSH concentration was altered. Since BSPGSH binds primarily to albumin with two classes of binding sites and does not distribute into red cells, Eq. (14.8) may be used to describe the hypothetical reference curve. With curve-fitting of the data, the major portion of the tracer outflow profile was found to be the throughput component (Figure 14.9), which decreases with increasing BSPGSH concentration (Geng et al., 1995a). Due to the convenient fact that BSPGSH is not metabolized, it was feasible to perform tissue-binding studies. These demonstrated two classes of binding sites with measurable affinities that resulted in very low tissue unbound fractions. Normally, Eq. (14.16) would be used, together with known values of f_u and f_t, to estimate the influx (k_1), efflux (k_{-1}), and sequestration (k_{seq}) constants in accordance with the previously developed barrier-limited models (Goresky et al., 1992). However, in this instance, extremely high concentrations of BSPGSH were found in bile and in the corresponding assayed tissue concentration (C_t) obtained upon homogenization of the same liver at

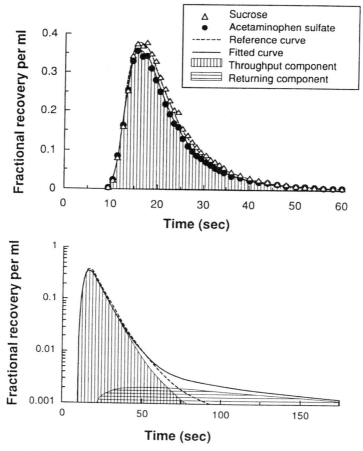

FIGURE 14.8. Outflow dilution profile of tracer [³H]acetaminophen sulfate conjugate in the red cell, albumin-perfused rat liver, presented in linear (upper panel) and logarithmic fashion (lower panel). Due to its binding to albumin, a hypothetic reference (based on albumin and sucrose curves) was used as its vascular reference to model the membrane-limited behavior of the tracer acetaminophen sulfate. Note the very large throughput component (upper and lower panels). The very small returning component was not visible in the linear plot (upper panel) but can be seen in the semilogarithmic plot (lower panel). [Taken from data of Goresky et al. (1992).]

the end of the single-pass perfusion study. When these were normalized to the logarithmic average concentration (based on total BSPGSH), similar concave-down trends were obtained (Figure 14.10). Due to the high BSPGSH concentrations in bile, liver homogenization will lead to C_t and therefore $C_{t,u}$ ($f_t C_t$) levels that are artificially too high due to inclusion of bile ductular elements containing BSPGSH (Scharschmidt and Schmid, 1978; Geng et al., 1995a, 1995b). Thus, $C_{t,u}$

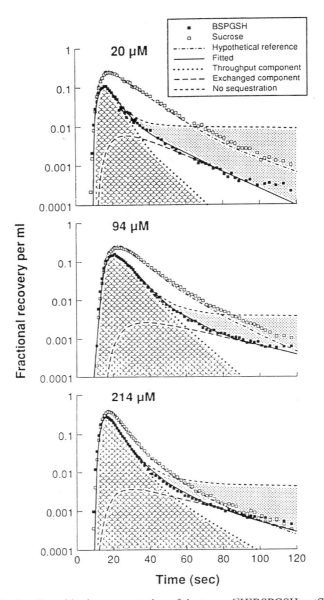

FIGURE 14.9. Semilogarithmic representation of the tracer [³H]BSPGSH outflow profiles for rat liver perfusion (12 ml · min⁻¹) studies conducted at various input BSPGSH concentrations. The steady state bulk extraction ratio equals [1 − integral of fractional recovery]. The outflow profiles are resolved into the throughput (scalloped) and returning (area under the long dashed line) components. The reference curve to which the BSPGSH curve is related is outlined by a dash–dot profile. Its space of distribution is slightly smaller than that for labeled sucrose. The form the labeled BSPGSH curve would have had in the absence of biliary excretion was calculated by setting $k_{seq} = 0$. The area between this curve and the observed points (the sequestered component) is outlined by dotted shading, the "sequestered component." Note that the throughput component increases in relation to the appropriate reference curve, with increasing concentration. [Reproduced from Geng et al. (1995a) with permission.]

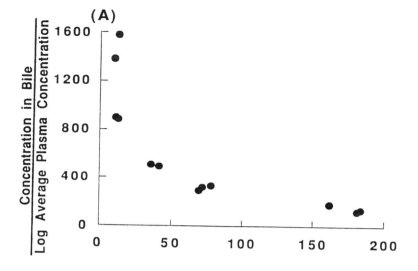

BSPGSH Log Average Plasma Concentration (μM)

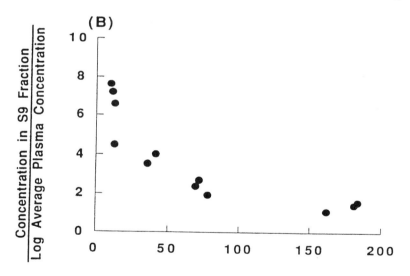

BSPGSH Log Average Plasma Concentration (μM)

FIGURE 14.10. Change in the bile to logarithmic average plasma concentration ratio (A) and S9 fraction to sinusoidal plasma concentration ratio (B), with increase in logarithmic average plasma concentration of BSPGSH. BSPGSH is highly concentrated in bile (A), and the concentration ratio diminishes with increase in sinusoidal plasma concentration. BSPGSH is also concentrated in the S9 supernatant, although to a lesser degree; the concentration ratio again decreases with increase in logarithmic average plasma concentration (B). [Reproduced from Geng et al. (1995a) with permission.]

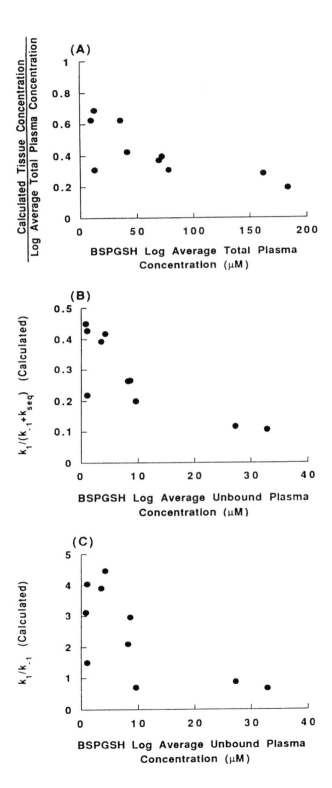

TABLE 14.3. Apparent kinetic parameters obtained from steady-state bulk experiments with BSPGSH compared with parameters from MID experiments with [³H]BSPGSH, obtained within the same livers.

	Steady state		MID		
	Plasma extraction	Biliary excretion	Sinusoidal influx	Sinusoidal efflux	Biliary excretion
Transport rate versus $\hat{C}_u$ [a]	3.3 ± 0.6[d]	3.6 ± 0.9[b]	3.7 ± 1.5		
K_m (μM)[c]					
V_{max} (nmol · min⁻¹ · g⁻¹)[b]	50 ± 3.2	50 ± 4.7[b]	83 ± 17		
Transport rate versus $\hat{C}_{t,u}$ [b]		3.6 ± 2.8		1.8 ± 1.9	1.8 ± 0.9
K_m (μM)[c]					
V_{max} (nmol · min⁻¹ · g⁻¹)[c]		97 ± 46		15 ± 11	94 ± 25

[a] $\hat{C}_{p,u}$ is the logarithmic average unbound plasma concentration of BSPGSH.
[b] $C_{t,u}$ is the calculated unbound concentration of BSPGSH in liver tissue.
[c] After nonlinear fitting with different weighting function, the sets of parameters were selected by comparing their coefficients of variation, residual sums of squares, and weighted residual plots.
[d] Standard deviation of parameter estimate.

was calculated instead. Since f_t remains low and relatively constant among the levels of the C_t's measured, substitution of f_t into the rate coefficients for efflux and sequestration (k_{-1} and k_{seq}) yields their corresponding rate constants; then $C_{t,u}$ may be estimated from Eq. (14.12) with foreknowledge of k_1.

With this method, the ratios of the calculated tissue to logarithmic average concentration are in fact considerably lower (about one-tenth) than those measured (cf. Figures 14.11A and 14.10B), likely due to the micellar sink phenomenon in bile (Scharschmidt and Schmid, 1978). The estimated apparent tissue to plasma partition ratio, or $k_1/(k_{-1} + k_{seq})$, displays a concave-down profile with increasing concentration (shown logarithmically) (Figure 14.11B), or in a form that is expected of carrier-mediated systems (cf. Figure 14.7). Regression of k_{-1} and k_{seq} against $V_{max}/(K_m + C_{t,u})$ will provide the corresponding parameters for efflux ($V_{max} = 15$ nmol · min⁻¹ · g⁻¹ and $K_m = 1.8$ μM) and excretion ($V_{max} = 94$ nmol · min⁻¹ · g⁻¹ and $K_m = 1.8$ μM) (Figures 14.11B and 14.11C). Regression of the influx constant (k_1) against ($V_{max}/(K_m + \hat{C}_u)$ provides those constants for influx ($V_{max} = 83$ nmol · min⁻¹ · g⁻¹ and $K_m = 3.7$ μM) (Table 14.3); the permeability surface area product ($P_{in}S$) for influx ($k_1 V_{cell}$) also displays concentration dependence (Figure 14.12). By comparing values of the kinetic constants, it is readily apparent that influx, the initial step for uptake, will become saturated first and

FIGURE 14.11. Variation of the theoretical tissue partition coefficients in relation to BSPGSH concentration in plasma or tissue. (A) The ratio of calculated tissue concentration to logarithmic average plasma concentration, related to logarithmic average total plasma concentration; and (B) $k_1/(k_{-1} + k_{seq})$, the theoretical value for $C_{t,u}/\hat{C}_u$ and (C) k_1/k_{-1}, the theoretical value of tissue partition coefficient when $k_{seq} = 0$, related to the logarithmic average plasma unbound concentration. [Reproduced from Geng et al. (1995a) with permission.]

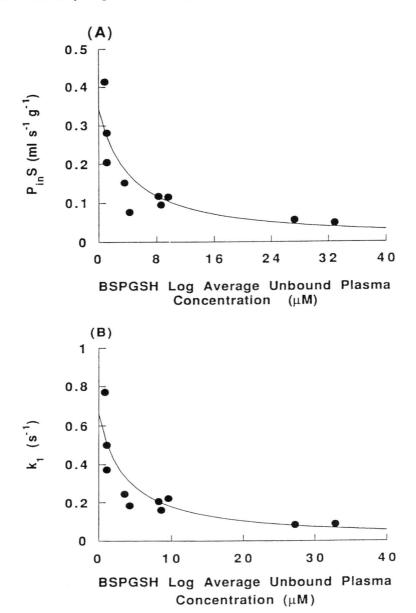

FIGURE 14.12. Variation of the influx permeability surface area product (A) and the influx rate constant (B) as a function of logarithmic average unbound concentration of BSPGSH. [Reproduced from Pang et al. (1995) with permission.]

that net uptake results in tissue concentrations that are too low to saturate the ATP-dependent transporter at the canalicular membrane.

This example of carrier-mediated influx may be used to illustrate an additional point. It is imperative to employ the steady-state unbound tissue concentration for the estimation of kinetic parameters for bulk BSPGSH excretion (Eq. (14.14)); this yielded a V_{max} of 97 nmol · min^{-1} · g^{-1} and a K_m of 3.6 μM, values that are similar to those obtained from MID (Table 14.3). Regression of the steady-state biliary excretion or net plasma disappearance rates against the logarithmic average unbound concentration $\hat{C}_u$ (Eq. (14.13)), however, resulted in values (V_{max} of 50 nmol · min^{-1} · g^{-1} and a K_m of 3.3 to 3.6 μM) that are different from those obtained with MID for either influx, efflux, or removal (Table 14.3); as expected, the two sets of values do not coincide. In this case, the intracellular BSPGSH levels do not reach sufficiently high values to saturate efflux and excretion, and the limiting feature of the transfer process at bulk concentrations of BSPGSH reached appears to be the influx process.

Flow-Limitation

Among the substrates examined in our laboratory, salicylamide (Pang et al., 1994b) and acetaminophen (Pang et al., 1995), both of which are neutral species, are found to display flow-limited distribution. The throughput component is absent for salicylamide, and Eq. (14.15) adequately described the behavior of this highly conjugated compound. Since the binding of acetaminophen to red blood cells has resulted in its slow efflux (or a red cell carriage effect), the flow-limited distribution characteristics of acetaminophen are shown only with erythrocyte-free liver preparation, for which the throughput component of acetaminophen is virtually nonexistent.

Precursor–Product Relationships and
Red Blood Cell Binding

Acetaminophen uptake and its conversion to the sulfate conjugate have been used to exemplify precursor–product relationships. When the red cell distribution of acetaminophen was first explored, an unremarkable value of 1.1 was found for λ_R, the distribution ratio in red blood cells, while acetaminophen was not bound to albumin (Pang et al., 1995). However, upon injection of pre-equilibrated MID doses containing the noneliminated references and [³H]acetaminophen into the portal vein of the single-pass perfused rat liver under varying conditions of hematocrit, the [³H]acetaminophen curve displays a high dependence of the hematocrit, exhibits an early high peak, paralleling that for red cells, followed by a prolonged decline (Figure 14.13); the peaks are, however, of much lower magnitude than those of the labeled RBC. After the rapid decline phase, a more prolonged decline follows, which exhibits a slope on the semilogarithmic plot that is almost parallel to that for D$_2$O. The rapid rise and then decline, previous to the later-in-time and

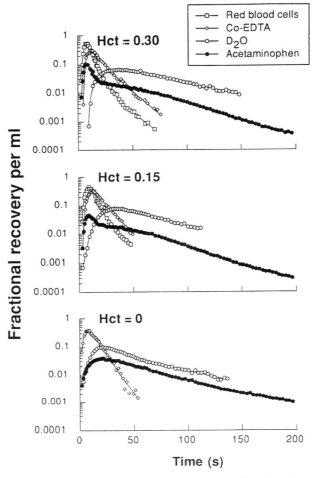

FIGURE 14.13. Semilogarithmic plots of the recovered outflow fractions of [³H]acetaminophen and the noneliminated reference indicators in multiple-indicator dilution experiments conducted at hematocrit = 0.3 (upper panel), 0.15 (middle panel), and 0 (bottom panel). [Reproduced from Pang et al. (1995) with permission.]

protracted decay, constitutes a kind of "cap" on the outflow profile for acetaminophen that varied with the hematocrit; in the absence of RBCs, the "cap" disappears. The changes in the outflow patterns exhibited by labeled acetaminophen with change in hematocrit are characteristic of a red cell carriage phenomenon due to red cell trapping of substrate, with slow efflux from red cells (Goresky et al., 1975; Pang et al., 1995).

The model that describes the binding of acetaminophen but slow efflux from

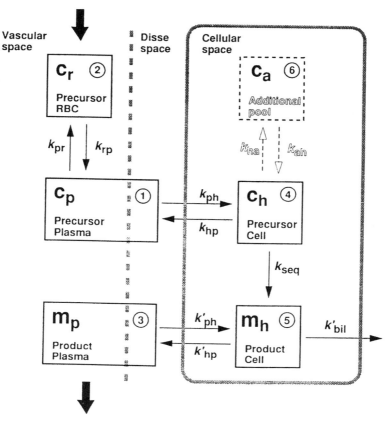

FIGURE 14.14. Schematic presentation of red cell carriage of acetaminophen (with the red cell and plasma as separate vascular components), its transfer into liver cell, distribution into an intracellular pool, and sulfation, with return to plasma and biliary excretion. [Reproduced from Pang et al. (1995) with permission.]

RBCs now requires the addition of a separate RBC pool in the vascular compartment (Figure 14.14). The model includes two pools of interchanging precursor material traveling along the sinusoids in the red cells and plasma at different velocities, the material in the red cells traveling with the velocity of flow and that in the plasma, like [^{58}Co]EDTA, in a delayed-wave fashion (Goresky, 1963). The model also describes the relationship between the plasma and red cell outflow profiles for the eliminated substance (c_p and c_r for acetaminophen in plasma and red cells), that for its metabolite in plasma (m_p for acetaminophen sulfate), and that of the reference substances, labeled red blood cells and [^{58}Co]EDTA. Acetaminophen in plasma then gains access to the hepatocellular compartment where acetaminophen c_h is converted to acetaminophen sulfate m_h, which in turn is released to plasma m_p. The additional parameters that describe the behavior of the eliminated acetaminophen tracer in this model include the rate constants for red cell efflux k_{rp}, for hepatocellular influx k_{ph}, for hepatocellular efflux k_{hp}, and for hepatocellular sequestration k_{seq}, as well as those concerning an additional pool, k_{ha} and k_{ah}, added to explain the "tail" of the curve. These rate constants, in contrast to those ordinarily used in Goresky's terminology, are associated with the compartment in which the flux originates. For modeling, the algorithms previously presented by Schwab (1984) for the description of L-lactate metabolism in the perfused rat liver (Schwab et al., 1985) were used.

The optimized parameters arising from the fit, when summarized (Table 14.4), show a slow efflux rate constant of acetaminophen from red cells (0.047 ml · s⁻¹ · ml⁻¹ RBC water, on average), with an acetaminophen turnover time in red cells ($1/k_{rp}$) of about 21 s, a value that is slightly larger than the transit time of red blood cells in the system. The hepatocyte influx rate constant for acetaminophen (k_{ph} and and the efflux rate constant, k_{hp}, were independent of the hematocrit, and values for the influx ($P_{ph}S$) and efflux ($P_{hp}S$) hepatocytic permeability surface area products were similar. The values are high in relation to the hepatic blood flow rate (0.019 ± 0.003 ml · s⁻¹ · g⁻¹). The calculated sequestration (or intrinsic) clearances for acetaminophen were similar at all three levels of hematocrit (Table 14.4), even though a much reduced sequestration clearance had been anticipated when there were no red cells inasmuch as the oxygen supply and oxygen consumption of the liver were much reduced.

TABLE 14.4. Fitted parameters from acetaminophen MID studies in perfused rat liver persusion experiments directed at acetaminophen transport and sulfation (flow 12 ml/min)

%RBC in perfusate	Permeability surface area products				Hepatocyte sequestration clearance
	Red cell	Hepatocyte			
	$P_{rp}S_r$ [a]	$P_{ph}S$	$P_{hp}S$	P_{ph}/P_{hp}	CL_{int} [b]
(v/v)	$\left(\dfrac{\text{ml/s}}{\text{ml RBC water}}\right)$	(ml s⁻¹ g⁻¹)	(ml s⁻¹ g⁻¹)		(ml s⁻¹ g⁻¹)
30%	0.049 ± 0.014	0.24 ± 0.10	0.23 ± 0.09	1.1 ± 0.1	0.018 ± 0.004
15%	0.044 ± 0.015	0.44 ± 0.20	0.33 ± 0.19	1.4 ± 0.2	0.021 ± 0.004
0%		0.38 ± 0.03	0.36 ± 0.09	1.1 ± 0.2	0.014 ± 0.002
_P_ values					
ANOVA		N.S.[c]	N.S.	N.S.	N.S.
Paired _t_ [d]	N.S.[c]				

[a] k_{rp} (or $P_{rp}S_r$), which is the rate constant for transfer into RBC, has been defined with respect to k_{pr} (see Pang et al., 1995). Values are normalized with respect to red cell water.

[b] CL_{int} equals $k_{seq}V_{cell}$.

[c] Not significant.

[d] 30% RBC vs. 15% RBC.

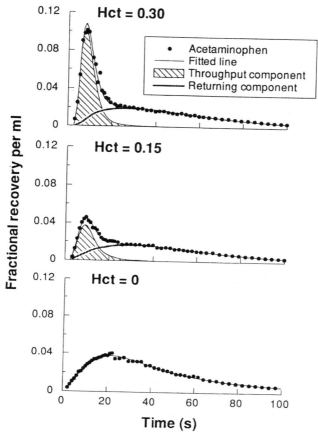

FIGURE 14.15. Plot of the fit of the acetaminophen curves. The acetaminophen in red cells that did not leave them during their passage along the vasculature (a red cell throughput) are in shaded area. The value of the underlying hematocrit (Hct) is given. Reproduced from Pang et al. (1995) with permission.

In the presence of red blood cells, the acetaminophen outflow profiles are made up of two components: one for acetaminophen in red cells and, the second, acetaminophen in plasma. After leaving the sinusoids, tracer acetaminophen will equilibrate between the two phases. The process would have been complete at the time of analysis of the samples. Nevertheless, the outflow concentration profiles in toto may be used to reconstruct events in the sinusoids and outflow profiles in the two phases, at the time when the blood bearing the tracer left the sinusoid. The fitted acetaminophen outflow profiles are illustrated in Figure 14.15. The component of the acetaminophen in the red cells, which never left them during their

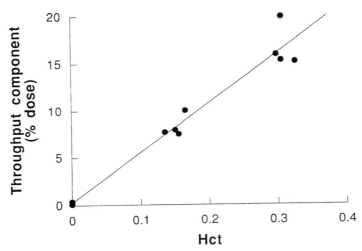

FIGURE 14.16. The throughput component (this time in both red cells and plasma), expressed as a percent of dose, versus hematocrit (Hct). Data are the same as in Figure 14.15. Reproduced from Pang et al. (1995) with permission.

passage along the vasculature, is shaded. The throughput component of the [³H]acetaminophen outflow curve, which has not entered the liver cells and which includes tracer in both the red cells and in plasma (the proportion of the latter is exceedingly small), correlates directly with the hematocrit (Figure 14.16). The decrease in the accessibility of the [³H]acetaminophen in red cells explains the lower steady-state hepatic extraction ratio for acetaminophen and the greater relative integrated outflow curve or area under the curve (AUC) of [³H]acetaminophen (also expressed as 1 − extraction ratio) at a hematocrit of 30%, in comparison with that at 15%.

In terms of modeling of the metabolite, acetaminophen sulfate conjugate, parameters obtained previously with preformed acetaminophen sulfate are necessary for linkage with the precursor parameters. With these, the product curve fits well and its AUC varies inversely with hematocrit (Figure 14.17). The overall analysis demonstrates a slow release of acetaminophen from the red blood cells and rapid liver cell entry, so that red cell binding is displayed as a red cell carriage effect which reduces the rate of liver cell entry and hence of sulfation of [³H]acetaminophen. The liver cells exhibit a concomitant, very low permeability to the product acetaminophen sulfate, leading to protracted product outflow curves. An inferred slow efflux-mediated storage phenomenon for product evolves, shown by the greater excretion of the acetaminophen sulfate when generated from acetaminophen than with administration of acetaminophen sulfate conjugate (Figure 14.18). A characteristic set of single sinusoidal concentration profiles described by the above for acetaminophen and acetaminophen sulfate is

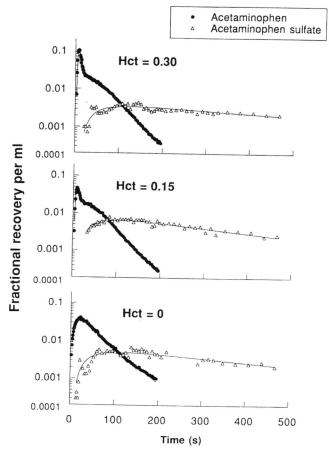

FIGURE 14.17. Fitted theoretical outflow profiles of [³H]acetaminophen and [³H]aceta-minophen sulfate (same data as Figure 14.13) calculated according to the scheme presented in Figure 14.14 for a hematocrit of 0.3 (upper panel), 0.15 (middle panel), and 0 (bottom panel). The profiles were calculated by use of the optimal fitted values for the parameters, derived for this particular set of profiles. Reproduced from Pang et al. (1995) with permission.

illustrated in Figure 14.19, with a vascular transit time of 10 s. A general pattern is seen. The red cell acetaminophen concentration decays slowly along the sinusoidal length whereas the plasma concentration decays much more quickly along the length since the membrane permeability is so high (almost flow-limited) that the plasma and liver cell water concentrations are virtually identical to one another. Since the permeability of the liver cell membrane to acetaminophen sulfate product is low, the steady-state concentration of generated acetaminophen sulfate in

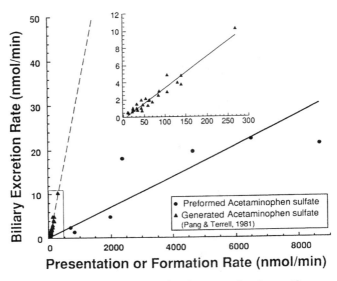

FIGURE 14.18. Differences in biliary excretion for acetaminophen sulfate, generated as a metabolite of acetaminophen (data of Pang and Terrell, 1981b) or when administered (data of Goresky et al., 1992) to the once-through perfused rat liver, versus the formation/presentation rate.

the liver cells will accumulate and will be much greater with respect to that for acetaminophen, and this decreases along the length because of biliary excretion. It is surmised that, of the generated acetaminophen sulfate which escapes to plasma, little reenters the liver cells. The concentration of the generated acetaminophen sulfate in plasma increases along the length and is highest in the blood leaving the sinusoid. From these, it is found that the proportion of tracer acetaminophen sulfate entering the liver cells is exceedingly small, of the order of 1% of the total, and the material does not enter red cells.

It must be noted that the value of the partition coefficient λ_R, while conveying the notion of the extent of red cell distribution, does not provide insight as to slow or fast equilibria. For drugs that have a high red cell distribution ($\lambda_R > 1$), a late smooth curve with no separable early component may be found when the tracer binds to red cells and leaves exceedingly quickly. This was observed for labeled xenon (Goresky et al., 1988), salicylamide sulfate (Xu et al., 1994), and salicylamide (Pang et al., 1994b), all compounds with substantial λ_R values. For these substrates, the conformation and transit time will nevertheless be influenced by hematocrit, a *red cell capacity effect* (Goresky et al., 1988). When the pre-equilibrated labeled tracer leaves red cells relatively slowly, the outflow dilution curves will exhibit an early component more or less coincident in time and symmetrical to the reference labeled red cell outflow curve, *a red cell carriage effect,* as shown for thiourea (Goresky et al., 1975) and acetaminophen (Pang et

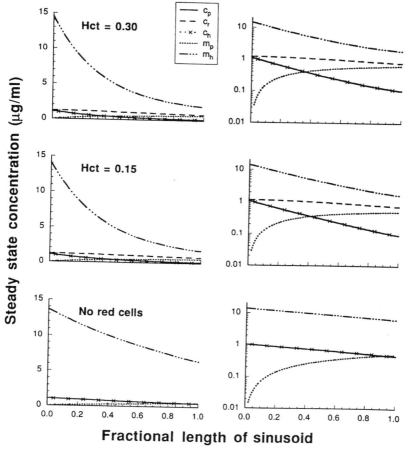

FIGURE 14.19. Underlying steady-state concentration profiles calculated with optimized parameters derived from fitting the transient profiles of the representative set of experiments. Values were calculated for a representative sinusoid, with a red cell transit time of 10 s. Reproduced from Pang et al. (1995) with permission.

al., 1995). Extensive binding of the carbonic anhydrase inhibitors (Wallace and Riegelman, 1977; Bayne et al., 1981; Lin et al., 1992) and of the immunosuppressive agents cyclosporin A (Lemaire et al., 1982) and tacrolimus (Piekoszewski et al., 1993) to red cells have been reported. Among these, reduced extraction ratios have been observed at higher hematocrits (Piekoszewski et al., 1993). For substrates that exhibit a red cell carriage effect, hepatic removal is expected to be higher when non–pre-equilibrated substrate in plasma is introduced than after pre-equilibration of the substrates within blood prior to its delivery to the liver, inferring that the red cell binding of drugs can play an important role in delimiting hepatic removal (Goresky et al., 1975; Lee and Chiou, 1989a, 1989b).

Concluding Remarks

We have adopted use of the MID technique for liver perfusion. Kinetic equivalents of the noneliminated reference indicators [^{58}Co]EDTA for labeled sucrose and D_2O for ^3H_2O are found, and this has allowed us free use of the [^3H]- and [^{14}C]-radiolabels for the tracer substrate or metabolite. The MID technique has aided in the validation of the prograde–retrograde perfusion technique in providing assurance of equal access of hepatocytes, albeit in reverse directions. The MID technique is a constitutive protocol in the examination of enzyme zonation in the HAPV and HAHV perfusion technique. In addition to the exploration of liver hemodynamics and microcirculation, integration of theoretical developments and the experimental procedures have further provided us with a quantitative description of drug and metabolite handling. The more recent and exciting finding is on the red cell carriage effect in delimiting entry and metabolism of a drug (acetaminophen). With the present tools in hand, other drug–metabolite conversion, futile cycling, and mechanistic levels of interaction may be readily explored.

Acknowledgment. This work was supported by the National Institutes of Health (GM-38250) and the Medical Research Council of Canada (MRC-9104).

References

Almond, N. E. and A. M. Wheatley. Measurement of hepatic perfusion in rats by laser Doppler flowmetry. *Am. J. Physiol.* 262(*Gastrointest. Liver Physiol.* 25):G203–G209, 1992.

Baron, J., R. A. Redick, and F. P. Guengerich. Effect of 3-methylcholanthrene, β-naphtho-flavone, and phenobarbital on the 3-methyl-cholanthrene inducible isozyme of cytochrome P-450 within centrilobular, midzonal, and periportal hepatocytes. *J. Biol. Chem.* 257:953–957, 1982.

Barnhart, J. R. and R. Clarenburg. Factors determining the clearance of bilirubin in perfused rat liver. *Am. J. Physiol.* 225:497–508, 1973.

Bass, L. Saturation kinetics in hepatic drug removal: A statistical approach to functional heterogeneity. *Am. J. Physiol.* 244:G583–G589, 1983.

Bass, L. Heterogeneity within observed regions: Physiologic basis and effects on estimation of rates of biodynamic processes. *Circulation* 72(Suppl IV):47–52, 1985.

Bass, L. and P. J. Robinson. Effects of capillary heterogeneity on rates of steady state uptake of substances by the intact liver. *Microvasc. Res.* 22:43–57, 1981.

Bass, N. M., M. E. Barker, J. A. Manning, A. L. Jones, and R. K. Ockner. Acinar heterogeneity of fatty acid binding protein expression in livers of male, female, and clofibrate-treated rats. *Hepatology* 9:12–21, 1989.

Bayne, W. F., F. T. Tao, G. Rogers, L. C. Chu and F. Theeuwes. Time course and disposition of methazolamide in human plasma and red blood cells. *J. Pharm. Sci.* 70:75–81, 1981.

Boyer, J. L., E. Elias, and T. J. Layden. The paracellular pathway and bile formation. *Yale J. Biol. Med.* 52:61–67, 1979.

Braakman, I., G. M. M. Groothuis, and D. K. F. Meijer. Zonal compartmentation of perfused rat liver: Plasma reappearance of rhodamine B explained. *J. Exp. Pharmacol. Ther.* 239:869–873, 1989.

Burger, H. J., R. Gebhardt, C. Mayer, and D. Mecke. Different capacities for amino acid transport in periportal and perivenous hepatocytes isolated by digitonin/collagenase perfusion. *Hepatology* 9:22–28, 1989.

Chiba, M. and K. S. Pang. Effect of protein binding on 4-methylumbelliferyl sulfate desulfation kinetics in perfused rat liver. *J. Pharmacol. Exp. Ther.* 266:492–499, 1993.

Chiba, M. and K. S., K. Poon, J. Hollands and K. S. Pang. Glycine conjugation of benzoic acid and its acinar localization in perfused rat liver. *J. Pharmacol. Exp. Ther.* 268:419–416, 1994.

Chiba, M. and K. S. Pang. Glutathione depletion kinetics with acetaminophen: A simulation study. *Drug Metab. Dispos.* 23:622–630, 1995.

Conway, J. G., F. C. Kauffman, T. Tsukada, and R. G. Thurman. Glucuronidation of 7-hydroxycoumarin in periportal and pericentral regions of the liver lobule. *Mol. Pharmacol.* 25:487–493, 1984.

Conway, J. G., F. C. Kauffman, and R. G. Thurman. Effect of glucose on 7-hydroxycoumarin glucuronide production in periportal and pericentral regions of the liver lobule. *Biochem. J.* 226:749–756, 1985.

Conway, J. G., F. C. Kauffman, T. Tsukada, and R. G. Thurman. Glucuronidation of 7-hydroxycoumarin in periportal and pericentral regions of the lobule in livers from untreated and 3-methyl-cholanthrene-treated rats. *Mol. Pharmacol.* 33:111–119, 1987.

DeBaun, J. R., J. Y. R. Smith, E. C. Miller and J. A. Miller. Reactivity in vivo of the carcinogen N-hydroxy-2-acetylaminofluorene: Increase by sulfate ion. *Science* 167:184–186, 1971.

de Lannoy, I. A. M. and K. S. Pang. A commentary. The presence of diffusional barriers on drug and metabolite kinetics. Enalaprilat as a generated *versus* preformed metabolite. *Drug Metab. Dispos.* 14:513–520, 1986.

de Lannoy, I. A. M. and K. S. Pang. Diffusional barriers on drug and metabolite kinetics. *Drug Metab. Dispos.* 15:51–58, 1987.

de Leeue, A. M. and D. L. Knook. The ultrastructure of sinusoidal liver cells in the intact rat at various ages. In: *Pharmacological, Morphological and Physiological Aspects of Aging,* edited by C. F. A. van Bezooijen.) Rijswik: Eurage, pp. 91–96, 1984.

Fleischner, G., D. K. F. Meijer, W. G. Levine, Z. Gatmaitan, R. Gluck, and I. M. Arias. Effect of hypolipidemic drugs, nafenopin and clofibrate, on the concentration of ligandin and Z protein in rat liver. *Biochem. Biophys. Res. Commun.* 67:1401–1407, 1975.

Froment, G. F. and K. B. Bischoff. Elements of reaction kinetics. In: *Chemical Reactor Analysis and Design.* New York: John Wiley and Sons, p. 10, 1976.

Gärtner, U., R. J. Stockert, W. G. Levine, and A. W. Wolkoff. Effect of nafenopin on the uptake of bilirubin and sulfobromophthalein by the isolated perfused rat liver. *Gastroenterology* 83:1163–1169, 1982.

Geng, W. P., A. J. Schwab, C. A. Goresky, and K. S. Pang. Carrier-mediated uptake and excretion of bromosulfophthalein-glutathione conjugate in perfused rat liver. A multiple indicator dilution study. *Hepatology* 22:1108–1207, 1995a.

Geng, W.-P., K. Poon, and K. S. Pang. Entry and removal of carrier-mediated *vs.* flow-limited substrates: A simulation study. *J. Pharmacokinet. Biopharm.* 23:347–378, 1995b.

Gillette, J. R. Overview of drug-protein binding. *Proc. N.Y. Acad. Sci.* (Washington DC) 226:6–17, 1973.

Gillette, J. R. and K. S. Pang. Theoretical aspects of pharmacokinetic drug interactions. *Clin. Pharmacol. Ther.* 22:623–639, 1977.

Gonzalez-Fernandez, J. M. and A. E. Atta. Maximal substrate transport in capillary networks. *Microvasc. Res.* 5:180–198, 1973.

Gooding, P. E., J. Chayen, B. Sawyer, and T. F. Slater. Cytochrome P-450 distribution in rat liver and the effect of sodium phenobarbitone administration. *Chem. Biol. Interact.* 20:299–310, 1978.

Goresky, C. A. A linear method for determining liver sinusoidal and extravascular volumes. *Am. J. Physiol.* 204:626–640, 1963.

Goresky, C. A. and A. C. Groom. Microcirculatory events in the liver and the spleen. In: *Handbook of Physiology—The Cardiovascular System IV*, edited by E. M. Renkin and C. C. Michel. Washington, DC: Am. Physiol. Society, pp. 689–780, 1984.

Goresky, C. A., G. G. Bach, and E. Nadeau. On the uptake of materials by the intact liver: The transport and net removal of galactose. *J. Clin. Invest.* 52:991–1009, 1973.

Goresky, C. A., G. G. Bach, and E. Nadeau. Red cell carriage of label. Its limiting effect on the exchange of materials in the liver. *Circ. Res.* 36:328–351, 1975.

Goresky, C. A., D. S. Daly, D. Mishkin, and I. M. Arias. Uptake of labeled palmitate by the intact liver: Role of intracellular binding sites. *Am. J. Physiol.* 234 (*Endocrinol. Metab. Gastrointest. Physiol.* 3):E542–E553, 1978.

Goresky, C. A., E. R. Gordon, and G. G. Bach. Uptake of monohydric alcohols by liver: Demonstration of a shared enzymic space. *Am. J. Physiol.* 244 (*Gastrointest. Liver Physiol.* 7):G198–G214, 1983.

Goresky, C. A., A. J. Schwab, and C. P. Rose. Xenon handling in the liver: Red cell capacity effect. *Circ. Res.* 63:767–778, 1988.

Goresky, C. A., K. S. Pang, A. J. Schwab, F. Barker III, W. F. Cherry, and G. G. Bach. Uptake of a protein bound polar compound, acetaminophen sulfate, by perfused rat liver. *Hepatology* 16:173–190, 1992.

Goresky, C. A., G. G. Bach, and A. J. Schwab. Distributed-in-space product formation in vivo: Linear kinetics. *Am. J. Physiol.* 264(*Heart Circ. Physiol.* 33):H2007–H2028, 1993a.

Goresky, C. A., G. G. Bach, and A. J. Schwab. Distributed-in-space product formation in vivo: Enzyme kinetics. *Am. J. Physiol.* 264(*Heart Circ. Physiol.* 33):H2029–H2050, 1993b.

Goresky, C. A., A. J. Schwab, and K. S. Pang. Flow, cell entry, metabolic disposal and product formation in the liver. In: *The Liver. Biology and Pathobiology, 3rd* edition, edited by I. A. Arias, J. Boyer, N. Fausto, W. Jakoby, D. Schachter, and D. Shafritz. New York: Raven Press, Chapter 58, pp. 1107–1141, 1994.

Goresky, C. A., G. G. Bach, A. J. Schwab, and K. S. Pang. Liver cell entry in vivo and enzymic conversion. In: *Whole Organ Approach to Cellular Metabolism: Capillary Permeation, Cellular Transport and Reaction Kinetics*, edited by J. Bassingthwaighte, C. A. Goresky, and J. N. Linehan. New York: Springer-Verlag, 1997.

Grausz, H. and R. Schmid. Reciprocal relation between plasma albumin level and hepatic sulfobromophthalein removal. *New Engl. J. Med.* 284:1403–1404, 1971.

Gumucio, J. J., C. Balabaud, D. L. Miller, L. F. Demason, H. D. Appleman, T. J. Stoecker, and D. R. Franzblau. Bile secretion and liver cell heterogeneity in the rat. *J. Lab. Clin. Med.* 91:350–362, 1978.

Gumucio, J. J., D. L. Miller, M. D. Krauss, and C. C. Zanolli. Transport of fluorescent compounds into hepatocytes and the resultant zonal labeling of the hepatic acinus in the rat. *Gastroenterology* 80:639–646, 1981.

Gumucio, D. L., J. J. Gumucio, J. A. P. Wilson, C. Cutter, M. Krauss, R. Caldwell, and E. Chen. Albumin influences sulfobromophthalein transport by hepatocytes of each acinar zone. *Am. J. Physiol.* 246:G86–G95, 1984.

Jones, A. L., G. T. Hreadek, R. H. Renston, K. Y. Wong, G. Karlagnais, and G. Paumgartner. Autoradiographic evidence for hepatic lobular concentration gradient of bile acid derivative. *Am. J. Physiol.* 238 (*Gastrointest. Liver Physiol.* 1):G233–G237, 1980.

Jungerman, K. and N. Katz. Functional hepatocellular heterogeneity. *Hepatology* 2:385–395, 1982.

Kitamura, T., P. Jansen, C. Hardenbrook, Y. Kamimoto, Z. Gatmaitan, and I. M. Arias. Defective ATP-dependent bile canalicular transport of organic anions in mutant (TR^{-1}) rats with conjugated hyperbilirubinemia. *Proc. Natl. Acad. Sci. U.S.A.* 87:3557–3561, 1990.

Knapp, S. A., M. D. Green, T. R. Tephly, and J. Baron. Immunohistochemical demonstration of isozyme- and strain-specific differences in the intralobular localizations and distributions of UDP-glucuronosyltransferases in livers of untreated rats. *Mol. Pharmacol.* 33:14–21, 1988.

Lee, H.-J. and W. L. Chiou. Erythrocytes as barriers for drug elimination in the isolated perfused rat liver. I. Doxorubicin. *Pharm. Res.* 10:833–839, 1989a.

Lee, H.-J. and W. L. Chiou. Erythrocytes as barriers for drug elimination in the isolated perfused rat liver. II. Propranolol. *Pharm. Res.* 10:840–843, 1989b.

Lemaire, M. and J. P. Tillement. Role of lipoproteins and erythrocytes in the in vivo binding and distribution of cyclosporin A in the blood. *J. Pharm. Pharmacol.* 34:715–718, 1982.

Levenspiel, O. Design for multiple reactions. In: *Chemical Reaction Engineering*, 2nd edition. New York: John Wiley and Sons Inc., pp. 182–185, 1970.

Levy, G. and A. Yacobi. Effect of protein binding on elimination of warfarin. *J. Pharm. Sci.* 63:805–806, 1974.

Lin, J. H., T.-H. Lin, and H. Cheung. Uptake and stereoselective binding of the enantiomers of MK-927, a potent carbonic anhydrase inhibitor by human erythrocytes in vitro. *Pharm. Res.* 9:339–344, 1992.

McFarlane, B. M., J. Spios, C. D. Gove, I. G. McFarlane, and R. Williams. Antibodies against the hepatic asialoglycoprotein receptor perfused in situ preferentially attach to periportal liver cells in the rat. *Hepatology* 11:408–415, 1990.

Meier, P. J. and B. Stieger. Canalicular membrane adenosine triphosphate-dependent transport systems. *Prog. Liver Dis.* 11:27–44, 1993.

Meijer, D. K. F. Current concepts on hepatic transport of drugs. *J. Hepatol.* 4:259–268, 1987.

Miller, D. L., C. S. Zanolli, and J. J. Gumucio. Quantitative morphology of the sinusoids in the hepatic acinus. *Gastroenterology* 76:965–969, 1979.

Miyauchi, S., Y. Sugiyama, T. Iga, and M. Hanano. Membrane-limited transport of the conjugative metabolites of 4-methylumbelliferone in rats. *J. Pharm. Sci.* 77:688–692, 1988.

Morris, M. E. and K. S. Pang. Competition between two enzymes for substrate removal in liver: Modulating effects of competitive pathways. *J. Pharmacokinet. Biopharm.* 15:473–496, 1987.

Morris, M. E., V. Yuen, and K. S. Pang. Competing pathways in drug metabolism. II. Competing pathways in drug metabolism. Enzymic systems for 2- and 5-sulfoconjugation are distributed anterior to 5-glucuronidation in the metabolism of gentisamide by the perfused rat liver. *J. Pharmacokinet. Biopharm.* 16:633–656, 1988a.

Morris, M. E., V. Yuen, B. K. Tang, and K. S. Pang. Competing pathways in drug metabolism.

I. Effect of varying input concentrations on gentisamide conjugation in the once-through in situ perfused rat liver preparation. *J. Pharmacol. Exp. Ther.* 245:614–652, 1988b.

Motta, P., M. Muto, and T. Fujita. *The Liver: An Atlas of Scanning Electron Microscopy.* New York: Igaku Shoin, plate 7.5, p. 14, 1978.

Murray, G. I., M. D. Burke, and S. W. B. Even. Glutathione localization by a novel *o*-phthaladehyde histofluorescence method. *Histochem. J.* 18:434–440, 1986.

Nathanson, M. H. and J. L. Boyer. Special article. Mechanisms and regulation of bile secretion. *Hepatology* 14:551–566, 1991.

Novikoff, A. B. Cell heterogeneity within the hepatic lobule of the rat (staining reactions). *J. Histochem. Cytochem.* 7:240–244, 1959.

Pang, K. S. and J. R. Gillette. Kinetics of metabolite formation and elimination in the perfused rat liver preparation: Differences between the elimination of preformed acetaminophen and acetaminophen formed from phenacetin. *J. Pharmacol. Exp. Ther.* 207:178–194, 1978.

Pang, K. S. and G. J. Mulder. A commentary: Effect of flow on formation of metabolites. *Drug Metab. Dispos.* 18:270–275, 1990.

Pang, K. S. and M. Rowland. Hepatic clearance of drugs. I. Theoretical consideration of a "well-stirred" model and a "parallel tube" model. Influence of hepatic blood flow, plasma and blood cells binding, and the hepatocellular activity on hepatic drug clearance. *J. Pharmacokinet. Biopharm.* 5:625–653, 1977.

Pang, K. S. and R. N. Stillwell. An understanding of the role of enzymic localization of the liver on metabolite kinetics: A computer simulation. *J. Pharmacokinet. Biopharm.* 11:451–468, 1983.

Pang, K. S. and J. A. Terrell. Retrograde perfusion to probe the heterogeneous distribution of hepatic drug metabolizing enzymes in rats. *J. Pharmacol. Exp. Ther.* 216:339–346, 1981a.

Pang, K. S. and J. A. Terrell. Conjugation kinetics of acetaminophen by the perfused liver preparation. *Biochem. Pharmacol.* 38:1959–1965, 1981b.

Pang, K. S. and X. Xu. Drug metabolism factors in drug discovery and design. In: *Pharmacokinetics: Regulatory-Industrial-Academic Perspectives,* edited by P. G. Welling and F.L.-S. Tse. New York: Marcel Dekker Inc., pp. 383–447, 1988.

Pang, K. S., H. Koster, I. C. M. Halsema, E. Scholtens, G. J. Mulder, and R. N. Stillwell. Normal and retrograde perfusion to probe the zonal distribution of sulfation and glucuronidation activities of harmol in the perfused rat liver preparation. *J. Pharmacol. Exp. Ther.* 224:647–653, 1983.

Pang, K. S., W. F. Cherry, J. A. Terrell, and E. H. Ulm. Disposition of enalapril and its diacid metabolite, enalaprilat, in a perfused rat liver preparation. Presence of a diffusional barrier into hepatocytes. *Drug Metab. Dispos.* 12:309–312, 1984.

Pang, K. S., X. Xu, M. E. Morris, and V. Yuen. Kinetic modeling of conjugations in liver. *Fed. Proc.* 46:2439–2441, 1987.

Pang, K. S., W. F. Cherry, J. Accaputo, A. J. Schwab, and C. A. Goresky. Combined hepatic arterial-portal venous or hepatic venous flows once-through the in situ perfused rat liver to probe the abundance of drug metabolizing activities. Perihepatic venous *O*-deethylation activity for phenacetin and periportal sulfation activity for acetaminophen. *J. Pharmacol. Exp. Ther.* 247:690–700, 1988a.

Pang, K. S., W. F. Lee, W. F. Cherry, V. Yuen, J. Accaputo, A. J. Schwab, and C. A. Goresky. Effects of perfusate flow rate on measured blood volume, Disse space, intracellular water spaces, and drug extraction in the perfused rat liver preparation: Characterization by the technique of multiple indicator dilution. *J. Pharmacokinet. Biopharm.* 16:595–605, 1988b.

Pang, K. S., F. Barker III, A. J. Schwab, and C. A. Goresky. ^{14}C-urea and [58]Co-EDTA as reference indicators in hepatic multiple indicator dilution studies. *Am. J. Physiol.* 259(*Gastrointest. Liver Physiol.* 22):G32–G40, 1990.

Pang, K. S., A. J. Schwab, and C. A. Goresky. Deterministic factors underlying drug and metabolite clearances in rat liver perfusion studies. In: *Perfused Liver: Clinical and Basic Applications,* edited by F. Ballet and R. G. Thurman. London: Les Editions INSERM and John Libbey Eurotext, pp. 259–302, 1991a.

Pang, K. S., N. Xu, and C. A. Goresky. D_2O as a substitute for ^3H_2O as a reference indicator in liver multiple indicator dilution studies. *Am. J. Physiol.* 261(*Gastrointest. Liver Physiol.* 24):G929–G936, 1991b.

Pang, K. S., W. F. Cherry, F. Barker, III, and C. A. Goresky. Esterases for enalapril hydrolysis is concentrated in the perihepatic venous region of the rat liver. *J. Pharmacol. Exp. Ther.* 257:294–301, 1991c.

Pang, K. S., I. A. Sherman, A. J. Schwab, W. Geng, F. Barker III, J. A. Dlugosz, G. Cuerrier, and C. A. Goresky. Role of the hepatic artery in the metabolism of phenacetin and acetaminophen: An intravital microscopic and multiple indicator dilution study in perfused rat liver. *Hepatology* 20:672–683, 1994a.

Pang, K. S., F. Barker III, A. J. Schwab, and C. A. Goresky. Demonstration of rapid entry and a cellular binding space for salicylamide in perfused rat liver: A multiple indicator dilution study. *J. Pharmacol. Exp. Ther.* 270:285–295, 1994b.

Pang, K. S., F. Barker III, A. J. Schwab, and C. A. Goresky. Sulfation of acetaminophen by the perfused rat liver: The effect of red blood cell carriage. *Hepatology* 22:267–282, 1995.

Perl, W. and F. P. Chinard. A convection-diffusion model of indicator transport through an organ. *Circ. Res.* 22:273–298, 1968.

Piekoszewski, W., F. S. Chow, and W. J. Jusko. Disposition of tacrolimus (FK506) in rabbits. Role of red cell binding in hepatic clearance. *Drug Metab. Dispos.* 21:690–698, 1993.

Rappaport, A. M. The structural and functional unit in the human liver (liver acinus). *Anat. Rec.* 130:673–689, 1958.

Rappaport, A. M. Hepatic blood flow: Morphologic aspects and physiologic regulation. In: *Liver and Biliary Tract Physiology. I International Review of Physiology, Vol. 21,* edited by N. B. Javitt. Baltimore, MD: University Park Press, pp. 1–63, 1980.

Ratna, S., M. Chiba, L. Bandyopadhyay, and K. S. Pang. Futile cycling between 4-methylumbelliferone and its conjugates in perfused rat liver. *Hepatology* 17:839–853, 1993.

Redick, J. A., W. B. Jakoby, and J. Baron. Immunohistochemical localization of glutathione-S-transferase in livers of untreated rats. *J. Biol. Chem.* 257:15200–15203, 1982.

Roberts, M. S. and M. Rowland. A dispersion model of hepatic elimination: 3. Application to metabolite formation and elimination kinetics. *J. Pharmacokinet. Biopharm.* 14:289–308, 1986.

Sato, H., Y. Sugiyama, S. Miyauchi, Y. Sawada, T. Iga, and M. Hanano. A simulation study on the effect of a uniform diffusional barrier across hepatocytes on drug metabolism by evenly or unevenly distributed uni-enzyme in the liver. *J. Pharm. Sci.* 75:3–8, 1986.

Scharschmidt, B. F. and R. Schmid. The micellar sink. *J. Clin. Invest.* 62:1122–1131, 1978.

Schwab, A. J. Extension of the theory of the multiple indicator dilution technique to variable systems with an arbitrary number of rate constants. *Math. Biosci.* 71:57–79, 1984.

Schwab, A. J. and C. A. Goresky. Hepatic uptake of protein-bound ligands: effect of an unstirred Disse space. *Am. J. Physiol.* 270(*Gastrointest. Liver Physiol.* 33):G869–G880, 1996.

Schwab, A. J., F. M. Zwiebel, A. Bracht, and R. Scholz. Transport and metabolism of L-lactate in perfused rat liver studied by multiple pulse labelling. In: *Carrier-Mediated Transport from Blood to Tissue*, edited by D. M. Yudilevich and G. E. Mann. New York: Longman, pp. 339–344, 1985.

Schwab, A. J., F. Barker III, C. A. Goresky, and K. S. Pang. Transfer of enalaprilat across rat liver cell membranes is barrier-limited. *Am. J. Physiol.* 258(*Gastrointest. Liver Physiol.* 21):G461–G475, 1990.

Sherman, I. A., S. C. Pappas, and M. M. Fisher. Hepatic microvascular changes associated with liver fibrosis and cirrhosis *Am. J. Physiol.* 258(*Heart Circ. Physiol.* 27):H460–H465, 1990.

Sherman, I. A., F. Barker III, J. Dugloz, F. M. Sadeghi and K. S. Pang. Dynamics of arterial and portal flow interactions in perfused rat liver: An intravital microscopic study. *Am. J. Physiol.* 271(*Gastrointest. Liver Physiol.* 34):G201–G210, 1996.

Smith, M. T., N. Loveridege, E. D. Wills, and J. Chayen. The distribution of glutathione content in the rat liver lobule *Biochem. J.* 182:103–108, 1979.

Sorrentino, D., R. B. Robinson, C.-L. Kiang, and P. D. Berk. At physiological albumin/oleate concentrations oleate uptake by isolated hepatocytes, cardiac myocytes, and adipocytes is a saturable function of the unbound oleate concentration. Uptake kinetics are consistent with the conventional theory. *J. Clin. Invest.* 84:1325–1333, 1989a.

Sorrentino, D., R. A. Weisiger, N. M. Bass, and V. Licko. The hepatocellular transport of sulfobromophthalein-glutathione by clofibrate treated-perfused rat liver. *Lipids* 24:438–442, 1989b.

St-Pierre, M. V., A. J. Schwab, C. A. Goresky, W.-F. Lee, and K. S. Pang. The multiple indicator dilution technique for characterization of normal and retrograde perfusions in the once-through rat liver preparation. *Hepatology* 9:285–296, 1989.

St-Pierre, M. V., P. I. Lee, and K. S. Pang. A comparative investigation of hepatic clearance models: predictions of metabolite formation and elimination. *J. Pharmacokinet. Biopharm.* 20:105–145, 1992.

Theilmann, L., Y. R. Stollman, I. M. Arias, and A. W. Wolkoff. Does Z-protein have a role in transport of bilirubin and bromosulfophthalein by isolated perfused rat liver? *Hepatology* 5:923–926, 1985.

Thurman, R. G., F. C. Kauffman, and K. Jungermann. *Regulation of Hepatic Metabolism. Intra- and Intercellular Compartmentation.* New York: Plenum Press, 1987.

Ullrich, D., G. Fisher, N. Katz, and K. W. Bock. Intralobular distribution of UDP-glucurono syltransferase in livers from untreated, 3-methylcholanthrene- and phenobarbital-treated rats. *Chem.-Biol. Interact.* 48:181–190, 1984.

Wallace, S. M. and S. Riegelman. Uptake of acetazolamide by human erythrocytes in vitro. *J. Pharm. Sci.* 66:729–731, 1977.

Watanabe, Y., G. P. Püschel, A. Gardemann, and K. Jungermann. Presinusoidal and proximal intrasinusoidal confluence of hepatic artery and portal vein in rat liver: Functional evidence by orthograde and retrograde bivascular perfusion. *Hepatology* 19:1198–1207, 1994.

Weisiger, R. A. Dissociation from albumin: A potentially rate-limiting step in the clearance of substances. *Proc. Natl. Acad. Sci. U.S.A.* 82:1563–1567, 1985.

Weisiger, R. A., C. A. Mendel, and R. R. Cavalieri. The hepatic sinusoid is not well-stirred: Estimation of the degree of axial mixing by analysis of lobular concentration gradients formed during uptake of thyroxine by the perfused rat liver. *J. Pharm. Sci.* 75:233–237, 1986.

Weisiger, R. A., S. M. Pond, and L. Bass. Albumin enhances unidirectional fluxes of fatty acid across a lipid-water interface: theory and experiments. *Am. J. Physiol.* 257(*Gastrointest. Liver Physiol.* 20):G904–G916, 1989.

Weisiger, R. A., S. Pond, and L. Bass. Hepatic uptake of protein-bound ligands: Extended sinusoidal perfusion model. *Am. J. Physiol.* 261(*Gastrointest. Liver Physiol.* 24):G872–G884, 1991.

Wilkinson, G. R. Clearance approaches in pharmacology. *Pharmacol. Rev.* 39:1–47, 1987.

Winkler, W., S. Keiding, and N. Tygstrup. Clearance as a quantitative measure of structure and function. In: *The liver: Quantitative Aspects of Structure and Function,* edited by P. Paumgartner and R. Preisig. Basel: Karger, pp. 144–155, 1973.

Wolkoff, A. W., C. A. Goresky, J. Sellin, S. Gatmaitan, and I. M. Arias. Role of ligandin in transfer of bilirubin from plasma into liver. *Am. J. Physiol.* 236:E638–G648, 1979.

Xu, X. and K. S. Pang. Hepatic modeling of metabolite kinetics in sequential and parallel pathways: salicylamide and gentisamide metabolism in perfused rat liver. *J. Pharmacokinet. Biopharm.* 17:645–671, 1989.

Xu, X., B. K. Tang, and K. S. Pang. Sequential metabolism of salicylamide exclusively to gentisamide-5-glucuronide and not gentisamide sulfate conjugates in the single pass in situ perfused rat liver. *J. Pharmacol. Exp. Ther.* 253:965–973, 1990a.

Xu, N., A. Chow, C. A. Goresky, and K. S. Pang. Effects of retrograde flow on measured blood volume, Disse space, intracellular water space, and drug extraction in the perfused rat liver: Characterization by the multiple indicator dilution technique. *J. Pharmacol. Exp. Ther.* 254:914–925, 1990b.

Xu, X., P. Selick, and K. S. Pang. Effects of nonlinear protein binding and heterogeneity of drug metabolizing enzymes on hepatic drug removal. *J. Pharmacokinet. Biopharm.* 21:43–76, 1993.

Xu, X., A. J. Schwab, F. Barker III, C. A. Goresky, and K. S. Pang. Salicylamide sulfate cell entry in perfused rat liver: A multiple indicator dilution study. *Hepatology* 19:229–244, 1994.

Yamamoto, K., I. A. Sherman, M. J. Phillips, and M. M. Fisher. Three-dimensional observations of the hepatic arterial terminations in rat, hamster and human liver by scanning electron microscopy of microvascular casts. *Hepatology* 5:452–456, 1985.

Zhao, Y., C. A. W. Snel, G. J. Mulder, and K. S. Pang. Glutathione conjugation of bromosulfophthalein in perfused rat liver: Studies with the multiple indicator dilution technique. *Drug Metab. Dispos.* 21:1070–1078, 1993.

Zierler, K. L. Equations for measuring blood flow by monitoring of radioisotopes. *Circ. Res.* 16:309–321, 1965.

15

A Generalized Mathematical Theory of the Multiple-Indicator Dilution Method

Andreas J. Schwab

Introduction

Tracer techniques have proven to be invaluable for the quantitative assessment of metabolic and biological transport processes. The multiple-indicator dilution technique as a special form of a tracer method is particularly suitable for the study of relatively rapid processes occurring in intact organs. In the previous chapters, the principles and some important applications of the multiple-indicator dilution technique have been addressed. In this chapter, a more detailed analysis of the mathematical foundations of this technique will be presented. This allows us to put this technique in a more general framework of tracer analysis, yielding a systematic approach to finding numerical solutions that apply in a more general way, whereas the analytical solutions presented in Chapter 13 are applicable only in simpler cases.

Compartmental Systems

Traditionally, mass transport within biological systems has been described by compartmental systems, where the material under investigation is assumed to be located in a finite number of discrete compartments. These may be physical compartments in space (such as a cell or the blood plasma) or representative of some chemical form of the material (such as asactaminophen and acetaminophen sulfate). The concentration (or, more accurately, the chemical potential) of the material in each compartment is assumed to be uniform over the whole space of the compartment. An important special case is the situation where a minimal amount of a labeled substance (a tracer) is introduced into a system that is otherwise in a steady state. In this case, the transfer of tracer from one compartment to another or its loss into the environment will be proportional to its amount in the source compartment (i.e., the compartment where the tracer comes from). The resulting class of linear systems has been widely investigated, and a number of theoretical aspects and applications are found in the literature (Jacquez, 1996;

Cobelli and DiStefano, 1980). In the following, only some aspects of the theory of compartmental systems are presented as a basis for a generalized theory of indicator dilution.

If there are n compartments, then the behavior of the system is described by the following system of linear differential equations:

$$\frac{dq_i}{dt} = \sum_{\substack{i=1 \\ i \neq j}}^{n} (k_{ij} q_j - k_{ji} q_i) - k_{0i} q_i + f_i, \quad i = 1 \ldots n, \quad (15.1)$$

where q_i is the amount of tracer in compartment number i, and t is time. The quantities k_{ij}, $i \neq j$ are called transfer coefficients for mass transfer from compartment j to compartment i, k_{0i} is the transfer coefficient for irreversible loss of material from compartment i into the environment, and f_i is the rate of influx of material into compartment i from the environment.[1] These equations can be written in a more concise manner by introducing vectors $\mathbf{q}$ and $\mathbf{f}$,

$$\mathbf{q} = \begin{pmatrix} q_1 \\ q_2 \\ \cdot \\ \cdot \\ \cdot \\ q_n \end{pmatrix} \qquad \mathbf{f} = \begin{pmatrix} f_1 \\ f_2 \\ \cdot \\ \cdot \\ \cdot \\ f_n \end{pmatrix}$$

and a square matrix $\mathbf{A}$ that contains the transfer coefficients:

$$\mathbf{A} = \begin{pmatrix} k_{11} & k_{12} & \ldots & k_{1n} \\ k_{21} & k_{22} & \ldots & k_{2n} \\ \cdot & \cdot & \cdot & \cdot \\ \cdot & \cdot & & \cdot \\ \cdot & \cdot & \cdot & \cdot \\ k_{n1} & k_{n2} & \cdots & k_{nn} \end{pmatrix},$$

where the diagonal elements are defined as

$$k_{jj} = -\sum_{\substack{i=0 \\ i \neq j}}^{n} k_{ij}$$

($-k_{jj}$ is the turnover rate of compartment j). With these definitions, the system (15.1) is written as

$$\frac{d\mathbf{q}}{dt} = \mathbf{A}\mathbf{q} + \mathbf{f}. \quad (15.2)$$

---

[1] In the pharmacokinetics literature, transfer coefficients take the form k_{ji}, where the first index corresponds to the source compartment and the second to the destination compartment.

This has the solution

$$\mathbf{q} = \int_0^t e^{t\mathbf{A}} \mathbf{f}(t - \tau) \, d\tau. \tag{15.3}$$

This equation contains a matrix as an exponent, a formulation explained in the Appendix. Equation (15.3) represents the convolution of the normalized fundamental matrix, $e^{t\mathbf{A}}$, with the input function $\mathbf{f}(t)$. We will, in this context, not go into the details of calculating the elements of $e^{t\mathbf{A}}$, a classical task that includes the evaluation of the eigenvalues and eigenvectors of matrix $\mathbf{A}$ (See Appendix 1).

If the system is initially empty, and tracer is injected instantaneously at time $t = 0$ into the system, then for each compartment i, we obtain $\mathbf{f} = \delta(t)\mathbf{q}_0$, where $\mathbf{q}_0$ is a vector with elements q_{0i}, which are the amounts injected into compartments $1 \ldots i$. In this case, Eq. (15.2) will become

$$\frac{d\mathbf{q}}{dt} = \mathbf{A}\mathbf{q} + \delta(t)\mathbf{q}_0. \tag{15.4}$$

The solution of Eq. (15.4)

$$\mathbf{q} = e^{t\mathbf{A}}\mathbf{q}_0 , \tag{15.5}$$

contains the impulse responses or transport functions of the compartmental system.

Extension of the Principles of Compartmental Analysis to Space-Distributed Systems

The multiple-indicator dilution technique was originally designed for investigating mass transport within capillaries or sinusoids. The basic concept and the mathematical principles of this technique have been outlined in Chapter 13. The strength of this concept is that it accommodates concentrations varying continuously with space. In particular, a metabolized substrate will have a higher concentration at the upstream end of a vascularized tissue than at its downstream end, whereas the contrary is the case for a product formed by the metabolism of that substrate (Goresky et al., 1993a, 1993b; see also Chapter 14). The compartmental theory cannot take this into account, since concentrations are always considered to be uniform. The concise notation used in compartmental theory, entailing the vectors and matrices described above, allows the formulation of quite complex situations in a parsimonious manner. It is the aim of the following development to formulate the theory for axially distributed systems in such a way that it acquires the generality of the compartmental systems theory, and satisfies the needs for analysis of high-resolution data acquired using the multiple-indicator dilution method. For this, partial differential equations are formulated using matrix notation, including a compartmental matrix as used for the description of compartmental systems.

Modeling the Events Within a Single Capillary

General Formulation

We shall start our exploration by first considering the events within a single capillary. An extension to the case of an array of many capillaries will follow later. For this, we shall modify the concepts developed for compartmental systems in such a way that concentration changes along the length of the capillary, in the direction of flow, are taken into account. We shall, however, neglect diffusional gradients within the flowing and nonflowing regions separated by membranes in concentrations in the directions perpendicular to flow. In doing this, we follow the original assumptions of Sangren and Sheppard (1953) and Goresky (1963, 1970), namely, that diffusion in the lateral direction (perpendicular to flow) is considered instantaneous, whereas diffusion in the direction of flow is considered to be so slow that it can be neglected. These assumptions represent approximations of the true events, and on the basis of the size and geometry of the microvasculature of well-perfused organs are felt not to be highly compromising for water-soluble solutes of molecular weights from hundreds to a few thousands of daltons. Transport of solutes along the vasculature is then governed by the average linear velocity within the capillary lumen; any velocity differences between the central and the lateral parts of the lumen, e.g., due to parabolic or Poisseuille flow, become ineffective since the resulting concentration differences are immediately dissipated by lateral diffusion. Substances that are poorly soluble in aqueous media are present within the cells bound to proteins or dissolved in cellular lipids. In this case, diffusion across the cells will be much slower, and lateral diffusion within the cells has to be taken into account, as discussed in Chapter 16.

In compartmental system analysis, "pool" usually designates a certain amount of a substance contained in uniform concentration within a subdivision of space. In the context of a distributed-in-space model, "pool" is understood in a local sense, that is, pools are considered to be located in infinitesimally small sections of the sinusoidal path within the liver. For each point along the flow path of a capillary, a space variable τ is defined as the time needed for an indicator confined to the vascular space to reach that point (which is equal to the ratio of the distance from the entry point to the linear flow velocity, or the ratio of cumulative capillary volume to capillary flow). The local fractional amount u_i is defined such that $u_i d\tau$ is the fraction of the injected tracer found within pool i between τ and $\tau + d\tau$. The material contained in each pool travels along the sinusoids with a relative velocity, w_i, which is defined to be 1 for an indicator confined to the vascular space and 0 for the part of a tracer contained within parenchymal cells; $\mathbf{W}$ is a diagonal matrix of the w_i. In analogy to compartmental systems, the transfer of label between pools is quantified by a set of transfer coefficients k_{ij}, defined as the amount of tracer transferred from pool j to pool i per unit time and per unit tracer amount in the source pool j. For simplicity, and because of lack of better knowledge, we assume that mass transfer coefficients are constant throughout the organ. A compartmental matrix $\mathbf{A}$ is defined similar to that above. These definitions are modifications of those used previously (Goresky, 1963, 1970; Schwab, 1984) in that linear flow velocities are normalized, and the values for the u_i are normalized by setting the flow to a capillary as well as the amount of tracer reaching a capillary equal to 1.

With these definitions, the system of partial differential equations describing the events within a single capillary can be written concisely as:

$$\frac{\partial \mathbf{u}}{\partial t} + \mathbf{W}\frac{\partial \mathbf{u}}{\delta \tau} = \mathbf{A}\mathbf{u} + \delta(\tau)\mathbf{f}(t),\qquad(15.6)$$

where $\mathbf{f}(t)$ is a function vector containing, for each mobile pool, the fraction of total tracer reaching the organ per unit time (the elements corresponding to stationary pools must all be zero). For example, two elements of $\mathbf{f}(t)$ might be $f_1(t)$ for tracer in plasma and $f_2(t)$ for tracer inside red blood cells. The shape of each of these functions will be governed by the dispersion of tracer within injection and collection catheters and the nonexchanging part of the vasculature, such as large vessels or, with some lung experiments, a cardiac ventricle (Clough et al., 1993). In general, the input functions $f_i(t)$ will be proportional to each other.

For instantaneous injection of tracer, the system becomes

$$\frac{\partial \mathbf{u}}{\partial t} + \mathbf{W}\frac{\partial \mathbf{u}}{\delta \tau} = \mathbf{A}\mathbf{u} + \delta(\tau)\delta(t)\mathbf{u}_0,\qquad(15.7)$$

where $\mathbf{u}_0$ is a vector containing the relative tracer amounts injected into each mobile pool. Again, the solution for continuous administration of tracer is obtained by convolution of the impulse response with the appropriate input function.

Solutions Using Laplace Transforms

We now perform Laplace transformation with respect to t or with respect to τ. First, with time, the Laplace transform is defined as follows:

$$\hat{\mathbf{u}}(s) = \int_0^\infty e^{-st}\mathbf{u}\, dt.$$

Equation (15.7) is then transformed into a system of ordinary differential equations:

$$s\hat{\mathbf{u}}(s) + \mathbf{W}\frac{d\hat{\mathbf{u}}(s)}{d\tau} = \mathbf{A}\hat{\mathbf{u}}(s) + \delta(\tau)\mathbf{u}_0.\qquad(15.8)$$

The solution of Eq. (15.8) is found in a way analogous to the derivation of Eq. (15.5) from Eq. (15.4):

$$\hat{\mathbf{u}}(s) = \mathbf{W}^{-1}e^{\tau(\mathbf{W}^{-1}\mathbf{A} - s\mathbf{W}^{-1})}\mathbf{u}_0 .\qquad(15.9)$$

In most cases of interest, the tracer will enter parenchymal cells where the velocity w_i is zero. Whenever this is the case, a solution of this Eq. (15.8) cannot be obtained in terms of the matrices $\mathbf{A}$ and $\mathbf{W}$, since $\mathbf{W}$ becomes singular and its inverse, $\mathbf{W}^{-1}$, is not defined. To overcome this problem, pools are renumbered in such a way that the mobile pools obtain the lowest numbers, and $\mathbf{u}$, $\mathbf{A}$, $\mathbf{W}$, and $\mathbf{u}_0$ are split as follows:

$$\mathbf{u} = \begin{pmatrix} \mathbf{u}_p \\ \mathbf{u}_q \end{pmatrix} \qquad \mathbf{A} = \begin{pmatrix} \mathbf{A}_{pp} & \mathbf{A}_{pq} \\ \mathbf{A}_{qp} & \mathbf{A}_{qq} \end{pmatrix} \qquad \mathbf{W} = \begin{pmatrix} \mathbf{W}_p & 0 \\ 0 & 0 \end{pmatrix} \qquad \mathbf{u}_0 = \begin{pmatrix} \mathbf{u}_{0p} \\ 0 \end{pmatrix},$$

where the indices p and q stand for the sets of mobile and stationary pools, respectively, and 0 denotes an all-zero vector or matrix. The matrices $\mathbf{A}_{pp}$ and $\mathbf{A}_{qq}$ contain coefficients for transfer within mobile and stationary pools, respectively; $\mathbf{A}_{qp}$ contains coefficients for the transfer of label from mobile to stationary pools, corresponding to cellular influx, and $\mathbf{A}_{pq}$ contains coefficients for the transfer of label from stationary to mobile pools, corresponding to cellular efflux. Equation (15.7) is split as follows:

$$\frac{\partial \mathbf{u}_p}{\partial t} + \mathbf{W} \frac{\partial \mathbf{u}_p}{\partial \tau} = \mathbf{A}_{pp}\mathbf{u}_p + \mathbf{A}_{pq}\mathbf{u}_q + \delta(\tau)\delta(t)\mathbf{u}_{0p} \tag{15.10}$$

$$\frac{\partial \mathbf{u}_q}{\partial t} = \mathbf{A}_{qp}\mathbf{u}_p + \mathbf{A}_{qq}\mathbf{u}_q . \tag{15.11}$$

Laplace transformation yields

$$s\hat{\mathbf{u}}_p(s) + \mathbf{W} \frac{\partial \hat{\mathbf{u}}_p(s)}{\partial \tau} = \mathbf{A}_{pp}\hat{\mathbf{u}}_p(s) + \mathbf{A}_{pq}\hat{\mathbf{u}}_q(s) + \delta(\tau)\mathbf{u}_{0p} \tag{15.12}$$

$$s\hat{\mathbf{u}}_q(s) = \mathbf{A}_{qp}\hat{\mathbf{u}}_p(s) + \mathbf{A}_{qq}\hat{\mathbf{u}}_q(s) . \tag{15.13}$$

$\hat{\mathbf{u}}_q(s)$ can be eliminated, yielding,

$$s\hat{\mathbf{u}}_p(s) + \mathbf{W}_p \frac{d\hat{\mathbf{u}}_p(s)}{d\tau} = \mathbf{A}_{pp}\hat{\mathbf{u}}_p(s) + \mathbf{A}_{pq}(s\mathbf{I} - \mathbf{A}_{qq})^{-1}\mathbf{A}_{qp}\hat{\mathbf{u}}_p(s) + \delta(\tau)\mathbf{u}_{0p} , \tag{15.14}$$

where $\mathbf{I}$ is a unit matrix of appropriate dimension, from which,

$$\frac{d\hat{\mathbf{u}}_p(s)}{d\tau} = \mathbf{W}_p^{-1}[\mathbf{A}_{pp} + \mathbf{A}_{pq}(s\mathbf{I} - \mathbf{A}_{qq})^{-1}\mathbf{A}_{qp} - s\mathbf{I}]\hat{\mathbf{u}}_p(s) + \mathbf{W}_p^{-1}\delta(\tau)\mathbf{u}_{0p} . \tag{15.15}$$

Since $\mathbf{W}_p$ is not singular, the solution of Eq. (15.14) exists and is given by

$$\hat{\mathbf{u}}_p(s) = \mathbf{W}_p^{-1} e^{\tau \mathbf{W}_p^{-1}[\mathbf{A}_{pp} - \mathbf{A}_{pq}(\mathbf{A}_{qq} - s\mathbf{I})^{-1}\mathbf{A}_{qp} - s\mathbf{I}]}\mathbf{u}_{0p}. \tag{15.16}$$

The original vector $\mathbf{u}$ will then be obtained by inverse Laplace transform of Eq. (15.16).

Alternatively, Laplace transformation with respect to the space variable τ may be used, as follows:

$$\tilde{\mathbf{u}}(p) = \int_0^\infty e^{-p\tau}\mathbf{u}\, d\tau. \tag{15.17}$$

Equation (15.7) is then transformed into a system of ordinary differential equations:

$$\frac{d\tilde{\mathbf{u}}(p)}{dt} + s\mathbf{W}\tilde{\mathbf{u}}(p) = \mathbf{A}\tilde{\mathbf{u}}(p) + \delta(t)\mathbf{u}_0 \tag{15.18}$$

that have the solutions

$$\tilde{\mathbf{u}}(p) = e^{t(\mathbf{A} - p\mathbf{W})}\mathbf{u}_0. \tag{15.19}$$

The original vector $\mathbf{u}$ may be obtained by inverse Laplace transform of Eq. (15.19).

According to the above development, there is a choice of two different strategies for the solution of these equations: Laplace transport with respect to time or to space. Strictly speaking, both strategies are mathematically equivalent and should in theory yield the same results. However, as we shall see below, in all but the simplest cases, only Laplace transformation with respect to time can be used to obtain algebraic solutions. On the other hand, we will use Laplace transformation with respect to τ to obtain some approximate solutions.

Analytical Solutions

According to the protocol of the multiple-indicator dilution method, the injected material usually includes a reference indicator, which, by definition, is confined to the vascular space and is not metabolized. In this case, only one pool has to be considered, and each of the resulting matrices and vectors will have only one single element. We set $\mathbf{A} = (0)$ (no metabolism or membrane transport), $\mathbf{W} = (1)$, and $\mathbf{u}_0 = (1)$. If we chose to use Laplace transformation with respect to time, substitution into Eq. (15.9) yields

$$\hat{u}_R(s) = e^{-s\tau}. \tag{15.20}$$

By applying the time-shift theorem, the solution of Eq. (15.20) is easily obtained as follows:

$$u_R = \delta(t - \tau). \tag{15.21}$$

The same result is obtained using Laplace transformation with respect to space, substituting into Eq. (15.19) instead of Eq. (15.9). This result is in accordance with the assumption of dispersionless plug or piston flow, and is equivalent to Eq. (1.1) in Chapter 13.

Similarly, for a tracer that equilibrates rapidly between the vascular and some extravascular space according to a "flow-limited distribution," the relative linear flow velocity will be $\mathbf{W} = (1/[1 + \gamma])$ (Goresky, 1963; see Chapter 13), where γ is the ratio of extravascular to vascular space. Thus, form Eq. (15.9),

$$\hat{u}_D(s) = (1 + \gamma)e^{-(1 + \gamma)s\tau}, \tag{15.22}$$

from which

$$u_D = (1 + \gamma)\delta(t - [1 + \gamma]\tau), \tag{15.23}$$

which is equivalent to Eq. (13.2). (Note that, per definition, u_D includes indicator within the extravascular space, hence the factor $1 + \gamma$.) Equation (15.23) may be derived similarly using Laplace transformation with respect to space from Eq. (15.19) instead of Eq. (15.9).

As a more complex example, consider the formation of a metabolic product behind a membrane barrier. The pools to consider are precursor (u_1, u_3) and product (u_2, u_4) in the vascular (u_1, u_2) and parenchymal (u_3, u_4) spaces. The compartmental matrix has the form

$$\mathbf{A} = \begin{pmatrix} -k_1\theta_1' & 0 & k_{-1} & 0 \\ 0 & -k_1'\theta_2' & 0 & k'_{-1} \\ k_1\theta_1' & 0 & -[k_{-1} + k_{enz}] & 0 \\ 0 & k_1'\theta_2' & k_{enz} & -k'_{-1} \end{pmatrix}, \tag{15.24}$$

where k_1 and k_{-1} are rate constants for influx and efflux of tracer precursor from parenchymal cells, k'_1, k'_{-1} are the corresponding rate constants for product, defined as the ratio of permeability–surface area products to the parenchymal cell volume (Goresky et al., 1970), $\theta'_1 = \theta_1 (1 + \gamma_1)$ and $\theta'_2 = \theta_2 (1 + \gamma_2)$ are the parenchymal to (interstitial plus vascular) space ratios for precursor and product, respectively, γ_1 and γ_2 are the interstitial to vascular space ratios, and θ_1 and θ_2 are the parenchymal to vascular space ratios for precursor and product, respectively. The velocity matrix will be

$$\mathbf{W} = \begin{pmatrix} \dfrac{1}{1 + \gamma_1} & 0 & 0 & 0 \\ 0 & \dfrac{1}{1 + \gamma_2} & 0 & 0 \\ 0 & 0 & 0 & 0 \\ 0 & 0 & 0 & 0 \end{pmatrix}. \tag{15.25}$$

The vector of relative injection amounts is

$$\mathbf{u}_0 = \begin{pmatrix} 1 \\ 0 \\ 0 \\ 0 \end{pmatrix}, \tag{15.26}$$

representing injection of the precursor, or

$$\mathbf{u}_0' = \begin{pmatrix} 0 \\ 1 \\ 0 \\ 0 \end{pmatrix}, \tag{15.27}$$

representing injection of product. Partitioning yields:

$$\mathbf{A}_{pp} = \begin{pmatrix} -k_1\theta_1' & 0 \\ 0 & -k_1'\theta_2' \end{pmatrix} \tag{15.28}$$

$$\mathbf{A}_{pq} = \begin{pmatrix} k_{-1} & 0 \\ 0 & k'_{-1} \end{pmatrix} \tag{15.29}$$

$$\mathbf{A}_{qp} = \begin{pmatrix} k_1\theta_1' & 0 \\ 0 & k_1'\theta_2' \end{pmatrix} \tag{15.30}$$

$$\mathbf{A}_{qq} = \begin{pmatrix} -[k_{-1} + k_{enz}] & 0 \\ k_{enz} & -k'_{-1} \end{pmatrix}. \tag{15.31}$$

The exponent in Eq. (15.16) obtains the form:

$$\tau(\mathbf{A}_{pp} - \mathbf{A}_{pq}(\mathbf{A}_{qq} - s\mathbf{I})^{-1}\mathbf{A}_{qp} - s\mathbf{I}) = \tau\begin{pmatrix} -M_1(s) & 0 \\ Q(s) & -M_2(s) \end{pmatrix}, \tag{15.32}$$

where

$$M_1(s) = (1 + \gamma_1)\left(s + k_1\theta_1' - \frac{k_1 k_{-1}\theta_1'}{s + k_{-1} + k_{enz}}\right)$$

$$M_2(s) = (1 + \gamma_2)\left(s + k_1'\theta_2' - \frac{k_1'k'_{-1}\theta_2'}{s + k'_{-1}}\right)$$

$$Q(s) = (1 + \gamma_2)k_{enz}\left(\frac{k_1\theta'_1}{s + k_{-1} + k_{enz}}\right)\left(\frac{k'_{-1}}{s + k'_{-1}}\right).$$

These expressions are analogous to those in Goresky et al. (1993a). Evaluation of the matrix exponential in Eq. (15.16) is obtained by finding the eigenvalues and eigenvectors of the exponent matrix (see Appendix 2). In this special case, where the exponent is a triangular matrix, the eigenvalues are equal to the diagonal elements, and the solution is:

$$\begin{pmatrix} \hat{u}_1 \\ \hat{u}_2 \end{pmatrix} = \begin{pmatrix} e^{-\tau M_1(s)} & 0 \\ \dfrac{Q(s)}{M_1(s) - M_2(s)}[e^{-\tau M_1(s)} + e^{-\tau M_2(s)}] & e^{-\tau M_2(s)} \end{pmatrix} u_{0p} \tag{15.33}$$

In the matrix at the right-hand side of Eq. (15.33), the left column represents the response to injection of the labeled precursor, whereas the right column represents the response to injection of the labeled product; the top row represents the response observed in the precursor leaving the capillary, and the bottom row represents that observed in the product leaving the capillary. Note that the top right element of the matrix is zero, since, with an irreversible enzymic conversion, no labeled precursor is observed after injection of the labeled product. In the case of a reversible enzymic reaction, or in that of a back-reaction leading to futile cycling (Ratna et al., 1982), all elements would be different from zero.

Equation (15.33) represents the Laplace transform of the capillary responses. To obtain the time-domain responses, one has to perform inverse Laplace transformation of each of the elements of the matrix in Eq. (15.33). For injection of the precursor and observation of the precursor, we obtain

$$\hat{u}_1(s) = (1 + \gamma_1) \exp\left(-[1 + \gamma_1] \tau \left[s + k_1\theta_1' - \frac{k_1 k_{-1}\theta_1'}{k_{-1} + k_{enz} + s}\right]\right) \quad (15.34)$$

This is equivalent to Eq. (5) in Goresky et al. (1973) or Eq. (A7) in Schwab et al. (1990). Inverse Laplace transformation of Eq. (15.34) will yield the response of a single capillary. Appropriate Taylor expansion of the exponential of the fraction in Eq. (15.34) yields

$$\hat{u}_1(s) = (1 + \gamma_1) \exp(-\tau'[k_1\theta_1' + s]) \left(1 + \sum_{n=1}^{\infty} \frac{[k_1\theta_1' k_{-1}\tau']^n}{n![k_{-1} + k_{enz} + s]^n}\right), \quad (15.35)$$

where $\tau' = (1 + \gamma_1) \tau$. Inverse Laplace transformation yields (Goresky et al., 1970, 1973; Schwab et al., 1990):

$$u_1 =$$

$$(1 + \gamma_1)e^{-k_1\theta_1'\tau'}\left[\delta(t - \tau') + S(t - \tau')e^{-(k_{-1} + k_{enz})(t - \tau')}\sum_{n=1}^{\infty} \frac{(k_1 k_{-1}\theta_1'\tau')^n(t - \tau')^{n-1}}{n!(n-1)!}\right],$$

$$(15.36)$$

where S is a step function, i.e., $S(x) = 0$ for $x < 0$ and $S(x) = 1$ for $x > 0$.

The sum in Eq. (15.36) also can be formulated in terms of a modified first-order Bessel function, I_1. The Taylor expansion of this function is

$$I_1(x) = \sum_{n=1}^{\infty} \frac{(x/2)^{2n-1}}{n!(n-1)!}. \quad (15.37)$$

After an appropriate algebraic manipulation and substitution of Eq. (15.37) (Goresky et al., 1970), one obtains

$$u_1 = (1 + \gamma_1)e^{-k_1\theta_1'\tau'}\left[\delta(t - \tau')\right.$$

$$\left. + S(t - \tau')e^{-(k_{-1} + k_{enz})(t - \tau')}\sqrt{\frac{k_1 k_{-1}\theta_1'\tau'}{t - \tau'}} I_1(2\sqrt{k_1 k_{-1}\theta_1'\tau')(t - \tau')})\right], \quad (15.38)$$

For injection of the precursor and observation of the product, we obtain

$$\hat{u}_2 = \frac{Q(s)}{M_1(s) - M_2(s)}\left[e^{-\tau M_1(s)} + e^{-\tau M_2(s)}\right].$$

Inverse Laplace transformation of this expression, for the case where $\gamma_1 = \gamma_2$, was developed in Goresky et al. (1993a, Appendix). The result is

$$u_2 = \frac{k_1 k_{enz} k'_{-1}\theta_1'}{(k_1\theta_1' - k_1'\theta_2')(r_1 - r_2)}$$

$$\left\{\left[e^{-k_1'\theta_2\tau} - e^{-k_1\theta_1\tau}\right]\left[e^{r_1(t - \tau')} - e^{r_2(t - \tau')}\right]\right.$$

$$+ \int_{\tau'}^{t}\left[e^{-\theta_2 k_1'\tau - k'_{-1}(\xi - \tau')}\sqrt{\frac{k_1'k'_{-1}\theta_2'\tau'}{\xi - \tau'}}I_1(2\sqrt{k_1'k'_{-1}\theta_2'\tau'[\xi - t']})\right.$$

$$-e^{-\theta_1 k_1\tau - (k_{-1} + k_{enz})(\xi - \tau')}\sqrt{\frac{k_1 k_{-1}\theta_1'\tau'}{\xi - \tau'}}I_1(2\sqrt{k_1'k_{-1}\theta_1'\tau'[\xi - t']})\right]$$

$$\left.[e^{r_1(t - \xi)} - e^{r_2(t - \xi)}]d\xi\right\}, \tag{15.39}$$

where r_1 and r_2 are the roots of the following quadratic equation in r:

$$(\theta_1'k_1 - k_1'\theta_1')r^2$$

$$+ (k'_{-1}\theta_1'k_1 - k_1'\theta_1'k_{-1} - k_1'\theta_1'k_{enz} + \theta_1'k_1 k_{enz})r + k'_{-1}\theta_1'k_1 k_{enz} = 0.$$

When labeled product is injected, no labeled precursor is observed; the equation for the labeled product is analogous to Eq. (15.36), with the enzymic conversion terms removed:

$$u_2' = (1 + \gamma_2)e^{-k_1'\theta_2'\tau''}$$

$$\left[\delta(t - \tau'') + S(t - \tau'')e^{-k_{-1}'(t - \tau'')}\sum_{n=1}^{\infty}\frac{(k_1'k_{-1}'\theta_2'\tau'')^n(t - \tau'')^{n - 1}}{n!(n - 1)!}\right] \tag{15.40}$$

or

$$u_2' = (1 + \gamma_2)e^{-k_1'\theta_2'\tau''}$$

$$\left[\delta(t - \tau'') + S(t - \tau'')e^{-k_{-1}'(t - \tau'')}\sqrt{\frac{k_1'k'_{-1}\theta_2'\tau''}{\tau - \tau''}}I_1(2\sqrt{(k_1'k'_{-1}\theta_2'\tau'')(\tau - \tau'')})\right] \tag{15.41}$$

where $\tau'' = (1 + \gamma_2)\tau$.

For the cases of two and three consecutive barriers, similar expressions for injection and sampling of the precursor have been found containing Bessel functions and an integral like Eq. (15.39) (two nested integrals in the case of three barriers) and the roots of a cubic equation. It is conjectured that, for an increasing number of barriers and for inclusion of products, polynomial equations of higher order will have to be solved, and several integrals will have to be nested to formulate closed solutions. Although these will be mathematically exact, is obvious that numerical procedures for their calculation will be impractical. Even in the cases demonstrated above, the numerical calculations may be sometimes very demanding in computer time usage.

An alternative to this is the use of numerical methods, such as finite difference methods. The idea for the latter is to divide the independent parameters, time and space, into small intervals and transform the differential equations into equations where the differentials are approximated by differences.

Finite Difference Methods

The Method of Lines

For the type of partial differential equations treated here (of the hyperbolic type), an appropriate method is the method of lines. For this, the capillary length is subdivided into segments, and a set of ordinary differential equations is formulated for each segment. These are then integrated with respect to time, using well-known numerical integration schemes for systems of ordinary differential equations, such as the Runge–Kutta method. Integration thus occurs along parallel lines in the time/space plane, hence the name of the method.

Whereas the methods involving Laplace transformations yield expressions for impulse responses directly, the assumption of an instantaneous injection will, in the case of the finite difference or line methods, lead to discontinuous boundary conditions that are difficult to handle numerically. In many cases, this problem may be handled using special procedures (Schwab, 1984). In the following, we will treat the case of continuous delivery of the tracers according to Eq. (15.6).

The principle of this method is the approximate calculation of a solution at time $t + \Delta t$ given the solution at time t. If the integration interval Δt is chosen small enough, the space derivative $\partial \mathbf{u}/\partial \tau$ may be considered independent of t within that interval. Therefore, for $\tau > 0$, Eq. (15.6) may be replaced by a system of ordinary differential equations:

$$\frac{d\mathbf{u}}{dt} = \mathbf{A}\mathbf{u} - \mathbf{W}\frac{\partial \mathbf{u}}{\partial \tau}\bigg|_t , \qquad t > 0. \qquad (15.42)$$

The term in Eq. (15.6) containing a delta function will impose the boundary condition

$$\mathbf{u} = \mathbf{f}(t), \qquad \tau = 0$$

The method of lines consists in approximating the space derivative $\partial \mathbf{u}/\partial \tau$ by a finite difference approximation, in the simplest case, by $\Delta \mathbf{u}/\Delta \tau$.

The Sliding Element Method

The sliding element method (Bassingthwaighte et al., 1989) is represented by the following solution of Eq. (15.42) (see Appendix 2):

$$\mathbf{u}(\tau, t + \Delta t) = e^{\Delta t \mathbf{A}} \mathbf{u}^{\text{slided}}(t), \qquad t > 0, \tag{15.43}$$

where the vector $\mathbf{u}^{\text{slided}}(t)$ is composed of elements $u_i^{\text{slided}}(t) = u_i(\tau - w_i \Delta t, t)$.

According to this, the solution after each time step Δt is obtained by first shifting ("sliding") those elements of the solution at the previous step (time t) that correspond to mobile pools for a distance $w_i \Delta t$ along the space axis (a distance equal to the time step if the velocity of the tracer is the same as that of the reference tracer), and then integrating with respect to time.

Modeling of the Response of a Whole Organ

General Considerations

The microcirculation of a whole organ includes a large number of capillaries and other microvessels that form a network. Following the pioneering work of Goresky (1963) for the liver, we consider the simple case of capillaries that are perfused in parallel. Capillaries are assumed to differ in their transit times for a vascular indicator such as labeled red cells, and to be otherwise identical in their properties. The feeding and draining of "large vessels," which include arteries, areterioles, veins, and venules, are considered to have a common, relatively short, transit time, t_0.

The response of the whole organ is derived from that of a single capillary by integrating over all flow paths with different transit times. This concept is straightforward whenever capillaries are independent and parallel.

If $n(\tau)$ is the distribution of transit times among flow paths [that is, $n(\tau)\,d\tau$ is the proportion of flow with transit times between τ and $\tau + d\tau$], then the concentrations at the outflow of the whole liver will be the flow-weighted average of the responses of single sinusoidal paths, according to the integrals

$$\mathbf{h}(t + t_0) = \mathbf{W} \int_0^\infty n(\tau) \mathbf{u}\, d\tau, \tag{15.44}$$

where $\mathbf{h}$ is a vector function whose elements are the fraction of injected tracer emerging from the liver per unit time, and t_0 is the time spent in large vessels where no dispersion or exchange occurs.

The above derivation is straightforward for the case where capillaries are perfused independently in parallel. However, it has been applied early on in cases such as in the liver, where capillaries form a network (Goresky et al., 1963, 1970). This extension of the scope of Eq. (15.44) is permissible if the behavior of the system is linear in the sense defined above. This has been proven for the steady-state single-barrier case by Forker and Cai (1993). More recently, an equation that is equivalent to Eq. (15.44) was derived using a stochastic approach, showing that the structure of the intravascular mixing processes in the capillary bed is irrelevant, provided that the sojourn time in the capillaries is the same as for a purely intravascular reference tracer (Weiss and Roberts, 1996). This will be the case if the tracer leaving the vascular space returns to it at the same location within the capillary bed.

Reference and Flow-Limited Tracers

The outflow profile for the intravascular plasma reference indicator is obtained by substituting Eq. (15.21) into Eq. (15.44), with $\mathbf{W} = (1)$:

$$h_R(t + t_0) = n(t). \tag{15.45}$$

For a tracer undergoing flow-limited distribution, one obtains, similarly,

$$h_D(t + t_0) = w\, n(wt) = \frac{1}{1 + \gamma}\, n\!\left(\frac{t}{1 + \gamma}\right) = \frac{1}{1 + \gamma}\, h_R\!\left(\frac{t - t_0}{1 + \gamma}\right). \tag{15.46}$$

Equation (15.46) relates the response h_D for the indicator undergoing flow-limited distribution to that of the reference indicator, h_R, according to a linear scaling of the time and relative tracer amount axes.

The General Case

In order to find the whole organ responses for the general case, one must substitute the inverse Laplace transform of Eq. (15.9) or Eq. (15.19) into Eq. (15.44). The use of numerical procedures for Laplace transform inversion and for the integral in Eq. (15.44) leads to problems of excessive computing time and insufficient accuracy. One may avoid Laplace inversion altogether by approximating the experimental and the calculated outflow profiles by the Laguerre function, and then compare the Laguerre coefficients instead of the outflow profiles to optimize transfer coefficients (Salcudean et al., 1981).

An approximate solution for the general case is based on a multiexponential approximation of the reference profile (Schwab, 1984; Schwab et al., 1985). For this, the transit time distribution is approximated by a sum of exponentials as follows:

$$n(\tau) = \sum_{i=1}^{n} \alpha_i e^{-\beta_i \tau}. \tag{15.47}$$

This is meant solely as a shape descriptor, with no physical meaning, and the (real or complex) parameters α_i and β_i are arbitrary. The vascular reference curve is then approximated by

$$h_R(t + t_0) = \sum_{i=1}^{n} \alpha_i e^{-\beta_i t}. \tag{15.48}$$

Thus, the parameters α_i and β_i may be determined from the reference curve (Schwab, 1984; Schwab et al., 1985). A suitable method for this is the method of moments (Dyson and Isenberg, 1971). In practice, three to six exponential terms are sufficient to approximate a measured reference curve; however, this is only possible if the upslope of the curves starts at t_0, a restriction that usually is met with outflow obtained from livers or lungs.

Substitution of Eq. (15.47) into Eq. (15.44) yields

$$\mathbf{h}(t + t_0) = \mathbf{W}\sum_{i=1}^{n}\alpha_i\int_0^{\infty}\mathbf{u}e^{-\beta_i\tau}\,d\tau \qquad (15.49)$$

Comparing this equation with Eq. (15.17) reveals that the integrals contained herein have the form of Laplace transforms, with β_i as the Laplace variable. Thus, Eq. (15.49) can be reformulated as follows:

$$\mathbf{h}(t + t_0) = \mathbf{W}\sum_{i=1}^{n}\alpha_i\tilde{\mathbf{u}}(\beta_i). \qquad (15.50)$$

Substitution of Eq. (15.19) yields

$$\mathbf{h}(t + t_0) = \mathbf{W}\sum_{i=1}^{n}\alpha_i e^{t(\mathbf{A} - \beta_i\mathbf{W})}\mathbf{u}_0. \qquad (15.51)$$

When the elements of the matrices $\mathbf{A}$ and $\mathbf{W}$ and the values for the α_i and β_i are known, the outflow profiles contained in $\mathbf{h}$ can be calculated. This method is very fast, since it involves mainly the calculation of eigensystems for matrices of limited size, for which rapid numerical algorithms are readily available.

Example Applications

Sulfate Conjugation of Acetaminophen in Perfused Rat Liver

When labeled acetaminophen is injected into an isolated perfused rat liver via the portal vein, it enters hepatocytes where it is metabolized to form acetaminophen sulfate. At the same time, it enters red cells from where it is released only slowly, giving rise to a red blood cell carriage phenomenon (Pang et al., 1995; see also Chapter 14). The structure of the model used to evaluate the outflow profiles is shown in Fig. 14 in that chapter. According to this structure, the following compartmental matrix is constructed:

$$\mathbf{A} = \begin{pmatrix} -[k_{pr} + k_{ph}] & k_{rp} & 0 & k_{hp} & 0 & 0 \\ k_{pr} & -k_{rp} & 0 & 0 & 0 & 0 \\ 0 & 0 & -k'_{ph} & 0 & k'_{hp} & 0 \\ k_{ph} & 0 & 0 & -[k_{ph} + k_{seq} + k_{ha}] & 0 & k_{ah} \\ 0 & 0 & k'_{ph} & k_{seq} & -[k'_{hp} + k'_{bil}] & 0 \\ 0 & 0 & 0 & k_{ha} & 0 & -k_{ah} \end{pmatrix}. \qquad (15.52)$$

The six columns and rows of matrix $\mathbf{A}$ correspond to:

1. tracer acetaminophen in plasma;
2. tracer acetaminophen within red blood cells;
3. tracer acetaminophen sulfate in plasma;
4. tracer acetaminophen within parenchymal cells (primary pool);
5. tracer acetaminophen sulfate within parenchymal cells;
6. tracer acetaminophen within parenchymal cells (additional pool).

The transfer coefficients k_{pr} and k_{rp} represent membrane transport of acetaminophen into and from red blood cells, k_{ph} and k_{hp} represent membrane transport of acetaminophen into and from parenchymal cells, and k'_{ph} and k'_{hp} represent membrane transport of acetaminophen sulfate into and from parenchymal cells.[2] Each of these transfer coefficients is defined as the ratio of the permeability–surface area product to the volume of the space where the flux originates. For example,

$$k_{hp} = \frac{P_{hp} S}{V_{\text{sin}}(1 - \text{Hct})(1 + \gamma)},$$
(15.53)

where V_{sin} is the sinusoidal volume, Hct is hematocrit, and γ is the interstitial to vascular plasma space ratios for acetaminophen. Moreover, k_{seq} represents the metabolic transformation of acetaminophen to acetaminophen sulfate within hepatocytes, k_{bil} represents the biliary excretion of acetaminophen sulfate, and k_{ha} and k_{ah} represent the exchange between the primary and the additional pools of parenchymal acetaminophen. The diagonal matrix of relative velocities is

$$\mathbf{W} = \begin{pmatrix} \dfrac{1}{1+\gamma} & 0 & 0 & 0 & 0 & 0 \\ 0 & 1 & 0 & 0 & 0 & 0 \\ 0 & 0 & \dfrac{1}{1+\gamma_{AS}} & 0 & 0 & 0 \\ 0 & 0 & 0 & 0 & 0 & 0 \\ 0 & 0 & 0 & 0 & 0 & 0 \\ 0 & 0 & 0 & 0 & 0 & 0 \end{pmatrix},$$
(15.54)

where γ_{AS} is the interstitial to vascular plasma space ratios for acetaminophen sulfate.

---

[2]We use here the nomenclature customary in the pharmokinetics literature, where the first index corresponds to the source compartment and the second to the destination compartment.

The vector of the relative amounts injected is

$$
\mathbf{u}_0 =
\begin{pmatrix}
\dfrac{1}{1 + \lambda_R \beta} \\[2mm]
\dfrac{\lambda_R \beta}{1 + \lambda_R \beta} \\[2mm]
0 \\
0 \\
0 \\
0
\end{pmatrix},
\tag{15.55}
$$

where β is defined as the ratio of red blood cells to total blood space, and λ_R is the red blood cell: plasma partition coefficient for acetaminophen.

The observable outflow profiles are those for acetaminophen and acetaminophen sulfate in blood. These were calculated using the approximation of the reference profile by a sum of exponentials, as described above. Since the red blood cells cannot be separated from plasma in the outflowing blood, the observed outflow profile for acetaminophen will be the sum of those for red blood cells and plasma. The whole blood profiles were then found as the sum of the first two elements of vector $\mathbf{h}$ in Eq. (15.51). In Figure 15 from Chapter 14, the fitted whole blood outflow profile for acetaminophen was resolved into the plasma and red cell components, corresponding to the first and second elements of $\mathbf{h}$. In contrast, since acetaminophen sulfate does not enter red blood cells, its outflow profile is obtained directly from Eq. (15.51) (Figure 17 in Chapter 14).

A similar mathematical method has been used for assessing lactate metabolism in the perfused rat liver, using lactate that was labeled with tritium attached to the carbon atom carrying the hydroxyl group (Bracht et al., 1980). It was shown that the hepatic metabolism of this tracer produces labeled water. Kinetic analysis of indicator dilution experiments revealed two intracellular pools of interconverting material, tentatively interpreted as an intracellular lactate pool and a combined pool of other NADH-dependent dehydrogenase substrates (Schwab, 1984; Schwab et al., 1985a, 1985b). However, quantitative interpretation of the results is hampered by a kinetic isotope effect of unknown extent in the metabolism of the tracer. In the lactate dehydrogenase reaction, a tritium–carbon bond is cleaved at a rate much slower than cleavage of the analogous hydrogen–carbon bond.

Concluding Remarks

For interpretation of multiple-indicator dilution experiments, a large number of mathematical methods have been developed in the past. In this chapter, an overview of some of these has been given. Each method has its advantages and disadvantages. Analytical solutions are most accurate, but do not exist for all cases. On the other hand, finite difference approximations are most general, but their accuracy may be unsatisfactory. The same is true for methods involving the numerical inversion of Laplace transforms. A method based on multiexponential approximations is very fast, although restrictions with regard to curve shapes exist. By using several different algorithms on the same problem, one will obtain much more confidence regarding the accuracy of the results.

References

Bassingthwaighte, J. B., C. Y. Wang, and I. S. Chan. Blood-tissue exchange via transport and transformation by capillary endothelial cells. *Circ. Res.* 65:997–1020, 1989.

Bracht, A., A. Kelmer-Bracht, A. Schwab, and R. Scholz. Transport of inorganic anions in perfused rat liver. *Eur. J. Biochem.* 114:471–479, 1981.

Bracht, A., A. J. Schwab, and R. Scholz. Untersuchungen von Flußgeschwindigkeiten in der isolierten perfundierten Rattenleber durch Pulsmarkierung mit radioaktiven Substraten und mathematischer Analyse der Auswaschkinetiken. *Hoppe-Seyle's Z. Physiol. Chem.* 361:357–377, 1980.

Clogh, A. V., D. Cui, J. H. Linehan, G. S. Krenz, C. A. Dawson, and M. B. Maron. Model-free nunerical deconvolution of recirculating indicator concentration curves. *J. Appl. Physiol.* 74:1444–1453, 1993.

Cobelli, C. and J. J. DiStefano. Parameter and structural identifiability concepts and ambiguities—A critical review and analysis. *Am. J. Physiol.* 239:R7–R24, 1980.

Dyson, R. D. and I. Isenberg. Analysis of exponential curves by a method of moments, with special attention to sedimentation equilibrium and fluorescence decay. *Biochemistry* 10:3233–3241, 1971.

Forker, E. L. and Z.-C. Cai. Mathematical modeling as strategy for understanding hepatic transport of organic solutes. In: *Hepatic Transport and Bile Secretion: Physiology and Pathophysiology,* edited by N. Tavaloni and P. D. Berk. New York: Raven Press, pp. 41–53, 1993.

Goresky, C. A. A linear method for determining liver sinusoidal and extravascular volumes. *Am. J. Physiol.* 204:626–640, 1963.

Goresky, C. A., W. H. Ziegler, and G. G. Bach. Capillary exchange modeling: Barrier-limited and flow-limited distribution. *Circ. Res.* 27:739–764, 1970.

Goresky, C. A., G. G. Bach, and B. E. Nadeau. On the uptake of material by intact liver. The transport and net removal of galactose. *J. Clin. Invest.* 52:975–990, 1973.

Goresky, C. A., G. G. Bach, and B. E. Nadeau. Red cell carriage of label. Its limiting effect on the exchange of materials in the liver. *Circ. Res.* 36:328–351, 1975.

Goresky, C. A., G. G. Bach, and A. J. Schwab. Distributed-in-space product formation in vivo: Linear kinetics. *Am. J. Physiol.* 264:H2007–H2028, 1993a.

Goresky, C. A., G. G. Bach, and A. J. Schwab. Distributed-in-space product formation in vivo: Enzymic kinetics. *Am. J. Physiol.* 264:H2029–H2050, 1993b.

Jacquez, J. A. *Compartmental Analysis in Biology and Medicine,* 3rd Edition. Ann Arbor, MI: BioMedware, 1996.

Pang, K. S., F. Barker, A. Simard, A. J. Schwab, and C. A. Goresky. Sulfation of acetaminophen by the perfused rat liver: The effect of red blood cell carriage. *Hepatology* 22:267–282, 1995.

Ratna, S., M. Chiba, L. Bandyopadhyay, and K. S. Pang. Futile cycling between 4-methylumbelliferone and its conjugates in perfused rat liver. *Hepatology* 17:838–853, 1993.

Salcudean, S. E., P. Bélanger, C. A. Goresky, and C. P. Rose. The use of Laguerre functions for parameter identification in a distributed biological system. *IEEE Transact. Biomed. Eng.* 28:767–775, 1981.

Sangren, W. C. and C. W. Sheppard. A mathematical derivation of the exchange of a labeled substance between a liquid flowing in a vessel and an external compartment. *Bull. Mathem. Biophys.* 15:387–394, 1953.

Schwab, A. J. Extension of the theory of the multiple indicator dilution technique to metabolic systems with an arbitrary number of rate constants. *Math. Biosc.* 71:57–79, 1984.

Schwab, A. J., F. M. Zwiebel, A. Bracht, and R. Scholz. Transport and metabolism of L-lactate in perfused rat liver studied by multiple pulse labelling. In: *Carrier-mediated transport from blood to tissue,* edited by D. M. Yudilevich and G. E. Mann, London: Longman, pp. 339–344, 1985.

Schwab, A. J., A. Bracht, and R. Scholz. Investigation of rapid metabolic reactions in whole organs by multiple pulse labelling. In: *Mathematics in biology and medicine,* edited by V. Capasso, E. Grosso, and S. L. Paveri-Fontana (Lecture Notes in Biomathematics 57), Berlin; New York: Springer-Verlag, pp. 348–353, 1985.

Schwab, A. J., A. Bracht, and R. Scholz. Transport of D-lactate in perfused rat liver. *Eur. J. Biochem.* 102:537–547, 1979.

Weiss, M. and M. S. Roberts. Tissue distribution kinetics as determinant of transit time dispersion of drugs in organs: Application of a stochastic model to the rat hindlimb. *J. Pharmacokin. Biopharm.* 24:173–196, 1996.

Appendix 1: Exponentials with Matrix-Valued Exponents

The usual definition for an exponential with a matrix-valued exponent $\mathbf{X}$ is

$$e^{\mathbf{X}} = \mathbf{I} + \frac{\mathbf{X}}{1!} + \frac{\mathbf{X}^2}{2!} + \frac{\mathbf{X}^3}{3!} + \cdots, \tag{15.56}$$

with an identity matrix $\mathbf{I}$ that has the same dimension as $\mathbf{X}$. This is analogous to a Taylor expansion of a scalar exponential. Assuming that all of the eigenvalues of $\mathbf{X}$ are distinct, the following relations hold:

$$\mathbf{X} = \mathbf{M}\Lambda\mathbf{M}^1 \tag{15.57}$$

and

$$e^{\mathbf{X}} = \mathbf{M}e^{\Lambda}\mathbf{M}^{-1}, \tag{15.58}$$

where Λ is a diagonal matrix containing the eigenvalues of $\mathbf{X}$, and $\mathbf{M}$ is a matrix whose columns are the eigenvectors of $\mathbf{X}$. These expressions can be evaluated after finding the eigenvalues and the eigenvectors of the matrix $\mathbf{X}$, using standardized numerical methods.

The following theorems are needed:

$$e^{(a+b)\mathbf{X}} = e^{a\mathbf{X}}e^{b\mathbf{X}} = e^{ab}e^{\mathbf{X}} \tag{15.59}$$

$$\mathbf{X}e^{\mathbf{X}} = e^{\mathbf{X}}\mathbf{X}. \tag{15.60}$$

Appendix 2: Sliding Element Method

The inhomogeneous system, Eq. (15.42), has the solution

$$\mathbf{u}(\tau, t + \Delta t) = e^{\Delta t \mathbf{A}} \mathbf{u}(\tau, t) - \int_t^{t + \Delta t} e^{(t + \Delta t - \xi)\mathbf{A}} \mathbf{W} \frac{\partial \mathbf{u}}{\partial \tau}\bigg|_{\xi} d\xi \qquad (15.61)$$

or, after collecting constant terms outside the integral, assuming $\partial \mathbf{u}/\partial \tau$ independent of t in the interval Δt:

$$\mathbf{u}(\tau, t + \Delta t) \approx e^{\Delta t \mathbf{A}} \left[\mathbf{u}(\tau, t) - e^t \mathbf{A} \left(\int_t^{t + \Delta t} e^{-t\mathbf{A}} d\xi \right) \mathbf{W} \frac{\partial \mathbf{u}}{\partial \tau}\bigg|_t \right]. \qquad (15.62)$$

Integration yields

$$\mathbf{u}(\tau, t + \Delta t) \approx e^{\Delta t \mathbf{A}} \left[\mathbf{u}(\tau, t) - \mathbf{A}^{-1}(e^{-\Delta t \mathbf{A}} - \mathbf{I}) \mathbf{W} \frac{\partial \mathbf{u}}{\partial \tau}\bigg|_t \right] \qquad (15.63)$$

or, by approximating $e^{\Delta t}\mathbf{A} \approx \mathbf{I} + \Delta t\, \mathbf{A}$ [i.e., keeping the first two terms in Eq. (15.56)],

$$\mathbf{u}(\tau, t + \Delta t) \approx e^{\Delta t \mathbf{A}} \left[\mathbf{u}(\tau, t) - \Delta t \mathbf{W} \frac{\partial \mathbf{u}}{\partial \tau}\bigg|_t \right]. \qquad (15.64)$$

The expression in square brackets above is a vector with elements

$$u_i^{\text{slided}}(t) = u_i(\tau, t) - \frac{\partial u_i}{\partial \tau}\bigg|_t w_i \Delta t, \qquad (15.65)$$

and, since

$$\frac{\partial u_i}{\partial \tau}\bigg|_t \Delta \tau \approx u_i(\tau, t) - u_i(\tau - \Delta \tau, t), \qquad (15.66)$$

setting $\Delta \tau = w_i \Delta t$ yields

$$u_i^{\text{slided}}(t) \approx u_i(\tau - w_i \Delta t, t). \qquad (15.67)$$

From this, Eq. (15.64) becomes

$$\mathbf{u}(\tau, t + \Delta t) = e^{\Delta t \mathbf{A}} \mathbf{u}^{\text{slided}}(t), \qquad t > 0, \qquad (15.43)$$

where the vector $\mathbf{u}^{\text{slided}}(t)$ is composed of elements $u_i^{\text{slided}}(t)$.

16

Impact of Extracellular and Intracellular Diffusion on Hepatic Uptake Kinetics

Richard A. Weisiger

The selective transport of small molecules is among the most fundamental functions of living cells. Selectivity is largely due to the plasma membrane, which contains carrier proteins that increase the permeability of many molecules. However, not all molecules require carriers. Molecules that are sufficiently hydrophobic can easily dissolve in the membrane core, but have much greater difficulty crossing the aqueous layers on either side of the plasma membrane (Figure 16.1). For many such molecules, transport across these intra- and extracellular water layers may limit the uptake rate under physiologic circumstances. In response to this limitation, organisms have evolved aqueous carrier systems (the soluble binding proteins) that catalyze the transport of poorly soluble molecules across these water layers. This chapter will discuss these transport processes and how soluble carrier systems may influence the observed uptake kinetics. It shall be argued that plasma and cytosolic binding proteins represent true carrier systems, producing all of the features of carrier-mediated kinetics. Failure to consider these carrier systems can lead to misinterpretation of uptake data.

Molecules fall into three basic categories: *hydrophilic compounds* (e.g., carbohydrates and inorganic ions), *hydrophobic compounds* (e.g., triglycerides and hydrocarbons), and *amphipathic compounds* (e.g., molecules with both hydrophilic and hydrophobic portions such as fatty acids, bile acids, bilirubin, thyroid, and steroid hormones).

Movement of molecules into and out of cells typically occurs by diffusion, a form of random motion powered by thermal vibrations. For hydrophilic molecules, the lipid core of the membrane represents a *free energy barrier* (Figure 16.1). In order to cross the membrane, energy must be available to get across the free energy "hump" to the other side. By providing aqueous channels through membranes, membrane carrier proteins reduce the free energy needed to cross the membrane until it is less than the energy available from thermal vibrations (160, 161), thus regulating the uptake rate (135).

In contrast, many amphipathic and hydrophobic molecules are more soluble in the lipid core of the membrane than in the aqueous layers on either side (Figure 16.1). For these molecules, transport across water layers may be much slower than

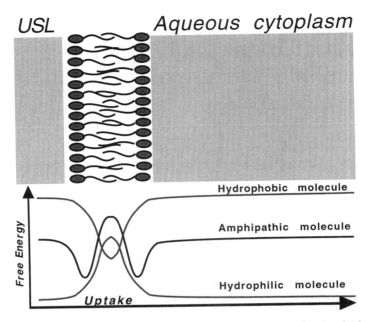

FIGURE 16.1. Barriers to uptake of small molecules by cells. Uptake of molecules is limited whenever they have insufficient thermal energy to climb a free energy barrier (portions of energy curves with positive slope). The membrane core represents a free energy barrier to uptake of hydrophilic molecules (bottom curve), but not to hydrophobic molecules (top curve), which are much more soluble in lipid. Amphipathic molecules prefer to bind to the membrane surface, where their lipid portion can interact with the membrane core and their hydrophilic portion can interact with solvent water. The major barrier to uptake of amphipathic and hydrophobic molecules is often dissociation from membrane or protein-binding sites and diffusion across aqueous layers. [Reprinted from reference (158) with permission.]

transport through the plasma membrane (4, 16, 155, 160). The height of the barrier generally reflects the number of hydrogen bonds in the solvent water that are broken when the molecule is dissolved. In other words, the surface tension of water tends to exclude amphipathic molecules on a submicroscopic scale just as oil drops are excluded from water on a visible scale. By reducing aqueous solubility, these energy barriers restrict the movement of amphipathic molecules across water layers on either side of the plasma membrane.

Many biologically important compounds are only sparingly soluble in water. For example, the solubility of long-chain fatty acids at physiologic pH is <10 μM (22, 102), while that of bilirubin is <1 nM (19, 21). Other amphipathic molecules with low solubilities include phospholipids (169), hydrophobic bile acids (120), retinoids (12), CoA-esters (25), cholesterol (78), thyroid and steroid hormones (17), and a wide variety of exogenous drugs and toxins that are collectively referred to as "organic anions." The liver maintains efficient fluxes of these compounds despite their low aqueous solubilities.

The energy required for an amphipathic or hydrophobic molecule to dissolve in water can be reduced by soluble binding proteins. These aqueous carriers provide mobile hydrophobic binding sites that allow amphipathic molecules to penetrate into and across aqueous layers that would otherwise block transport (155). This function has been compared to "boats" (the binding proteins) carrying nonswimming passengers (the amphipaths) across a river (the water layer) (99). This is a primary function of plasma albumin, which binds long-chain fatty acids and most organic anions (69, 112). Plasma also contains binding proteins for steroid (123) and thyroid (97) hormones, heme (133), retinoids (105), and a variety of other poorly soluble metabolites. Likewise, cytoplasm contains a large number of soluble binding proteins. These include fatty acid binding protein (FABP) (13), bile acid binding proteins (139), retinol binding proteins (12), phospholipid binding proteins (169), thyroid hormone binding proteins (9, 76), sterol binding proteins (126), heme binding protein (77), tocopherol binding protein (168), and a diverse group of glutathione S-transferases with overlapping specificities (65).

Extracellular Barriers

Every cell is surrounded by a thin layer of stagnant fluid, often referred to as the unstirred layer (USL) or barrier layer (10). Within this layer, diffusion is the primary transport mechanism. In experimental systems such as cultured cells, the thickness of this layer typically ranges from 30 to several hundred micrometers and cannot be reduced to less than about 7 μm without such vigorous stirring that the cells are disrupted (10). If the permeability of the USL is less than the permeability of the plasma membrane ($P_{usl} < P_m$), aqueous diffusion rather than membrane transport will determine the influx rate (155, 160). The diffusion rates for water-soluble molecules such as glucose are large enough (and their membrane permeabilities low enough) that extracellular diffusion barriers are rarely important in vivo. However, the much larger USL that is typically present in experimental systems (e.g., tissue culture and suspended cells) often influences uptake rates even for highly soluble molecules (10).

As discussed by others in this book, vascular endothelial cells tend to block access of small molecules in plasma to tissues. Molecules must either take a paracellular route around the endothelial cells or a transcellular route through them. Amphipathic molecules may cross endothelial cells with the aid of albumin receptors that form part of a transcytotic pathway (see Chapter 2). Liver endothelium, however, contains multiple openings that permit direct movement of molecules and small lipoproteins across the endothelium. Within the vascular sinusoids of the liver, plasma is stirred by the motion of erythrocytes, which partially disrupt the laminar layers that form near the sinusoidal margins. In addition, erythrocytes may promote convection within the subendothelial space by a process known as "endothelial massage" (165). The physical thickness of the USL adjacent to the plasma membrane of the liver is unknown, but it seems unlikely to be more than ~1 μm (161).

Although the permeability of such a small USL is relatively large (8, 10), the permeability of the plasma membrane for many amphipathic molecules is still larger. For example, the permeability of the liver plasma membrane for unbound long-chain fatty acids is ~0.2 cm/s (161), or about one million times greater than for chloride (28). The extremely high membrane permeability of fatty acids and organic anions presumably reflects the presence of specific membrane carriers for these molecules (116, 140, 148), although passive permeation of the membrane may also contribute (14, 35, 106, 141, 142, 151, 164). Because of this high permeability, unstirred layers as small as 0.5 μm may significantly limit the uptake rate for fatty acids in the perfused liver (161).

The *effective* thickness of the USL will be larger if the diffusion rate is reduced by matrix proteins (e.g., collagen) within the subendothelial space (117). Reichen et al. have shown that the reduced hepatic transport of bile acids and aminopyrene by cirrhotic rat liver is primarily caused by the larger extracellular diffusion barrier (117). Similar diffusion barriers have been found in human cirrhotics (55), and are a likely cause of the defects in hepatic clearance and secretion that characterize this disorder. Extracellular matrix proteins are also found in the normal liver (121), where their effect on diffusion rates is unknown. Sinusoidal lining cells further increase the magnitude of extracellular diffusion barriers (14, 15). Although endothelial cells are fenestrated, the openings cover only 6–8% of the sinusoidal surface (165). The reduced surface area available for diffusion may represent a "bottleneck" to uptake (165).

Role of Plasma-Binding Proteins

It has been known for many years that the diffusive flux of poorly soluble molecules can be enhanced by adding a second "bound" flux (Figure 16.2). This mechanism is known as *codiffusion*. It explains how bile acid micelles augment the diffusion of bile acids across intestinal unstirred layers (162) and how myoglobin enhances the diffusive flux of oxygen within muscle cells (50). Most formulations of codiffusion assume rapid equilibrium between bound and unbound species. However, dissociation rates of amphipathic molecules such as fatty acids and bilirubin from their binding proteins can be quite slow (154). Thus, this assumption is not always valid.

Bass and Pond extended the codiffusion model to include slow dissociation of the amphipath from its binding protein (11). Their model produced kinetic patterns similar to those seen experimentally and thought to reflect a saturable interaction of the albumin–fatty acid complex with an albumin receptor on the cell surface (Figure 16.3). Later work showed that unstirred water layers as small

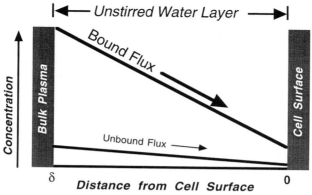

FIGURE 16.2. Equilibrium codiffusion. The maximum diffusional flux of amphipaths across the extracellular USL is limited by their low solubility in plasma water. Plasma-binding proteins such as albumin provide a second diffusional flux across the USL, thus increasing the total flux. Because rapid binding and dissociation are assumed, the concentration gradients are linear. Compare with Figure 16.4. [Modified from reference (155) with permission.]

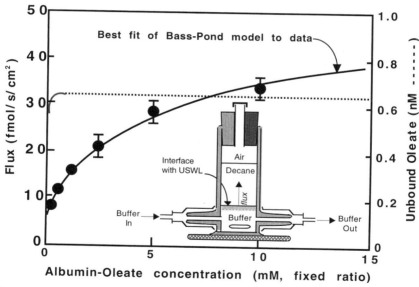

FIGURE 16.3. Diffusion across an unstirred layer showing "albumin receptor" kinetics. Experimental data are shown for the flux of oleate, an amphipathic molecule, from a buffer solution into decane (a hydrocarbon) as the albumin and oleate concentrations were varied using a simple experimental system (inset). The primary rate-limiting step was shown to be diffusion across a 48-μm USL adjacent to the decane interface (160). Added albumin increased the flux both by codiffusion (see later) and by reducing the size of a layer of binding disequilibrium adjacent to the interface surface. Curve shows the best-fit line of the Bass–Pond diffusion model (11) to the data. [Modified from reference (160) with permission.]

as 0.5 μm could restrict uptake of protein-bound amphipaths if they have a very large membrane permeability (161).

An important finding by these investigators is that there is always a thin layer of binding disequilibrium at the cell surface known as the *disequilibrium atmosphere* (shaded area to the right of the broken line in Figure 16.4), which reflects the fact that the cell takes up ligand only from the unbound pool. Within this layer, codiffusion fails (because dissociation cannot keep up with uptake). Transport across this layer is therefore dependent solely on the diffusion of unbound amphipath. The depth of this layer depends on the albumin concentration

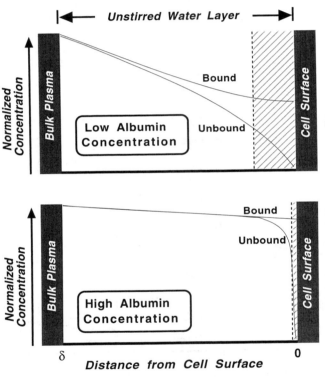

FIGURE 16.4. Effect of albumin concentration on codiffusion gradients within the USL. At low albumin concentrations, unbound ligand must diffuse across a larger disequilibrium atmosphere thickness (shaded area on right of top panel). Higher albumin concentrations reduce the thickness of the disequilibrium atmosphere, increasing the unbound concentration at the cell surface and thus the uptake flux. In this way, higher binding protein concentrations increase the permeability of the USL for unbound ligand. Note that bound and unbound concentrations have been normalized to their values in the bulk plasma. Compare with Figure 16.2, in which the disequilibrium atmosphere thickness is zero. [Reprinted from reference (155) with permission.]

(compare the top and bottom panels in Figure 16.4) (11). Albumin and other binding proteins increase the permeability of the USL by decreasing the thickness of this layer.

The Bass–Pond codiffusion model has now been tested in an artificial transport system (160), suspended cells (114, 115), perfused rat liver (128, 161), intact pigs (108, 109) and in humans (66). It appears to account for both the absolute rate of uptake and the uptake kinetics, although some discrepancies remain (114).

The cycle in which the albumin–ligand complex diffuses to the cell surface, dissociates, and albumin diffuses back to the bulk plasma may be viewed as a carrier system (Figure 16.5). If dissociation of the complex is very rapid, it is kinetically identical to a membrane carrier that diffuses between the two sides of a membrane (138, 159). If dissociation is slower, the kinetics become more complex (160), but they still show saturation, mutual competition, and countertransport (138). The only difference is that membrane carriers typically carry

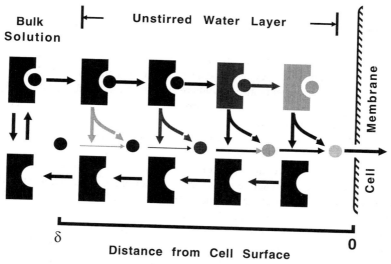

FIGURE 16.5. Diagram of codiffusion mechanism. Albumin and other plasma-binding proteins act as carriers to shuttle amphipathic molecules across the extracellular USL. Albumin–ligand complexes diffuse down a concentration gradient toward the membrane. Net dissociation of ligand occurs during this process due to uptake of unbound ligand at the membrane surface. Uncomplexed albumin is then available to diffuse back to the bulk plasma and begin the cycle again. The carrier cycle is kinetically equivalent to that of a membrane carrier. The size and intensity of the arrows reflect the magnitude of the net flux. Note that most of the flux is due to the diffusion of bound ligand near the bulk solution, while all of the flux is carried by the unbound ligand at the cell surface. This gives rise to the disequilibrium layer. [Reprinted from reference (155) with permission.]

hydrophilic molecules across the hydrophobic core of the membrane, while plasma-binding proteins carry hydrophobic molecules across the hydrophilic USL.

If a method for catalyzing the dissociation of ligands from albumin existed at or near the cell surface, the thickness of the disequilibrium layer could be reduced and the uptake rate increased. The existence of such *surface-mediated dissociation* mechanisms frequently has been postulated (155) but never proven. Catalysis of dissociation due to albumin receptors, collision of the albumin complex with the cell surface, or local changes in solvent properties should produce a kinetic pattern which is very similar to that due to slow dissociation (154) and/or slow diffusion (160). Consequently, such phenomena are very difficult to either prove or disprove. We have taken the position that additional uptake mechanisms should not be assumed unless it is shown that known phenomena (including diffusion, dissociation, and membrane transport) are insufficient to account for experimental results (161).

Intracellular Barriers

The living cell is often viewed as a minute bag of water containing dissolved molecules and suspended organelles. Cytoplasmic cell water (cytosol) is typically treated as a single "well-stirred" compartment. Thus, we speak of "the" cytoplasmic pH and "the" cytosolic concentration of calcium, glucose, or ATP as if these values are identical everywhere within the cytoplasm. However, recent data suggest that cells are not uniform in composition, but often contain substantial concentration gradients. Moreover, these gradients have, over evolutionary time, helped determine the size, shape, and structure of living cells.

Small molecules move through cells by three basic transport mechanisms: convection, diffusion, and active transport. Convection, also known as bulk flow, is important only in certain cell types. These include cells with very active cytoplasmic streaming (e.g., amoebas and other cells that move by amoeboid motion) and cells with a substantial flow of cell water from one pole to another. An example of the latter is the renal tubular cell (52, 71, 149). Because this flow is not turbulent, it is unable to mix the contents of the cell. Instead, bulk water flow displaces any existing concentration gradients and mobile organelles in the direction of the flow (71). Active transport may occur when a "molecular motor" (kinesin) attached to a membrane vesicle or organelle moves through the cytoplasm following a cytoplasmic filament using a ratchet mechanism driven by the hydrolysis of ATP (34, 37). The latter mechanism is well developed in neurons, whose axons are far too long to rely on simple diffusion for transport (1, 3), and is also found in liver, where it is involved in delivering vesicles to surface membranes (92).

Bulk Transport Rates

Diffusion is the primary mechanism of cytoplasmic transport in most cells. It occurs by random motion of molecules (i.e., Brownian motion), which is driven by thermal vibrations. According to Fick's law of diffusion, the diffusional flux J is dependent on the concentration difference ΔC, the diffusion constant D, and the distance x:

$$J = \frac{\Delta CD}{x} . \tag{16.1}$$

The diffusion constant D is a measure of the mobility of molecules in solution. It is determined by the properties of both the molecule and the medium. Thus, the diffusion "constant" for a molecule will be different in different media or for different conditions. The molecular theory of diffusion was worked out by Einstein shortly before he developed his special theory of relativity. According to the Stokes–Einstein equation[1]:

1. D is inversely related to the molecular diameter.
2. D is proportional to the absolute temperature.
3. D is inversely proportional to the viscosity.

As will be discussed later, the diffusion constant of a molecule in cytoplasm is always lower than the comparable value in water due to greater viscosity, tortuosity, and binding to cytoplasmic organelles. Because these properties are poorly defined in most cases, we often speak of an "effective" or "apparent" diffusion constant, D_{eff}, that incorporates these effects. Thus, D_{eff} is the diffusion constant that the molecule appears to display if one assumes that the cytoplasm is homogeneous.

Like membranes, cytoplasm has a permeability, P_c. The permeability of the cytoplasm is inversely related to the distance x that the transported molecule must cross to reach its target:

$$P_c = \frac{D_{eff}}{x} . \tag{16.2}$$

We are used to thinking of the plasma membrane as the primary barrier to transport of molecules into and across cells. This is true, however, *only when the permeability of the plasma membrane for that molecule (P_m) is less than the permeability of the cytoplasm ($P_m \ll P_c$)*. Consider a liver cell transporting bile

---

[1] The Stokes–Einstein equation is $D = RT/6\pi r\eta Å$. It defines the diffusion constant D of a molecule as a function of the gas constant R, the absolute temperature T, its radius of gyration r, the viscosity of the solvent η, and Avogadro's number $Å$. For globular molecules, r is proportional to the cube root of the molecular weight.

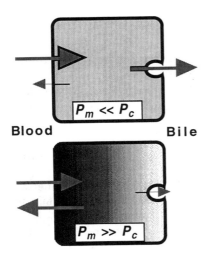

Blood

Bile

FIGURE 16.6. Effect of cytoplasmic transport rate on bile acid secretion. When membrane transport is rate-limiting (top panel), efflux from the cell is minimal (small arrow) and most molecules reach the bile canaliculus (right) before they efflux from the cell. When cytoplasmic transport is slow, however, efflux nearly equals influx and less bile acid reaches the canaliculus before effluxing. Biliary excretion removes bile acid near the canaliculus faster than it can be replenished by transport, causing cytoplasmic concentration gradients. Compare these steady-state gradients with the transient gradients produced by a short pulse of uptake in Figure 16.17. [Reprinted from reference (159) with permission.]

acids from blood to bile. If the permeability of the plasma membrane is much less than that of the cytoplasm, then transport across the plasma membrane will be rate-limiting and no cytoplasmic concentration gradient will form (Figure 16.6, top). If, however, $P_m \gg P_c$, then a large concentration gradient will form a steady state in the presence of an effective biliary secretion mechanism (Figure 16.6, bottom) or during the "initial rate" phase of uptake (see Figure 16.17). In such cases, the concentration of the transported molecule in the layer of cytoplasm adjacent to the plasma membrane will increase until the rate of uptake into the cell (influx) approaches the back flux from the cell (efflux). Thus, one effect of slow cytoplasmic transport is that molecules tend to efflux from the cell before they can be metabolized or excreted (arrows).

Cytoplasmic Information Flux

These statements apply to the so-called "steady-state" condition, in which cellular concentration gradients have had time to develop into profiles that remain stable with time. However, cells are not steady-state machines. They sense changes in their environment through a variety of signal transduction mechanisms and typically respond by changing the concentrations of specific molecules in the cytoplasm (e.g., calcium, cyclic AMP, messenger RNA, transcription, and translation factors). A critical feature of this control loop is *how long* it takes for a signal to travel from one part of the cell to another. For example, the hormone tri-iodothyronine (T_3) takes on average about 1 min to move from the surface of the liver cell to the nucleus where it acts (86). Thus, T_3 is unable to convey information to the nucleus about environmental changes that occur on time

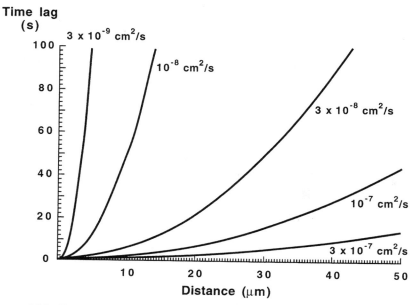

FIGURE 16.7. Time required for information carried by a diffusing molecule to cross a cell. The time lag is plotted as a function of distance and the effective diffusion constant (D_{eff}) of the intracellular messenger. Most cells larger than 10–20 µm are unable to respond to their environments in a timely fashion (see text).

scales of less than several minutes. To attempt to do so would be like trying to use smoke signals to send a play-by-play description of a basketball game—most information content would be lost. Calcium, on the other hand, crosses most cells within seconds (7, 103) and is much better suited to carrying information rapidly.

A critical feature of diffusion is that the time lag required for diffusion gradients to form is proportional to *the square* of the distance[2] (130). Thus, it takes four times as long for a signal carried by an intracellular messenger such as cyclic AMP to cross a 20-µm cell than a 10-µm cell, and nine times as long to cross a 30-µm cell (Figure 16.7). This lag time limits the size of living cells. Thus, another important feature of diffusion is that:

4. Information signals are delayed by the *square* of the distance.

Delayed delivery of information might be tolerable if the information content were retained. Thus, a large amount of information can be recovered from a

---

[2]The time *t* required for a diffusional front to cross a distance *x* is approximately $t = x^2/2D$, where D is the effective diffusion constant (31).

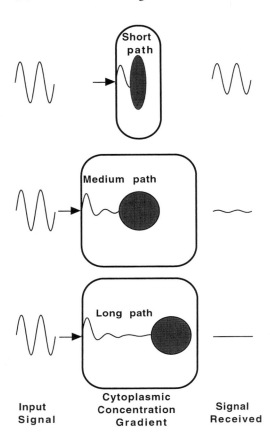

Input
Signal

**Cytoplasmic
Concentration
Gradient**

**Signal
Received**

FIGURE 16.8. How the nucleus senses changes in the environment. Information (input signal with characteristic frequency f) is carried to the nucleus by diffusion of messenger molecules. If the diffusional path x is longer than the distance, the molecule can diffuse in a single on–off cycle (i.e., if $x > \sqrt{D/f}$), the signal dissipates before it reaches the nucleus, and no information is transmitted (bottom panel). A typical baud rate for a small messenger molecule ($D_{eff} = 10^{-6}$ cm^2/s) over a distance of 10 μm is one bit per second. Because thyroid hormone (T_3) has a lower diffusion constant [$D_{eff} \approx 3 \times 10^{-8}$ cm^2/s (89)], the comparable baud rate for this hormone is only about 0.03 bits per second.

distant spacecraft despite a delay of many hours required for the radio waves to reach earth. Likewise, the information coded in messenger RNA does not degrade during the time required for it to diffuse to the ribosome providing the molecule remains intact. Unlike radio waves, however, diffusional signals rapidly dissipate as they travel through a medium (Figure 16.8). Thus, the information "receiver" (e.g., cell nucleus) cannot be located further from the "transmitter" (e.g., plasma membrane) than the distance that the information-carrying molecule (e.g., cyclic AMP) can diffuse in the characteristic time required to switch the signal from "off" to "on" and back again. The maximum rate of switching is known as the *baud rate* (bits per second). For diffusion of a neurotransmitter across a synapse (distance <0.1 mm) the maximum baud rate is very large ($\sim 10^4$ bits per second; Figure 16.9). For diffusion of cyclic AMP throughout a cell (10 μm), the baud rate is much lower (~ 1 bit per second). The final important feature of diffusion is thus:

5. The maximum rate of information transfer is inversely related to the *square* of the distance (i.e., the baud rate $\approx D/x^2$).

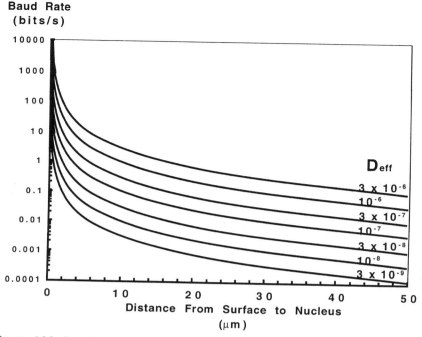

FIGURE 16.9. Rate that nucleus receives information about its environment depends on cell size. Plot shows number of binary bits of information per second that can be carried from the plasma membrane to the nucleus by diffusion of a messenger molecule with cytoplasmic diffusion constant D_{eff} (cm²/s). Seven bits are required to specify a signal level to an accuracy of less than 1%. Information transfer limitations may limit the size of most cells to about 10 μm (see text).

Why Are Cells Small?

Mammalian cells typically measure 5–20 μm in diameter, while most living cells are less than about 100 μm (10^{-2} cm). Aqueous diffusional constants for biological molecules in cytoplasm range over three orders of magnitude: from about 5×10^{-6} cm²/s for water and inorganic ions to 3×10^{-9} cm²/s for substances that bind extensively to cytoplasmic membranes (Table 16.1).

Exceptions to this size rule occur in three cases. The first is when convective mixing is substantial. Thus, amoebas and cellular macrophages can be much larger, ranging up to several hundred micrometers in diameter. Such large cells remain viable because cytoplasmic streaming, which is a form of convective stirring, helps mix their cytoplasm. Note, however, that amoeba cytoplasm also allows much larger values of D_{eff} than are typical for mammalian cells (Table 16.1), suggesting that convection alone is not sufficient to allow such large cells to exist. Another exception is cells with multiple nuclei (e.g., striated muscle

TABLE 16.1. Effective diffusion rates in water and tissues.*

Molecule	D_{eff} cm²/s × 10⁻⁸	Tissue	Method	Ref.
Actin	0.3	Chicken gizzard cell	FRAP	(156)
Actin	10–50	Amoeba cytoplasm	FRAP	(157)
Albumin	0.6	Chicken gizzard cell	FRAP	(156)
Albumin	1	Fibroblast	FRAP	(166)
Albumin	1.7	Fibroblast	FRAP	(59)
Albumin	9.2	Sea urchin oocyte	FRAP	(158)
Albumin	29	15% Albumin solution	modeling	(36)
Albumin	40	Amoeba cytoplasm	FRAP	(157)
Albumin	51	5% Albumin solution	modeling	(36)
Albumin	60	Rabbit ear interstitium	FRAP	(26)
Albumin	63	Dilute water solution	modeling	(36)
Albumin	66	Dilute water solution	lag time	(79)
Albumin	88	Dilute water solution	light scattering	(42)
Apoferritin	1.6	Fibroblast	FRAP	(59)
Carboxyfluorescein	1	Fibroblast	FRAP	(59)
Carboxyfluorescein	2	Liver	FRAP	(75)
Dextran (68 kDa)	2	Liver	FRAP	(111)
Fluorescein	0.33	Liver	FRAP	(75)
Hemoglobin	6	Erythrocyte	NMR	(70)
Insulin	0.9	Fibroblast	FRAP	(59)
Myoglobin	50	Myocyte	spectroscopy	(63)
Ovalbumin	3.5	Human macrophage	FRAP	(157)
Palmitate	0.2–0.6	Liver	MID	(53)
Protons (hydronium)	100–200	Muscle	modeling	(57)
Protons (hydronium)	140	Neuron	pH lag time	(2)
Protons (hydronium)	9500	Dilute water solution	lag time	(79)
Stearate-NBD	0.3–0.5	Liver	FRAP	(74)
Tri-iodothyronine	3	Liver	MID	(89)
Tubulin	7.5	Sea urchin oocyte	FRAP	(158)
Water (self diffusion)	570	Duck embryo	NMR	(27)
Water (self diffusion)	1940	Dilute water solution	NMR	(27)

*Most studies were done at 25°C, but range from 18 to 37°C. Consult original references for details.

cells and true slime molds) that have adopted a distributed approach to information processing. In these cells, the response to external stimuli is determined by whichever nucleus is closest. Even mammalian liver cells, which are only about 20 μm in diameter (114), tend to be multinuclear, suggesting that they are too large to be governed by a single nucleus. Finally, concentration gradients may be *desirable* in certain cells. For example, oocytes encode information in cytoplasmic concentration gradients that direct embryonic differentiation after fertilization (95). Accordingly, the largest single cells are oocytes, ranging up to 8 cm in diameter for ostrich egg yolks. Collectively, these data suggest that the maximum size of cells may be limited by the need for rapid cytoplasmic transport of metabolic substrates and information.

Inserting typical values from Table 16.1 into Figure 16.7 shows that the time required for information about the environment to reach the nucleus of cells and

TABLE 16.2. Types of cytoplasmic transport.*

Type	Mechanism	Rate	Role for binding proteins	Effect of cytoskeletal inhibitors	Examples
I	Aqueous diffusion of unbound form	Very fast $D_{eff} \approx 10^{-6}$ cm²/s $t_{diff} \approx 1$ s	—	—	Glucose ATP Potassium
II	Aqueous diffusion of protein-bound form	Fast $D_{eff} < 10^{-7}$ cm²/s $t_{diff} > 10$ s	Yes	—	Bilirubin Fatty acids Calcium
III	Vesicular transport	Slow $D_{eff} \approx 10^{-9}$ cm²/s $t_{diff} \approx 1000$ s	—	Inhibition if driven by kinesin	Bile acids ICG High-conc. bile acids
IV	Lateral diffusion of membrane-bound form	Very slow $D_{eff} \approx 10^{-10}$ cm²/s $t_{diff} \approx 10,000$ s	—	—	ICG plus colchicine
V	Convection	Variable	Yes	Inhibition	—

*The five major forms of cytoplasmic transport are listed along with order-of-magnitude estimates of their diffusion rates (D_{eff}) and the approximate time required to diffuse across a 10-μm liver cell.

for the processed response to that information to reach the appropriate effector sites within the cytoplasm grows dramatically with the size of the cell. Thus, cells larger than about 10–20 μm in diameter may be unable to respond in a timely manner to their environments.

Varieties of Cytoplasmic Diffusion

Depending on their physical and chemical properties, small molecules may diffuse through the cytoplasm while dissolved in the aqueous phase, while bound to soluble proteins, or while partially or completely bound to membranes. Diffusion of the membrane-bound form may occur by lateral diffusion of the molecule within the membrane (99, 101, 129) or by the diffusion of membrane vesicles with their bound ligands (39, 124). Vesicles may also be actively transported through the cytoplasm by molecular motors that are driven by the hydrolysis of ATP (37). The effective cytoplasmic diffusion constant, D_{eff}, is the sum of the diffusion constants of all of these constants after weighting for the fraction of the molecule in each phase.[3] Table 16.2 lists the major mechanisms of cytoplasmic transport with their approximate rates.

Factors that Influence Cytoplasmic Diffusion

A number of factors modulate the diffusional flux of molecules within cells. These include the tortuosity of the diffusional path, cytosolic viscosity, binding of the molecule to membranes and proteins, molecular crowding, and size-dependent sieving by cytoskeletal filaments.

Tortuosity

Cells contain many obstructions that increase the diffusional path and thus reduce the diffusional flux (Figure 16.10). For example, liver cytoplasm contains ~10,000 cm² of membranes (one square meter) per milliliter of cytoplasm (153).

Most cytosolic molecules are unable to pass through these membranes but must follow aqueous channels around them. This huge membrane density (Figure 16.11) greatly increases the tortuosity (length) of the diffusional path, further reducing the cytoplasmic diffusion rates (27, 111, 122). The effect of tortuosity can be incorporated into Eq. (16.1) as a larger value of the distance x, but is more commonly used to reduce the value of D, which becomes an *effective* diffusion constant D_{eff}. This value is thus not a constant at all, but varies according to the composition of the cytoplasm. This approach is used because D_{eff} is an experimentally measurable value, while other features of cytoplasm such as tortuosity and viscosity are more difficult to quantitate.

---

[3]Mathematically, $D_{eff} = \Sigma_1^n X_i D_i$, where X_i is the fraction of the molecule in that phase (e.g., bound to membranes) and D_i is the diffusion constant of that form of the molecule within the cytoplasm (90).

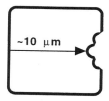

**Direct diffusional path
(no   tortuosity)**

**Minimum diffusional path
is   longer due to tortuosity**

FIGURE 16.10. Diffusion of bile acid from blood to bile canaliculus. The minimum diffusional path (right panel) is much longer than the direct path (left panel) due to the presence of cytoplasmic structures such as membranes and organelles.

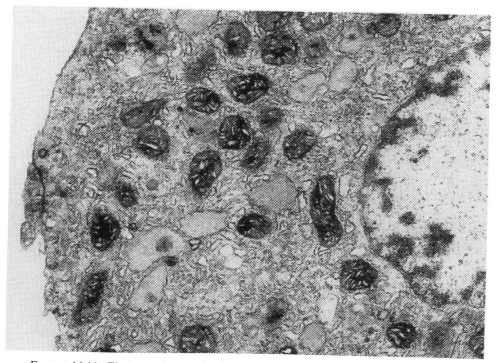

FIGURE 16.11. Electron micrograph of liver cell. Cytoplasmic membranes increase the tortuosity of the diffusional path, thus reducing the effective diffusion constant of molecules within the cell.

Restricted Mobility Due to Cytoskeleton

Cytoplasm contains a lattice of cytoskeletal filaments that significantly alters the solvent properties of cytoplasmic water (29, 81, 83). These filaments make up 16–21% of the volume of a typical cell and have a surface area of ~19,000 cm²/ml (43). This network further restricts the diffusional mobility of large molecules by molecular sieving (80, 81) and of many soluble molecules by reversible (e.g., Van der Waals) binding interactions (43, 59). In addition, size-dependent molecular sieving is important for larger molecules, completely immobilizing molecules larger than about 520 Å in diameter (80).

Viscosity and Molecular Crowding

Overall, cytosol has the consistency of a viscoelastic gel whose aqueous phase has a viscosity which is two to six times that of water (81, 84, 93), although lower viscosities have been reported (64, 82). The viscosity of this gel appears regulated in part by pH and intracellular Ca^{2+}, with maximum viscosity at physiologic values (81), although measurements in purified cytosol may not be representative of living tissue (29, 82). A major portion of cell water is tightly bound to cytoplasmic membranes and proteins, thus limiting its mobility and further reducing cytoplasmic diffusion rates (27, 81, 83). The cytosolic protein concentration is 15–26% (41), suggesting that there should be significant hydrodynamic, steric, and electrostatic interactions among dissolved proteins that should further limit their mobility (83). Because little water is free, small changes in cytoplasmic water content can produce dramatic changes in cytoplasmic viscosity and D_{eff} values (56, 59, 64, 93, 110).

These data suggest that the diffusion of aqueous molecules should be dramatically slower in cytoplasm than in free solution. Using fluorescent dextrans, Peters et al. estimated that cytoplasmic diffusion of hydrophilic molecules in liver cells is approximately 20 times slower than in water (111). In other words, a liver cell that is 20 μm across is diffusionally equivalent to a 400-μm sphere of unstirred water. Extracellular water layers much smaller than this have been shown to limit the uptake of rapidly cleared compounds such as fatty acids and bile acids (10, 114, 161). Because amphipaths often bind extensively to intracellular membranes (20, 33, 78, 90, 139), their diffusion through cytoplasm should be slower than for comparably sized dextrans. Although energy-dependent vesicular transport pathways also exist (5, 38), vesicular transport is relatively slow and may be important only at higher amphipath concentrations (39).

Cytoplasmic Diffusion of Amphipathic Molecules

Amphipaths are molecules with detergent properties, containing both hydrophobic and hydrophilic domains. As such, they tend to bind to membranes (146) and to various soluble binding proteins and have only limited solubility in water. Amphipaths are a very broad class that includes not only physiologic molecules such as long-chain fatty acids, cholesterol, bilirubin, thyroid, and steroid hormones, but also exogenous drugs and toxins that are collectively referred to as *organic anions* and *organic cations*. Although convection and vesicular transport may contribute significantly to the cytoplasmic transport of some amphipathic molecules (5, 32, 38, 51), most cytoplasmic transport is believed to occur by diffusion of the molecule while it is bound to cytoplasmic binding proteins such as ligandin and fatty acid binding protein (44, 100, 146).

A role for soluble binding proteins in transport has long been postulated (39, 44, 45, 94, 96, 98–100, 131, 132, 134, 139, 144, 146, 147, 150). Although these basic ideas were first proposed more than 15 years ago (100) and have gained wide support since then (39, 45, 94, 132, 134, 139), no adequate methods were available to investigate cytoplasmic transport in living cells until recently.

The idea that the binding of a small molecule to a larger one can increase the diffusion rate in cytoplasm is counterintuitive. Normally, increasing the size of the diffusing species should decrease diffusion. However, amphipathic molecules spend only a small part of the time in the cytosol, often binding extensively to intracellular membranes that are themselves essentially immobile (146). Because lateral diffusion within membranes is very slow (58), membrane-bound organic anions contribute relatively little to the diffusional flux in most cases (Figure 16.12). Diffusion rates should thus be increased by cytoplasmic binding proteins, which reduce the fraction of the organic anion in the relatively immobile membrane-bound pool (146).

These predictions have been confirmed by recent studies showing that organic anions and other amphipaths diffuse through hepatic cytoplasm very slowly. Effective cytoplasmic diffusion constants range from $\sim 3 \times 10^{-8}$ cm$^2 \cdot$ s^{-1} for triiodothyronine (89) to $\sim 3 \times 10^{-9}$ cm$^2 \cdot$ s^{-1} for a stearate analog (87) and for fluorescein (75). These values are two to three orders of magnitude smaller than for the unbound anion in free water, and correspond to half-times for cytoplasmic equilibration of more than 1 min. Hydrophobic molecules take longer to reach the bile than their more hydrophilic brethren (6), presumably reflecting more extensive binding to cytoplasmic membranes. Higher levels of binding proteins help prevent membrane binding. As shown in Table 16.3, diffusion constants for a fluorescent fatty acid were nearly twice as fast in female than in male hepatocytes (88), as expected from the fact that female cells contain substantially more fatty acid binding protein than male cells (107).

Driving Forces for Cytoplasmic Diffusion

Concentration gradients are inherently unstable.[4] In the absence of energy input, they decay toward equilibrium. Thus, diffusive fluxes must, in all cases, reflect input of free energy. Driving forces may be generated within the same cell, such as

---

[4]More precisely, it is *activity* gradients that are unstable. The activity of a molecule in solution is related to its thermodynamic free energy. In most cases, this value is determined by the concentration. However, binding of the molecule to membranes or proteins may alter activity, as discussed in reference (155).

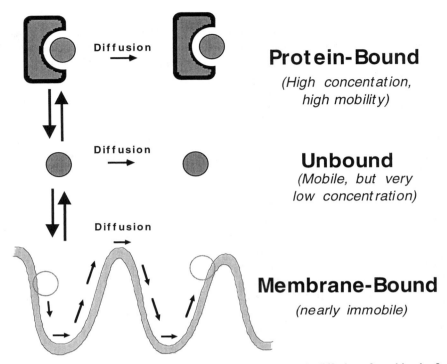

FIGURE 16.12. How binding proteins stimulate the cytoplasmic diffusion of amphipaths. In the absence of binding protein, most of the amphipathic molecules (circles) would be bound to membranes, where they are relatively immobile. The tiny unbound pool, although highly mobile, is too small to contribute significantly to the diffusional flux. Protein binding increases the soluble (and thus diffusible) pool by competing with membranes for the amphipathic molecule. [Reprinted from reference (159) with permission.]

TABLE 16.3. Effect of binding protein level on membrane binding and mobility.*

	[FABP]	Cytosolic fraction %	D_{eff} (cm^2 s^{-1}) × 10^{-9}
Male	100%	18.2 ± 2.7	3.05 ± 0.21
Female	200%	35.1 ± 7	5.03 ± 0.37

*Data from reference (88).

by the active transport of a molecule across the cell plasma membrane. For example, bile acids are actively pumped into liver cells across the basolateral membrane (125), while many organic anions and cations are actively pumped out of liver cells across the bile canalicular membrane (170). In each case, cytoplasmic concentration gradients may be created and sustained by the pumping mechanism if it is sufficiently rapid (see the discussion of Figure 16.2). In addition, gradients may be created within a cell by metabolic utilization of a substrate within the same cell (50) or in a different cell that may be adjacent or located remotely. Thus, rapid utilization of long-chain fatty acids by heart muscle may generate diffusion gradients across the endothelial cells that line the cardiac capillaries (47), and excretion of calcium into urine is ultimately responsible for the concentration gradient across the intestinal epithelial cell that controls the rate of calcium absorption (23, 24, 40).

Methods for Viewing Cytoplasmic Transport

Direct Observation

Molecules with intrinsic fluorescence or molecules that alter the fluorescence of other probes may be observed directly by several techniques, all of which are based on observing the relaxation of an experimentally created concentration gradient over time. Confocal microscopy has been used to follow the rate of diffusion of calcium into xenopus oocytes after stimulating uptake using vasopressin (104). A concentric ring of calcium fluorescence was observed that required several minutes to reach the center of the cell at 25°C. These calcium waves reflect not only the diffusion of calcium through the cytoplasm, but also the release and reuptake of calcium from cytoplasmic storage points triggered by phosphatidyl inositol (103). Direct microinjection of fluorescent probes into cells can be used (166), although this may create local anomalies in cytoplasmic composition. Jürgens and co-workers used heme absorbence to show that the diffusion constant of myoglobin in rat diaphragm is $\sim 5 \times 10^{-7}$ cm²/s (63). However, the most commonly used method of direct observation is fluorescence repolarization after photobleaching (FRAP).

FRAP uses a highly collimated laser beam to generate transient concentration gradients within cells by causing irreversible bleaching (Figure 16.13), and then measures cytoplasmic mobility by determining how rapidly the gradients dissipate (18, 59, 60, 75, 80, 90, 111, 166). In most cases, cells have been previously loaded with the fluorescent molecule by preincubation or microinjection. With the use of appropriate filters on the laser and photomultiplier, most common fluorescent molecules are suitable. Temperature and CO_2 levels are typically physiologic.

FRAP has numerous *advantages* for studying intracellular transport:

1. It is direct and model-independent (i.e., requires few assumptions).
2. The portion of the cell cytoplasm to be studied can be selected.
3. Studies are performed in a single cell rather than averaging results over many cells, allowing cell heterogeneity to be investigated.
4. Multiple studies can be performed on a single coverslip of cells, greatly facilitating data acquisition.
5. Convection can be distinguished from diffusion.
6. Multiple intracellular pools can be detected (if present), providing that their diffusion rate constants are sufficiently different and they do not rapidly interconvert.

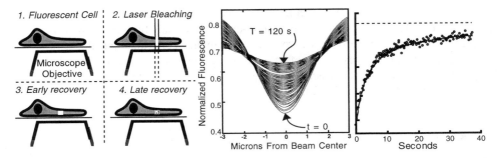

FIGURE 16.13. Laser photobleaching method (FRAP). A cultured cell is preloaded with a fluorescent probe molecule and positioned on a microscope stage. After selecting a suitable portion of the cytoplasm, a strong laser blast is used to irreversibly bleach probe molecules within a limited area. Care is used not to heat or otherwise injure the cell. The rate at which unbleached probe molecules diffuse back into the bleached area provides a measure of D_{eff}. [Modified from reference (90) with permission.]

FRAP's major *limitations* are:

1. The molecules to be studied must have suitable fluorescence properties.
2. A method for getting the molecule into the cytoplasm must be found.
3. Once in the cell, the probe must not be metabolized or excreted before the photobleaching study is complete.
4. Addition of the fluorescent side group may alter the properties of the molecule.

FRAP has been used primarily to measure the lateral mobility of molecules within membranes (167). However, it has recently been used to measure the intracellular mobility of proteins (59, 61, 166), calcium (18), dextrans (111), ficols (80, 83), fatty acids (61, 68, 85, 88), fluorescein (75), carboxyfluorescein (75), and other molecules (59, 64). In all of these studies, it was found that cytoplasmic diffusion is much slower than comparable diffusion rates in free solution, and amphipathic and hydrophobic molecules diffuse through cytoplasm less rapidly than hydrophilic molecules.

Indirect Determination

Indirect methods infer cytoplasmic transport rates by measuring the rates of component steps in the process and then using models to calculate how much slower the rate is expected to be in living tissues due to tortuosity, viscosity, and binding effects (30, 40, 54, 113, 122, 137, 147). Conversely, rates of component steps can be calculated from the measurement of overall rates for an experimental system (48, 49, 62, 136, 152). As an example of the latter, we consider the rate of efflux of a molecule from a tissue preloaded with a radioactive tracer. If cytoplasmic transport is rapid and perfusion even, the concentration in the effluent

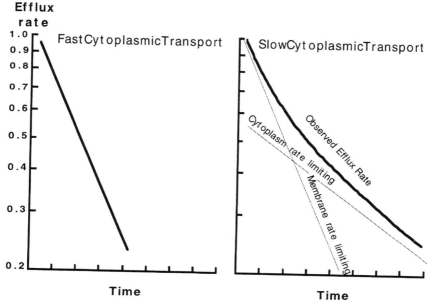

FIGURE 16.14. Effect of slow cytoplasmic transport on efflux curves. The curves show the concentration in the effluent of a tissue perfused single-pass with buffer after being pre-loaded with a radioactive indicator. (A) Fast cytoplasmic transport. A single exponential is seen that reflects the permeability of the plasma membrane. (B) Slow cytoplasmic transport. Initial efflux rates are limited by the plasma membrane, but the efflux rate slows later as cytoplasmic gradients develop. The final efflux rate reflects both membrane and cytoplasmic permeabilities. The rate of cytoplasmic transport can be calculated from fitting an appropriate model to the data. [Reprinted from reference (159) with permission.]

will be a simple exponential with time (Figure 16.14A). In contrast, if cytoplasmic transport is slow,[5] the curve will be nonlinear (Figure 16.14B). The nonlinearity develops when the rate of efflux from surface layers of cytoplasm exceeds the rate at which the radioactive molecule can be replenished by diffusion from a deeper layer of cytoplasm.

Unfortunately, this approach is sensitive to uneven perfusion of the tissue: Poorly perfused regions can simulate slow cytoplasmic transport by releasing their radioactivity back into the perfusate slowly, thus invalidating the approach.

This problem can be largely overcome by using the multiple-indicator dilution (MID) technique. This method is similar to that shown in Figure 16.14, except that the tissue is not uniformly loaded at the start of the experiment but is instead

---

[5]In this context, "slow" means that the characteristic time for equilibration (x^2/D) is longer than or comparable to the characteristic time for removal by all other transport processes (membrane transport, plasma flow, etc). These times are simply the inverse of the rate constants for each of these steps, and are more fully discussed in reference (154).

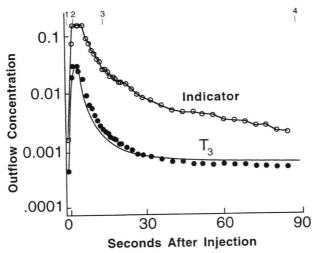

FIGURE 16.15. Results of MID study with tri-iodothyronine (T_3). A bolus of labeled T_3 and a nontransported indicator (albumin) was injected into the portal vein of a perfused rat liver. The best-fit outflow profile for T_3 estimated by treating the cytoplasm as a single compartment (lower curve) underestimates the concentration at early times and overestimates it a later times. This error can be eliminated by incorporating slow cytoplasmic diffusion into the model (see Figure 16.16). Cytoplasmic transport rates can thus be measured in this manner (89, 91). Numbers at the top refer to diagrammatic panels in Figure 16.17. [Redrawn from reference (89) with permission.]

perfused with a bolus of the radioactive tracer. Effluent radioactivity is expressed relative to a nontransported indicator, which is used as a reference. Because poorly perfused regions of the tissue take up little radioactivity from the bolus, this method is relatively insensitive to the uniformity of tissue perfusion within the physiological range.

This approach was devised by Luxon and Weisiger (91) as an extension of earlier work by Goresky (46). Although it is less direct, the MID method has the distinct advantage that it can be used with any molecule that can be obtained in radioactive form. Unlike FRAP, the labeling of the molecule does not significantly change its physical properties. Using this approach, the values for D_{eff} of tri-iodothyronine (89) and palmitate (53) were found to be 3×10^{-8} cm²/s and $2–6 \times 10^{-9}$ cm²/s, respectively. More importantly, the values obtained for palmitate by the MID method closely match the values obtained for NBD-stearate by the FRAP approach, thus helping to validate both methods. Luxon has further demonstrated that D_{eff} can be markedly reduced by displacing fatty acids from their cytosolic binding proteins (85), in agreement with the concepts presented in Figure 16.14. A reduction in D_{eff} may explain why indomethacin, which blocks the binding of taurocholate to its cytoplasmic binding protein, both delays and reduces biliary bile acid excretion in the perfused rat liver (145). The MID approach has also been

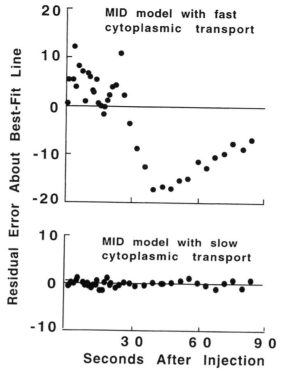

FIGURE 16.16. MID model comparison. Quality of fits of standard MID model that assumes instantaneous cytoplasmic transport (top panel) and diffusion MID model, which includes slow cytoplasmic transport (bottom panel). Both models had the same number of unknown (fitted) parameters. The diffusion MID model replaces the metabolism rate constant with a diffusion rate constant. [Redrawn from reference (89) with permission.]

used by Rivory and co-workers to estimate the cytoplasmic diffusion rate of water and certain lipophilic drugs in the perfused liver (118, 119).

Cytoplasmic transport can also be inferred from the fixation of tissues after brief exposure to labeled molecules (143). Cytoplasmic transport of hydronium ions (hydrated protons) has been measured from the lag between the changes in pH at two different sites in a giant neuron (2), while transport of water has been estimated by NMR techniques (67, 73, 122, 163).

Summary

Transport models that neglect extracellular diffusion barriers or treat the cytoplasm of cells as a "well-stirred" compartment are valid only when the membrane permeability is much less than the permeability of the extracellular USL and of the cytoplasm. This is most likely to be true for steady-state studies of small, hydro-

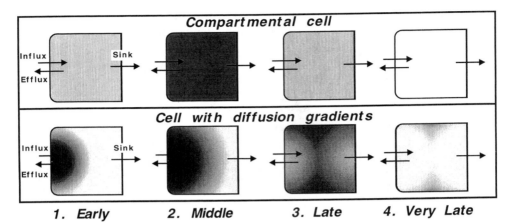

FIGURE 16.17. Schematic view of cytoplasmic concentration gradients during an MID experiment. Intracellular concentration is represented by the intensity of shading from white (none) to dark (high) for four phases of a typical MID experiment. Cells with slow cytoplasmic diffusion initially have more efflux than a compartmental cell at early and middle periods, less efflux for the late period, and more efflux for the very late period. Proper analysis of the outflow curves allows estimation of intracellular transport rates. Numbers (1–4) correspond to time points labeled in Figure 16.15.

philic molecules with modest metabolism or elimination rates. Conversely, such models are most likely to fail when the experiment is very short (e.g., the multiple-indicator dilution method) or the rate of metabolism or excretion is very rapid (e.g., fatty acid metabolism by the liver). The binding of amphipathic and hydro-phobic molecules to intracellular structures may greatly impede their mobility. Albumin and intracellular binding proteins act as true carrier systems to shuttle these molecules across intracellular water layers on either side of the plasma membrane. These aqueous carriers are potential sites of metabolic regulation. Additional study is needed to determine the degree to which these transport systems limit and regulate the metabolism of amphipathic molecules under phys-iological conditions.

References

1. Aizawa, H., Y. Sekine, R. Takemura, Z. Zhang, M. Nangaku, and N. Hirokawa. Kinesin family in murine central nervous system. *J. Cell. Biol.* 119:1287–1296, 1992.
2. Al-Baldawi, N.F. and R.F. Abercrombie. Cytoplasmic hydrogen ion diffusion coeffi-cient. *Biophys. J.* 61:1470–1479, 1992.
3. Amaratunga, A., S.E. Leeman, K.S. Kosik, and R.E. Fine. Inhibition of kinesin synthesis in vivo inhibits the rapid transport of representative proteins for three transport vesicle classes into the axon. *J. Neurochem.* 64:2374–2376, 1995.

4. Andersen, O. and M. Fuchs. Potential energy barriers to ion transport within lipid bilayers: Studies with tetraphenylborate. *Biophys. J.* 15:795–830, 1975.

5. Aoyama, N., T. Ohya, K. Chandler, S. Gresky, and R.T. Holzbach. Transcellular transport of organic anions in the isolated perfused rat liver: The differential effects of monensin and colchicine. *Hepatology* 14:1–9, 1991.

6. Aoyama, N., H. Tokumo, T. Ohya, K. Chandler, and R.T. Holzbach. A novel transcellular transport pathway for non-bile salt cholephilic organic anions. *Am. J. Physiol.* 261:G305–G311, 1991.

7. Atri, A., J. Amundson, D. Clapham, and J. Sneyd. A single-pool model for intracellular calcium oscillations and waves in the Xenopus laevis oocyte. *Biophys. J.* 65:1727–1739, 1993.

8. Baker, K.J. and S.E. Bradley. Binding of sulfobromophthalein (BSP) sodium by plasma albumin. Its role in hepatic BSP extraction. *J. Clin. Invest.* 45:281–287, 1966.

9. Barlow, J.W., L.E. Raggatt, C.F. Lim, D.J. Topliss, and J.R. Stockigt. Characterization of cytoplasmic T3 binding sites by adsorption to hydroxyapatite: Effects of drug inhibitors of T3 and relationship to glutathione-*S*-transferases. *Thyroid* 2:39–44, 1992.

10. Barry, P.H. and J.M. Diamond. Effects of unstirred layers on membrane phenomena. *Physiol. Rev.* 64:763–872, 1984.

11. Bass, L. and S.M. Pond. The puzzle of rates of cellular uptake of protein-bound ligands. In: *Pharmacokinetics: Mathematical and Statistical Approaches to Metabolism and Distribution of Chemicals and Drugs,* edited by A. Pecile and A. Rescigno. London: Plenum Press, 1988, pp. 241–265.

12. Bass, N.M. Cellular binding proteins for fatty acids and retinoids: Similar or specialized functions? *Mol. Cell. Biochem.* 123:191–202, 1993.

13. Bass, N.M., R.M. Kaikaus, and R.K. Ockner. Physiology and molecular biology of hepatic cytosolic fatty acid-binding protein. In: *Hepatic Transport and Bile Secretion: Physiology and Pathophysiology,* edited by N. Tavoloni and P.D. Berk. New York: Raven Press, 1993, pp. 421–446.

14. Bassingthwaighte, J.B., L. Noodleman, G.J. Van der Vusse, and J.F. Glatz. Modeling of palmitate transport in the heart. *Mol. Cell. Biochem.* 88:51–58, 1989.

15. Bassingthwaighte, J.B., C.Y. Wang, and I.S. Chan. Blood-tissue exchange via transport and transformation by capillary endothelial cells. *Circ. Res.* 65:997–1020, 1989.

16. Benz, R., P. Lauger, and K. Janko. Transport kinetics of hydrophobic ions in lipid bilayer membranes: Charge-pulse relaxation studies. *Biochem. Biophys. Acta.* 455:701–720, 1976.

17. Billheimer, J.T. and J.L. Gaylor. Effect of lipid composition on the transfer of sterols mediated by non-specific lipid transfer protein (sterol carrier protein₂). *Biochem. Biophys. Acta Lipid. Metab.* 1046:136–143, 1990.

18. Blatter, L.A. and W.G. Wier. Intracellular diffusion, binding, and compartmentalization of the fluorescent calcium indicators indo-1 and fura-2. *Biophys. J.* 58:1491–1499, 1990.

19. Brodersen, R. Bilirubin. Solubility and interaction with albumin and phospholipid. *J. Biol. Chem.* 254:2364–2369, 1979.

20. Brodersen, R. Binding of bilirubin to albumin. *CRC Crit. Rev. Clin. Lab. Sci.* 11:305–399, 1980.

21. Brodersen, R. and J. Theilgaard. Bilirubin colloid formation in neutral aqueous solution. *Scand. J. Clin. Lab. Invest.* 24:395–398, 1969.

22. Brodersen, R., H. Vorum, E. Skriver, and A.O. Pedersen. Serum albumin binding of

palmitate and stearate. Multiple binding theory for insoluble ligands. *Eur. J. Biochem.* 182:19–25, 1989.

23. Bronner, F. Intestinal calcium transport: the cellular pathway. *Miner Electrolyte Metab.* 16:94–100, 1990.

24. Bronner, F., D. Pansu, and W.D. Stein. An analysis of intestinal calcium transport across the rat intestine. *Am. J. Physiol.* 250:G561–G569, 1986.

25. Catalá, A. Interaction of fatty acids, acyl-CoA derivatives and retinoids with microsomal membranes: Effect of cytosolic proteins. *Mol. Cell. Biochem.* 120:89–94, 1993.

26. Chary, S.R. and R.K. Jain. Direct measurement of interstitial convection and diffusion of albumin in normal and neoplastic tissues by fluorescence photobleaching. *Proc. Natl. Acad. Sci. USA* 86:5385–5389, 1989.

27. Cheng, K.H. Quantitation of non-Einstein diffusion behavior of water in biological tissues by proton MR diffusion imaging: Synthetic image calculations. *Magn. Reson. Imaging* 11:569–583, 1993.

28. Claret, M. and J.L. Mazet. Ionic fluxes and permeabilities of cell membranes in rat liver. *J. Physiol. (Lond).* 223:279–295, 1972.

29. Clegg, J.S. Properties and metabolism of the aqueous cytoplasm and its boundaries. *Am. J. Physiol.* 246:R133–R151, 1984.

30. Cooper, R., N. Noy, and D. Zakim. A physical-chemical model for cellular uptake of fatty acids: prediction of intracellular pool sizes. *Biochemistry* 26:5890–5896, 1987.

31. Crank, J. *The Mathematics of Diffusion,* 2nd ed. New York: Oxford University Press, 1989, pp. 326–337.

32. Crawford, J.M. and J.L. Gollan. Transcellular transport of organic anions in hepatocytes: Still a long way to go. *Hepatology* 14:192–197, 1991.

33. Crawford, J.M. and J.L. Gollan. Hepatocellular transport of bilirubin: The role of membranes and microtubules. In: *Hepatic Transport and Bile Secretion: Physiology and Pathophysiology,* edited by N. Tavoloni and P.D. Berk. New York: Raven Press, 1993, pp. 447–466.

34. Cyr, J.L. and S.T. Brady. Molecular motors in axonal transport. Cellular and molecular biology of kinesin. *Mol. Neurobiol.* 6:137–155, 1992.

35. Daniels, C., N. Noy, and D. Zakim. Rates of hydration of fatty acids bound to unilamellar vesicles of phosphatidylcholine or to albumin. *Biochemistry* 24:3286–3292, 1985.

36. Dwyer, J.D. and V.A. Bloomfield. Brownian dynamics simulations of probe and self-diffusion in concentrated protein and DNA solutions. *Biophys. J.* 65:1810–1816, 1993.

37. Endow, S.A. The emerging kinesin family of microtubule motor proteins. *Trends Biochem. Sci.* 16:221–225, 1991.

38. Erlinger, S. Role of intracellular organelles in the hepatic transport of bile acids. *Biomed. Pharmacother.* 44:409–416, 1990.

39. Erlinger, S. Intracellular events in bile acid transport by the liver. In: *Hepatic Transport and Bile Secretion: Physiology and Pathophysiology,* edited by N. Tavoloni and P.D. Berk. New York: Raven Press, 1993, pp. 467–476.

40. Feher, J.J., C.S. Fullmer, and R.H. Wasserman. Role of facilitated diffusion of calcium by calbindin in intestinal calcium absorption. *Am. J. Physiol.* 262:C517–C526, 1992.

41. Fulton, A. How crowded is the cytoplasm? *Cell* 30:375–378, 1985.

42. Gaigalas, A.K., J.B. Hubbard, M. McCurley, and S. Woo. Diffusion of bovine serum albumin in aqueous solutions. *J. Phys. Chem.* 96:2355–2359, 1992.

43. Gershon, N.D., K.R. Porter, and B.L. Trus. The cytoplasmic matrix: its volume and surface area and the diffusion of molecules through it. *Proc. Natl. Acad. Sci. USA* 82:5030–5034, 1985.

44. Glatz, J.F. and G.J. Van der Vusse. Intracellular transport of lipids. *Mol. Cell. Biochem.* 88:37–44, 1989.

45. Glatz, J.F.C. and G.J. Van der Vusse. Cellular fatty acid-binding proteins: Current concepts and future directions. *Mol. Cell. Biochem.* 98:237–251, 1990.

46. Goresky, C.A. Uptake in the liver: The nature of the process. *Can. J. Physiol. Pharmacol.* 21:65–101, 1980.

47. Goresky, C.A., W. Stremmel, C.P. Rose, S. Guirguis, A.J. Schwab, H.E. Diede, and E. Ibrihim. The capillary transport system for free fatty acids in the heart. *Circ. Res.* 74:1015–1026, 1994.

48. Groebe, K. and G. Thews. Role of geometry and anisotropic diffusion for modeling PO_2 profiles in working red muscle. *Respir. Physiol.* 79:255–278, 1990.

49. Groebe, K. and G. Thews. Calculated intra- and extracellular PO_2 gradients in heavily working red muscle. *Am. J. Physiol.* 259:H84–H92, 1990.

50. Groebe, K. and G. Thews. Basic mechanisms of diffusive and diffusion-related oxygen transport in biological systems: A review. *Adv. Exp. Med. Biol.* 317:21–33, 1992.

51. Hayakawa, T., O. Cheng, A. Ma, and J.L. Boyer. Taurocholate stimulates transcytotic vesicular pathways labeled by horseradish peroxidase in the isolated perfused rat liver. *Gastroenterology* 99:216–228, 1990.

52. Hebert, S.C., J.A. Schafer, and T.E. Andreoli. The effects of antidiuretic hormone (ADH) on solute and water transport in the mammalian nephron. *J. Membr. Biol.* 58:1–19, 1981.

53. Luxon, B.A., D.C. Holly, M.T. Milliano, and R.A. Weisiger. Sex differences in membrane and intracellular hepatic transport of palmitate support a balanced uptake mechanism. *Am. J. Physiol.* (in press), 1998.

54. Hou, L., F. Lanni, and K. Luby-Phelps. Tracer diffusion in F-actin and Ficoll mixtures. Toward a model for cytoplasm. *Biophys. J.* 58:31–43, 1990.

55. Huet, P.M., C.A. Goresky, J.P. Villeneuve, D. Marleau, and J.O. Lough. Assessment of liver microcirculation in human cirrhosis. *J. Clin. Invest.* 70:1234–1244, 1982.

56. Häussinger, D., N. Saha, C. Hallbrucker, F. Lang, and W. Gerok. Involvement of microtubules in the swelling-induced stimulation of transcellular taurocholate transport in perfused rat liver. *Biochem. J.* 291:355–360, 1993.

57. Irving, M., J. Maylie, N.L. Sizto, and W.K. Chandler. Intracellular diffusion in the presence of mobile buffers. Application to proton movement in muscle. *Biophys. J.* 57:717–721, 1990.

58. Jacobson, K., Z. Derzko, E.S. Wu, Y. Hou, and G. Poste. Measurement of the lateral mobility of cell surface components in single, living cells by fluorescence recovery after photobleaching. *J. Supramol. Struct.* 5:565–576, 1976.

59. Jacobson, K. and J. Wojcieszyn. The translational mobility of substances within the cytoplasmic matrix. *Proc. Natl. Acad. Sci. USA* 81:6747–6751, 1984.

60. Jacobson, K. and J. Wojcieszyn. The translational mobility of substances within the cytoplasmic matrix. *Proc. Natl. Acad. Sci. USA* 81:6747–6751, 1984.

61. Jans, D.A., R. Peters, P. Jans, and F. Fahrenholz. Vasopressin V2-receptor mobile fraction and ligand-dependent adenylate cyclase activity are directly correlated in LLC-PK1 renal epithelial cells. *J. Cell. Biol.* 114:53–60, 1991.

62. Jones, D.P., T.Y. Aw, and A.H. Sillau. Defining the resistance to oxygen transfer in tissue hypoxia. *Experientia* 46:1180–1185, 1990.

63. Jürgens, K.D., T. Peters, and G. Gros. A method to measure the diffusion coefficient of myoglobin in intact skeletal muscle cells. *Adv. Exp. Med. Biol.* 277:137–143, 1990.

64. Kao, H.P., J.R. Abney, and A.S. Verkman. Determinants of the translational mobility of a small solute in cell cytoplasm. *J. Cell. Biol.* 120:175–184, 1993.

65. Kaplowitz, N. Physiological significance of glutathione S-transferases. *Am. J. Physiol.* 239:G439–G444, 1980.

66. Keiding, S., P. Ott, and L. Bass. Enhancement of unbound clearance of ICG by plasma proteins, demonstrated in human subjects and interpreted without assumption of facilitating structures. *J. Hepatol.* 19:327–344, 1993.

67. Kimmich, R., T. Gneiting, K. Kotitschke, and G. Schnur. Fluctuations, exchange processes, and water diffusion in aqueous protein systems. A study of bovine serum albumin by diverse NMR techniques. *Biophys. J.* 58:1183–1197, 1990.

68. Kolega, J. and D.L. Taylor. Gradients in the concentration and assembly of myosin II in living fibroblasts during locomotion and fiber transport. *Mol. Biol. Cell.* 4:819–836, 1993.

69. Kragh-Hansen, U. Structure and ligand binding properties of human serum albumin. *Dan. Med. J.* 37:57–84, 1990.

70. Kuchel, P.W. and B.E. Chapman. Translational diffusion of hemoglobin in human erythrocytes and hemolysates. *J. Magn. Reson.* 94:574–580, 1991.

71. Kuwahara, M., L.B. Shi, F. Marumo, and A.S. Verkman. Transcellular water flow modulates water channel exocytosis and endocytosis in kidney collecting tubule. *J. Clin. Invest.* 88:423–429, 1991.

72. Lake, J.R., V. Licko, R.W. Van Dyke, and B.F. Scharschmidt. Biliary secretion of fluid-phase markers by the isolated perfused rat liver. Role of transcellular vesicular transport. *J. Clin. Invest.* 76:676–684, 1985.

73. Latour, L.L., K. Svoboda, P.P. Mitra, and C.H. Sotak. Time-dependent diffusion of water in a biological model system. *Proc. Natl. Acad. Sci. USA* 91:1229–1233, 1994.

74. Lenzen, R., F. Tarseti, R. Salvi, E. Schuler, R. Dembitzer, and N. Tavoloni. Physiology of canalicular bile formation. In: *Hepatic Transport and Bile Secretion: Physiology and Pathophysiology,* edited by N. Tavoloni and P.D. Berk. New York: Raven Press, 1993, pp. 539–552.

75. LeSage, G.D., W.E. Robertson, J.L. Phinizy, and A. Dominquez. Cytoplasmic and membrane-based diffusion of organic anions in hepatocyte couplets and isolated endoplasmic reticulum vesicles. *Gastroenterology* 102:A841 (abstract), 1992.

76. Lichter, M., G. Fleischner, R. Kirsch, J. Levi, K. Kamisaka, and I.M. Arias. Ligandin and Z protein in binding of thyroid hormones by the liver. *Am. J. Physiol.* 230:113–1155, 1976.

77. Liem, H.H., J.A. Grasso, S.H. Vincent, and U. Muller Eberhard. Protein-mediated efflux of heme from isolated rat liver mitochondria. *Biochem. Biophys. Res. Commun.* 167:528–534, 1990.

78. Liscum, L. and N.K. Dahl. Intracellular cholesterol transport. *J. Lipid. Res.* 33:1239–1254, 1992.

79. Longsworth, L. Temperature dependence of diffusion in aqueous solutions. *J. Phys. Chem.* 58:770–773, 1954.

80. Luby-Phelps, K., P.E. Castle, D.L. Taylor, and F. Lanni. Hindered diffusion of inert tracer particles in the cytoplasm of mouse 3T3 cells. *Proc. Natl. Acad. Sci. USA* 84:4910–4913, 1987.

81. Luby-Phelps, K., F. Lanni, and D.L. Taylor. The submicroscopic properties of

cytoplasm as a determinant of cellular function. *Ann. Rev. Biophys. Biophys. Chem.* 17:369–396, 1988.

82. Luby-Phelps, K., S. Mujumdar, R.B. Mujumdar, L.A. Ernst, W. Galbraith, and A.S. Waggoner. A novel fluorescence ratiometric method confirms the low solvent viscosity of the cytoplasm. *Biophys. J.* 65:236–242, 1993.

83. Luby-Phelps, K. and D.L. Taylor. Subcellular compartmentalization by local differentiation of cytoplasmic structure. *Cell. Motil. Cytoskel.* 10:28–37, 1988.

84. Luby-Phelps, K. and R.A. Weisiger. Role of cytoarchitecture in cytoplasmic transport. *Comp. Biochem. Physiol.* 115B:295–306, 1996.

85. Luxon, B.A. Inhibition of binding to fatty acid binding protein reduces the intracellular transport of a fatty acid analog: Further evidence for a transport function for FABP. *Gastroenterology* 104:A936 (abstract), 1993.

86. Luxon, B.A., R.R. Cavalieri, and R.A. Weisiger. A new method for measuring cytoplasmic transport: Application to 3,5,3'-triiodothyronine (T3). *Clin. Res.* 39:460A (abstract), 1991.

87. Luxon, B.A. and R.A. Weisiger. Cytoplasmic transport: A potentially rate-limiting step in the hepatic utilization of fatty acids. *Hepatology* 14:255A (abstract), 1991.

88. Luxon, B.A. and R.A. Weisiger. Sex differences in cytoplasmic transport of a fatty acid analog: Evidence for a transport function for fatty acid binding protein (FABP). *Hepatology* 16:144A (abstract), 1992.

89. Luxon, B.A. and R.A. Weisiger. A new method for measuring cytoplasmic transport: Application to 3,5,3'-triiodothyronine (T3). *Am. J. Physiol.* 263:G733–G741, 1992.

90. Luxon, B.A. and R.A. Weisiger. Sex differences in intracellular fatty acid transport: Role of cytoplasmic binding proteins. *Am. J. Physiol.* 265:G831–G841, 1993.

91. Luxon, B.A. and R.A. Weisiger. Extending the multiple indicator dilution method to include slow cytoplasmic diffusion. *Math. Biosci.* 113:211–230, 1993.

92. Marks, D.L., N.F. LaRusso, and M.A. McNiven. Isolation of the microtubule-vesicle motor kinesin from rat liver: Selective inhibition by cholestatic bile acids. *Gastroenterology* 108:824–833, 1995.

93. Mastro, A.M., M.A. Babich, W.D. Taylor, and A.D. Keith. Diffusion of a small molecule in the cytoplasm of mammalian cells. *Proc. Natl. Acad. Sci. USA* 81:3414–3418, 1984.

94. Matarese, V., R.L. Stone, D.W. Waggoner, and D.A. Bernlohr. Intracellular fatty acid trafficking and the role of cytosolic lipid binding proteins. *Prog. Lipid. Res.* 28:245–272, 1990.

95. McGinnis, W. and M. Kuziora. The molecular architects of body design. *Sci. Am.* 270:58–61, 64–66, 1994.

96. Meijer, D.K.F. and G.M.M. Groothuis. Hepatic transport of drugs and proteins. In: *Oxford Textbook of Clinical Hepatology,* edited by N. McIntyre, J.P. Benhamou, J. Bircher, M. Rizzetto, and J. Rodes. New York: Oxford University Press, 1991, pp. 40–78.

97. Mendel, C.M., R.A. Weisiger, A.L. Jones, and R.R. Cavalieri. Thyroid hormone-binding proteins in plasma facilitate uniform distribution of thyroxine within tissues: A perfused rat liver study. *Endocrinology* 120:1742–1749, 1987.

98. Meuwissen, J.A.T.P. and K.P.M. Heirwegh. Binding proteins in plasma and liver cytosol, and transport of bilirubin. In *Transport by Proteins,* edited by G. Blauer and H. Sund. New York: W. de Gruyter, 1978, pp. 387–403.

99. Meuwissen, J.A.T.P. and K.P.M. Heirwegh. Aspects of bilirubin transport. In:

Bilirubin, Volume II, edited by K.P.M. Heirwegh and S.B. Brown. Boca Raton, FL: CRC Press, 1982, pp. 39–83.

100. Meuwissen, J.A.T.P., B. Ketterer, and K.P.M. Heirwegh. Role of soluble binding proteins in overall hepatic transport of bilirubin. In *Chemistry and Physiology of Bile Pigments,* edited by P. Berk and N. Berlin. Bethesda, MD: National Institutes of Health, 1977, pp. 323–337.

101. Morre, D.J., W.D. Merritt, and C.A. Lembi. Connections between mitochondria and endoplasmic reticulum in rat liver and onion stem. *Protoplasma* 73:43–49, 1971.

102. Murkerjee, P. Dimerization of anions of long-chain fatty acids in aqueous solutions and the hydrophobic properties of the acids. *J. Phys. Chem.* 69:2821–2827, 1965.

103. Nathanson, M.H. Cellular and subcellular calcium signaling in gastrointestinal epithelium. *Gastroenterology* 106:1349–1364, 1994.

104. Nathanson, M.H., M.S. Moyer, A.D. Burgstahler, A.M. O'Carroll, M.J. Brownstein, and S.J. Lolait. Mechanisms of subcellular cytosolic Ca^{2+} signaling evoked by stimulation of the vasopressin V1a receptor. *J. Biol. Chem.* 267:23282–23289, 1992.

105. Noy, N. and Z.-J. Xu. Interactions of retinol with binding proteins: Implications for the mechanism of uptake by cells. *Biochemistry* 29:3878–3883, 1990.

106. Noy, N. and D. Zakim. Physical chemical basis for the uptake of organic compounds by cells. In: *Hepatic Transport and Bile Secretion: Physiology and Pathophysiology,* edited by N. Tavoloni and P.D. Berk. New York: Raven Press, 1993, pp. 313–336.

107. Ockner, R.K., D.A. Burnett, N. Lysenko, and J.A. Manning. Sex differences in long chain fatty acid utilization and fatty acid binding protein concentration in rat liver. *J. Clin. Invest.* 64:172–181, 1979.

108. Ott, P., S. Keiding, and L. Bass. Intrinsic hepatic clearance of indocyanine green in the pig: Dependence on plasma protein concentration. *Eur. J. Clin. Invest.* 22:347–357, 1992.

109. Ott, P., S. Keiding, A.H. Johnsen, and L. Bass. Hepatic removal of two fractions of indocyanine green after bolus injection in anesthetized pigs. *Am. J. Physiol. Gastrointest. Liver. Physiol.* 266:G1108–G1122, 1994.

110. Periasamy, N., H.P. Kao, K. Fushimi, and A.S. Verkman. Organic osmolytes increase cytoplasmic viscosity in kidney cells. *Am. J. Physiol. Cell. Physiol.* 263: C901–C907, 1992.

111. Peters, R. Nucleo-cytoplasmic flux and intracellular mobility in single hepatocytes measured by fluorescence microphotolysis. *EMBO J.* 3:1831–1836, 1984.

112. Peters, T., Jr. Serum albumin. *Adv. Prot. Chem.* 37:161–245, 1985.

113. Phillips, M.C., W.J. Johnson, and G.H. Rothblat. Mechanisms and consequences of cellular cholesterol exchange and transfer. *Biochim. Biophys. Acta* 906:223–276, 1987.

114. Pond, S.M., C.K.C. Davis, M.A. Bogoyevitch, R.A. Gordon, R.A. Weisiger, and L. Bass. Uptake of palmitate by hepatocyte suspensions: Facilitation by albumin? *Am. J. Physiol.* 262:G883–G894, 1992.

115. Pond, S.M., R.A. Gordon, Z.-Y. Wu, R.A. Weisiger, and L. Bass. Effects of gender and pregnancy on hepatocellular uptake of palmitic acid: Facilitation by albumin. *Am. J. Physiol. Gastrointest. Liver. Physiol.* 267:G656–G662, 1994.

116. Potter, B.J. and P.D. Berk. Liver plasma membrane fatty acid binding protein. In: *Hepatic Transport and Bile Secretion: Physiology and Pathophysiology,* edited by N. Tavoloni and P.D. Berk. New York: Raven Press, 1993, pp. 253–268.

117. Reichen, J., B. Egger, N. Ohara, T.B. Zeltner, T. Zysset, and A. Zimmermann. Deter-

minants of hepatic function in liver cirrhosis in the rat Multivariate analysis. *J. Clin. Invest.* 82:2069–2076, 1988.

118. Rivory, L.P. Probing hepatic structure and function with the multiple indicator dilution technique. Ph.D. Thesis. University of Queensland, St. Lucia, Australia, 1991.

119. Rivory, L.P., M.S. Roberts, and S.M. Pond. Axial tissue diffusion can account for the disparity between current models of hepatic elimination for lipophilic drugs. *J. Pharmacokinet. Biopharm.* 20:19–61, 1992.

120. Roda, A., A. Minutello, M.A. Angellotti, and A. Fini. Bile acid structure-activity relationship: Evaluation of bile acid lipophilicity using 1-octanol/water partition coefficient and reverse phase HPLC. *J. Lipid Res.* 31:1433–1443, 1990.

121. Rojkind, M. Extracellular matrix. In: *The Liver: Biology and Pathobiology, 2nd ed.,* edited by I.M. Arias, W.B. Jakoby, H. Popper, D. Schachter, and D.A. Shafritz. New York: Raven Press, 1988, pp. 707–716.

122. Rorschach, H.E., C. Lin, and C.F. Hazlewood. Diffusion of water in biological tissues. *Scanning Micros.* 5:S1–S10, 1991.

123. Rosner, W. Plasma steroid-binding proteins. *Endocrinol. Metabol. Clin. North Am.* 20:697–720, 1991.

124. Rothman, J.E. Mechanisms of intracellular protein transport. *Nature* 372:55–63, 1994.

125. Ruifrok, P.G. and D.K. Meijer. Sodium ion-coupled uptake of taurocholate by rat-liver plasma membrane vesicles. *Liver* 2:28–34, 1982.

126. Scallen, T.J., A. Pastuszyn, B.J. Noland, R. Chanderbhan, A. Kharroubi, and G.V. Vahouny. Sterol carrier and lipid transfer proteins. *Chem. Phys. Lipids* 38:239–261, 1985.

127. Scharschmidt, B.F., J.R. Lake, E.L. Renner, V. Licko, and R.W. Van Dyke. Fluid phase endocytosis by cultured rat hepatocytes and perfused rat liver: Implications for plasma membrane turnover and vesicular trafficking of fluid phase markers. *Proc. Natl. Acad. Sci. USA* 83:9488–9492, 1986.

128. Schwab, A.J. and C.A. Goresky. Hepatic uptake of protein-bound ligands: Effect of an unstirred Disse space. *Am. J. Physiol.* 270:G869–G880, 1996.

129. Scow, R.O., E.J. Blanchette-Mackie, M.G. Wetzel, and A. Reinila. Lipid transport in tissue by lateral movement in cell membranes. In: *The Adipocyte and Obesity: Cellular and Molecular Mechanisms,* edited by A. Angel and C.H. Hollenberg. New York: Raven Press, 1983, pp. 165–169.

130. Shah, J.C. Analysis of permeation data: Evaluation of the lag time method. *Int. J. Pharm.* 90:161–169, 1993.

131. Simion, F.R., B. Fleischer, and S. Fleischer. Two distinct mechanisms for taurocholate uptake in subcellular fractions from rat liver. *J. Biol. Chem.* 259:10814–10822, 1984.

132. Sleight, R.G. Intracellular lipid transport in eukaryotes. *Ann. Rev. Physiol.* 49:193–208, 1987.

133. Smith, A. and W.T. Morgan. Hemopexin-mediated heme transport to the liver. Evidence for a heme-binding protein in liver plasma membranes. *J. Biol. Chem.* 260:8325–8329, 1985.

134. Spener, F., T. Borchers, and M. Mukherjea. On the role of fatty acid binding proteins in fatty acid transport and metabolism. *FEBS. Lett.* 244:1–5, 1989.

135. Stein, W.D. Concepts of mediated transport. In: *Membrane Transport,* edited by S.L. Bonting and J.J. de Pont. Amsterdam: Elsevier Press, 1981, pp. 123–157.

422   Richard A. Weisiger

136. Stein, W.D. Facilitated diffusion of calcium across the rat intestinal epithelial cell. *J. Nutr.* 122:651–656, 1992.

137. Stewart, J.M., W.R. Driedzic, and J.A. Berkelaar. Fatty-acid-binding protein facilitates the diffusion of oleate in a model cytosol system. *Biochem. J.* 275:569–573, 1991.

138. Weisiger, R.A. When is a carrier not a membrane carrier? The cytoplasmic transport of amphipathic molecules. *Hepatology* 24:1288–1295, 1996.

139. Stolz, A., H. Takikawa, M. Ookhtens, and N. Kaplowitz. The role of cytoplasmic proteins in hepatic bile acid transport. *Ann. Rev. Physiol.* 51:161–176, 1989.

140. Stremmel, W., C. Tiribelli, and K. Vyska. The multiplicity of sinusoidal membrane carrier systems of organic anions. In: *Hepatic Transport and Bile Secretion: Physiology and Pathophysiology*, edited by N. Tavoloni and P.D. Berk. New York: Raven Press, 1993, pp. 225–234.

141. Stump, D.D., R.M. Nunes, D. Sorrentino, L.M. Isola, and P.D. Berk. Characterization of two distinct components of hepatic oleate uptake. *Hepatology* 16:865A, 1992.

142. Stump, D.D., R.M. Nunes, D. Sorrentino, L.M. Isola, and P.D. Berk. Characteristics of oleate binding to liver plasma membranes and its uptake by isolated hepatocytes. *J. Hepatol.* 16:304–315, 1992.

143. Suchy, F.J., W.F. Balistreri, J. Hung, P. Miller, and S.A. Garfield. Intracellular bile acid transport in rat liver as visualized by electron microscope autoradiography using a bile acid analogue. *Am. J. Physiol.* 245:G681–G689, 1983.

144. Sweetser, D.A., R.O. Heuckeroth, and J.I. Gordon. The metabolic significance of mammalian fatty-acid-binding proteins: Abundant proteins in search of a function. *Ann. Rev. Nutr.* 7:337–359, 1987.

145. Takikawa, H., J.C. Fernandez-Checa, J. Kuhlenkamp, A. Stolz, M. Ookhtens, and N. Kaplowitz. Effect of indomethacin on the uptake, metabolism and excretion of 3-oxocholic acid: Studies in isolated hepatocytes and perfused rat liver. *Biochim. Biophys. Acta* 1084:247–250, 1991.

146. Tipping, E. and B. Ketterer. The influence of soluble binding proteins on lipophile transport and metabolism in hepatocytes. *Biochem. J.* 195:441–452, 1981.

147. Tipping, E., B. Ketterer, and L. Christodoulides. Interactions of small molecules with phospholipid bilayers. Binding to egg phosphatidylcholine of some organic anions (bromosulphophthalein, oestrone sulphate, haem and bilirubin) that bind to ligandin and aminoazo-dye-binding protein A. *Biochem. J.* 180:327–337, 1979.

148. Tiribelli, C. Determinants in the hepatic uptake of organic anions. *J. Hepatol.* 14:385–390, 1992.

149. Verkman, A.S., J.A. Dix, and J.L. Seifter. Water and urea transport in renal microvillus membrane vesicles. *Am. J. Physiol.* 248:F650–F655, 1985.

150. Voelker, D.R. Organelle biogenesis and intracellular lipid transport in eukaryotes. *Microbiol. Rev.* 55:543–560, 1991.

151. Von Dippe, P. and D. Levy. Characterization of the bile acid transport system in normal and transformed hepatocytes. Photoaffinity labeling of the taurocholate carrier protein. *J. Biol. Chem.* 258:8896–8901, 1983.

152. Vork, M.M., J.F. Glatz, and G.J. Van der Vusse. On the mechanism of long chain fatty acid transport in cardiomyocytes as facilitated by cytoplasmic fatty acid-binding protein. *J. Theor. Biol.* 160:207–222, 1993.

153. Weibel, E.R., W. Stäubli, H.R. Gnägi, and F.A. Hess. Correlated morphometric and biochemical studies on the liver cell. *J. Cell. Biol.* 68:91, 1969.

154. Weisiger, R.A. Dissociation from albumin: A potentially rate-limiting step in the clearance of substances by the liver. *Proc. Natl. Acad. Sci. USA* 82:1563–1567, 1985.

155. Weisiger, R.A. The role of albumin binding in hepatic organic anion transport. In: *Hepatic Transport and Bile Secretion: Physiology and Pathophysiology,* edited by N. Tavoloni and P.D. Berk. New York: Raven Press, 1993, pp. 171–196.

156. Wang, Y.-L., F. Lanni, P. McNeil, B. Ware, and L. Taylor. Mobility of cytoplasmic and membrane-associated actin in living cells. *Proc. Natl. Acad. Sci. USA* 79:4660–4664, 1982.

157. Kreis, T., B. Geiger, and J. Schlessinger. Mobility of microinjected rhodamine actin within living chicken gizzard cells determined by fluorescence photobleaching recovery. *Cell* 29:835–845, 1982.

158. Salmon, E., W. Saxton, R. Leslie, M. Karow, and J. McIntosh. Measurements of spindle microtubule dynamics by fluorescence redistribution after photobleaching. *J. Cell. Biol.* 7:253A-1 (abstract), 1983.

159. Weisiger, R.A. Cytoplasmic transport of lipids: Role of binding proteins. *Comp. Biochem. Physiol.* 115B:319–331, 1996.

160. Weisiger, R.A., S.M. Pond, and L. Bass. Albumin enhances unidirectional fluxes of fatty acid across a lipid-water interface: Theory and experiments. *Am. J. Physiol.* 257:G904–G916, 1989.

161. Weisiger, R.A., S.M. Pond, and L. Bass. Hepatic uptake of protein-bound ligands: Extended sinusoidal perfusion model. *Am. J. Physiol.* 261:G872–G884, 1991.

162. Westergaard, H. and J.M. Dietschy. The mechanism whereby bile acid micelles increase the rate of fatty acid and cholesterol uptake into the intestinal mucosal cell. *J. Clin. Invest.* 58:97–108, 1976.

163. Wheatley, D.N., A. Redfern, and R.P. Johnson. Heat-induced disturbances of intracellular movement and the consistency of the aqueous cytoplasm in HeLa S-3 cells: A laser-Doppler and proton NMR study. *Physiol. Chem. Phys. Med. NMR* 23:199–216, 1991.

164. Wieland, T., M. Nassal, W. Kramer, G. Fricker, U. Bickel, and G. Kurz. Identity of hepatic membrane transport systems for bile salts, phalloidin, and antamanide by photoaffinity labeling. *Proc. Natl. Acad. Sci. USA* 81:5232–5236, 1984.

165. Wisse, E., R.B. De Zanger, K. Charrels, P. Van Der Smissen, and R.S. McCuskey. The liver sieve: Considerations concerning the structure and function of endothelial fenestrae, the sinusoidal wall and the space of Disse. *Hepatology* 5:683–692, 1985.

166. Wojcieszyn, J.W., R.A. Schlegel, E.S. Wu, and K.A. Jacobson. Diffusion of injected macromolecules within the cytoplasm of living cells. *Proc. Natl. Acad. Sci. USA* 78:4407–4410, 1981.

167. Yguerabide, J., J.A. Schmidt, and E.E. Yguerabide. Lateral mobility in membranes as detected by fluorescence recovery after photobleaching. *Biophys. J.* 40:69–75, 1982.

168. Yoshida, H., M. Yusin, I. Ren, J. Kuhlenkamp, T. Hirano, A. Stolz, and N. Kaplowitz. Identification, purification, and immunochemical characterization of a tocopherol-binding protein in rat liver cytosol. *J. Lipid Res.* 33:343–350, 1992.

169. Zilversmit, D.B. Lipid transfer proteins. *J. Lipid Res.* 25:1563–1569, 1984.

170. Zimniak, P. and Y.C. Awasthi. ATP-dependent transport systems for organic anions. *Hepatology* 17:330–339, 1993.

5
Metabolism in the Lung

17

The Uptake and Metabolism of Substrates by Endothelium in the Lung

John H. Linehan, Said H. Audi, and Christopher A. Dawson

Introduction

The work of Vane (16) and others (8, 9, 15) showed that a number of vasoactive substances carried in venous blood are removed from the blood and/or chemically modified by the pulmonary endothelial cells during passage through the pulmonary circulation. This stimulated interest in developing the means for evaluating these "nonrespiratory functions" of the lung. A further motivation for attempting to understand the metabolic functions of the pulmonary endothelium is the concept that, if these metabolic functions of the endothelium could be measured in vivo, they might provide information about the physiological or pathophysiological status of the endothelial cells.

Nondestructive in vivo evaluation of the metabolic functions of cells within an organ—"in vivo cell biology"—can be accomplished by: 1) measuring the metabolic signatures embedded in the blood, or other accessible body fluid, in the form of concentrations of substances either produced or removed by cells of the intact organ; 2) indicator dilution using, as at least one indicator, a substrate for some metabolic process carried out by the cells of the organ; or by 3) functional imaging using substrates or ligands that are sequestered by some process of interest (e.g., as in nuclear medicine) or using a detectable signature (e.g., as in fMRI). In terms of the variety of substrates (both naturally occurring and synthetic) that can be used, the kinds of detection systems (taking advantage of a wide range of substrate physical-chemical properties for the on-line and off-line blood analysis and external imaging modalities) that can be used, and the information content of the data, the indicator dilution methods are the most versatile methods for studying the in vivo function of the cells of the internal organs of the body without damaging the cells.

Indicator dilution methods can take different forms depending, at least in part, on the time frame of observation necessary to detect the influence or existence of the metabolic process. The sudden injection, multiple-indicator dilution (MID) method with inflow and/or outflow detection is particularly applicable to the study of in vivo cell functions having reaction rates whose characteristic time constants are of the order of the transit time for the blood flowing through the organ. The MID method is based on explaining, both qualitatively and quantitatively, the separation in time and concentration that occurs between multiple indicators as they pass through the organ. The information content of the data can be complex because there are many factors that can influence the amount of a substance that is removed and/or modified on passage through the organ in addition to the uptake and/or metabolism processes occurring on or within the cells themselves. These include organ perfusion (blood flow, perfused surface area, capillary mean transit time, and the distribution of these transit times) and reactions taking place in the blood (plasma-protein binding, metabolism by blood-borne enzymes, uptake and/ or binding to formed elements, and substrate conformational rearrangements). With the appropriate experimental design and/or indicator probes, the MID data can contain information about all of these processes. The MID method is particularly appealing for studying pulmonary endothelial cell functions in vivo because of the relative easy access available for the indicator injection and sampling. However, the pulmonary endothelium is normally provided with a rather high blood flow with respect to the metabolic requirements of the endothelial cells themselves. One would not expect to find substantial differences between a reference indicator and tracer concentration–time curves for the typical substrates and products for intermediary metabolism of the cells. Instead, the processes that can be detected and quantified tend to be those wherein the pulmonary endothelium is involved in processing the systemic venous blood before it passes to the arterial system. The indicators used most extensively for studying the metabolic functions of the pulmonary endothelium are substrates for carrier-mediated uptake processes or substrates or ligands for endothelial surface enzymes. Generally, their uptake/metabolism process results in significant differences in the reference and tracer concentration–time curves. These differences provide the data set useful for quantitative elucidation of the processes controlling the fate of the substrate. The finite number of uptake/metabolism sites on the endothelium is an extensive property of the process, and the affinity of the sites for the substrate is an intensive property of the process, for each of which the possibility exists for quantification. We have developed what has become to be referred to as the "bolus-sweep" method to this end. In this method, sufficient substrate is included in the bolus such that, during the passage of the bolus through the pulmonary capillaries, the time-varying substrate concentration within the capillaries first increases and then decreases, sweeping through a range of concentrations such that saturation kinetics applies to the uptake/metabolism of the substrate. In this way, the data contain sufficient information to enable the estimation, via a kinetic model, of both the intensive and extensive properties of the uptake/metabolism process.

This chapter will discuss the implementation of the bolus-sweep multiple-indicator dilution technique wherein a bolus containing the indicators is rapidly injected into the pulmonary artery. The bolus is assumed to contain a reference indicator as well as an indicator for the substrate of interest. The reference indicator in this case is assumed to be convected with and confined to the blood as the blood flows through the lung. Thereby it traces the behavior of the substrate within the vascular space without being altered itself. As a result of the injection process and dispersion upstream from the lungs (e.g., within systemic veins or heart), the reference indicator concentration–time curve arriving at the capillary inlet is dispersed. That is, at the capillary inlet, the concentrations rise fairly rapidly to a peak and then more slowly decrease. At the capillary inlet, both the reference and substrate indicator concentration–time curves are congruent. As the indicators pass through the capillary bed, they undergo further dispersion. The dispersion processes are, undoubtedly, due to flow and path length variability, although plug flow is assumed within a given capillary segment. The net result is that the reference indicator curve in the venules is dispersed more than at the capillary inlet. The substrate is subjected to the same dispersive processes, but, in addition, it is taken up or metabolized by the pulmonary endothelium. If the range of concentrations of the substrate at the capillary inlet are of the order of the Michaelis–Menten concentration K_m, that is, encompassing the concentration resulting in a reaction velocity equal to one-half the maximum velocity, V_{max}, then saturation kinetics with respect to the uptake/metabolism process are to be expected. Assuming that the uptake/metabolism kinetic processes at the endothelium and dispersion processes within the vessels dominate the fate of the substrate during passage through the capillaries, and that the reference indicator provides information for separating the impact of the dispersive (kinematic) processes from that of the kinetic processes, a mathematical model can be constructed that can be used to decode the data, yielding quantitative parameters that are descriptive of the kinetic processes.

The development of the organ model begins with the consideration of events within a single capillary. The quantitative basis of the development is the species (mass) balance that expresses the relationship among the various chemical-physical processes in which the substrate participates. A key feature of the analysis is the identification of the processes (one or more) that dominate the fate of the substrate. This identification ultimately leads to methods of data analysis that are the means of obtaining parameters that are potentially useful for interpreting changes in the endothelial cell function within the intact organ. In this chapter we will introduce these concepts by consideration of two different substrates, namely, benzoyl-phenylalanyl-alanyl-proline (BPAP) and serotonin [5-hydroxytryptamine (5-HT)], BPAP is a substrate for angiotensin-converting enzyme (ACE) that is expressed at the luminal surface of the endothelial cell. Serotonin is taken up by the endothelial cells via a carrier-mediated process.

Model and Theory

Single-Capillary Model

We begin with the following description of the models for 5-HT and BPAP. 1) Each capillary element includes a luminal volume (vascular region), a surrounding extravascular (cellular) volume, and a surface separating the two. The vascular reference indicator is confined to the capillary lumen. 2) Diffusion of the vascular indicator and substrate(s) in the direction of flow is negligible compared with axial convective transport; that is, indicators are transported from inflow to outflow only by convection. 3) Flow is restricted to the capillary lumen.

5-HT

Within each capillary element, it is assumed that the endothelial uptake of 5-HT occurs via a saturable transport mechanism. A simple representation of this process is the Michaelis–Menten equation (10). Within the endothelial cells, 5-HT can be sequestered and/or chemically altered [e.g., deamination to 5-hydroxy-indole acetic acid (5-HIAA)] (5). For mathematical simplicity, the 5-HT metabolism-sequestration process is assumed to be irreversible and obeys first-order kinetics. For such a capillary element, the concentrations of the reference indicator and the substrate are described by the following species balance equations (10):

$$\frac{\partial [R]}{\partial t} + W\frac{\partial [R]}{\partial x} = 0 \tag{17.1}$$

$$\frac{\partial [S]}{\partial t} + W\frac{\partial [S]}{\partial x} = -\frac{1}{Q_c}\left(\frac{V_{max}[S]}{K_m + [S]} - \frac{PS}{Q_e}F_e\right) \tag{17.2}$$

$$\frac{\partial F_e}{\partial t} = \left(\frac{V_{max}[S]}{K_m + [S]} - \frac{PS}{Q_e}F_e\right) - k_{met}F_e \tag{17.3}$$

$$F_e = [S_e]Q_e ,$$

where W is the average velocity within the vascular region Q_c; $[R](x,t)$ and $[S](x,t)$ are the vascular concentrations of the reference indicator and the substrate at distance x from the capillary inlet ($x = 0$) and time t, respectively; $[S_e](x,t)$ is the substrate concentration within the endothelial cells at distance x from the capillary inlet and time t; V_{max} represents the maximum rate of substrate uptake; K_m is the concentration at which the rate of 5-HT uptake is equal to $V_{max}/2$; PS/Q_e is the tissue parameter that controls the back-diffusion of 5-HT into the vascular region; and k_{met} is the rate of 5-HT sequestration/metabolism within the extravascular volume, Q_e.

To model bolus injections, the solution to Eqs. (17.1)–(17.3) is constrained by the initial ($t = 0$) conditions $[R](x,0) = [S](x,0) = [S_e](x,0) = 0$, and boundary ($x = 0$) conditions $[S_e](0,t) = 0$, and $[R](0,t) = [S](0,t) = C_a(t)$, where $C_a(t)$ is the capillary inlet function. The identifiable parameters in Eqs. (17.1)–(17.3) are V_{max}, K_m, PS/Q_e, and k_{met}.

BPAP

BPAP exists in two forms, which are *cis* and *trans*, with respect to the alanyl-proline bond (11). Only the *trans* form is an ACE substrate (11). The *cis–trans* conversion rate is slow enough relative to the pulmonary capillary mean transit time $\overline{t}_c$, such that, during the single pass through the capillary, the *cis* form behaves as a nonhydrolyzed "impurity" (11). Thus, in the species balance equations below, the *cis* form of BPAP is assumed to be convected through the lung capillary. The inlet concentration of the *cis* form is assumed to be its equilibrium fraction, θ, times the inlet BPAP concentration. In addition, BPAP associates with plasma albumin. The association and dissociation rates are rapid relative to $\overline{t}_c$. Thus, the fraction of BPAP associated with albumin is assumed to be constant throughout the capillary. Under these assumptions, the BPAP concentrations are described by the following species balance equations:

$$\frac{\partial [T]}{\partial t} + W\frac{\partial [T]}{\partial x} = -\frac{1}{Q_c\left(1 + \dfrac{[P]}{K_d}\right)}\left(\frac{V_{\max}[T]}{K_m + [T]}\right) \tag{17.4}$$

$$\frac{\partial [C]}{\partial t} + W\frac{\partial [C]}{\partial x} = 0 \tag{17.5}$$

$$[B](L,t) = [T](L,t) + [C](L,t),$$

where $[T](x,t)$ and $[C](x,t)$ are the concentrations of the *cis* and *trans* forms at distance x from the capillary inlet and time t, respectively; $[B](L,t)$ is the substrate concentration at the capillary outlet ($x = L$) at time t; and L is the capillary length. The saturable hydrolysis process of the *trans* form is represented by the Michaelis–Menten equation, where $V_{\max}$ is the maximum rate of hydrolysis, and K_m is the concentration at which the rate of hydrolysis is equal to $V_{\max}/2$. $[P]$ is the plasma protein concentration, and K_d is the plasma protein equilibrium dissociation constant.

To model bolus injections, the solution to Eqs. (17.4)–(17.5) is constrained by the initial ($t = 0$) conditions $[T](x,0) = [C](x,0) = 0$, and boundary ($x = 0$) conditions $[T](0,t) = (1 - θ)\ C_a(t)$, and $[C](0,t) = θ\ C_a(t)$, where $C_a(t)$ and θ are the capillary inlet function and the substrate fraction in the Cis form, respectively. In Eqs. (17.4)–(17.5), the identifiable parameters are $V_{\max}/(1 + ([P]/K_d))$, K_m, and θ.

Organ Model

To construct an organ model from Eqs. (17.1)–(17.3) or (17.4)–(17.5) for a single-capillary element, the capillary inlet concentration, $C_a(t)$, and the distribution of pulmonary capillary transit times, $h_c(t)$, are needed (1, 2, 3, 10).

As described by Audi et al. in Chapter 22 of this book and in reference (2), we have estimated that, for a normal dog lung lobe, the pulmonary capillary mean transit time, $\overline{t}_c$, was approximately 48% of the total vascular mean transit time, $\overline{t}$; the relative dispersion of $h_c(t)$, $RD_c = σ_c/\overline{t}_c$, was approximately 0.75; and the skewness coefficient of $h_c(t)$, m_c^3, was approximately 1.8, where m_c^3 and $σ_c$ are the third central moment and standard deviation of $h_c(t)$, respectively. These quantifiers of $h_c(t)$ can be exploited using a functional form such as a shifted random walk function, $RF(t)$, which is a probability density function whose functional values are determined by its first three moments (1–3).

Let $C_a(t) = (q/F)\, h_a(t)$ be a function representing the time–concentration curve of the vascular reference indicator and the substrate in the precapillary portion of the system (injection system and arterial bed), where q is the mass of the injected indicator, F is the total flow through the organ, and $h_a(t)$ is the transport function for the precapillary portion of the system. Let $h_v(t)$ be the transport function of the postcapillary portion of the system (venous bed and sampling system). Assuming random coupling conditions (1–3, 10) between the capillary and noncapillary transit time distributions, the organ reference indicator outflow curve $C_R(t)$ is

$$C_R(t) = C_a(t) * h_c(t) * h_v(t), \qquad (17.6)$$

where $*$ is the convolution operator (1–3, 10).

$h_a(t)$ and $h_v(t)$ can also be represented by a shifted random walk function. Assuming that $h_a(t) = h_v(t)$, the parameters of the shifted random walk function representing $C_a(t)$ and $h_v(t)$ can be specified using a least squares optimization procedure that minimizes the difference between $C_R(t)$ and $C_a(t) * h_c(t) * h_v(t)$ (1–3, 10).

For given initial and boundary conditions, Eqs. (17.1)–(17.3) for 5-HT or Eqs. (17.4)–(17.5) for BPAP can be solved numerically, for example, by a finite difference method (7). By virtue of the form of the governing partial differential equations, the solution for the heterogeneous capillary bed is obtained from the solution of a single-capillary element with $C_a(t)$ as the capillary input concentration curve. It should be appreciated that, by solving the model at any location along a capillary element having the maximum capillary transit time, the single-capillary solution also provides along its instantaneous length the output for all capillary transit times between the minimum and maximum capillary transit times (1, 10). This derives from the random-coupling concept that an aliquot of blood entering a capillary at $t = 0$ will pass by all longitudinal capillary elements on its course to a capillary of maximum length (1–3). To provide the whole organ output for the reference indicator, $C_R(t)$, and the substrate, $C_S(t)$, a mass balance is written at the capillary outflows in which the outputs for all transit times are each weighted according to $h_c(t)$ (1, 10).

Experimental Methods

For the examples in this chapter, the experiments were carried out using isolated dog lung lobes as previously described (11, 14). The lung lobe was perfused at a constant flow rate with blood (in the case of 5-HT) or with an artificial perfusate [Krebs–Ringer bicarbonate solution containing 5% bovine serum albumin (BSA)] in the case of BPAP. Each bolus contained a vascular reference indicator (indocynanine green dye), a trace dose of labeled substrate (either ^{3}H-BPAP or ^{14}C-5-HT), and varying amounts of the unlabeled substrate (either BPAP or 5-HT). Just prior to injection, the venous outflow was directed to a sample collector, and samples were collected at equal time intervals. The samples were analyzed for total $^3H/^{14}C$ by liquid scintillation counting methods and for dye concentration spectrophometrically. The ^3H in BPAP and its labeled metabolite ^3H-benzoyl-phenylalanine in each sample were separated by thin-layer chromatography and quantified by radiochromatogram scanning (11).

The top panel of Figure 17.1 shows an example of the concentration versus time outflow curves of $[^{14}C]$-5-HT obtained from a dog lung lobe following three bolus injections, each with a different amount of unlabeled 5-HT. The lower panel shows the resulting instantaneous extraction ratio, $E(t)$, where $E(t) = 1 - [C_S(t)/C_R(t)]$, $C_S(t)$ is the 5-HT concentration in the venous effluent, and $C_R(t)$ is the concentration of the reference indicator times the mass of the injected unlabeled serotonin divided by the mass of the injected reference indicator.

The $E(t)$ curves are shown because they reveal differences in the relative shapes of the $C_R(t)$ and $C_S(t)$ curves that are not readily apparent from the curves themselves. When the injected bolus contains a trace dose (10 nmoles) of unlabeled serotonin (small relative to K_m), $E(t)$ tends to be concave downward, rising early in time followed by a decreasing trend later in time as $C_R(t)$ falls. This extraction pattern has been interpreted as reflecting the influence of both the heterogeneous perfusion of the capillary bed on the uptake of 5-HT (rising trend early in time) and the returning flux of unmetabolized 5-HT from inside the endothelial cell to the vascular space (decreasing trend later in time) (5, 10). Thus, the rising portion of $E(t)$ is due to the average concentration of indicators in the early samples being weighted by indicators coming from capillaries with transit times that are predominantly shorter than the mean capillary transit time. Thereby, as time progresses, capillaries with longer transit (residence) times contribute to the samples, and the extraction increases. The falling portion of $E(t)$ reflects the intracellular accumulation of 5-HT, which, along with the falling 5-HT vascular concentration, provides a steep gradient for returning flux (5, 10). The returning indicator later in time results in the appearance of a decrease in the uptake of 5-HT.

As the dose of 5-HT in the injected bolus increases, the saturable nature of the uptake process is revealed in Figure 17.1 by the decrease in the extraction (area difference between $C_R(t)$ and $C_S(t)$) and an apparently upward concavity of the extraction ratio at early times. This upward concavity in $E(t)$ has been interpreted as follows (5, 10). The dispersed $C_a(t)$ results in time-varying substrate concentrations at the capillary inlet. As the capillary concentration of the substrate rises on the ascending part of the $C_a(t)$ curve, becoming of the order of K_m, the instantaneous values of $E(t)$ do not rise in proportion to the increasing capillary transit time because of the saturation effect, whereas, on the descending portion of $C_a(t)$, the capillary concentration of the substrate falls relative to K_m in addition to the increasing transit times. The concavity of $E(t)$ depends on the amount of substrate injected, on the heterogeneity of capillary perfusion, $h_c(t)$, and on the dispersion of the capillary input function, $C_a(t)$, relative to $h_c(t)$ (3, 5, 10). Again, the falling portion of $E(t)$ at longer times reflects the impact of returning flux of unaltered substrate.

The model fits in Figure 17.1 were obtained by simultaneously fitting Eqs. (17.1)–(17.3) to the substrate outflow curves from the three bolus injections with

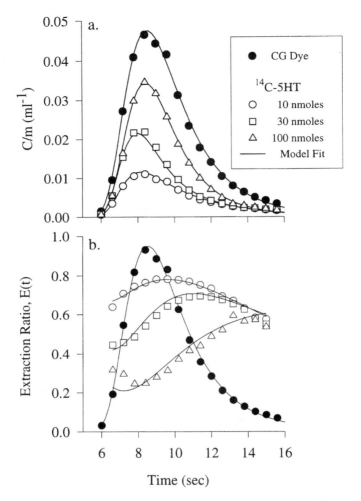

FIGURE 17.1. (a) Example of concentration vs. time data obtained with three tracer bolus injections each with a different amount of unlabeled 5-HT. Ordinate is normalized concentration, where C is either concentration of substrate, 5-HT, or vascular reference indicator, indocynanine green dye (CG), and m is the injected dose of respective indicators. Only one reference indicator curve is shown because all three reference indicator curves are virtually superimposible. (b) Instantaneous extraction ratio, $E(t)$, data for concentration curves in (a). The CG dye concentration curve is plotted without a scale for timing perspective. Solid lines superimposed on the data represent the simultaneous model fit to all three substrate outflow curves with the same set of model parameters. The estimated parameter values were V_{max} = 35.4 nmol/s, K_m = 1.22 nmol/ml, k_{met} = 0.33 s^{-1}, and PS/Q_e = 0.53 s^{-1}. Coefficient of variations between data and model fit was 7.0%.

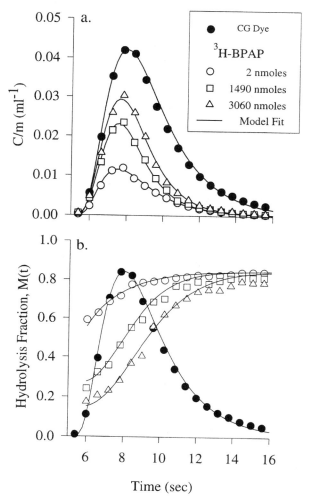

FIGURE 17.2. (a) Example of concentration vs. time data obtained with three tracer bolus injections each with a different amount of unlabeled BPAP. Ordinate is normalized concentration, where C is either concentration of substrate, BPAP, or vascular reference indicator, indocynanine green dye (CG), and m is the injected dose of respective indicators. Only one reference indicator curve is shown because all three reference indicator curves are virtually superimposible. (b) Instantaneous hydrolysis, $M(t)$, data for concentration curves in (a). The CG dye concentration curve is plotted without a scale for timing perspective. Solid lines superimposed on the data represent the simultaneous model fit to all three substrate outflow curves with the same set of model parameters. The estimated model parameters were $V_{max}/(1 + ([P]/K_d)) = 2466$ nmol/s, $K_m = 47.8$ nmol/ml, and $\theta = 0.154$. Coefficient of variations between data and model fit was 7.5%.

one set of parameter values. Parameter optimization was carried out using a modified Levenberg–Marquardt algorithm (12). Theoretically, if the actual capillary input function, $C_a(t)$, and the capillary transport function, $h_c(t)$, were known, the injection of a single bolus with a sufficient amount of substrate would provide the range of concentrations needed to calculate V_{max} and K_m. However, as a practical matter, the use of multiple bolus injections, each with a different amount of unlabeled substrate, helps to avoid sensitivity to inaccuracies in $C_a(t)$ and $h_c(t)$ estimates and to reduce correlations between model parameters, thus increasing the robustness of the parameter estimation procedure.

An example for ^{3}H-BPAP is shown in Figure 17.2(a). The instantaneous hydrolysis fraction, $M(t)$, where $M(t) = 1 - [C_S(t)/C_R(t)]$, $C_S(t)$ is the BPAP concentration in the venous effluent, and $C_R(t)$ is the concentration of the reference indicator times the mass of injected unlabeled BPAP divided by the mass of injected reference indicator, is shown in Figure 17.2(b). The explanation for the shapes of the $M(t)$ curves for BPAP and the $E(t)$ curves for 5-HT are similar, except that the BPAP hydrolysis is irreversible (11). Therefore, there is no returning flux, that is, no decrease in $M(t)$ at longer times. $M(t)$ plateaus at a value that is less than 1.0 (~ 0.85), revealing the *cis* fraction, which is not hydrolyzed on the time course of a single pass through the lungs. When the injected dose is high enough, the portions of the curves with higher substrate concentrations (relative to K_m) reveal the saturation of the hydrolysis process as with 5-HT.

The ability to measure these kinetic parameters for these and other substrates within the intact organ represents a tool for studying cells within their organ environment and in vivo (4–6, 8, 13, 14). One advantage of the estimation of both V_{max} and K_m is that, as an extensive property, V_{max} has the potential for reflecting the quantity of endothelium that is perfused. This can be an important variable in studies wherein organ perfusion is altered by experimental or pathophysiological conditions. On the other hand, K_m, as an intensive property, is more likely to reflect chemical changes in the transport/metabolic process, as opposed to simply the number of sites available to plasma passing through the lungs.

Acknowledgment. This study was supported by the Department of Veterans Affairs and the National Heart, Lung, and Blood Institute Grant HL-24349.

References

1. Audi, S.H., C.A. Dawson, J.H. Linehan, G.S. Krenz, S.B. Ahlf, and D.L. Roerig. An interpretation of ^{14}C-urea and ^{14}C-primidone extraction in isolated rabbit lungs. *Ann. Biomed. Engrg.* 24:337–351, 1996.
2. Audi, S.H., G.S. Krenz, J.H. Linehan, D.A. Rickaby, and C.A. Dawson. Pulmonary capillary transport function from flow-limited indicators. *J. Appl. Physiol.* 77:332–351, 1994.
3. Bronikowski, T.A., C.A. Dawson, and J.H. Linehan. On indicator dilution heterogeneity: A stochastic model. *Math. Biosci.* 83:199–225, 1987.

4. Dawson, C.A., C.W. Christensen, D.A. Rickaby, J.H. Linehan, and M.R. Johnson. Lung damage and pulmonary uptake of serotonin in intact dogs. *J. Appl. Physiol.* 58:1761–1766, 1985.

5. Dawson, C.A., J.H. Linehan, D.A. Rickaby, and T.A. Bronikowski. Kinetics of serotonin uptake in the intact lung. *Ann. Biomed. Engrg.* 15:217–227, 1987.

6. Dawson, C.A., D.L. Roerig, and J.H. Linehan. In: *Clinics in Chest Medicine,* edited by S.G. Jenkinson. Philadelphia: Saunders, Vol. 10, 1989.

7. Finlayson, B.A. *Numerical Methods For Problems with Moving Fronts.* Seattle, WA: Ravenna Park Publishing, 1992.

8. Gillis, C.N. and B.R. Pitt. The fate of circulatory amines within the pulmonary circulation. *Ann. Rev. Physiol.* 44:269–281, 1982.

9. Janod, A.F. Metabolism, production and release of hormones and mediators in the lung. *Am. Rev. Resp. Dis.* 112:93–108, 1975.

10. Linehan, J.H., T.A. Bronikowski, and C.A. Dawson. Kinetics of uptake and metabolism by the endothelial cell from indicator dilution data. *Ann. Biomed. Engrg.* 15:210–215, 1987.

11. Linehan, J.H., T.A. Bronikowski, D.A. Rickaby, and C.A. Dawson. Hydrolysis of a synthetic angiotensin-converting enzyme substrate in dog lungs. *Am. J. Physiol.* 257:H2006–H2016, 1989.

12. Marquardt, D. An algorithm for least-squares estimation of nonlinear parameters. *SIAM J. Appl. Math.* 11:431–441, 1963.

13. Peeters, F.A., T.A. Bronikowski, C.A. Dawson, J.H. Linehan, H. Bult, and A.G. Herman. Kinetics of serotonin uptake in isolated rabbit lungs. *J. Appl. Physiol.* 66:2328–2337, 1989.

14. Rickaby, D.A., J.H. Linehan, T.A. Bronikowski, and C.A. Dawson. Kinetics of serotonin uptake in the dog lung. *J. Appl. Physiol.* 51:405–414, 1981.

15. Said, S.I. Metabolic functions of the pulmonary circulation. *Circ. Res.* 50:325–333, 1982.

16. Vane, J.R. The release and fate of vasoactive hormones in the circulation. *Br. J. Pharmacol.* 35:209–242, 1969.

18

Pulmonary Endothelial Surface Reductase Kinetics

Christopher A. Dawson, Robert D. Bongard, David L. Roerig,
Marilyn P. Merker, Yoshiyuki Okamoto, Said H. Audi, Lars E. Olson,
Gary S. Krenz, and John H. Linehan

The role of the endothelium as a metabolically active organ having a number of regulatory functions is well established. The in vivo evaluation of these functions tends to be a difficult problem, and much of the research in this area is being carried out using simpler systems such as cultured endothelial cells. The results from such studies provide increased motivation for understanding how the various endothelial functions operate in vivo. The multiple-indicator dilution method (MID) is an approach for studying in vivo endothelial cell biology. The lungs are unique with regard to in vivo application of the MID for the study of capillary permeation, cellular transport, and reaction kinetics in that access to a single inlet (e.g., a systemic vein or the pulmonary artery) and single outlet (e.g., a peripheral systemic artery) is more readily available than in any other organ, and the MID has been applied to the in vivo study of these functions of the pulmonary endothelium (13–16, 18, 21, 22). The MID method is suited for studying those processes that occur rapidly enough that their effects can be observed in the time frame of a single pass through the lungs. In general, capillary blood flow tends to be so high in comparison to the rates of endothelial cell utilization of typical substrates for intermediary metabolism that the MID is not applicable to such substrates. However, the pulmonary endothelium also carries out a number of metabolic functions that appear to be directed at modulating blood concentrations of certain substances rather than at the metabolic requirements of the endothelial cells themselves. These include the control of levels of certain vasoactive hormones reaching the arterial circulation after release from various organs into the venous blood (14, 16, 17). To process the blood, these metabolic functions must operate at very high rates, which also makes them amenable to the MID method. There has been considerable interest in the application of the MID method using these hormones as potential probes for the in vivo evaluation of pulmonary endothelial cell biology. Recently, we became aware of another type of rapid metabolic function carried out by the pulmonary endothelium, namely, the reduction of certain electron acceptors as they pass through the lungs (5, 6, 25).

The plasma membranes of various cell types have been found to transport electrons from internal donors to external acceptors via transplasma membrane oxidoreductases (8–10, 26, 27, 31–34). In macrophages and neutrophils, such a phenomenon has evolved to provide for the transfer of electrons from intracellular NADPH to molecular oxygen to produce superoxide on the external surface (12, 29, 30), and the physiological significance of this transfer is well established (4). However, the function of transmembrane electron transfer in other mammalian cell types that have been studied, including erythrocytes, hepatocytes, and some tumor cell lines (8, 9), is less well understood. In hepatocytes, cytosolic NADH, NADPH, glutathione, xanthine, and ascorbate have been found to be electron donors for the electron transfer (8, 9). Monodehydroascorbate, oxygen, ferric compounds, and lipid hydroperoxides have been identified as naturally occurring substrates for various cell types (9, 27), but whether these are the physiologically important electron acceptors is not clear. These transmembrane electron transport systems have been implicated in the regulation of growth and development, antioxidant defense, protection of receptor protein sulfhydryl groups on the plasma membrane surface, and membrane transport of various metabolites, iron, and ions (9, 19, 27). Stimulation and inhibition of the electron transport by various hormones, such as glucagon and insulin, and by α-adrenergic receptor agonists have been observed (9, 11), suggesting that the cell surface reductase activity is regulated.

While several other cell types have been studied (9, 10), there is little information regarding possible plasma membrane electron transport in vascular endothelial cells (6, 34). However, in a study in which the thiazine dye methylene blue (MB) was added to the blood recirculating through an isolated lung, we noticed that after the arterial and venous cannulas had been washed the venous cannula was blue, while the arterial cannula, made of the same plastic material, remained colorless (5). Since both cannulas had been exposed to approximately the same dye concentration, this suggested that passage through the lung had some transient effect on the dye that increased its affinity for the plastic. Methylene blue is a redox dye that is blue in the oxidized form and colorless in the reduced form. In addition, the blue oxidized form, MB^+, is hydrophilic, while the reduced leuko form, MBH, is hydrophobic. We guessed that the MB^+ might have been reduced on passage through the lungs, and then the less water soluble MBH form precipitated on the cannula. Since the MBH form autooxidizes in the presence of oxygen, we surmised that the blue color on the cannula may have resulted from the autooxidation of the precipitated MBH.

To provide a more quantitative approach for examining the sequence of events, we perfused a group of rabbit lungs with a recirculating artificial salt solution containing methylene blue and 5% bovine serum albumin. The presence of the albumin turned out to be important, as will be discussed below. We then measured the optical absorbance at the absorption peak for the MB^+ (665 nm) in samples obtained from the arterial inlet and the venous outlet. The samples were transferred to the spectrophotometer as rapidly as possible, and the absorbance was followed for several minutes. The results from one sample pair are shown in Figure 18.1. Initially, the absorbance was lower in the venous sample than in the arterial sample. The absorbance increased with time in both samples, but it increased much more in the venous sample, so that when the absorbances finally

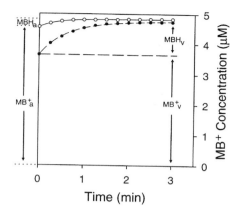

FIGURE 18.1. Autooxidation of methylene blue in arterial (open circles) and venous (closed circles) samples of perfusate. The ordinate is the concentration of the MB⁺ form of the dye in the cuvette. The abscissa is the time following removal from the perfusion system. The subscripts *a* and *v* refer to the arterial and venous concentrations, respectively. The final *a–v* difference in MB⁺ concentration, after full oxidation, represents the fraction sequestered by the lungs. [Reprinted from reference (5) with the permission of the American Physiological Society.]

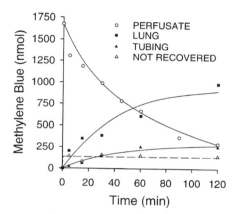

FIGURE 18.2. Disposition of methylene blue during recirculation of the dye through the lungs. The ordinate is the nmole of dye found in each phase: the perfusate, the lungs, and the tubing. There was also a small fraction that could not be unaccounted for ("not recovered"). The abscissa is the time following the beginning of the MB⁺ recirculation. [Reprinted from reference (5) with the permission of the American Physiological Society.]

reached constant values there was only a small difference between the arterial and venous samples. The explanation for this observation is that a fraction of the MB^+ was reduced as it passed through the lungs. In the spectrophotometer cuvette, the MBH autooxidized back to MB^+, accounting for the increase in absorbance with time. The MBH also autooxidized within the recirculating perfusate, so that by the time it returned to the arterial inlet there was only a small amount of the MBH left. The difference between the final arterial and venous concentrations of MB^+ reflects the uptake of a small fraction of the dye into the lungs. Thus, as the recirculation of the dye continued, the dye concentration in the perfusate decreased. The time course for this decrease can be seen in Figure 18.2. Figure 18.2 also shows that most of the dye lost from the perfusate could be found by extracting the dye from the lungs. There was also a fraction that precipitated on the tubing, and a smaller fraction that could not be accounted for. Thus, the results of these experiments were consistent with the hypothesized reduction of the dye during passage through the lungs.

Our ability to detect MBH within the perfusate, and thus to detect that the dye had been reduced within the lung, was the result of the fact that the perfusate contained albumin. When we repeated the experiments using a protein-free perfusate, the uptake of the dye into the lungs was much faster and no MBH could be detected in the venous effluent. This difference is due to the fact that the lipophilic MBH has a high affinity for albumin as well as for the tissue. Thus, in the absence of albumin in the perfusate, the MBH produced in the lungs stayed in the lungs rather than associating with the perfusate albumin for transport to our sampling site.

We also carried out experiments in which we injected a bolus of either MB^+ or MBH into the perfusate flowing into the rabbit lung and took rapid samples of the venous effluent. We observed that the fraction taken up into the lungs was much larger for MBH than for MB^+ (Figure 18.3). This was consistent with the much higher lipophilicity of the MBH, and it led us to consider the possibility that, in the recirculation experiments, it may have been the MBH form of the dye that was actually taken up by the lungs. This implies that, rather than the MB^+ being taken up, reduced, and then released back into the perfusate as MBH, the MB^+ was reduced outside the cells in the perfusate or on the endothelial surface. This hypothesis further implies two possibilities, one that a short-lived reducing agent was released into the perfusate, and the other that reduction took place on the endothelial surface. We considered superoxide as a possible candidate for a short-lived reducing agent released into the perfusate, but we found that the effect of superoxide dismutase was to speed the autooxidation rather than to inhibit reduction. Furthermore, the effluent perfusate did not reduce the dye.

To further evaluate the possibility that reduction occurred on the endothelial surface, we studied cultured bovine pulmonary arterial endothelial cells. With the cells grown on microcarrier beds, it was possible to place a fairly large endothelial surface area (about 70 cm²) in a spectrophotometer cuvette in such a way that, when the cell-coated beads settled to the bottom of the cuvette, they were below the light path. Thus, the beds were mixed in the cuvette filled with methylene blue–containing medium, and at intervals the beds were allowed to settle to the bottom so the dye concentration of the medium could be measured. The cultured cells took up the dye (Figure 18.4), and, when albumin was present in the medium, the autooxidation curve, similar to that for the venous samples obtained from the lungs, could be observed (Figure 18.5). Thus, the endothelial cells were apparently capable of carrying out the same process that occurred within the lungs.

To examine the hypothesis that the reduction occurred on the endothelial surface, we took advantage of another thiazine dye, toluidine blue O (TBO) (6). The blue oxidized form, TBO^+, is reduced at an even faster rate by the endothelial cells than is MB^+ (Figure 18.4), but, more importantly for this purpose, its primary amine side group facilitates covalent bonding of the TBO to other molecules. Thus, we synthesized a toluidine blue O-polyacrylamide polymer (TBOP) that was too large (36 kD) to enter the cells in either the oxidized or reduced form.

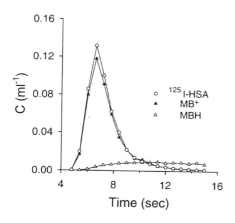

FIGURE 18.3. The concentrations (normalized to the injected amount) of ^{125}I-HSA and methylene blue obtained following the injection of a bolus containing either MB$^+$ or MBH with ^{125}I-HSA as the reference indicator. Results obtained from one lung perfused with a perfusate containing 5% bovine serum albumin. [Reprinted from reference (5) with the permission of the American Physiological Society.]

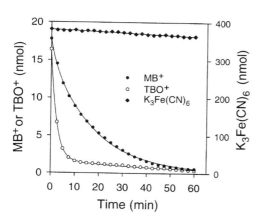

FIGURE 18.4. Examples of the time courses observed when MB$^+$, TBO$^+$, or ferricyanide was added to the medium bathing the endothelial cell–covered beads. The endothelial cell surface areas were 70.9, 78.6, and 73.5 cm^2 for the MB$^+$, TBO$^+$, and ferricyanide experiments, respectively. [Reprinted from reference (6) with the permission of the American Physiological Society.]

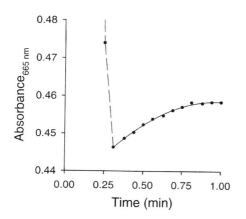

FIGURE 18.5. The autooxidation curve for methylene blue in the medium of the cultured endothelial cells. The initial fall in absorbance (dashed line) represents the decrease in light scattering with time that occurred as the beads fell through the medium, thus clearing the light path. Then the absorbance due to the MB$^+$ increased with time as the MBH produced by the cells was oxidized back to MB$^+$. [Reprinted from reference (5) with the permission of the American Physiological Society.]

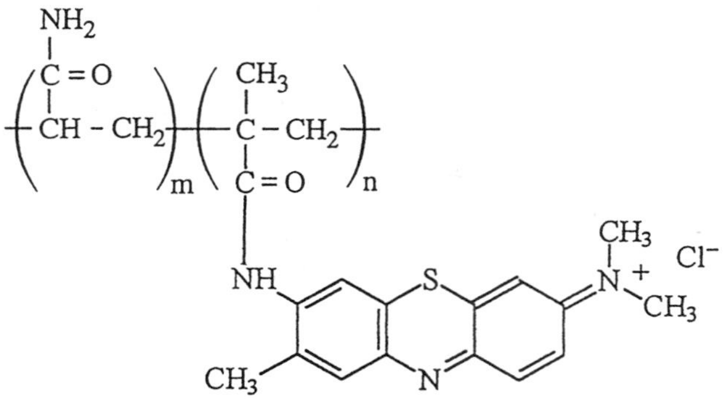

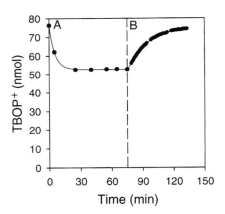

FIGURE 18.6. The structure of the toluidine blue O-polyacrylamide polymer (TBOP⁺).

FIGURE 18.7. The time course observed when TBOP⁺ (equivalent nmoles of TBO⁺) was added to the medium bathing the endothelial cell–covered beads. The TBOP⁺ was measured by absorbance difference between 590 and 750 nm. During time period A, the medium was bathing the cells between absorbance readings. During time period B, beginning at the dashed line, the medium was removed from the cells and allowed to autooxidize. [Reprinted from reference (6) with the permission of the American Physiological Society.]

Figure 18.6 shows the structure of the TBOP⁺, and Figure 18.7 shows the results obtained when the TBOP⁺ was added to the medium surrounding the cell-covered beads. There was net reduction of the polymer until an equilibrium was attained between reduction by the cells and autooxidation by the oxygen in the medium. When the medium was removed from the cells, autooxidation back to the original TBOP⁺ concentration confirmed that the polymer had not entered the cells.

To gain insight into the possibility that the smaller dye monomers (about 0.3 kD) might also be reduced on the endothelial surface, we took advantage of the fact that MB⁺ and TBOP⁺ have sufficiently different absorption spectra that they can be measured simultaneously. Thus, we examined the ability of the TBOP to inhibit the reduction of the MB⁺ uptake by the cultured cells. When both the polymer and MB⁺ were added to the medium surrounding the cells, the polymer inhibited the uptake of methylene blue, either by competition for the reducing process or by oxidizing any MBH formed (Figure 18.8). This inhibition by the

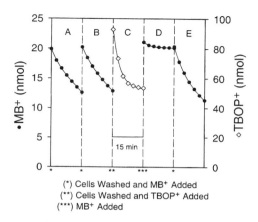

(*) Cells Washed and MB⁺ Added
(**) Cells Washed and TBOP⁺ Added
(***) MB⁺ Added

FIGURE 18.8. The effect of endothelial cells on amount of MB⁺ and
TBOP⁺ in the medium surrounding
the cells. Five consecutive medium
changes are shown. The decreases
with time represent reduction of MB⁺,
uptake of MBH, and reduction of
TBOP⁺. In panels A and B, the
decrease in MB⁺ can be seen during
two consecutive periods, during
which only MB⁺ was added to the medium. In panel C, the decrease in
TBOP⁺ can be seen when only
TBOP⁺ was added to the medium. In
panel D, MB⁺ was added in the pres-
ence of TBOP, and the uptake and reduction of MB were almost completely blocked. In
panel E, only MB was present, showing that the effect of TBOP was reversible.

external TBOP is consistent with the concept that the MB⁺ reduction also occurred on the surface of the cells.

Ferricyanide is another electron acceptor that has been used extensively for
studying transplasma membrane electron transport in other cell types (8, 9). It also
has the advantage that it does not permeate cell membranes in either the oxidized
ferricyanide or reduced ferrocyanide form. However, it is reduced much more
slowly by the endothelial cells than the thiazine dyes (Figure 18.4). On the other
hand, ferricyanide rapidly oxidizes MBH. When both ferricyanide and methylene
blue were added to the medium, the ferricyanide inhibited the methylene blue
uptake by the cells until the ferricyanide was depleted (Figure 18.9). This was
interpreted as indicating that the ferricyanide reduction was primarily the result of
its oxidation of MBH formed at the cell surface, and that once the ferricyanide had
been depleted the MBH entered the cells instead.

Another feature of the overall fate of these redox dyes is that the lungs turn blue
as they accumulate the dye. The question arises as to how the lungs can be blue if
the dye had been taken up in the reduced leuko form. The cell interior tends to be a
reducing environment, and the dye is quickly reduced by lung homogenate. To
examine this question, we added dye to cultured endothelial cells grown on flat
coverslips so that we could observe the intracellular distribution of the dye by
microscopic observation (25). Accumulation in the cells was clearly associated
with the blue staining of a group of organelles having the superficial appearance
and distribution of lysosomes or peroxisomes (25). Thus, it appears that the dye is
sequestered by organelles having a relatively high oxidative potential. Sequestration apparently occurs because the highly cationic, oxidized form cannot readily
permeate the organelle membrane, and thus it is trapped inside.

These observations have led us to a tentative list (Table 18.1) of reactions taking
place within the capillary, on the endothelial membrane, and within the endothelial cells as a means of explaining the fate of the dye within the lungs.

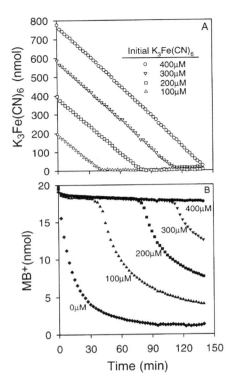

FIGURE 18.9. Reduction of ferricyanide (A) and uptake of methylene blue (B) by endothelial cells when both ferricyanide and methylene blue were present in the medium. In panel A, the ferricyanide concentrations are the initial concentrations. In panel B, the concentration designated beside each curve refers to the same initial ferricyanide concentration as in panel A. The initial methylene blue concentration was 10.2 mM in each case. (Reprinted from reference (6) with the permission of the American Physiological Society)

Even assuming that this list of reactions includes those playing a dominant role in the disposition of the dyes, many questions can be asked. For example, can the reduction process be attributed to a specific reductase associated with the plasma membrane? What is the intracellular source(s) of electrons? What are the natural electron acceptors? Are any of the steps regulated in response to physiological or pathophysiological stimuli? The later question implies that in vivo measurement of the dye disposition might reveal regulatory or pathophysiological changes in the processing of the dyes. In vivo measurement would require the tools for study of in vivo cell biology. Therefore, we have begun to evaluate the potential use of MID with thiazine dyes as in vivo probes for these processes. Since there are a number of variables that are likely to impact on the overall fate of the dye as it passes through the lungs, we have begun with studies of isolated perfused lungs wherein we have control over some of these variables, including flow rate, vascular pressures, and perfusate composition.

The in vivo disposition of the dyes is likely to be complicated by the binding to plasma proteins, uptake by red blood cells, variations in perfusion, and so on. To simplify the system for the initial evaluation of the information obtainable from the MID approach, we perfused isolated rabbit lungs with a protein-free solution.

TABLE 18.1. Reactions involving thiazine dyes.

Capillary	Endothelial membrane	Intracellular
$BH + 2O_2 \underset{k_{-3}}{\overset{k_3}{\rightleftharpoons}} B^+ + H^+ + 2O_2^-$	$B^+ + E \underset{k_{-1}}{\overset{k_1}{\rightleftharpoons}} B^+E$	$BH + 2O_2 \underset{k_{-3}}{\overset{k_3}{\rightleftharpoons}} B^+ + H^+ + 2O_2^-$
$2O_2^- + 2H^+ \underset{k_{-4}}{\overset{k_4}{\rightleftharpoons}} H_2O_2 + O_2$	$B^+E + NH \underset{k_{-2}}{\overset{k_2}{\rightleftharpoons}} BH + N^+ + E$	$2O_2^- + 2H^+ \underset{k_{-4}}{\overset{k_4}{\rightleftharpoons}} H_2O_2 + O_2$
$BH + H^+ + H_2O_2 \underset{k_{-5}}{\overset{k_5}{\rightleftharpoons}} B^+ + 2H_2O$		$BH + H^+ + H_2O \underset{k_{-5}}{\overset{k_5}{\rightleftharpoons}} B^+ + 2H_2O$
$2H_2O_2 \underset{k_{-6}}{\overset{k_6}{\rightleftharpoons}} 2H_2O + O_2$		$2H_2O_2 \underset{k_{-6}}{\overset{k_6}{\rightleftharpoons}} 2H_2O + y\,O_2$
$BH + P \underset{k_{-7}}{\overset{k_7}{\rightleftharpoons}} BHP$		$BH + A \underset{k_{-9}}{\overset{k_9}{\rightleftharpoons}} BHA$
$B^+ + P \underset{k_{-8}}{\overset{k_8}{\rightleftharpoons}} B^+P$		

BH denotes the reduced form and B⁺ the oxidized form of the thiazine dye under study; P is plasma protein; E is the luminal surface thiazine reductase; NH and N⁺ represent reduced and oxidized forms, respectively, of the electron donor (N is classified under the endothelial membrane heading because of its association with E, but, by analogy with similar surface reductases in other cells, it is actually the intracellular electron donor); and A represents intracellular sites of association with BH.

Under these conditions, when the dye is reduced it is immediately taken up into the lung, so there is virtually no MBH in the venous effluent, and the difference between the concentrations of the reference indicator and the MB⁺ in the venous effluent samples reflects both the amount reduced and the amount taken up by the lung. In this initial stage in the development of an interpretation of the MID data, it was necessary to determine whether the rate of MB⁺ reduction or the rate of MB⁺ supply was the rate-limiting process in this simple system. We did this by determining how the disappearance of the dye was influenced by the amount of time the dye was in contact with the endothelium by varying the flow rate, $\dot{Q}$. Capillary volume, Q_c, was maintained approximately constant by adjusting the venous pressure. The results of one such study are shown in Figure 18.10. The following species balance was assumed to describe the dominant aspects of the disposition of the reference indicator and the dye within a lung capillary under these conditions:

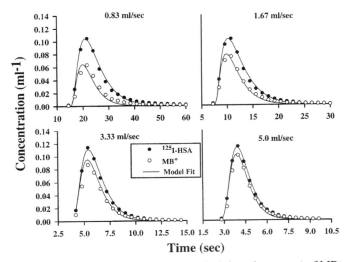

FIGURE 18.10. The concentrations (normalized to the injected amounts) of MB+ and reference indicator (^{125}I-HSA) versus time in the venous effluent of an isolated rabbit lung at four flow rates and the model equations (18.1a,b) fit to the data. $k_1E = 0.86$ ml/s. The perfusate was protein-free, and no MBH could be detected in the venous effluent.

$$\frac{\partial[R]}{\partial t} + w\frac{\partial[R]}{\partial x} = 0; \qquad \frac{\partial[B^+]}{\partial t} + w\frac{\partial[B^+]}{\partial x} = -k_1[E][B^+] \qquad (18.1a,b)$$

initial conditions ($t = 0$): $[B^+] = [R] = 0$; boundary conditions ($x = 0$): $[B^+] = [R] = h_n(t)$; $[B^+E] = 0$, where w is the flow velocity, R is the reference indicator (radiolabeled albumin in this case), $h_n(t)$ is the normalized input concentration function, and $Q_c = \int_0^\infty t h_c(t)\, dt$. Under these experimental conditions, the hypothesis was that the process represented by k_1E (i.e., dye association with the reductase, $E = [E]Q_c$) is the rate-limiting process so that it was the sole determinant of the disappearance of the dye from the perfusate.

To implement this model for the data in Figure 18.10, $h_c(t)$ was a random walk function with a relative dispersion ($\sqrt{\sigma_c^2}/\bar{t}_c$) of 0.9 and a mean transit time ($\bar{t}_c$) of 0.48 times the reference indicator mean transit time, where σ_c^2 is the variance of $h_c(t)$. These values were established for the rabbit lung in this preparation in reference (3). The $h_n(t)$ was also a random walk function with mean and relative dispersion that, when convolved with $h_c(t)$, gave the best fit to the reference indicator curve. Given the respective transit time functions, the model was fit to the MB+ and reference indicator outflow concentration curves from all four boluses simultaneously. The fitting procedure involved the numerical solution of the model at each iteration of a modified Levenberg–Marquardt algorithm. The solution is for a single capillary with $h_n(t)$ used as the input concentration curve. Since the dye concentration was well within the linear range with respect to the processes involved, neither the actual magnitude of the input nor the arrangement of the total noncapillary bolus dispersion upstream and downstream of the capillaries needed to be specifically addressed. The single-capillary model was solved to obtain the output for all capillary transit times between the minimum and maximum capillary transit times, and then the output for each transit time was weighted according to $h_c(t)$ to provide the whole organ output to be compared with the measured outflow concentrations during the optimization (2).

The model fits are also shown in Figure 18.10. The model fit the data from the four flows with only the one free parameter, $k_1E = 0.86$ ml/s, reasonably well, such that there are no systematic flow-dependent deviations between the model and the data. Conceptually, k_1E is the parameter that reflects the rate of TBO reduction and is independent of time spent in the capillaries per se. It is analogous to the permeability surface area product in the Crone model (11). Thus, the results appear consistent with the concept that the reduction process rather than the rate of dye delivery was the rate-limiting process within the whole organ.

For the study of a more complicated system, for example, with plasma albumin added to the perfusate, the association of the methylene blue with the albumin results in a very small fractional reduction and uptake (Figure 18.1). Thus, methylene blue does not provide much of a signal for kinetic analysis. We addressed this problem by using toluidine blue O, which is more rapidly reduced (Figure 18.4). Also, when albumin is present in the perfusate, the venous effluent contains both oxidized and reduced dye. Thus, the autooxidation of the reduced dye in

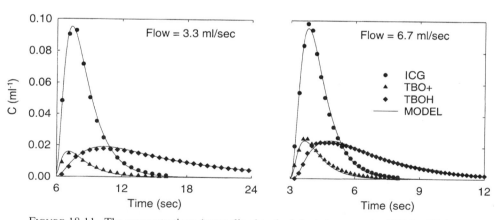

FIGURE 18.11. The concentrations (normalized to the injected amounts) of TBO+, TBOH, and reference indicator, indocyanine green dye (ICG), versus time in the venous effluent of an isolated rabbit lung at two flow rates (6.7 and 3.3 ml/s) and the model equations (18.2a–f) fit to the data, resulting in: $k_1E = 24.1$ ml/s, $k_8[P] = 0.78$ s^{-1}, $k_2[NH] = 6.2$ s^{-1}, $k_0[A] = 4.3$ s^{-1}, $k_{-9} = 2.2$ s^{-1}, and $Q_i = 11.2$ ml. The perfusate contained 5% bovine serum albumin, and, to put the rate constants in perspective, the capillary transit time was estimated to be 1.35 and 0.67 s at the low and high flow rates, respectively.

venous effluent samples prior to spectrophotometric analysis has the potential to confound the analysis. In theory, it would be optimal to measure the blue form of the dye as it exits from the capillaries. In practice, the autooxidation is slow enough that it is negligible during the passage from capillary outlet to venous outlet. Thus, the problem introduced by the presence of TBOH in the samples was mitigated by using an on-line flowthrough photodetector to sample the venous effluent oxidized dye concentration before it was enhanced by autooxidation of the TBOH (15). The individual samples of venous effluent were also collected and the TBOH allowed to oxidize so that the total effluent concentration TBO$^+$ plus TBOH could be measured. The TBOH was then determined by difference. We used indocyanine green dye, which binds rapidly and tightly to the albumin as the reference indicator. Figure 18.11 is an example of the results.

The following species balance, based on the chemical equations above, was assumed to describe the dominant aspects of the disposition of the reference indicator dye, R, and the redox dye, B, within a lung capillary under these experimental conditions:

$$\frac{\partial[R]}{\partial t} + w\frac{\partial[R]}{\partial x} = 0; \qquad \frac{\partial[B^+P]}{\partial t} + w\frac{\partial[B^+P]}{\partial x} = k_8[P][B^+];$$

$$\frac{\partial[B^+]}{\partial t} + w\frac{\partial[B^+]}{\partial x} = -k_1[E][B^+] - k_8[P][B^+] ;$$

$$\frac{\partial[BH]}{\partial t} + w\left(\frac{Q_c}{Q_c+Q_i}\right)\frac{\partial[BH]}{\partial x} =$$

$$\left(\frac{Q_c}{Q_c+Q_i}\right)k_2[NH][B^+E] + \left(\frac{Q_i}{Q_c+Q_i}\right)k_{-9}[BHA] - k_9[A][BH]);$$

$$\frac{\partial[B^+E]}{\partial t} = k_1[E][B^+] - k_2[B^+E][NH] ;$$

$$\frac{\partial[BHA]}{\partial t} = -k_{-9}[BHA] + k_9[A][BH] ; \qquad (18.2a\text{--}f)$$

initial conditions ($t = 0$): [B$^+$] = [BH] = [B$^+$E] = [BHA] = [B$^+$P] = [R] = 0; boundary conditions ($x = 0$): [B$^+$] = [R] = $h_n(t)$; [BH] = [B$^+$E] = [BHA] = [B$^+$P] = 0.

The model was fit simultaneously to the TBO$^+$ and TBOH data for both flow rates using the same method as described for Eqs. (18.1a,b) and the data in Figure 18.10. The identifiable model parameters under these conditions are $k_8[P]$, k_1E, $k_2[NH]$, $k_9[A]$, k_{-9}, and Q_i, where $Q_i = Q_e k_{-7}/(k_{-7} + [P]k_7)$ and Q_e is the extravascular volume accessible to the TBOH, and it is assumed on the basis of the data shown in Figure 18.3 that, as with MBH, the perfusate albumin binding of the TBOH is rapidly equilibrating relative to the transit time through the lungs. In addition, the binding of the TBO$^+$ to the perfusate albumin is assumed to be virtually irreversible in the same time frame. To put the results in perspective, the model estimate of k_1E from the data in Figure 18.11 of 24.1 ml/s implies that at the

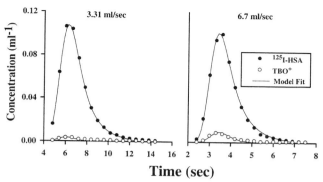

FIGURE 18.12. The concentrations (normalized to the injected amounts) of TBO+ and reference indicator (^{125}I-HSA) versus time in the venous effluent of an isolated rabbit lung at two flow rates. The perfusate was protein-free, and no TBOH was detectable in the venous effluent.

flow rates studied very little of the unbound TBO+ escaped reduction on passage through the lungs. This prediction is confirmed by the data shown in Figure 18.12, which were obtained in another rabbit lung at the same flow rates as in Figure 18.11 but with the albumin in the perfusate replaced by 5% 70-kD dextran. Thus, the presence of both TBO+ and TBOH in the venous effluent was primarily the result of their binding to the albumin in the perfusate. In addition, TBO+ reduction was much faster than MB+ reduction in the lungs, as was the case with the cultured endothelial cells.

These are preliminary observations, but they suggest that the quantitative evaluation of the roles of various endothelial cell processes in intact lungs and in vivo is potentially complex. However, the large difference in the estimated k_1E for TBO+ and MB+, and the manipulation of the physical chemical properties of the dye as in the case of the TBOP, suggest the potential for designing redox dyes with specific properties that will be useful for dissecting kinetic mechanisms and for providing different sensitivities to changes in blood and endothelial cell processes. With the appropriate probes, the role(s) of the endothelial cell reducing capacity in vivo might be determined.

There are some interesting observations which suggest that the ability to measure the kinetics of the endothelial transplasma membrane electron transport in the lungs in vivo might be useful. Increased activity of this transport may explain the increased reduction of nitroblue tetrazolium by the lungs following ischemia and reperfusion (1, 23), and it may be involved in the extracellular redox cycling of agents such as paraquate (7), which are particularly toxic to the lungs. The pulmonary endothelium is apparently capable of reducing the thiazine dyes at rates that are comparable to the normal rate of reduction of oxygen by the whole lung (5), and while its role in normal physiology of the lungs is unknown, one could imagine that the endogenous electron acceptors might be blood or plasma membrane constituents for which the reduced form is required for optimal function. Monodehydroascorbate, which is a known electron acceptor for hepatocyte trans-plasma membrane electron transfer (27), might be an example, but other readily oxidizable lipids, thiols, and hydroxyls might be protected from oxidation to phenoxyl radicals, hydroperoxides, disulfides, and carbonyls if any of these ox-idized forms were electron acceptors for the transplasma membrane electron transport. Because the reduction of the thiazine dyes by the pulmonary endo-thelium is sufficiently rapid to be detected during a single pass of the dyes through intact lungs (5), these dyes have potential as probes for the phenomenon in the intact organ, and for studying its role in the physiology of the lungs.

Acknowledgments. This work has been supported by National Heart, Lung, and Blood Institute Grants HL24349, HL52108, and the Department of Veterans' Affairs.

References

1. Auclair, C., E. Voisin, and H. Banoun. Superoxide dismutase-inhibitable NBT and cytochrome C reduction as probe of superoxide anion production: A reappraisal. In: *Oxy Radicals and Their Scavenger Systems, Molecular Aspects,* edited by G. Cohen and R.A. Greenwald. New York: Elsevier Science Publishing Co., Inc., Vol. I, 1983, pp. 312–315.
2. Audi, S.H., C.A. Dawson, J.H. Linehan, G.S. Krenz, S.B. Ahlf, and D.L. Roerig. An interpretation of ^{14}C-urea and ^{14}C-primidone extraction in isolated rabbit lungs. *Ann Biomed. Eng.* 24:337–351, 1996.
3. Audi, S.H., J.H. Linehan, G.S. Krenz, C.A. Dawson, S.B. Ahlf, and D.L. Roerig. Estimation of the pulmonary capillary transport function in isolated rabbit lungs. *J. Appl. Physiol.* 78(3):1004–1014, 1995.
4. Baggiolini, M., F. Boulay, J.A. Badwey, and J.T. Curnutte. Activation of neutrophil leukocytes: Chemoattractant receptors and respiratory burst. *FASEB J.* 7:1004–1010, 1993.
5. Bongard, R.D., G.S. Krenz, J.H. Linehan, D.L. Roerig, M.P. Merker, J.L. Widell, and C.A. Dawson. Reduction and accumulation of methylene blue by the lung. *J. Appl. Physiol.* 77:1480–1491, 1994.
6. Bongard, R.D., M.P. Merker, R. Shundo, Y. Okamoto, D.L. Roerig, J.H. Linehan, and C.A. Dawson. Reduction of thiazine dyes by bovine pulmonary arterial endothelial cells in culture. *Am. J. Physiol. 269 (Lung Cell. Mol. Physiol. 13):* L78–L84, 1995.
7. Britigan, B.E., T.L. Roeder, and D.M. Shasby. Insight into the nature and site of oxygen-centered free radical generation by endothelial cell monolayers using a novel spin trapping technique. *Blood* 79:699–707, 1992.
8. Crane, F.L., H. Low, and M.G. Clark. Plasma membrane redox enzymes. In: *The Enzymes of Biological Membranes, Second Edition,* edited by A.N. Martonosi. New York, London: Plenum Press, Vol. 4, 1985, pp. 465–510.
9. Crane, F.L., I.L. Sun, R. Barr, and H. Low. Electron and proton transport across the plasma membrane. *J. Bioenerget. Biomem.* 23:773–803, 1991.

10. Crane, F.L., I.L. Sun, M.G. Clark, C. Grebing, and H. Low. Transplasma-membrane redox systems in growth and development. *Biochim. Biophys. Acta.* 811:233–264, 1985.

11. Crone, C. The permeability of capillaries in various organs as determined by the use of the indicator diffusion method. *Acta Physiol. Scand.* 58:292–305, 1963.

12. Cross, A.R., O.T.G. Jones, A.M. Harper, and A.W. Segal. Oxidation-reduction properties of the cytochrome b found in the plasma-membrane fraction of human neutrophils. *Biochem. Int.* 194:599–606, 1981.

13. Dawson, C.A., C.W. Christiansen, D.A. Rickaby, J.H. Linehan, and M.R. Johnston. Lung damage and pulmonary uptake of serotonin in intact dogs. *J. Appl. Physiol.* 58:1761–1766, 1985, 1985.

14. Dawson, C.A. and J.H. Linehan. Biogenic amines. In: *Lung Biology in Health and Disease,* edited by D. Massaro. New York: Marcel Dekker, Inc., Vol. 41—Lung Cell Biology, 1989, pp. 1091–1139.

15. Dawson, C.A., D.L. Roerig, and J.H. Linehan. Evaluation of endothelial injury in the human lung. *Chest* 10:13–24, 1989.

16. Dupuis, J., C. Goresky, and D.J. Stewart. Pulmonary removal and production of endothelin in the anesthetized dog. *J. Appl. Physiol.* 76:694–700, 1994.

17. Gillis, C.N. Pharmacological aspects of metabolic processes in the pulmonary microcirculation. *Ann. Rev. Pharmacol. Toxicol.* 26:183–200, 1986.

18. Gillis, C.N. Pulmonary extraction of PGE1 in the adult respiratory distress syndrome. *Am. Rev. Respir. Dis.* 137:1–2, 1988.

19. Goldenberg, H. Plasma membrane redox activities. *Biochim. Biophys. Acta* 694:203–223, 1982.

20. Goldenberg, H., F.L. Crane, and J. Morre. NADH oxidoreductase of mouse liver plasma membranes. *J. Biol. Chem.* 254:2491–2498, 1979.

21. Goresky, C.A., J.W. Warnica, J.H. Gurgess, and B.E. Nadeau. Effect of exercise on dilution estimates of extravascular lung water and on the carbon monoxide diffusing capacity in normal adults. *Circulation* 37:379–389, 1975.

22. Harris, T.R., R.J. Roselli, C.R. Maurer, R.E. Parker, and N.A. Pou. Comparison of labeled propanediol and urea as markers of lung vascular injury. *J. Appl. Physiol.* 62:1852–1859, 1987.

23. Kennedy, T.P., N.V. Rao, C. Hopkins, L. Pennington, E. Tolley, and J.R. Hoidal. Role of reactive oxygen species in reperfusion injury of the rabbit lung. *J. Clin. Invest.* 83:1326–1335, 1989.

24. Low, H., F.L. Crane, E.J. Patrick, G.S. Patten, and M.G. Clark. Properties and regulation of trans-plasma membrane redox system of perfused rat heart. *Biochim. Biophys. Acta* 804:253–260, 1984.

25. Merker, M., B. Bongard, J. Linehan, Y. Okamoto, D. Vyprachticky, B.M. Brantmeier, D.L. Roerig and C. Dawson. Pulmonary endothelial thiazine uptake: Separation of cell surface reduction from intracellular reoxidation. *Am. J. Physiol. 272(Lung Cell. Mol. Physiol. 16):* L673–L680 (in press), 1997.

26. Morre, D.J., M. Davidson, C. Geilen, J. Lawrence, G. Flesher, R. Crowe, and F.L. Crane. NADH oxidase activity of rat liver plasma membrane activated by guanine nucleotides. *Biochem. Int.* 292:647–653, 1993.

27. Navas, P., J.M. Villalba, and F. Cordoba. Ascorbate function at the plasma membrane. *Biochim. Biophys. Acta* 1197:1–13, 1994.

28. Olson, L.E., R.D. Bongard, C.A. Dawson, and J.H. Linehan. On-line detection of reduction and sequestration of thiazine dyes by the lung. *FASEB J.* 8:A916, 1994.

29. Ravel, P. and F. Lederer. Affinity-labeling of an NADPH-binding site on the heavy subunit of flavocytrochrome b558 in particulate NADPH oxidase from activated human neutrophils. *Biochem. Biophys. Res. Commun.* 196:543–552, 1993.

30. Segal, A.W. and A. Abo. The biochemical basis of the NADPH oxidase of phagocytes. *Trends Biochem. Sci.* 18:43–47, 1993.

31. Sun, I.L., E.E. Sun, F.L. Crane, D.J. Morre, A. Lindgren, and H. Low. Requirement for coenzyme Q in plasma membrane electron transport. *Proc. Natl. Acad. Sci.* 89:11126–11130, 1992.

32. Toole-Simms, W., I.L. Sun, D.J. Morre, and F.L. Crane. Transplasma membrane electron and proton transport is inhibited by chloroquine. *Biochemistry* 4:761–769, 1990.

33. Villalba, J.M., A. Canalejo, M.I. Buron, F. Cordoba, and P. Navas. Thiol groups are involved in NADH-ascorbate free radical reductase activity of rat liver plasma membrane. *Biochem. Biophys. Res. Commun.* 192:707–713, 1993.

34. Zulueta, J.J., F.-S. Yu, I.A. Hertig, and V.J. Thannickal. Release of hydrogen peroxide in response to hypoxia-reoxygenation: Role of an NAD(P)H oxidase-like enzyme in endothelial cell plasma membrane. *Am. J. Respir. Cell Mol. Biol.* 12:41–49, 1995.

35. Zurbriggen, R. and J.L. Dreyer. An NADH-diaphorase is located at the cell plasma membrane in a mouse neuroblastoma cell line NB41A3. *Biochim. Biophys. Acta* 1183:513–520, 1994.

19
Water and Small Solute Exchanges in the Lungs

Francis P. Chinard

Introduction

The multiple indicator dilution technique was introduced some years ago (1, 2) as a means of obtaining information about phenomena and events occurring in intact organs, either in vivo or in isolated perfused preparations. Information obtained in this manner could then be integrated with the results of studies of isolated organelles or cultured cells in which the usual anatomical relationships could not be expected to survive. There is thus complementarity of these experimental approaches, the results of which may require sophisticated mathematical modeling to obtain valid and coherent values of various transport and metabolic coefficients. I have not attempted a complete review of the field. Apologies are extended to those who might feel neglected.

Water: Exchanges and Bulk Movement in the Lungs

Water is the single major chemical constituent of the lungs; sodium and chloride ions are the secondary major constituents of the extracellular fluid, and potassium ions are, after water, the single major constituent of the intracellular domain. The movement of water among the several compartments, red cells, plasma, and interstitial fluid, and between the intra- and extracellular domains, is considered to be regulated by simple physico-chemical factors and processes. The issue considered here is whether the water movement in the lungs is diffusive or convective ("in bulk," with solutes), with the understanding that both mechanisms could be operative sequentially or simultaneously and possibly in different regions of the pulmonary microvasculature.

Willard Gibbs and Ernest Starling

Gibbs was responsible for the developments of the concepts of free energy and chemical potentials (20). He clearly stated that the necessary condition for equi-

455

libration between two phases with respect to a given substance was equality of the chemical potentials of that substance, μ, in the two phases:

$$\mu'' = \mu', \tag{19.1}$$

where the primes denote the phases and the subsequent subscripts the specific substances involved.

In the case of water, w, the chemical potentials at a given temperature are determined by pressure and solute concentrations, so that the differences of the chemical potentials of water in two phases or compartments can be expressed as:

$$\Delta\mu_w = \mu_w'' - \mu_w' = \bar{V}_w \Delta P + RT \ln a_w''/a_w' \tag{19.2}$$

where $\bar{V}_w$ is the partial molal volume of water, P is pressure, R is the gas constant, T is absolute temperature, and a is the activity of water. The activity of water is determined by the other substances present in the compartment or phase.

Similar relationships hold for other neutral substances, while somewhat more complex relationships hold for the electrochemical potentials of ions. What is solvent and what is solute is irrelevant: the terms are used as a nomenclatural convenience. "Solvent" denotes the constituent present in the highest concentration. Expression (19.1) readily reduces to the more familiar van't Hoff expression

$$\Pi = cRT, \tag{19.3}$$

where Π is osmotic pressure and c is solute (not solvent) concentration (6, 8). It is understood that the solvent is present in both phases and the solute in only one.

There was much interest in osmometers and osmotic pressure toward the end of the nineteenth century, and it was at about this time that the brilliant English physiologist, Ernest Starling, through astute observations and perceptive thinking, established the relationship that can be formulated, at equilibrium, as:

$$P'' - \Pi'' = P' - \Pi', \tag{19.4}$$

where the double primes refer to the vascular side and the single primes to the interstitial fluid side of the endothelium (6, 9). The Πs, however, are understood to refer only to the effects of macromolecules and do not take into account the effects of small solutes on the properties of water. They are usually referred to as the oncotic pressures. In effect, the small solute effects cancel out.

A Note on Osmotic Pressure

In passing, it must be noted that the meaning of osmotic pressure has often been misunderstood, particularly by some physiologists (6, 9).

Osmotic pressure is simply one of the several colligative properties of mixtures in which the interactions of the constituents are measured. In dilute aqueous systems, where water is the dominant constituent, four colligative properties are measured (which refer to the effects on the properties of water): freezing point depression, vapor pressure lowering, boiling point elevation, and lowering of the escaping tendency of water across a barrier permeable to it but not to any other constituent of the system. This fourth colligative property is usually measured in a two-compartment isothermal system with pure water in one compartment and a solution in the other, the compartments being separated by a barrier permeable to water but not to solutes. Water will move from the pure water compartment to the solute-containing compartment unless a specific hydrostatic pressure, related to the solute concentration, is imposed on the solute-containing compartment. The pressure required to annul the movement of water is called the *osmotic pressure*. But the osmotic pressure is not an intrinsic property of the solution. Indeed, in the type of two-compartment system just described but with equal pressures on the two sides, the movement of water from the pure water compartment to the solution compartment can be prevented by introducing a specific concentration of solutes into the pure water compartment without there being at any time a pressure difference across the barrier. Further, in the same two-compartment system, but this time with pure water on both sides, an imposition of a pressure difference across the barrier will move water from the high pressure compartment to that with the lower pressure. This movement, or increased escaping tendency of water induced by the pressure difference, can be negated by introducing a specific concentration of solutes into the higher pressure compartment. Thus, solute concentrations and hydrostatic pressure have related but inverse effects on the escaping tendency of water. Osmotic pressure is not an intrinsic property of solutions any more than the other colligative properties are. It is simply a means of quantifying the effects of solutes on the properties of water. This concept underlies the interpretations of some of the experimental results that follow.

Classical Representations

A suggested interpretation of the relationship indicated by Eq. (19.4) is shown in Figure 19.1 where, rather than equilibrium in the whole system, equilibrium is maintained because outward filtration at the arteriolar end of the capillary is matched by inward return at the venular end. The movement of fluid is considered to be in bulk, that is, without separation of water and solutes.

In more recent years, this simultaneous bidirectional bulk movement has featured less prominently in publications with focus on the development of tissue edema. Thus, in the lungs, as indicated in Figure 19.2, it is considered that there is normally slight outward filtration along the length of the capillaries but at a sufficiently low enough rate for the lymph to drain off the net filtration occurring at the endothelial cell junctions. If the capacity of the lymphatic system is exceeded, fluid accumulation, i.e., edema, occurs.

In early physico-chemical experiments with artificial membranes shown to be strictly semipermeable, i.e., permeable to water but not to solutes, it was found that the rate of passage of water was directly proportional to the extent of the disequilibrium with respect to water as measured by the solute concentration. In a

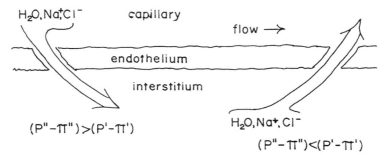

FIGURE 19.1. Early version of capillary–interstitium steady state. Hydrostatic pressure is assumed to fall along the length of the capillary. The balance of the Starling factors favors outward movement of water and solutes at the arteriolar end and reabsorption at the venular end.

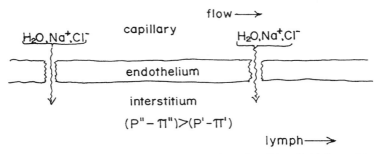

FIGURE 19.2. A current but incomplete representation of capillary fluid–interstitium relationships. There is slight imbalance of the Starling factors along the length of the capillary, so that there is net outward translocation of water and solutes. This takes place "in bulk," without separation of water and hydrophilic solutes, across the endothelial cell junctions and possibly by way of transvesicular pathways. This net movement outward of fluid is removed by way of the lymphatics.

similar manner, the rate of formation of glomerular fluid (the glomerular filtration rate, GFR) was related to the Starling factors P and Π by an expression of the type:

$$GFR = k(\Delta P - \Delta \Pi), \tag{19.5}$$

where k is a proportionality coefficient incorporating the characteristics (permeabilities, thickness, surface area) of the barrier separating plasma or blood from glomerular fluid, and $\Delta \Pi$ refers to the *osmotic* pressure difference. Similar relationships were considered to obtain in other tissues and organs besides the kidneys.

In the chemical potential approach (effectively the thermodynamics of irreversible processes), an additional factor, the reflection coefficient, σ, was introduced by Staverman (31) when there was leakage of solute across a less than perfect barrier. The values of σ could be between 1, for a perfect barrier, and 0, for a barrier offering no hindrance to the passage of solute, each solute having a specific value of σ at a given barrier.

In the Starling approach, a similar factor is introduced as a lumped coefficient involving all of the proteins of plasma (but none of the small solutes) for which the barrier might not be an absolute barrier. The values of σ for small solutes such as Na$^+$ and Cl$^-$ are thus considered to be 0. The expression for the GFR then becomes:

$$GFR = k(\Delta P - \sigma \Delta \Pi). \tag{19.6}$$

This type of formulation is usually considered to apply in the lungs in the development of the pulmonary edema of the adult respiratory distress syndrome (ARDS) where permeability of the capillary barrier to proteins is considered to be increased.

In the thermodynamic approach, the volume flux, J_v, can be expressed, in essence, as:

$$J_v = L_p(\Delta P - \sigma RT \Delta c_s) \tag{19.7}$$

and the solute flux, θ_s, as:

$$\phi_\sigma = \omega RTc_s + (1 - \sigma_s) \bar{c}_s J_v, \tag{19.8}$$

where ω is a coefficient relating to permeability and surface area, and $\bar{c}_s$ is the mean solute concentration in the membrane. The similarities of the expressions for J_v and for GFR are evident. A filtration coefficient, P_f, can be extracted from the experiments in which either hydrostatic or osmotic (i.e., solute concentration) gradients are imposed across a barrier. Hydrostatic and osmotic effects are here considered equivalent. Similarly, a diffusional permeability coefficient, P_d, can be extracted from ω in experiments in which either isotopic or nonperturbing substance concentration gradients are imposed.

The question to be considered is whether the ratio P_f/P_d, of the permeabilities of water measured by the use of isotopes, P_d, and by osmotic and hydrostatic pressure differences, P_f, is less than, equal to, or greater than unity. If $P_f = P_d$, a diffusive mechanism is involved. But if $P_f/P_d \gg 1$, then pores and convective (bulk flow) mechanisms must be considered (18). Hydrostatic and osmotic pressure effects are considered to be equivalent.

The generally accepted concept, particularly in textbooks, is of bulk movement of water without separation of small solutes through discrete pores at the endothelial cell junctions (see Figures 19.1 and 19.2). Part of the data cited as evidence in support of this concept of bulk movement is that interstitial and glomerular fluids have essentially the same compositions as the plasma from which they were apparently derived except for protein concentration and the constraints of the Gibbs–Donnan equilibrium. In other words, the argument offered is that because the fluids have the same composition as that of an ultrafiltrate of plasma, they must have been formed by ultrafiltration, in bulk, without separation of small solutes such as Na^+ and Cl^- from water. The small solutes would have zero reflection coefficients at the "pores" through which the water went.

The argument is fallacious. Comparison of the characteristics of two phases of a system cannot provide information on the mechanism or pathways involved in the transport of materials from one phase to another. The near identity of composition is not necessarily an indication of the pathways involved in achieving that identity.

What can be found concerning the ratio of P_f to P_d?

Filtration and Diffusion Permeability Coefficients

A traditional method of obtaining values for filtration permeability coefficients, P_f, has been to use isolated perfused organ preparations and to determine the rate of weight gain as related to an increase of imposed microvascular pressure. Unfortunately, in the lungs, no single value can be ascribed to P_f. The values range from 2.5×10^{-5} to 3.6×10^{-2} cm · s^{-1} (23). From permeability–surface area products in indicator dilution experiments of the type described below, values are obtained for the diffusional permeability coefficient, P_d, of water up to nearly 100 $\times 10^{-5}$ cm · s^{-1} (24). Both experimental approaches depend on values assigned to the surface area and reflect the properties of the endothelium, not just of the constituent endothelial cells.

A quite different approach has been used by Garrick et al. (18–20) to examine the properties of endothelial cells. Using the linear diffusion technique (28), they have obtained values for P_d at 37°C, average 311×10^{-5} cm · s^{-1} (18). Thus, the ratio of P_f/P_d is very close to unity with respect to these cells cultured from pulmonary artery endothelial plucks.

These results are interpreted as indicating that water movement through the endothelial cell membrane is by diffusion, probably through lipid areas. With respect to the endothelium proper, the calculated ratios P_f/P_d for the values cited range from 2.5×10^{-2} to 3.6×10^1. The appropriateness of calculating P_f/P_d ratios from such different experimental systems can be questioned. However, a diffusive mechanism for the passage of water across the pulmonary endothelium cannot be excluded.

Extraction of Tracer Water and Sodium in the Lungs

Some of the earliest indicator dilution experiments involved the use of T-1824 or Evans blue as a vascular (plasma) reference and deuterium oxide, D_2O, tritium oxide, THO, or oxygen-18 labeled water as tracers for water (3, 4, 7). Extractions of the order of 0.5 or 50% were routinely found. It was deduced that at least one-half of the water molecules entering the exchange bed left the vascular system and returned during the transit of the injected bolus through the lungs. The essentially identical results obtained with the three different water labels indicated that there was no isotope effect such as might have been observed if the permeabilities of the

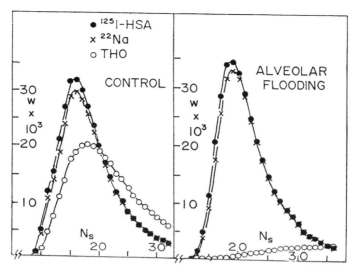

FIGURE 19.3. Outflow patterns of ^{125}I-human serum albumin (HSA), ^{22}Na, and tritium oxide (THO) in an isolated perfused dog lung. N_s denotes sample number. Left panel shows results under control conditions. Right panel shows results after alveolar flooding. With flooding there is a marked increase in the extraction of THO but virtually no change in the extraction of ^{22}Na. [From (11), with permission.]

different species were different and limiting the passage to the extravascularcompartment. In contrast, the extraction of ^{22}Na was only of the order of 0.05 to 0.10 (or 5 to 10%) in dogs (25). (The extraction of ^{22}Na is significantly higher in rats.) Further, with the introduction of "isotonic" fluid into the airways (Figure 19.3), the extraction of water increased in proportion to the amount of fluid introduced and could reach 0.99 or 99% without significant increase of the extraction of ^{22}Na (11). Essentially all of the water molecules entering the exchange regions of the vascular compartment left that compartment and returned in the time of the transit of the bolus—a matter of a relatively few seconds. There could not be much of a barrier to water at either the endothelial or the epithelial boundary, and the distribution of the tracer water was limited only by the volume available or accessible to it. Sodium, in sharp contrast, was barrier-limited, presumably at the endothelium (24). Clearly, at the endothelium, water and sodium ions do not always move together.

Further evidence that water and small solutes can move separately across the endothelium has been provided by Effros (12). In elegant experiments, he demonstrated that solute-free water entered the vascular compartment on presentation to the endothelium of a bolus of a hypertonic solution of poorly permeating solutes.

There is no experimental evidence that sodium and water must move together in the lungs. There is evidence that water and sodium can move separately, as illustrated. However, as will be shown below, the net movement of water depends on the net movement of sodium and other small solutes, but not necessarily by the same pathways.

Studies with Albumin

It had been shown earlier, in somewhat primitive experiments, that the injection of a bolus containing a high concentration of albumin along permeating nonradioactive indicators resulted in water movement from the tissues to the vascular compartment, while the indicators moved in the opposite direction from the vascular compartment to the tissues (7). This finding was in contrast to what would be expected if water and solutes moved together in bulk. The rate of water movement is estimated to have been of the order of $1 \text{ cm} \cdot \text{s}^{-1}$.

These experiments were suggested by studies that the author had participated in a few years earlier in which the expansion of the vascular volume was followed in edematous patients with the nephrotic syndrome who had received rapid injections of concentrated human serum albumin into the blood stream (5). Calculated rates of initial water movement were of the order of $10 \text{ cm}^3 \cdot \text{min}^{-1}$, but the periods of observation encompassed minutes and hours. The considerable changes in plasma volume were not associated with corresponding changes of small solute concentrations.

The role of albumin, that is, of "oncotic pressure," in affecting the distribution of water between the vascular and extravascular compartments in accordance with the Starling relationship and with the more fundamental thermodynamic relationships is clear. But in the multiple-indicator dilution experiments, the time scale is of the order of seconds to minutes; in the clinical studies, the time scale is of the order of minutes to hours. In the former, water without solutes moves across the barrier. In the latter, the data do not indicate that there is separation of water and small solutes. Are there real contradictions here? Can the apparently divergent results—movement of water alone in the short term and water with solutes in the longer term—be reconciled?

These questions are not new, nor is the attempt at resolution that follows, although it is based on recent data. A fascinating discussion was published some 60 years ago (14). In this, August Krogh stated that "fluid movements by filtration and osmosis take place in bulk and comprise all substances which will pass the wall; they are essentially different from diffusion of single particles." This statement was in response to the views expressed by Ancel Keys that there were different rates of diffusion across capillary walls for different solutes and that "whenever functional or experimental states alter osmotic balance between blood and tissue, readjustment tends to take place by a shift of water and only secondarily by a shift of solutes." Keys used the term "osmotic buffering" in this connection. Subsequently, Louis Flexner and his colleagues emphasized the role of diffusion in the transcapillary exchanges of isotopically labeled water and small solutes (16) and thus provided a basis for the development of the multiple-indicator dilution technique (1) and the extension of the diffusion hypothesis as contrasted to the convective "bulk" concept (7).

An Attempt at a Unified Approach

The simplest way to bring the experimental data and the derived interpretations into a coherent whole is to consider the substances (solvent and solutes) in terms of their chemical potentials and the effects of these substances on each others' properties such as activities and chemical potentials. (The designations "solvent" and "solute" are arbitrary and refer to greater or lesser relative concentrations; they have no intrinsic thermodynamic significance.) The Le Chatelier–Braun principle of moderation is usefully invoked here. In its simplest form, it indicates that modification of one of the factors determining an equilibrium and so tending to shift the equilibrium in a given direction will be accompanied by a modification of one of the other determining factors, which, acting alone, would result in a shift of the equilibrium in the opposite direction. The displacement of the equilibrium is thus moderated. This principle of moderation was extended by Prigogine and Defay (26, 27) to systems in a steady state. Thus, in a steady-state system, modification of one of the factors determining the steady state, for example, a rate, will result in a change of the steady state. But this change will be associated with modification of one of the other determining factors such that, if occurring alone, the steady state would have been shifted in the opposite direction with a change of rate opposite to that initially produced.

In the lungs, the permeability of the endothelium to water is relatively very high and to sodium very low. The surface of the endothelium consists of endothelial cells and their junctions, with the surface area of the former being far greater than that of the latter. The main pathway for water is the endothelial cells, and a minor pathway is at the junctions. Sodium ion is probably restricted to pathways across the endothelium at the junctions or possibly at other extracellular hydrophilic pathways, which may include transendothelial vesicles (see Figure 19.4). [This model also was suggested by Keys (14).]

The imposition of an increase of hydrostatic pressure on the vascular side results in an increase of the chemical potential of water (and of all solutes): some water crosses the endothelium, mostly across the endothelial cells. A quantity of sodium corresponding to its concentration in plasma cannot follow because of the low permeability of sodium at the endothelial cells, so a slight increase in sodium concentration occurs, decreasing the difference of chemical potentials of water across the barrier and thus moderating its rate of passage. However, the rate of passage of sodium will increase because of the increase in its concentration at the extracellular pathways. If the increase in hydrostatic pressure difference is maintained, the result will be the accumulation of fluid in the interstitium, with the composition of an ultrafiltrate of plasma but with its major constituents, water and sodium, having taken different pathways. The rate of passage of sodium might also be enhanced by the slight dilution of interstitial sodium that would occur with the passage of "pure" water across the endothelial cells. Thus, the determinant of the passage water across the pulmonary capillaries can be considered to be sodium permeability and transport. (Chloride, with the second highest molar concentrations, probably crosses the endothelial barrier under essentially the same conditions and restraints as sodium. There are fewer data available for chloride, so emphasis has been placed on sodium.)

Thus, the normal endothelium is similar to a slightly imperfect desalting membrane (reverse osmosis). Major outpouring of water is prevented by the low permeability to sodium, so that the effects of increases of hydrostatic pressure are moderated by osmotic buffering. If the permeability to sodium is increased, as may occur in certain clinical situations (possibly in ARDS), then the efficiency of the osmotic buffering will be less and the tendency to edema formation will be increased.

In a similar fashion, albumin and the other macromolecules would increase in concentration with translocation of water and small solutes across the endothelium. However, because of their much smaller molar concentrations, their moderating effects would be less pronounced than those of small solutes and would not come into play as effectively in the short term. The solutes with the lower reflection coefficients and higher permeabilities, sodium and chloride ions, are present at much higher concentrations (140 and 100 mM, respectively) than solutes with higher reflection coefficients and lower permeabilities (for example, albumin at 0.6 mM). A pure water movement of 1% following the imposition of an increase of hydrostatic pressure would be associated with an increase of 1.4 mM for a nonpermeating solute present at 140 mM while the increase would be only 0.006 mM for a solute such as albumin.

This model, as illustrated in Figure 19.4, implicitly includes a barrier resistance to sodium and other small solutes at the junctions and at the possible transcellular vesicular pathways. At these sites, the reflection coefficients for these small solutes would be less than at the endothelial cell surfaces but not 0. The reflection coefficients of the proteins would ordinarily be 1 at these surfaces and only slightly less than 1 at the junctional and other pathways.

Thus, the small solutes moderate the translocation of water that would occur by diffusion in response to an increase of its chemical potential brought about by an increase of hydrostatic pressure. Short-term, transient increases of hydrostatic pressure, such as those that may occur as part of the cardiac cycle, could thus be accommodated. What little translocation of water and solutes occurred would be disposed of without accumulation by the lymphatic and pleural systems.

Sustained increases of hydrostatic pressure that occur in left ventricular failure would be associated with accumulation of fluid in the pulmonary tissues as the run-off capacity was exceeded, even though the run-off rate might be increased as tissue pressure increased.

In the event of endothelial injury, as is considered in ARDS, several stages could be postulated. In a first stage, the barrier resistance to sodium and small solutes at the junctions could be decreased, so that permeability to the small

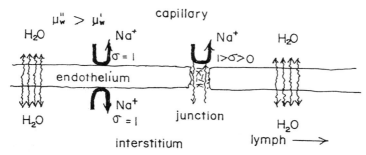

FIGURE 19.4. Schematic representation of pathways for water and sodium ions (and other small, hydrophilic solutes) across the pulmonary endothelium. Water exchanges freely across the endothelium, while Na^+ has a reflection coefficient close to 1.0 at the endothelial cell and less than 1.0 but greater than 0 at the junctions. Pinocytotic vesicles are considered a possible part of the extracellular hydrophilic pathways. The properties of the endothelial cell surfaces relative to water and small hydrophilic solutes are considered to be the same on the two sides of the endothelial cells and at the junctions.

solutes would be increased and their reflection coefficients decreased. However, the permeability and reflection coefficients of proteins would not necessarily be altered. In such a situation, accumulation of fluid in the lungs could be rapid but without much, if any, additional protein translocation. Repair of the injury at the junctions would be followed with fairly rapid removal of the accumulated edema fluid.

A second stage would involve more extensive injury at the junctions, with increased permeability and decreased reflection coefficients for albumin. This would necessarily result in additional decreases of the resistance to sodium and small solutes. Accumulation of fluid would be more rapid and its removal slower than with lesser degrees of injury.

A third stage could be with injury sufficient to allow the larger protein molecules easier translocation. The consequences indicated for the second stage would be even more pronounced.

Currently, emphasis has been given to permeability to proteins as an indicator of the junctional barrier in various clinical situations. If the model proposed here has validity, examination of changes of permeability to sodium (or chloride) could be informative. As indicated earlier, the extraction of sodium in dog lungs in vivo is about 0.05 or 5%. It is not yet known whether this extraction is limited to some extent by the smallness of the interstitial volume of distribution (or dilution space) for sodium or is due primarily to the diffusive or permeability-limited loss. The significantly low recovery of tracer Na^+ in multiple-indicator dilution studies of 1.0 or 100% has been noted (10, 11, 25), and may be explained by a combination of late, slow return from interstitial fluid to the blood and a small degree of lymphatic drainage.

Finally, it should be recognized that the vascular system of the lungs is really quite robust. Patients in intensive care units are frequently the recipients of very heavy fluid overloads without developing pulmonary edema, although they may have large amounts of peripheral, abdominal, and even chest wall edema. In experimental animals, administration of large amounts of fluid is associated with the production of anasarca (widespread subcutaneous edema) before there is a measurable increase in extravascular lung water (personal observations). Patients with low plasma albumin concentrations, as in the nephrotic syndrome, frequently have marked peripheral edema and even ascites (peritoneal fluid accumulation) without evidence of pulmonary edema.

A "tight" barrier at the junctions may be the basis for this highly effective defense system, its loosening the basis for its failure.

Summary and Conclusions

The question reviewed is whether water and solutes cross the capillary endothelium of the lungs in bulk by convective flow, without separation of water and solutes, or by diffusion of the individual constituents.

A physico-chemical approach is used to define osmotic pressure and the effects of solutes on the properties of water. Water and solute translocation across a more or less solute-permeable barrier are due to differences in chemical potentials across the barrier.

Water may be driven across an endothelial barrier by either convection with conductivity defined by a filtration coefficient, P_f, or by molecular diffusion with conductivity defined by a permeability coefficient P_d. For a convective mechanism, the ratio P_f/P_d should be greater than 1.0; in a diffusive mechanism, we should have $P_f = P_d$. Values for P_f have too great a range to be very reliable, but values for P_d from multiple-indicator dilution experiments are within that range. For endothelial cells, $P_f = P_d$: a diffusive mechanism is an adequate explanation.

In multiple-indicator dilution experiments, the apparent unidirectional extraction of tracer water is ordinarily 50%, but in the presence of alveolar flooding it can exceed 99%. At a minimum, therefore, all of the water traversing the pulmonary vasculature exchanges at least once with extravascular water. In contrast, the extraction of sodium remains at about 0.1 or 10%, even in the presence of massive alveolar flooding. This and other evidence indicates that water and small solutes such as sodium do not move in bulk.

The role of albumin, the dominant solute in the Starling relationship, is considered to be no different in principle than the role of other solutes.

The earlier, pioneering concepts of transcapillary exchanges and movement developed by August Krogh, Ancel Keys, Louis Flexner, and others illustrated the importance of osmotic buffering.

The huge diffusional exchanges and the much smaller net translocations of water and small solutes are integrated into a tentative physical model. Net translocation of water is determined by the permeability to and passage of sodium and other small solutes in the short term of seconds to minutes. Albumin plays a less important role in the long term of minutes to hours. The Le Chatelier–Braun principle of moderation as adapted to states by Prigogine and Defay is incorporated into the model along with the concept of osmotic buffering.

The model is applied to clinical situations such as heart failure, fluid overload, and lung injury. It is suggested that changes of permeability to sodium and small solutes may be as important as changes of permeability to proteins in the development of the adult respiratory distress syndrome.

References

1. Chinard, F.P., G.H. Vosburgh, and L.B. Flexner. The mechanism of passage of water and of various substances from blood to interstitial spaces in the legs, liver and head of the dog. *Abstract. XVIII Inter. Physiol. Congress. Copenhagen.* pp. 155–156, 1950.
2. Chinard, F.P. and L.B. Flexner. Capillary permeability. *Bull. Johns Hopkins Hosp.* 88:489, 1951.
3. Chinard, F.P. and T. Enns. Transcapillary pulmonary exchange of water in the dog. *Am. J. Physiol.* 178:197–202, 1954.
4. Chinard, F.P. and T. Enns. The relative rates of passage of deuterium and tritium oxides across capillary walls in the dog. *Am. J. Physiol.* 178:203–205, 1954.
5. Chinard, F.P., H.D. Lauson, H.A. Eder, and R.L. Greif. Plasma volume changes following the administration of albumin to patients with the nephrotic syndrome. *J. Clin. Invest.* 33:629–635, 1954.
6. Chinard, F.P. The definition of osmotic pressure. *J. Chem. Ed.* 31:66–69, 1954.
7. Chinard, F.P., G.H. Vosburgh, and T. Enns. Transcapillary exchange of water and of other substances in certain organs of the dog. *Am. J. Physiol.* 183:221–234, 1955.
8. Chinard, F.P. Colligative properties. *J. Chem. Ed.* 32:377–380, 1955.
9. Chinard, F.P. Osmotic pressure. *Science* 124:472–474, 1956.
10. Chinard, F.P. and W.O. Cua. Endothelial extraction of tracer water varies with extravascular water in dog lungs. *Am. J. Physiol.* 252(*Heart Circ. Physiol.* 21):H340–H348, 1987.
11. Cua, W.O., V. Bower, C. Tice, and F.P. Chinard. Pulmonary vascular extraction and distribution of antipyrine with alveolar flooding. *Am. J. Physiol.* 269(*Heart Circ. Physiol.* 38):H1811–H1819, 1995.
12. Effros, R.M. Osmotic extraction of hypotonic fluid from the lungs. *J. Clin. Invest.* 54:935–947, 1974.
13. Enns, T., F.P. Chinard. Relative rates of passage of $H^1H^3O^{16}$ and $H^1_2O^{18}$ across pulmonary capillary vessels in the dog. *Am. J. Physiol.* 195:133–136, 1956.
14. Faraday Society. *The Properties and Functions of Membranes, Natural and Artificial. A General Discussion.* London: Gurney and Jackson, 1937.
15. Finkelstein, A. *Water Movement Through Lipid Bilayers, Pores, and Plasma Membranes.* New York: John Wiley & Sons, 1987.
16. Flexner, L.B., A. Gellhorn, and M. Merrell. Studies on rates of exchange of substances between the blood and extravascular fluid. *J. Biol. Chem.* 144:35–40, 1942.
17. Garrick, R.A., U.S. Ryan, and F.P. Chinard. Endothelial cell permeability to water. *Biochim. Biophys. Acta.* 862:227–230, 1986.

18. Garrick, R.A., D.J. DiRisio, R. Gianuzzi, W.O. Cua, U.S. Ryan, and F.P. Chinard. The osmotic permeability of isolated calf pulmonary artery endothelial cells. *Biochim. Biophys. Acta.* 939:343–348, 1988.

19. Garrick, R.A., U.S. Ryan, V. Bower, W.O. Cua, and F.P. Chinard. The diffusional transport of water and small solutes inisolated lung endothelial cells and erythrocytes. *Biochim. Biophys. Acta.* 1148:108–116, 1993.

20. Gibbs, J.W. On the equilibrium of heterogeneous substances. *Trans. Conn. Acad.* 3:108–248, 1876; 343–524, 1878.

21. Guller, B., T. Yipintsoi, A.L. Orvis, and J.B. Bassingthwaighte. Myocardial sodium extraction at varied coronary flows in the dog. Estimation of capillary permeability by residue and outflow detection. *Circ. Res.* 37:359–378, 1975.

22. Michel, C.C. One hundred years of Starling's hypothesis. *News Physiol. Res.* 11:229–237, 1996.

23. Parker, J.C., A.C. Guyton, and A.E. Taylor. Pulmonary transcapillary exchange and pulmonary edema. *Intnl. Rev. Physiol. Cardiovasc. Physiol. III.* 18:261–315, 1979.

24. Perl, W., P. Choudhury, and F.P. Chinard. Reflection coefficients of dog lung endothelium to small hydrophilic solutes. *Am. J. Physiol.* 228:197–809, 1975.

25. Perl, W., F. Silverman, A.C. Delea, and F.P. Chinard. Permeability of dog lung endothelium to sodium, diols, amides and water. *Am. J. Physiol.* 230:1708–1721, 1976.

26. Prigogine, I. and R. Defay. *Thermodynamique Chimique conformément aux Méthodes de Gibbs et de De Donder.* Paris: Dunod, 1944.

27. Prigogine, I. *Etude Thermodynamique des Phénomènes irreversibles.* Paris: Dunod, 1947.

28. Redwood, W.R., E. Rall, and W. Perl. Red cell membrane permeability deduced from bulk diffusion coefficients. *J. Gen. Physiol.* 64:706–729, 1974.

29. Safford, R.E., E.A. Bassingthwaighte, and J.B. Bassingthwaighte. Diffusion of water in cat ventricular myocardium. *J. Gen. Physiol.* 72:513–538, 1978.

30. Starling, E.H. On the absorption of fluid from connective tissue spaces. *J. Physiol. London.* 19:312–326, 1896.

31. Staverman, A.J. The theory of measurement of osmotic pressure. *Rec. Trav. Chim.* 70:344–352, 1951.

20
Pulmonary Perfusion and the Exchange of Water and Acid in the Lungs

Richard M. Effros, Julie Biller, Elizabeth Jacobs, and Gary S. Krenz

Studies of transport across the membranes that separate the pulmonary vasculature from the alveoli have been associated with theoretical and practical problems which have challenged the ingenuity of investigators for decades. It is difficult to distinguish between alveolar and airway transport in the intact lung, or to define the relative roles of different cells that line the alveoli. Monolayers of type II pneumocytes have been used to characterize solute transport in the distal lungs, but uncertainties persist regarding the identity and properties of these cells (Mason et al., 1982; Crandall et al., 1982). The morphology of isolated alveolar cells changes in culture from that of type II pneumocytes to flatter cells, which appear to be similar to type I pneumocytes. Since the former cover less than 5% of the surface of the lungs, they may not be representative of the normal barrier that separates the gaseous phase from the blood flowing through the lungs. Whether the cells that have been in culture for longer intervals have transport properties similar to those of the type I pneumocytes will remain uncertain unless the function of these cells can be compared with that of cells in the intact lung.

Transport studies in the intact lung are complicated by difficulties in studying transport between the vasculature and the small amount of fluid that covers the pulmonary epithelium. Two general approaches have been used to study indicator movement across the distal pulmonary epithelium: 1) Radioactive indicators are introduced into the lungs in aerosols, and subsequent losses from the airspaces are followed by external scanning. We introduced a radioaerosol approach in the mid-1970s that continues to be used in many laboratories to assess the permeability of the lungs in both clinical and animal studies (Krauthammer et al., 1977; Rinderknecht et al., 1980). Although this approach has proven to be useful in detecting changes in the permeability of the lungs to a variety of substances, sampling has been restricted to the vascular compartment, and only one indicator has been used at a time in most of these studies. 2) Indicators can be instilled in a solution into the airspaces, an approach that has been used by investigators for several decades (Chinard et al., 1962). Although it is obviously unphysiological to fill the airspaces with solutions, flooding of the airspaces from the vasculature or mouth is unfortunately a common clinical event and therefore warrants study.

The present chapter reviews several older and newer indicator dilution approaches that have been used to study solute and water exchange in intact lungs. Emphasis is placed on indicator dilution studies of water and acid-base equilibration. The chapter ends with a discussion of constant infusion methods of detecting changes in the capillary surface area of perfused lungs.

Aquaporins and Water Transport Across the Alveolar–Capillary Barrier

Early studies of water transport in the lungs investigated the manner in which the lung becomes dehydrated when small volumes of hypertonic sucrose, NaCl, or urea are injected into systemic veins of anesthetized dogs (Effros, 1974). Blood was collected from the carotid artery in serial sample tubes, and the volume and solute concentration of the fluid that had been extracted from the lung tissues was determined by measuring the concentrations of hemoglobin, protein, Na^+, and K^+ concentrations in the pulmonary venous outflow. As illustrated in Figure 20.1, injections of hypertonic urea resulted in proportionate changes in the concentrations of each of these substances. This observation could only be explained if the water that had been transferred from the tissues contained neither Na^+ nor K^+. The study appears to represent the first use of what we referred to as a *baseline dilution* approach to characterize fluid transport between the pulmonary tissues and the vasculature. Rather than introducing an indicator into the system, the concentration of an endogenous solute is monitored. As indicated below, this approach has proven particularly useful in subsequent studies of water transport.

It was hypothesized that bolus injections of hypertonic solutions result in cellular rather than interstitial contraction, a conclusion that was subsequently confirmed with steady-state infusion and ultrastructural studies (Effros and Chang, 1979; Wangensteen et al., 1981). As shown in Figure 20.2, the action of these solutes is primarily at the cell membrane, which allows the movement of water but does not permit rapid solute transport. In effect, the solute and solvent drag reflection coefficients (σ_f) of the endothelial cell membranes to small solutes are very close to 1.0. As indicated below, it is quite likely that water lost from the cells moved through aquaporins that are now known to be present in the cell membranes. The failure of fluid to be extracted from the interstitium was attributed to the relatively large size of the interendothelial junctions, which presumably have solute reflection coefficients (σ_d) to each of the small solutes that are close to 0. The studies of Grabowski and Bassingthwaighte (1976) in the heart suggest that there may be a transient reduction in the interstitial volume, which lasts about 1 min after osmotic transients.

These osmotic experiments also showed a serious flaw in the isogravimetric approach used by Pappenheimer et al. (1951), who made estimates of the size of pores in the endothelium of cat legs with hypertonic infusions. These investigators believed that they could halt fluid flow between the interstitium and blood of legs perfused with concentrated solutions of inulin, glucose, or urea by increasing

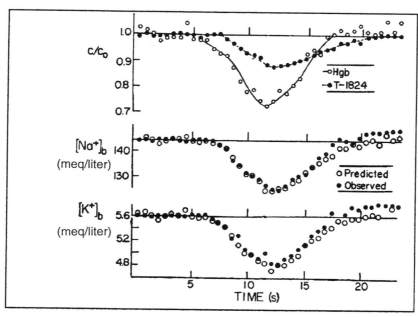

FIGURE 20.1. Osmotic transient studies. Following injection of hypertonic urea into the pulmonary inflow (via a systemic vein), concentrations of hemoglobin, T-1824–labeled albumin, Na+, and K+ decrease, indicating flow of fluid from the lung tissues to the perfusate. Changes in the concentrations of each of these indicators are proportionate to one another, and concentrations of both Na+ and K+ can be predicted on the basis of the corresponding changes in hemoglobin and T-1824 albumin. These observations indicated that there was very little Na+ or K+ in the fluid that had been extracted from the lungs. (Reproduced from Effros, R. M. Osmotic extraction of hypotonic fluid from the lungs. *J. Clin. Invest.* 54:935–947, 1974, Fig. 4 by copyright permission of The American Society for Clinical Investigation.)

intravascular pressures to prevent weight changes. They calculated values for the permeability of the capillaries to a variety of substances, assuming that each of these solutes moved through the same channels, and extrapolated their data to what they thought would be the permeability of the endothelium to water molecules. Utilizing this approach, they calculated that the "pores" that permit water and solute transport out of the capillaries of the leg are 3.0 nm in diameter. Although the weight of these legs could be kept constant when infused with hypertonic solutions by raising intravascular pressures, it is by no means clear that there were no movements of fluids in the preparation. The absence of solutes in the water extracted from the lungs with hypertonic injections suggests that, in any isogravimetric experiment in which the organ is perfused with hypertonic solutions of small solutes, isotonic fluid is forced into the interstitium by the hydrostatic pressure difference between the vascular and interstitium at the very same time that solute-free water is lost from the cells (Figure 20.2). Equivalent cellular contraction and interstitial expansion are responsible for the fact that the weight of

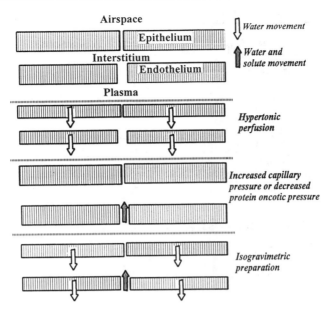

FIGURE 20.2. Schematic figure of water transport between the lungs and tissues under various experimental interventions. Infusions of the lungs with hypertonic urea results in cellular contractions with very little change in the intersitial volume. Increases in vascular pressures (or decreases in plasma protein) increase the size of the intersitial compartment but have little effect on cellular volume. In the isogravimvetric preparation, interstitial distention should be increased by the same amount as cellular contraction.

the organ can be kept unchanged. The lung experiments also showed that comparisons of fluid movement in the lungs in response to osmotic and hydrostatic gradients would be difficult to interpret. Lightfoot et al. (1976) reanalyzed gravimetric and osmotic experiments of Pappenheimer with a multicomponent model. The calculation of vascular reflection coefficients from gravimetric experiments by comparing changes in lung weight caused by increasing vascular pressure or the osmolality of the perfusate (Taylor and Gaar, 1970; Perl et al., 1975) must also take into account the fact that increases in vascular pressures will tend to increase interstitial volumes with little effect upon cellular volumes whereas increases in osmolality will decrease cellular volumes but have relatively little effect on the interstitial volumes.

Because there is normally very little fluid in the airspaces, experiments in air-filled lungs provide no information regarding the properties of the pulmonary epithelial barrier that separates the alveolar and interstitial compartments of the lungs. More recently, we have conducted experiments in which isotonic saline was initially instilled into the airspaces of perfused isolated rat lungs (Effros et al., 1988). The lungs were then perfused with hypertonic solutions of glucose or sucrose. Once again, a baseline approach was used to study the movement of

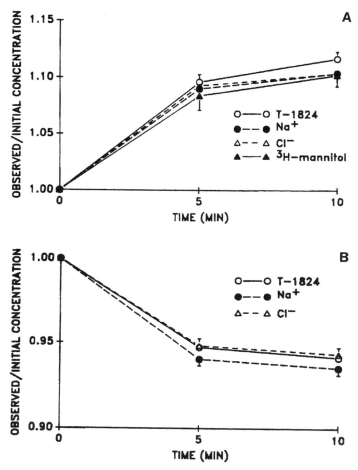

FIGURE 20.3. Perfusion of fluid-filled lungs with hypertonic glucose resulted in proportional increases in airspace concentrations of small and large molecules (top graph). Decreases in perfusate concentrations are also proportionate. These observations are consistent with the conclusion that movement of water across the pulmonary epithelium is not associated with the movement of measurable amounts of solute. (Reprinted with permission from Effros, R. M., G. R. Mason, K. Sietsema, J. Hukkanen, and P. Silverman. Pulmonary epithelial sieving of small solutes in rat lungs *J. Appl. Physiol.* 65:640–648, 1988, Fig. 7.)

water out of the airspaces and its appearance in the vascular space. Samples were collected from both the perfusate and airspace compartments. As noted in Figure 20.3, concentrations of Na+ and Cl− in the airspaces increased proportionately with those of albumin labeled with Evans Blue, which was also incorporated in the airspace fluid, and proportionate decreases were observed in the perfusate. These experiments indicated that the solvent drag reflection coefficient of the epithelial membranes to these small solutes is very close to 1.0. It could be argued that the failure of solutes to accompany the water extracted from the airspaces was related to the fact that the dimensions of the interendothelial junctions are much larger than those of the sugar molecules. However, the relatively slow rate at which salts and sugars cross the epithelium would seem to argue against a large number of such "leaky" pores in the normal epithelial membrane.

Using a conceptual approach analogous to that of Pappenheimer et al. (1951), Solomon and his associates conducted osmotic and tracer studies to determine if there were "pores" present in the cell membranes of red cells (Paganelli and Solomon, 1957; Sidel and Solomon, 1957). They found that the permeability of cell membranes to water determined in the presence of osmotic gradients is from three to four times as great as the corresponding movement of labeled water and calculated that these pores must have dimensions of about 0.35 nm in diameter. Subsequent studies, in which the effects of unstirred layers were more precisely accounted for, have suggested that osmotic permeability exceeds tracer permeability by a factor of as much as 11 (Moura et al. 1984), leading to an unrealistic estimate that the "pores" are 2.5 nm in diameter, or larger than the interendothelial gap junctions (about 1 nm).

Clearly there must be something radically wrong with the model of cell membrane pores postulated by these investigators. The correct answer to this "puzzle" appears to have been provided by Lea (1963), who postulated the presence of "single-file" diffusion, in which water molecules line up in a row within a very confined channel that connects the aqueous solutions on the outside and inside of the cells. It can be calculated that the number of molecules of water that would be present in this channel would equal the ratio between osmotic and diffusional permeabilities. In other words, if P_f is 11 times P_d, then there are 11 molecules of water in the channel at any one time. Considerable effort was devoted to the identification of the putative water channel. Although a variety of known channels were postulated to be responsible for water transport in red cells, they were subsequently deemed to play a relatively minor role in water transport (Verkman, 1993). Utilizing radiation inactivation analysis, Van Hoek et al. (1991, 1992) concluded that the molecular weights of the proteins that are responsible for water transport across the cell membranes of proximal tubules and red cells are approximately 30,000 daltons. In the course of a study of red cell Rh proteins, Agre (1992) discovered a protein that was abundant in red cell membranes and that has a molecular weight of 28,000. Insertion of mRNA for this protein into Xenopus ova cell membranes resulted in marked increases in the P_f of the cell membranes. This protein was initially designated as the Channel-Forming Integral Protein (CHIP28) or, more recently, as aquaporin-1, since it represents the first in a family of membrane proteins, most of which conduct water molecules. The aquaporin-1 channels are extremely narrow, and no molecules other than water have been found that traverse these structures. They are best characterized in terms of the model of single-file diffusion. More recently, Preston et al. (1993) found several familial cohorts who lack the blood group type that corresponds to the aquaporin-1. As expected, water movement in their red cells in response to osmotic gradients was no more rapid than that which would be expected from tracer movement of water, in other words, $P_f = P_d$. What was particularly remarkable was the fact that none of these patients appeared to have suffered ill effects from this genetic and physiological deficiency! Although they have less efficient water transport in their red cells, it is not known whether water transport across other membranes, for example, the endothelium, is also abnormal. The extreme rarity of this condition suggests that there may be additional mutations present in these individuals which enhance water transport across other membranes. Mutations in other aquaporins can result in serious disorders. Aquaporin-2 is an inducible protein that is recruited to the surface of the collecting duct cells when they are stimulated with antidiuretic hormone. Abnormalities of this protein are responsible for a form of inheritable diabetes insipidus (Van Lieburg et al., 1994).

Evidence that aquaporin-1 might be present in the lungs was first inferred in the experiments of Hasegawa et al. (1994a), who detected the presence of mRNA of aquaporin-1 in lung tissues. Subsequently, Folkesson et al. (1994) reported that they could slow water transport across type II cells with $HgCl_2$, one of the few inhibitors of aquaporins that have been reported to date. Since type II cells normally cover a very small fraction of the epithelial surface area, additional experiments were conducted in intact lungs, in which most of the epithelium is covered by type I pneumocytes. These investigators found that, following instillation of hypertonic solutions into dog lungs, the flow of fluid into the airways was decreased by prior infusions of $HgCl_2$. They concluded that aquaporin-1 channels were present in the alveolar epithelium.

Interpretation of studies of water movement in whole lungs is complicated by the possible effects of flow on exchange. The observation of Folkesson et al. (1994) that $HgCl_2$ slowed water transport could be due to effects of this agent on tissue perfusion rather than epithelial permeability. Chinard and Cua (1994) reported that infusions of mercurials increase pulmonary vascular resistance. It is possible to distinguish between effect on permeability and tissue perfusion by incorporating a "flow-limited" indicator in the fluid introduced into the airspaces or vasculature. Evidence has been obtained that equilibration of molecules which rapidly cross the cellular membranes of the lung is virtually complete by the time the perfusate has traversed the pulmonary capillaries. This is presumably responsible for the fact that losses of ^3HOH, $^{14}CO_2$, [^{14}C]-butanol, and [^{14}C]-ethanol from the airspaces into the perfusate occur at very similar rates (Effros et al., 1981, 1985).

In order to determine if the effects of $HgCl_2$ on water transport were due to changes in perfusion, we incorporated ^3HOH in the airspace as a "flow-limited" indicator of exchange (Effros et al., 1997). Five milliliters of water with ^3HOH were instilled into the airspaces of perfused rat lungs, and a direct comparison was made of the rates at which labeled and unlabeled molecules were lost from the airspaces and entered the vasculature of perfused rat lungs. The net movement of water into the perfusate was followed by measuring the dilution of Evans Blue labeled albumin (baseline dilution technique). Clearances of both ^3HOH and HOH were calculated by the equations indicated in Appendix 1 at the end of this chapter.

As illustrated in a typical experiment in Figure 20.4, the rate at which ^3HOH

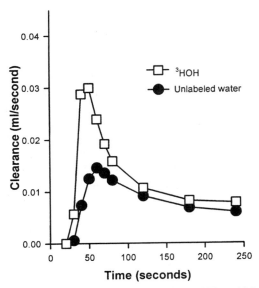

FIGURE 20.4. Following instillation of unlabeled water (HOH) and labeled water (^{3}HOH) into the airspaces of isolated, perfused rat lungs, the clearance of ^{3}HOH exceeds that of unlabeled water (HOH). (Adapted from Fig. 1 in Effros, R. M., C. Darin, E. R. Jacobs, R. A. Rogers, G. Krenz, and E. E. Schneeberger. Water transport and the distribution of aquaporin-1 in pulmonary air spaces. *J. Appl. Physiol.* 83:1002–1016, 1997.)

entered the perfusate exceeded the rate of net transfer of unlabeled water into the perfusate. This difference was presumably unrelated to any difference in the manner in which labeled and unlabeled water diffused through membranes. It merely reflected the fact that there was initially no ^{3}HOH in the perfusate and consequently no back-diffusion (see Figure 20.5). In contrast, concentrations of HOH in the airspace were 55.5 moles per liter, whereas those in the perfusate were 55.2 moles per liter, and considerable back-diffusion of unlabeled water would be expected under these circumstances. These observations indicate that, under normal circumstances, the net movement of water between the airspaces and perfusate in response to realistic osmotic gradients is limited by the permeability of the epithelium. Had there been equilibration between the airspaces and the perfusate during transit through the lungs, then the outflow patterns of the labeled and unlabeled water would have been the same. Perfusion of these lungs with $HgCl_2$ did slow the clearance of ^{3}HOH from the airspaces, but this effect was relatively small. A mathematical model of water exchange between the airspaces and vasculature is provided in Appendix 2 (Krenz and Effros, 1995).

In a subsequent group of experiments, 0.6 ml of water with ^{3}HOH was injected into the pulmonary arteries of perfused, air-filled rat lungs, and samples of the outflow were collected. As indicated in Figure 20.6, initially both HOH and ^{3}HOH were lost from the perfusate and then, as isotonic, unlabeled perfusate entered the lungs, both HOH and ^{3}HOH returned to the pulmonary vasculature. Clear-

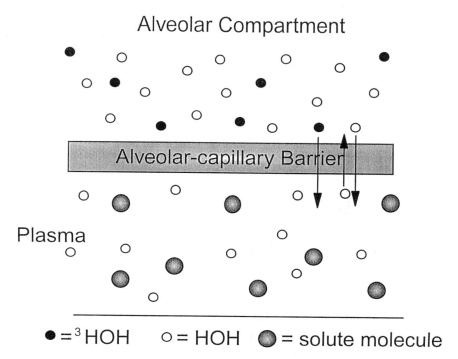

FIGURE 20.5. Schematic diagram that illustrates the relative concentration differences of labeled and unlabeled water in the lungs after instillation into the airspaces. The initial absence of back-diffusion of ^{3}HOH initially enhances its clearance relative to HOH.

ances of HOH were much closer to those of ^{3}HOH than in the studies in which these indicators were introduced into the airspaces and collected from the vasculature. This observation is consistent with the hypothesis that the endothelium is more permeable to the osmotic movement of water than the epithelium. Incorporation of HgCl$_2$ in the perfusate markedly decreased the clearance of HOH relative to ^{3}HOH, suggesting that the barriers which separate the pulmonary vasculature from the tissue do have aquaporins that can be inhibited by mercurials.

The simplest way to explain the differences observed in the experiments with fluid-filled and air-filled lungs is to assume that aquaporin-1 is asymmetrically distributed in the alveolar–capillary barrier (see Figure 20.7). Schneeberger (personal communication) used immunogold labeling to show that the aquaporin-1 molecules are abundant in the endothelium but are absent in type I pneumocytes. Evidence that aquaporin-1 is present on many endothelial beds has been reported by Schnitzer (1996), although it may be functionally absent in some endothelial cell preparations (Garrick, 1993). On the pulmonary epithelium, aquaporin-1 appears to be restricted to the type II cells and adjoining cells (Folkesson et al., 1994; King et al., 1996), which represent a small fraction of the epithelial lining area. Aquaporin-5, another water channel that is inhibited by HgCl$_2$, is present on

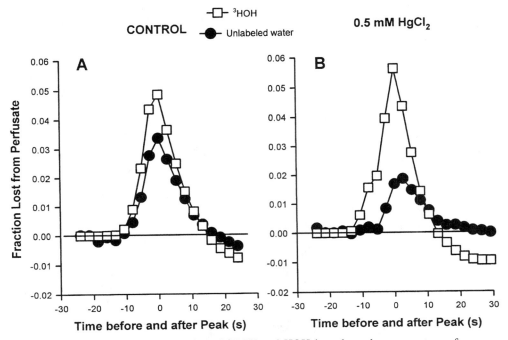

FIGURE 20.6. Following injection of ³HOH and HOH into the pulmonary artery of an isolated rat lung, clearance of ³HOH exceeds that of HOH, but this difference is much less than that observed after intratracheal injections (compare with Figure 20.4). Perfusion with HgCl₂ slowed net movement of unlabeled water relative to that of labeled water. (From Effros et al. 1997, fig. 6, with permission.)

the apical surfaces of type I cells (Raina et al., 1995; Lee et al., 1996), but the present functional studies suggest that the net transport of water through these channels may be slower than through the aquaporin-1 channels that are on the endothelium. Aquaporin-4 is a third water channel that has been found in the lung (Hasegawa, 1994b). It is not inhibited by HgCl₂ and is found with aquaporin-3 on the basolateral surfaces of the airway epithelial cells.

We also found that changes in temperature from 37 to 10°C have relatively little effect on the rate of water transfer from the airspaces to the perfusate, implying that the activation energy for water across the epithelium is very low. This would be consistent with a very low Arrhenius energy of activation and with passage through water channels.

The presence of flow-limitation of labeled water in the lung must always be considered in whole organ studies. Schnitzer and Oh (1996) found that the uptake of ³HOH by perfused rat lungs was limited after the lungs were perfused with HgCl₂. If the uptake of ³HOH is limited by flow rather than diffusion, then the decreased uptake that they observed could reflect decreased perfusion rather than an alteration in the permeability of the capillaries to ³HOH. In a recent study by

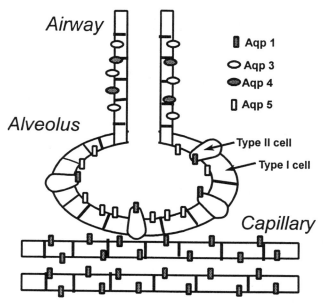

FIGURE 20.7. Schematic diagram of distribution of the aquaporins (Aqp) in the alveolar–capillary barrier. Aqp1 is on endothelium and Type II cells, Aqp 5 is on apical surfaces of Type I cells, Aqp 3 and 4 are on basolateral surfaces of airway cells.

Carter et al. (1996), the diffusion of D_2O into the airspaces of mouse lungs was monitored with an intraalveolar fluorescence technique. They reported that increasing flow rates from 1.7 to 2 ml/min accelerated the movement of D_2O into the airspaces, but a further increase to 2.3 ml/min was without effect; they concluded that exchange was diffusion-rather than flow-limited at these higher flows. However, increases in flow in the latter experiments may have been accomplished by vascular recruitment rather than accelerating the velocity of flow in the monitored vessels.

Understanding the function of pulmonary aquaporins will presumably depend in part on additional indicator dilution studies that are conducted in intact lungs. However, several conclusions and speculations can be made on the basis of our current knowledge:

1. It is likely that perfusion of the lungs with hypertonic or hypotonic solutions of small molecules and eletrolytes results in the flow of water through aquaporin-1 channels in the endothelium. As indicated in our early osmotic transient experiments, it is primarily the cellular rather than the interstitial compartment which is dehydrated by these solutions. However, when concentrations of large molecules are increased, there may be some extracellular movement of water between the interstitium of the lungs and the vasculature, since the reflection coefficients of the junctions to such macromolecules may be sufficiently high to yield a flow of fluid from the interstitium. Increases in

hydrostatic pressure may also drive water through either the aquaporins or the interendothelial junctions.

2. The presence of channels that are specifically permeable to water but no other molecules casts doubt on the validity of extrapolating the molecular permeability of "pores" to water molecules, as suggested by Pappenheimer et al. (1951) for the endothelium of the leg and by Berg et al. (1989) for the pulmonary epithelium.

3. It has been known for more than 100 years that large volumes of water can be aspirated with relatively little harm to the lungs (Colin, 1873). Some water aspiration is inevitable during drinking, and of course water may enter the lungs while swimming. If, as suggested in these experiments, epithelial permeability is less than endothelial permeability to water transport, then it seems plausible that the osmolality of the pulmonary parenchymal cells would be more closely linked to that of the plasma than that of any fresh or salt water that might be aspirated. This would tend to minimize rapid changes in cell volume that might compromise the integrity of the alveolar epithelial barrier. It is also possible that a paucity of water channels in the epithelium may help reduce water losses from the lungs.

4. In principle, it should be possible to calculate the relative permeabilities of the endothelium and epithelium of the lungs to water. We are currently involved in developing mathematical models for this purpose (see Appendix 2). The concomitant use of intraarterial injections of ^3HOH or some other flow-limited indicator in hypertonic or hypotonic solutions with baseline dilution analysis should make it possible to analyze the permeability of any vascular bed to water.

5. Aquaporin-1 channels are present in abundance in continuous endothelial beds, but they are not found in fenestrated and sinusoidal exchange vessels (Schnitzer, 1996). This suggests that the aquaporins normally provide an important route for the transudation of fluid from capillaries when hydrostatic pressures are increased. However, it must be recognized that, once water leaves the vasculature and dilutes the interstitium, concentration of the plasma and dilution of the interstitial electrolytes will result in osmotic gradients that effectively inhibit any further edema formation. For example, a difference in Na^+ concentration of 1 meq/L (and associated anions of 1 meq/L) will result in an osmotic force equivalent to 40 mmHg, countering any further transudation. Since solutes cannot traverse aquaporin-1 channels, which are exquisitely specific for water, electrolytes and other solutes must move through other routes before additional movement of water can occur. Although it is quite likely that these solutes diffuse through the intercellular junctions, it is also possible that they cross the cell membranes, for example, through sodium channels and transporters, and that transudation might be regulated by inhibiting these routes of solute transfer. Aquaporin-1 is also abundant in the cell membranes of the proximal tubule, where it serves to conduct water from the tubular lumen to the interstitium of the kidney in response to the active transport of Na^+. It is conceivable that reabsorption of water from the interstitium to

the lumen of the capillaries occurs in parallel with the active transport of Na^+ by the endothelial cells.

Neutralization of Aspirated HCl in the Lungs and the Pathogenesis of Bronchiectasis in Cystic Fibrosis

The peculiar intersection of the respiratory and gastrointestinal tracts in the pharynx can be readily explained on a phylogenetic basis but can hardly be deemed an optimal arrangement. A complex set of reflexes have evolved to minimize aspiration of gastric fluid, which may have a pH as low as 1.0, but such aspiration is a common event, particularly among the very young, the very old, and those with swallowing disorders (Titjen et al., 1994). Acute aspiration of acid typically results in bronchospasm and may induce life-threatening, noncardiogenic pulmonary edema (adult respiratory distress syndrome). Chronic aspiration is associated with bronchiectasis, which tends to be localized in the lower lobes of the lungs, because they are more likely to fill with fluid after aspiration in adults (Beakey et al., 1951; Kennedy, 1962). Although research has been conducted concerning the pH threshold at which lung injury occurs in various species (Titjen, 1994), there appears to be no information regarding the manner in which acids, or, for that matter, alkaline solutions, are buffered in the lungs.

A major incentive for studying acid neutralization in the lungs is the recent observation that the cystic fibrosis transmembrane conductance regulator (CFTR) normally transports both HCO_3^- and Cl^- (Poulsen et al., 1994; Smith and Welsh, 1992; Cucchiara et al., 1991). This prompted us to hypothesize that defects in this anionic transport channel may impair the neutralization of gastric acid aspirated by young children with cystic fibrosis (CF) and may be responsible for the characteristic bronchiectasis that they develop (Effros and Darin, 1995a). It is the pulmonary complications of CF that are responsible for most of the morbidity and mortality sustained by these patients. It has been generally assumed that most of the airway pathology associated with CF is due to inspissation of secretions caused by decreased secretion of Cl^- by the airway mucosa. For this reason the disease was referred to as "mucoviscidosis" (Farber, 1945). However, inspissation of secretions is also seen in asthmatics, who seldom develop bronchiectasis. This suggests that there may be other factors, such as inadequate buffering of acid, which contribute to the development of bronchiectasis in CF. Up to 80% of CF children have acid reflux from the stomach as documented by pH electrodes placed in the esophagus, esophageal manometry, and gastro-esophageal scintigraphy (Malfroot and Dab, 1991). Furthermore, the administration of cisapride, a drug that enhances motility and diminishes reflux, appears to reduce the cough and wheezing experienced by CF children (Malfroot and Dab, 1991). This response was attributed to decreased bronchospasm associated with acid exposure but could also be related to a reduction in the destructive effects of acid upon the bronchial walls. We have proposed that defective HCO_3^- and Cl^- transport due to inadequate CFTR function in CF patients plays an important role in the develop-

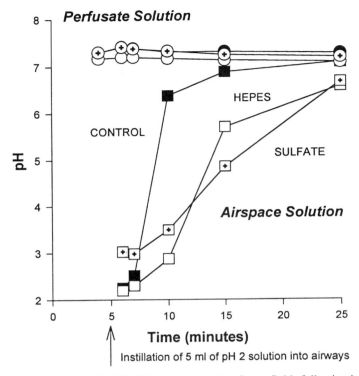

FIGURE 20.8. Changes in the pH of the airspace and perfusate fluids following instillation of isotonic saline with HCl at pH 2.0 into the trachea (at 5 min) in isolated rat lung. Note that when HCO_3^- in the perfusate is replaced by HEPES, buffering of the airspace fluid is slowed. Similarly, replacement of Cl^- with SO_4^{2-} also slowed buffering.

ment of bronchiectasis. By the age of 6 months, approximately 25% of children with CF have bronchiectasis, and by the time they reach 2 years, virtually all have bronchiectasis (Bedrossian, 1976). The characteristic upper lobe distribution of bronchiectasis in these children may be due to the fact that they are recumbent much of the day. In normal individuals, HCO_3^- secreted by the nonparietal cells of the stomach (Flamstrom, 1994), the esophagus (Brown et al., 1993), and the salivary glands (Helm et al., 1987) helps to neutralize HCl and protect the mucosa of the stomach and esophagus. It is possible that abnormal CFTR transport of HCO_3^- in these organs is responsible for much of gastric hyperacidity (Cox, 1987) and peptic esophageal disease that is observed in these children (Bendig et al., 1982).

   The multiple-indicator dilution approach has proven to be ideally suited for investigating the neutralization of acid in lungs. As indicated in Figure 20.8, aspirated acid is rapidly neutralized in isolated rat lungs. In these experiments, 5 ml of 10-mM HCl in 154-mM NaCl were instilled into the tracheas of isolated, perfused rat lungs. However, when HCO_3^- in the perfusate was replaced with the relatively impermeant buffer (HEPES), the rate of buffering was slowed significantly. Buffering was also slowed when Cl^- in the airspace was replaced with SO_4^{2-}. These observations suggest that the transport of both HCO_3^- and Cl^- is essential for prompt neutralization of aspirated acid. However, it was important to be sure that the effects of replacing these ions was not related to factors other than anion transport, such as the rate of fluid reabsorption from the airspaces, the permeability of the epithelium, or the rate of tissue perfusion. These experiments are preliminary, but they are worth mentioning since they emphasize the value of the multiple-indicator dilution approach.

The rate of fluid reabsorption from the airspaces was monitored by incorporating fluorescein labeled dextran (m.w. 54,000) in the airspace fluid. Concentrations of this indicator increased by approximately 10% during the course of these experiments, indicating reabsorption of fluid regardless of whether isotonic saline at pH 7, pH 4 (buffered with citrate), or pH 2 was instilled into the lungs. Replacement of HCO_3^- in the perfusate with HEPES had no effect on their rate of fluid reabsorption when saline at pH 2 was instilled into the lungs. However, reabsorption from the lungs was completely abolished when Cl^- in the airspaces was replaced with SO_4^{2-}, and some fluid actually entered the airspaces.

The permeability of the epithelium to ^3H-mannitol instilled into the airspaces was also evaluated. The ratio of the concentration of ^3H-mannitol to that of the dextran was used as an index of pulmonary epithelial permeability. Instillation of saline at pH 4 or pH 2 did increase the ratio of ^3H-mannitol to dextran concentrations in the airspace relative to that found in experiments in which saline at pH 7 was instilled into the lungs. Permeability was increased further when Cl^- in the airspaces was replaced with SO_4^{2-} but not when HCO_3^- in the perfusate was replaced with HEPES. It would be difficult to explain the *slowing* of neutralization that was observed after replacement of Cl^- by SO_4^{2-} to the increases in epithelial permeability that were observed in these experiments.

The rate at which ^3HOH instilled into the airspaces equilibrated with the perfusate was used for two purposes: (a) Since loss of this indicator from the airspaces is limited by perfusion rather than by permeability, the rate of equilibration between the airspaces and the vasculature could be used to detect alterations in lung perfusion. (b) By monitoring equilibration, it was possible to determine the maximal delay that might have resulted from poor mixing of the airspace fluid during the sampling process. Rapid equilibration was observed regardless of the perfusate and airspace solutions that were used.

$^{36}Cl^-$ was incorporated into the airspace fluid, and concentrations of both labeled and unlabeled chloride were measured in the airspace and perfusate to determine bidirectional fluxes of this ion between the airspaces and perfusate. Permeability of the alveolar-capillary to $^{36}Cl^-$ as well as ^3H-mannitol was increased by the presence of SO_4^{2-} in the airspaces, but, presumably because there were only tracer concentrations of Cl^- in the airspaces, neutralization of the airspace solutions was slowed in these experiments. Measurements were also made of Na^+ and K^+ concentrations in the airspace and perfusate solutions. These showed release of K^+ into the airspaces after the instillation of acid, suggesting that the exchange of K^+ for H^+ might also contribute to airspace buffering.

These experiments suggest that the exchange of HCO_3^- and Cl^- normally plays an important role in the rate at which acid is buffered in the lungs. Whether abnormalities of CFTR encountered in CF actually slow the neutralization of acid has yet to be determined. Mice in which the CFTR function has been eliminated by knock-out technology have relatively normal Cl^- transport in the airways because of the presence of Ca^{2+}-stimulated Cl^- transport that appears to be much more abundant in this species than in man (Clarke et al., 1994). Nor do these mice develop much in the way of lung disease. It might be possible to test this hypothesis in the nasal mucosa of patients with CF, but interpretation of data obtained in this model would be difficult to interpret in the presence of chronic infection.

The best test of the acid hypothesis of lung injury would be to evaluate the effect of suppressing acid secretion in the stomach with a histamine (H2) blocker or an agent that blocks H^+ secretion in the stomach upon the progression of bronchiectasis in children with CF. Ideally, such a trial would be best conducted from the time of birth to suppress the development of bronchiectasis. Once bronchiectasis has occurred, chronic infections result in progressively severe damage to the lungs.

Continuous Infusion Measurements of PS Products in the Lungs

Many of the unusual properties of the pulmonary circulation are related to its unique position between the venous and arterial circulations. Because virtually the entire circulation must traverse the pulmonary vasculature, it acts as an ideal filter that prevents the delivery of clots in the venous blood from reaching the systemic vessels. The remarkable compliance of the pulmonary vasculature permits it to accommodate small emboli or rapid increases in systemic blood flow with only minimal elevations in pulmonary artery pressure. In addition, the pulmonary vasculature is responsible for the metabolism of a wide variety of substances returning to the heart, in effect processing the blood before it is delivered to the extrapulmonary tissues.

Both the surface area and the metabolic activities of the pulmonary vasculature can be altered by pulmonary disorders, thereby affecting both pulmonary vascular pressures and metabolic activity. Distinction between effects on the surface area and metabolic function or permeability of the pulmonary vasculature has proven to be a particularly difficult problem. In the early 1950s, Chinard et al. (1955) found that extraction of extracellular solutes from blood traversing the pulmonary vasculature following bolus injections was much less than that in skeletal muscle. Furthermore, although extraction of these solutes should increase as blood flow is decreased, these changes are relatively small in magnitude and difficult to measure accurately. The tendency for extractions to remain relatively constant with increases in blood flow has been used to argue that the surface area of the pulmonary vasculature continues to be recruited at all attainable flows (Dupuis et al., 1990).

Because the volume of distribution of extracellular solutes in the lung is quite limited, extraction of these indicators rapidly decreases during constant infusions, and they must be administered as bolus injections, making it impossible to obtain continuous measurements of changes in the clearance of indicators from the blood stream. There would be an obvious advantage to alternative indicators with much larger volumes of distribution that could be used in constant infusions to detect alterations in the surface area of the pulmonary vasculature. This was the original strategy of Renkin (Renkin, 1959a, 1959b; Renkin and Rosell, 1962), who used both ^{42}K and ^{86}Rb to measure changes in permeability–surface area (PS) products in skeletal muscle. For reasons that are probably related to a fundamental interest in the permeability of the pulmonary vasculature to extracellular solutes rather than to the surface area of these vessels, this approach does not appear to have been tried until recently, when it was inadvertently found that increased pulmonary uptake of ^{201}Tl, a K^+ analog used in studies of myocardial perfusion, was associated with reduced survival in patients with coronary artery disease (Gill et al., 1987). Bingham et al. (1980) found that pulmonary uptake of ^{201}Tl could be increased by either raising left atrial pressures or inducing asystole and bradycardia with acetylcholine in experimental preparations. Each of these maneuvers decreases blood flow, and the investigators concluded that increased uptake was due to more prolonged exposure of the vasculature to the radionuclide.

In our studies, we perfused isolated rat lungs with an unlabeled solution and then abruptly changed the solution to a labeled solution (step increase in inflow concentrations) and collected the outflow over periods up to 5 min in duration. We chose to modify the Renkin protocol, incorporating a vascular indicator (^{125}I-albumin) as well as a potassium analog (^{201}Tl) in the labeled perfusion solution. This permitted us to determine when the unlabeled solution had been flushed from the vasculature. Much like Renkin, we found that the initial extraction of ^{201}Tl is high and then falls rapidly, and have attributed this to distribution into some portion of the interstitial compartment of the lungs. Again like Renkin, we intentionally avoided using this brief portion of the outflow curve to detect changes in PS products.

Utilizing this approach, we have studied the effects of three hemodynamic changes in these lungs. In the first, we either increased or decreased flows in Zone II lungs. This resulted in reciprocal changes in extraction (see Figure 20.9). PS products were clearly less at low flows than at high flows, and when flows were reduced abruptly, PS products also fell. However, it was difficult to show that PS products were increased when flows were increased (see Figure 20.10). In a second series of experiments, we were able to show that PS products of ^{201}Tl$^+$ obtained at high flows exceeded those at lower flows (not shown). In a third set of experiments, we abruptly decreased pulmonary arterial pressures and increased pulmonary venous pressures by swivelling the arterial and venous reservoirs, which had been mounted on a single bar. This maneuver served to decrease flows through the lungs but kept pressures in the capillaries essentially unchanged. It did

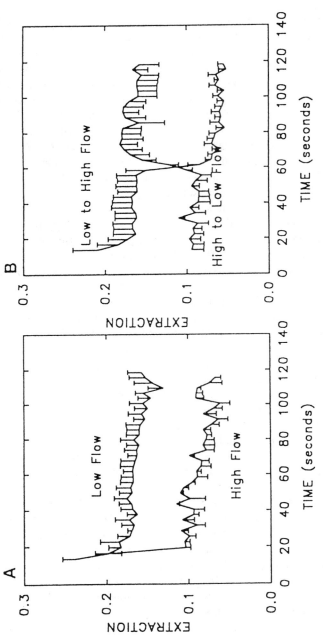

FIGURE 20.9. Relationship of the extraction of ^{201}Tl from the pulmonary vasculature of isolated rat lungs to perfusion rate. Note that extraction is reduced at high flows (From Effros et al. 1994, reprinted with permission from *J. Appl. Physiol.*)

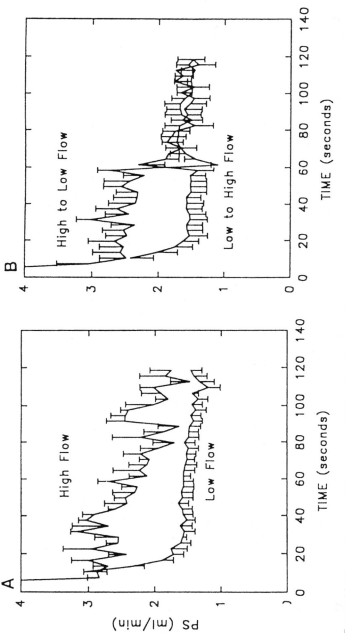

FIGURE 20.10. Relationship of *PS* products of ²⁰¹Tl to flow. Note that *PS* products are greater at high than at low flow. Abrupt decreases in flow resulted in decreases in *PS*, but increases in *PS* with increases in flow were not significant (From Effros et al. 1994, reprinted with permission from *J. Appl. Physiol.*.)

not result in any significant changes in the *PS* product of the lungs. Changes in flow did not result in any change in serum Na^+ or K^+ concentrations, suggesting that there had been no significant changes in the manner in which these solutes were exchanged between the plasma and tissues of the lungs. These observations suggest that constant infusion techniques can be used to measure alterations in surface area when the lungs are challenged by experimental protocols that may alter pulmonary hemodynamics.

Appendix 1: Calculation of Clearances of Labeled and Unlabeled Water

By following the movement of both labeled and unlabeled water in the perfused lungs, it is possible to distinguish between flow-limited and perfusion-limited exchange. Useful equations for calculating clearances of labeled and unlabeled water are indicated in this Appendix. A model for interpreting the data in terms of permeability–surface area products is described in Appendix 2.

Instillation of Water Into the Airspaces

Following instillation of *hypotonic* solutions into the airspaces, the entry of water into the pulmonary exchange vessels increases the flow leaving these vessels, and decreased Evans Blue–labeled albumin is calculated from the arterial inflow (F_a) and the Evans Blue concentrations in the arterial perfusate, $[EB]_a$, and venous perfusate, $[EB]_{ven}$, with the equation:

$$F_{ven} = F_a \frac{[EB]_a}{[EB]_{ven}} . \qquad (20.1)$$

The clearance (C_{HOH}) of water from the airspaces to the perfusate is calculated from the difference between the venous and arterial flows:

$$C_{HOH} = F_{ven} - F_a = F_a \frac{[EB]_a - [EB]_{ven}}{[EB]_{ven}} . \qquad (20.2)$$

The volume of water, $V_{A,i}$, remaining in the airspaces at time, i, is assumed equal to the amount originally instilled into the lungs (5 ml) minus the volume of water that was transferred from the airspaces to the perfusate prior to that time:

$$V_{A,i} = 5 - \sum_{n=0}^{i} C_{HOH} \Delta t_i , \qquad (20.3)$$

where Δt_i represents the duration of the time interval during which the ith sample was collected.

The rate of loss, $L_{A,i}$, of ^3HOH from the airspaces during the ith interval is calculated with the equation:

$$L_{A,i} = F_{ven}[^3HOH]_{ven,i} .$$ (20.4)

The amount, $(^3HOH)_{A,i}$, of ^3HOH remaining in the airspaces during the collection of the ith sample is assumed equal to the original amount instilled into the airspaces, $(^3HOH)_{A,0}$, minus the quantity that left in the venous outflow:

$$(^3HOH)_{A,i} = (^3HOH)_{A,0} - \sum_{n=0}^{i} L_{A,i}\Delta t_i .$$ (20.5)

The concentration of ^3HOH in the airspace fluid at the time that the ith sample of perfusate is collected (designated as $[^3HOH]_{A,i}$) is calculated from Eqs. (20.3) and (20.5):

$$[^3HOH]_{A,i} = \frac{(^3HOH)_{A,i}}{V_{A,i}} .$$ (20.6)

The clearance, C_{THO}, of ^3HOH from the airspaces during each collection interval is calculated with the equation:

$$C_{THO} = \frac{L_{A,i}}{[^3HOH]_{A,i}} .$$ (20.7)

Pulmonary Artery Injections

In order to study the permeability of the endothelium, lungs are perfused with the Evans Blue–labeled albumin solution for 2 min and 0.7 ml of water is injected into the arterial inflow. The injectate contains ^3HOH and $[^{14}C]$-dextran, which are dissolved in water containing the same concentration of Evans Blue as that present in the perfusion solution and 0.1 g/dl albumin. As the hypotonic injection bolus traverses the pulmonary capillaries, unlabeled water is lost from the vessels, resulting in an increase in the concentrations of Evans Blue–labeled albumin. The rate at which unlabeled water is lost from the vascular compartment to the tissues (which is equivalent to its clearance, C) was calculated with the equation:

$$C_{HOH} = F_a \frac{[EB]_{ven} - [EB]_a}{[EB]_{ven}} .$$ (20.8)

The fraction (D_{HOH}) of the injected volume of water (0.7 ml) that has been lost from the pulmonary venous outflow is calculated by dividing C_{HOH} by 0.7 (see Figure 20.8).

The fraction (D_{THO}) of labeled water lost from the perfusate samples is calculated in the following fashion: It is assumed that the labeled dextran (which was injected into the pulmonary artery with 0.7 ml of water containing ^3HOH) does not leave the vasculature during transit through the lungs. The concentration of ^3HOH (designated as the calculated value, $[^3HOH]_a$), that would have been present in each of the venous samples had there been no diffusion of ^3HOH out of that sample in transit through the lungs was determined with the equation:

$$[^3HOH]_a = \frac{[dex]_{ven}}{[dex]_{inj}} [^3HOH]_{inj}, \tag{20.9}$$

where $[dex]_{ven}$ represents the observed concentration of $[^{14}C]$-dextran in the venous sample and $[dex]_{inj}$ designates the corresponding concentration in the injected bolus. The fraction, D_{THO}, of the injected ^3HOH that is lost from each sample because of diffusion into the tissues is therefore:

$$D_{THO} = \frac{([^3HOH]_a - [^3HOH]_{ven})F_{ven}\Delta t}{[^3HOH]_{inj}V_{inj}}$$

$$= \frac{\left(\dfrac{[dex]_{ven}}{[dex]_{inj}} [^3HOH]_{inj} - [^3HOH]_{ven}\right)F_{ven}\Delta t}{[^3HOH]_{inj}V_{inj}}, \tag{20.10}$$

where Δt is the collection interval. The hypotonic bolus is followed by isotonic perfusate containing no ^3HOH, and consequently there was return of both labeled and unlabeled water to the vasculature at later times.

Appendix 2: Steady-State Equations of Water Absorption from the Airspaces

Let $c(t,x)$ be the radially averaged vascular concentraion of salt and $F(t,x)$ the radially averaged vascular flow at time t, at a position x within the pulmonary capillary. Following instillation of water into the airspaces, the net influx of water, V_w, into the vasculature is modeled as $P_f\Delta x\Delta tc(t,x)$ (Krenz and Effros, 1995). Assuming an initial salt concentration of c_0 and inflow F_0, simple mass balance results in:

$$\frac{\partial c}{\partial t} = \frac{-kP_f}{A} c^2 - \frac{1}{A} F \frac{\partial c}{\partial x} \qquad c(t,0) = c_0 \qquad c(0,x) = c \tag{20.11}$$

$$\frac{\partial F}{\partial x} = kP_f c \qquad F(t,0) = F_0 \qquad F(0,x) = F_0. \tag{20.12}$$

Under steady-state assumptions:

$$F(x) = \sqrt{F_0 + KP_f F_0 c_0 x} \qquad c(x) = F_0 c_0 / \sqrt{F_0^2 + kP_f F_0 c}. \tag{20.13}$$

Utilizing a simple model of exchange in which capillary length, flow, and permeability are assumed constant and the distribution of transit times is related to heterogeneous transit times through the venous circulation, the average concentration that is responsible for the observed rate of unlabeled water diffusion into the vessels from the airspace can be calculated.

References

Agre, P., G.M. Preston, B.L. Smith, J.S. Jung, S. Raina, C. Monon, W.B. Guggino, and S. Nielsen. Aquaporin CHIP: The archetypal molecular water channel. *Am. J. Physiol.* 265:F463–F476, 1993.

Bedrossian, C.W., S.D. Greenberg, D.B. Singer, and J.J. Hansen. The lung in cystic fibrosis. A quantitative study including prevalence of pathologic findings among different age groups. *Human Pathology* 7:195–204, 1976.

Bendig, D.W., D.K. Jeilheimer, M.L. Wagner, G.D. Ferry, and G.M. Harrisson. Complications of gastroesophageal reflux in patients with cystic fibrosis. *J. Pediatr.* 100:536–540, 1982.

Berg, M.M., K.J. Kim, R.L. Lubman, and E.D. Crandall. Hydrophilic solute transport across rat alveolar epithelium. *J. Appl. Physiol.* 66:2320–2327, 1989.

Bingham, J.B., K.A. McKusick, H.W. Strauss, C.A. Boucher, and G.M. Pohost. Influence of coronary artery disease on pulmonary uptake of thallium-201. *Am. J. Cardiol.* 46:821–826, 1980.

Breakey, A.S., C.T. Dotter, and I. Steinberg. Pulmonary complications of cardiospasm. *N. Eng. J. Med.* 245:441–447, 1951.

Brown, C.M., C.F. Snowdon, B. Slee, L.N. Sandle, and W.D.W. Rees. Measurement of bicarbonate output from the intact human oesophagus. *Gut* 34:872–880, 1993.

Carter, E.P., M.A. Matthay, J. Farinas, and A.S. Verkman. Transalveolar osmotic and diffusional water permeability in intact mouse lung measured by a novel surface fluorescence method. *J. Gen. Physiol.* 108:133–142, 1996.

Chinard, F.P., W.O. Cua, and V. Bower. Response of lung microvasculature to sulfhydryl reagents. *FASEB J.* 8:A1036, 1994.

Chinard, F.P., T. Enns, and M.F. Nolan. The permeability characteristics of the alveolar capillary barrier. *Trans. Assoc. Am. Physicians* 75:253–261, 1962.

Chinard, F.P., G.H. Vosburgh, and T. Enns. Transcapillary exchange of water and of other substances in certain organs of the dog. *Am. J. Physiol.* 183:221–234, 1955.

Clarke, L.L., B.R. Grub, J.R. Yankaskas, C.U. Cotton, A. McKenzie, and R.C. Boucher. Relationship of a non-cystic fibrosis transmembrane conductance regulator-mediated chloride conductance to organ-level disease in Cftr(-/-) mice. *Proc. Natl. Acad. Sci. USA* 91:479–483, 1994.

Colin, G. (1873) De l'absorption dans les voies aeriennes. *Physiol. Compares des Animaux.* Paris: Bailliere et Fils.

Cox, K.L., J.N. Isenberg, and M.E. Ament. Gastric acid hypersecretion in cystic fibrosis. *J. Pediatr. Gastroenterol. Nutr.* 1:559–565, 1982.

Cua, W.O., V. Bower, C. Tice, and T.P. Chinard. Pulmonary vascular extraction and distribution of antipyrine with alveolar flooding. *Am. J. Physiol.* 269(*Heart, Circ Physiol. 38*):H1811–H1819, 1995.

Cucchiara, S., F. Santamaria, M.R. Andreotti, R. Minella, P. Ercolini, V. Oggero, and G de Ritis. Mechanisms of gastro-oesophageal reflux in cystic fibrosis. *Arch. Dis. in Childhood* 66:617–622, 1991.

Dupuis, J., C.A. Goresky, C. Juneau, A. Calderone, J.L. Rouleau, C.P. Rose, and C. Goresky. Use of norepinephrine uptake to measure lung capillary recruitment with exercise. *J. Appl. Physiol.* 68:700–713, 1990.

Effros, R.M. Osmotic extraction of hypotonic fluid from the lungs. *J. Clin. Invest.* 54:935–947, 1974.

Effros, R.M. and C. Darin. Efficient anion transport is essential for prompt neutralization of inspired acid: A possible model of lung injury in cystic fibrosis. *Am. J. Resp. Crit. Care Med.* p. A741, 1995.

Effros, R.M. and R.S.Y. Chang. Distribution of water and proteins in the lungs in pulmonary edema. In: *Pulmonary Edema,* edited by A.P. Fishman and E.M. Renkin. Bethesda, MD: American Physiological Society, pp. 137–144, 1979.

Effros, R.M., G.R. Mason, and P. Silverman. Role of perfusion and diffusion in $^{14}CO_2$ exchange in the rabbit lung. *J. Appl. Physiol.* 51:1136–1144, 1981.

Effros, R.M., G.R. Mason, E. Reid, L. Graham, and P. Silverman. Diffusion of labeled water and lipophilic solutes in the lung. *Microvasc. Res.* 29:45–55, 1985.

Effros, R.M., G.R. Mason, K. Sietsema, J. Hukkanen, and P. Silverman. Pulmonary epithelial sieving of small solutes in rat lungs. *J. Appl. Physiol.* 65:640–648, 1988.

Effros, R.M., et al. Continuous measurements of changes in pulmonary capillary surface area with ^{201}Tl infusions. *J. Appl. Physiol.* 77:2093–2103, 1994.

Effros, R.M., C. Darin, E.R. Jacobs, R.A. Rogers, G. Krenz, and E.E. Schneeberger. Water transport and the distribution of aquaporin-1 in pulmonary air spaces. *J. Appl. Physiol.* 83:1002–1016, 1997.

Farber, S. Some organic digestive diseases in early life. *J. Mich. Med. Soc.* 44:587–594, 1945.

Fiegelson, J., F. Girault, and Y. Pecaue. Short communication: Gastro-oesophageal reflux and esophagitis in cystic fibrosis. *Acta Paediatr. Scand.* 76:989–990, 1987.

Flemström, G. Gastric and duodenal mucosal secretion of bicarbonate. In: *Physiology of the Gastrointestinal Tract,* 3rd edition, edited by L.R. Johnson. New York: Raven Press, pp. 1285–1309, 1984.

Folkesson, H., M.A. Matthay, H. Hasegawa, F. Kheradmand, and A.S. Verkman. Transcellular water transport in lung alveolar epithelium through mercurial-sensitive water channels. *Proc. Natl. Acad. Sci.* 91:4970–4974, 1994.

Garrick, R.A., U.S. Ryan, V. Bower, W.O. Cua, and F.P. Chinard. The diffusional transport of water and small solutes in isolated endothelial cells and erythrocytes. *Biochim. Biophys. Acta* 1148(1):108–116, 1993.

Gill, J.B., T.D. Ruddy, J.B. Newell, D.M. Finkelstein, H.W. Strauss, and A. Boucher. Prognostic importance of thallium uptake by the lungs during exercise in coronary artery disease. *N. Engl. J. Med.* 317:1485–1489, 1987.

Goodman, B.E. and E.D. Crandall. Dome formation in primary cultured monolayers of alveolar epithelial cells. *Am. J. Physiol.* 243:C96–C100, 1982.

Hasegawa, H., S.C. Lian, W.E. Finkbeiner, and A.S. Verkman. Extrarenal tissue distribution of CHIP28 water channels by in situ hybridization and antibody staining. *Am. J. Physiol.* 266:C893–C903, 1994a.

Hasegawa, H., T. Ma, W. Skach, M.A. Matthay, and A.S. Verkman. Molecular cloning of a mercurial-insensitive water channel expressed in selected water-transporting tissues. *J. Biol. Chem.* 269:5497–5500, 1994b

Helm, J.F., W.J. Dodds, and W.J. Hogan. Salivary response to esophageal acid in normal subjects and patients with reflux esophagitis. *Gastroenterology* 92:1393–1397, 1987.

Kennedy, H.H. "Silent" gastroesophageal reflux: An important but little known cause of pulmonary complications. *Dis. of the Chest* 42:42–45, 1962.

King, L.S. and P. Agre. Pathophysiology of the aquaporin water channels. [Review] *Ann. Rev. Physiol.* 58:619–648, 1996.

King, L.S., S. Nielsen, and P. Agre. Aquaporin-1 water channel protein in lung: Ontogeny, steroid-induced expression, and distribution in rat. *J. Clin. Invest.* 97:2183–2191, 1996.

Krauthammer, M.J., J. Rinderknecht, K. Taplin, K. Wasserman, J.M. Uszler, and R.M. Effros. Enhanced diffusion of small solutes across the pulmonary epithelium in pulmonary fibrosis. *Chest* 72:403, 1977.

Krenz, G. and R.M. Effros. Mathematical models of clearance of airway water. *FASEB J.* 9:A570, 1995.

Lea, E.J.A. Permeation through long narrow pores. *J. Theor. Biol.* 5:102–107, 1963.

Lee, M.D., K.Y. Bhakta, S. Raina, R. Yonescu, C.A. Griffin, N.G. Copeland, D.J. Gilbert, N.A. Jenkins, G.M. Preston, and P. Agre. The human Aquaporin-5 gene. Molecular characterization and chromosomal localization. *J. Biol. Chem.* 271(15):8599–8604, 1996.

Lightfoot, E.N., J.B. Bassingthwaighte, and E.F. Grabowski. Hydrodynamic models for diffusion in microporous membranes. *Ann. Biomed. Engin.* 4:78–90, 1976.

Malfroot, A. and I. Dab. New insights on gastro-oesophageal reflux in cystic fibrosis by longitudinal follow up. *Dis. in Childhood,* 66:617–622, 1991.

Mason, R.J., M.C. Williams, J.H. Widdicombe, M.J. Sanders, D.S. Misfeldt, and L.C. Berry, Jr. Transepithelial transport by pulmonary alveolar type II cells in primary culture. *Proc. Natl. Acad. Sci. USA* 79:6033–6037, 1982.

Moura, T.R., R.I. Macey, D.Y. Chien, D. Daran, and H. Santos. Thermodynamics of all-or-nothing channels closure in red cells. *J. Membr. Biol.* 81:105–111, 1984.

Paganelli, C.V. and A.K. Solomon. The rate of exchange of tritiated water across the human red cell membrane. *J. Gen. Physiol.* 41:259–277, 1957.

Pappenheimer, J.R., E.M. Renkin, and L.M. Borrero. Filtration, diffusion and molecular sieving through peripheral capillary membranes. A contribution to the pore theory of capillary permeability. *Am. J. Physiol.* 167:13–46, 1951.

Perl, W., P. Chowdhury, and P.P. Chinard. Reflection coefficients of dog lung endothelium to small hydrophilic solutes. *Am. J. Physiol.* 228:797–809, 1975.

Poulsen, J.H., H. Fischer, B. Illek, and T.E. Machen. Bicarbonate conductance and pH regulatory capability of cystic fibrosis transmembrane conductance regulator. *Proc. Natl. Acad. Sci. USA* 91:5340–5344, 1994.

Preston, G.M., B.L. Smith, M.L. Zeidel, J.J. Moulds, and P. Agre. Mutations in aquaporin-1 in phenothypically normal humans without functional CHIP water channels. *Science* 265:1585–1587, 1994.

Raina, S., G.M. Preston, W.B. Guggino, and P. Agre. Molecular cloning and characterization of an aquaporin cDNA from salivary, lacrimal, and respiratory tissues. *J. Biol. Chem.* 270:1908–1912, 1995.

Renkin, E.M. Transport of potassium-42 from blood to tissue in isolated mammalian skeletal muscles. *Am. J. Physiol.* 197:1205–1210, 1959a.

Renkin, E.M. Exchangeability of tissue potassium in skeletal muscle. *Am. J. Physiol.* 197:1211–1215, 1959b.

Renkin, E.M. and S. Rosell. Effects of different types of vasodilator mechanisms on vascular tonus and on transcapillary exchange of diffusible material in skeletal muscle. *Acta. Physiol. Scand.* 54:241–251, 1962.

Rinderknecht, J., L. Shapiro, M. Krauthammer, G. Taplin, K. Wasserman, J.M. Uszler, and R.M. Effros. Accelerated clearance of small solutes from the lungs in interstitial lung disease. *Am. Rev. Resp. Dis.* 121:105–117, 1980.

Schnitzer, J. and P. Oh. Aquaporin-1 in plasma membrane and caveolae provides mercury-sensitive water channels across lung endothelium. *Am. J. Physiol.* 270 (*Heart Circ. Physiol.*):H416–H422, 1996.

Sidel, V.W. and A.K. Solomon. Entrance of water into human red cells under an osmotic pressure gradient. *J. Gen. Physiol.* 41:243–257, 1957.

Smith, J.J. and M.J. Welsh. cAMP stimulates bicarbonate secretion across normal, but not cystic fibrosis airway epithelia. *J. Clin. Invest.* 89:1148–1153, 1992.

Taylor, A.E. and K.A. Gaar. Estimation of equivalent pore radii of pulmonary capillary and alveolar membranes. *Am. J. Physiol.* 218:1133–1140, 1970.

Titjen, P.A., R.J. Kaner, and C.E. Quinn. . Aspiration emergencies. *Clin. Chest Med.* 15:117–135, 1994.

Van Hoek, A.N., M.L. Horn, L.H. Luthjens, M.D. DeJong, J.A. Dempster, and C.H. van Os. Functional unit of 30 kDa for proximal tubule water channels as revealed by radiation inactivation. *J. Biol. Chem.* 226:1633–16635, 1991.

Van Lieburg, A.F., M.A. Verdijk, V.V. Knoers, A.J. van Essen, W. Proesmans, R. Mallmann, L.A. Monnens, B.A. van Oost, C.H. van Os, and P.M. Deen. Patients with autosomal nephrogenic diabetes insipidus homozygous for mutations in the aquaporin 2 water-channel gene. *Am. J. Human Genetics* 55:648–652, 1994.

Van Hoek, A.N., L.H. Luthjens, M.L. Horn, C.H. Van Os, and J.A. Dempster. A 30kDa functional size for the erythrocyte water channel determined in situ by radiation inactivation. *Biochem. Biophys. Res. Commun.* 184:1331–1338, 1992.

Verkman, A. *Water Channels.* Austin, TX: R.G. Landes Co, 1993.

Wangensteen, O.D., H. Bachofen, and E.R. Weibel. Lung tissue volume changes induced by hypertonic NaCl: Morphometric evaluation. *J. Appl. Physiol.* 51:1443–1450, 1981.

21

The Transport of Small Molecules Across the Microvascular Barrier as a Measure of Permeability and Functioning Exchange Area in the Normal and Acutely Injured Lung

Thomas R. Harris

Introduction

Most studies of capillary exchange in the lung have been aimed at the issue of fluid balance and an examination of factors that influence pulmonary edema. The physiology of capillary fluid exchange was established in quantitative form by Pappenheimer et al. (1951), who did their original work in peripheral systemic capillaries. Chinard and Enns (1954) pioneered the application of trace injections of radioisotopes for the estimation of extravascular water in the lung and may be said to have originated the use of the indicator dilution method in the lung. Crone (1963) and Renkin (1959) were the first to develop methods for the computation of permeability–surface area (*PS*) in capillaries from multiple tracer studies.

The 1970s saw renewed interest in fluid balance and permeability in the lung as the clinical entity of Adult Respiratory Distress Syndrome (ARDS) was defined by Ashbaugh et al. (1967). It was soon recognized that this syndrome involved pulmonary edema caused by excessive permeability in the lung capillaries and epithelial cells, as differentiated from classical pulmonary edema stemming from left heart failure and the resulting increase in pulmonary capillary pressure. A number of investigators began long-term research into the basic physiology of lung fluid balance. These studies have been reviewed by Taylor and Parker (1985), Bernard and Brigham (1984), Roselli and Harris (1989), and Harris and Roselli (1989).

The mortality associated with ARDS has remained at 50–60% since it was first described (Canonico and Brigham, 1997). However, there has been some change in criteria for the disease. The reported incidences range from 50 per 100,000 population to 5 per 100,000, depending on exclusion criteria. ARDS is defined as hypoxemia, patchy diffuse infiltrates on lung X-ray, low pulmonary compliance, and normal left ventricular filling pressure (Bernard and Brigham, 1984). The name of the syndrome has evolved to Acute Respiratory Distress Syndrome to differentiate its features from a more chronic form of fibrotic lung injury. It is associated with sepsis, trauma, complications of surgery, pancreatitis, fat embolism from long bone trauma, blood-borne and inhaled toxins, and other incidents that activate the immune system (Canonico and Brigham, 1997). The most serious and refractory cases are associated with sepsis syndrome and multiple organ failure (MOF), which appears to be a generalized microcirculatory disorder of vital organs. Numerous experimental studies in animals and isolated endothelial cell systems have generated a long list of candidate mediators expressed by the immune systems (Canonico and Brigham, 1997). Numerous blockers of the actions of these materials have been proposed and studied, but definitive therapy does not exist. However, increased capillary permeability is recognized as an important feature in both ARDS and MOF. This has generated interest in methods for the measurement and monitoring of this quantity in experimental animals and patients.

In addition to capillary permeability, precise characterization of ARDS benefits from knowledge of lung blood flow, intravascular and extravascular volumes of blood and edema fluid, and capillary surface area. These quantities and others can interact in complex ways to limit precise definition of the site of action of a candidate therapy. In the last 5 years, three methods have emerged for measuring major lung vascular functions, principally defined as pulmonary blood flow, exchange surface area, extravascular and intravascular volumes in the lung, and lung vascular permeability. These are the gamma-emitter scanning (GES) technique, which uses labeled macromolecules and blood markers (Byrne and Sugerman, 1988; Gorin et al., 1978; Roselli and Riddle, 1989), positron emission tomography (PET; Schuster, 1989), and indicator dilution (ID; Harris and Roselli, 1989; Harris et al., 1990; Rickaby, et al., 1981; Gillis and Catravas, 1982).

The GES method allows estimates of the pulmonary transcapillary escape rate (PTCER) constants for macromolecules. It requires sampling of the venous blood. Disadvantages include a limited availability of labeled molecules (since gamma-emitters are required), relatively high doses of radioactivity, poor spatial image quality, a limitation to the study of the transport of large molecules for the measurement of capillary permeability, relatively long counting times, and the existence of motion artifacts, which is related to these long counting periods.

PET has several significant advantages. It can provide spatially specific measures of extravascular lung water, vascular volume, PTCER, and blood flow. Further, a great variety of metabolites can be labeled which are short-lived. In spite of this power, there are some disadvantages to PET. It is a complex, virtually nonportable technology requiring an on-site cyclotron. Only a single tracer probe can be used for each analysis, that is, simultaneous multiple-tracer studies are not possible. The counting times for an entire scan are relatively long.

The use of indicator dilution methods in the lung is the primary subject of this chapter. Its main features are discussed in the following section.

The Use of the Indicator Dilution Method in the Lung

Experimental Features

Typical lung indicator dilution curves exhibit separation between red cell markers, plasma markers, diffusing tracers, and labeled water, as shown in Figure 21.1. A number of radioisotope ID studies of the lung have used the following tracer mixture (Harris et al., 1990; Olson et al., 1991): reference tracers of ^{51}Cr-labeled erythrocytes (25 µCi) and ^{125}I-labeled ovine serum albumin (5 µCi), a diffusing tracer ($[^{14}C]$-urea, -1,4-butanediol, or -1,2-propanediol, 8 µCi), and tritiated water (^{3}HOH, 50 µCi). Each study consists of a radioactive multiple-tracer bolus injection into the arterial side of the lung and the withdrawal of venous blood samples from a left atrial cannula or a peripheral artery in intact preparations. Radiolabeled tracer volumes vary from 1.5 to 3 ml.

After the isotope is injected into the arterial catheter, 40 sequential venous blood samples (1 sample/s) are collected into heparinized tubes on a rotating wheel. A 0.5-ml sample is pipetted from each aliquot, 3 ml of alcohol are added, and the samples are then centrifuged to isolate the beta radiation emitters. From a supernatant containing these beta emitters, 2 ml are pipetted into vials, which are then counted in a liquid scintillation counter for ^{14}C and ^{3}H activity. Radiation is detected from 0.5-ml gamma samples from each tube and from injectate diluted in the subject animal's or patient's blood. Corrections are made for isotope overlap and quench.

Reconstruction of the MID curves consists of first normalizing counts to injectate concentrations and then fitting each curve to a monoexponential decay. Areas under all tracer concentration–time curves are computed and inspected for similarities in areas. Occasional discrepancies traceable to systematic sampling errors are corrected by altering concentrations by a constant factor, restricting curve areas to match the ^{51}Cr-erythrocyte curve area.

Research has established that the indicator dilution method has a number of distinct advantages for evaluating lung vascular function. These are:

1. The separation of a diffusing tracer curve from its appropriate intravascular reference is a direct observation of microvascular transport and therefore must contain information about the state of the microcirculation at the time of indicator passage (Harris and Brigham, 1982; Bassingthwaighte and Goresky, 1985).
2. It is a rapid measurement, taking only 30–40 s, and therefore has the potential to track dynamic changes in the microcirculation (Harris et al., 1985).
3. It has been shown to quantitatively measure extravascular lung water, microvascular permeability–surface area product, and parameters that characterize saturable uptake by the endothelium (Rickaby et al., 1981; Harris and Brigham, 1982; Lewis et al., 1982; Syrota et al., 1982).
4. Parameters derived from indicator curves are altered by lung vascular damage

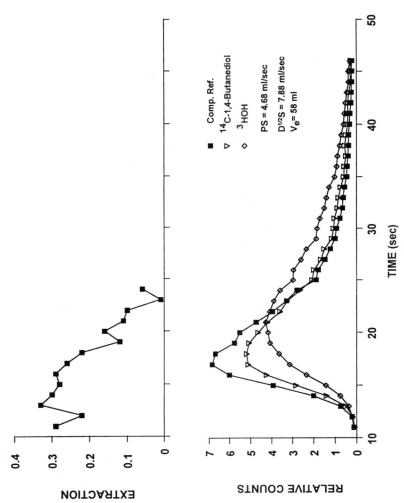

FIGURE 21.1. Typical indicator curves from the isolated perfused dog lung. The reference curve is a composite of ^{51}Cr-labeled red blood cells and ^{125}I-labelled albumin computed from Eq. (21.5) (Olson et al., 1991).

in animal experiments (Harris et al., 1976, 1978; Zelter et al., 1984; Bradley et al., 1988).

5. It can be performed in patients under intensive care and provides measures of microvascular function that alter with severity of respiratory distress (Brigham et al., 1979, 1983; Rinaldo et al., 1986; Harris et al., 1990).

6. Some tracers can be used as nonradioactive markers and, when appropriate instrumentation is used, can provide rapid readings of pulmonary blood flow, extravascular water volume, and capillary PS (Basset et al., 1981; Leksen et al., 1990; Olson et al., 1994, 1995).

Mathematical Theory

Capillary Transport Models

Three mathematical models of transcapillary exchange of small nonmetabolized solutes across the lung microvascular barrier have been used extensively. The easily applied Crone–Renkin (CR) model assumes that back-diffusion from the extravascular space is negligible before the appearance of the curve peak (Crone, 1963; Renkin, 1959). The Sangren–Sheppard (SS) model allows for back-diffusion, but after transcapillary movement the tracer is assumed to equilibrate in the extravascular space along a direction perpendicular to flow (Sangren and Sheppard, 1953). However, the one-parameter Effective Diffusivity (ED) model does not assume that the material is perfectly mixed in the extravascular region and incorporates the capillary permeability and extravascular diffusivity into an ED (Haselton et al., 1984a, 1984b).

For the CR, SS, and ED models, the differential equation describing the movement of indicator dilution tracers within the lung capillaries is as follows:

$$\frac{\partial C_D}{\partial t} + \frac{F_c}{V_c}\frac{\partial C_D}{\partial x'} = -N_D S', \qquad (21.1)$$

where C_D is the diffusing tracer concentration within whole blood in the capillary, F_c is capillary flow of the phases containing the tracer, $N_D S'$ is the rate of loss of material from the vascular phase, t is time, V_c is the intracapillary volume of the tracer-containing phases, and x' is the distance from the capillary entrance normalized to capillary length. This intravascular equation is based on the assumption that the diffusing tracer is equilibrated between the red cell and plasma phases. The following expressions characterize the diffusional properties assumed by each model: for the CR model,

$$N_D S' = \frac{PS_{\text{cap}}}{V_c}(C_D); \qquad (21.2)$$

for the SS model,

$$N_D S' = \frac{PS_{\text{cap}}}{V_c}(C_D - C'_{ID}); \qquad (21.3)$$

and for the ED model,

$$N_D S' = -\frac{DS}{V_c} \frac{\partial C'_{ID}}{\partial z}\bigg|_{z=0} \qquad (21.4)$$

where PS_{cap} is the permeability–surface area product, C'_{ID} is the concentration of diffusing tracer in the extravascular space, S is the surface area, and z represents the interstitial distance perpendicular to the direction of flow that is accessible to the diffusing tracer.

When tracers equilibrate rapidly between red cells and plasma, the red cell markers and the plasma markers can be combined to constitute a single composite reference. This behavior has been established for [14C]-urea (Parker et al., 1986), [14C]-1,3-propanediol (Harris et al., 1987), and [14C]-1,4-butanediol (Olson et al., 1991). Thus, the equation suggested by Goresky et al. (1969) can be used:

$$C_{comp} = \frac{[0.7(Hct)C_{RBC} + 0.94(1 - Hct)C_{PL}]}{0.7[Hct] + 0.94[1 - Hct]}, \qquad (21.5)$$

where Hct is the hematocrit, and C_{RBC} and C_{PL} represent the concentration of the reference indicator in the red cell and plasma phases, respectively. For reference indicators (C_R) including composite references, Eq. (21.1) holds, but since by definition there is no loss of material from the vascular space, $N_D S'$ is zero:

$$\frac{\partial C_R}{\partial t} + \frac{F_c}{V_c} \frac{\partial C_R}{\partial x'} = 0, \qquad (21.6)$$

where x' is the capillary length normalized to total capillary length.

The CR model (Crone, 1963) equation for computing PS_{cap} is based on the initial and boundary conditions of the negligible back-diffusion assumption and the integration of Eq. (21.1). That expression is given as:

$$PS_{cap} = -F_c \log_e[1 - E_i], \qquad (21.7)$$

where E_i is the integral extraction of C_D from C_R as shown below:

$$E_i = \frac{\int_0^{t_p}(C_R - C_D)\, dt}{\int_0^{t_p} C_R\, dt}, \qquad (21.8)$$

and t_p is the appearance time of the reference curve peak.

F_c is pulmonary water flow and is calculated from Goresky et al. (1969) as follows:

$$F_c = [0.7(Hct) + 0.94(1 - Hct)]\ (\text{pulmonary blood flow}). \qquad (21.9)$$

The SS model requires Eqs. (21.1) and (21.3) and an equation for the concentration of tracer in the extravascular space, C'_{ID}:

$$\frac{\partial C'_{ID}}{\partial t} = \frac{PS_{cap}}{V_1}(C_D - C'_{ID}). \qquad (21.10)$$

Here, V_I is the volume of distribution of the tracer in the extravascular space.

The ED model, like the SS model, allows for back-diffusion, but it also combines microvascular permeability (P), the thickness of the interstitial space (Z), and extravascular diffusivity (D_e) into a single equivalent diffusivity (D) parameter as follows (Haselton, 1989a, 1989b):

$$\frac{1}{D} = \frac{1}{D_e} + \frac{3}{ZP}.$$
(21.11)

The solution of this model in the Laplace domain at $x' = 1$ results in the following equation:

$$\bar{C}_D(s,1) = \bar{C}_R(s,1) \exp\left[-s^{1/2} \frac{D^{1/2}S}{F_c}\right],$$
(21.12)

where s is the Laplace transform variable. The analysis of experimental indicator curves yields $D^{1/2}S$. This model has the attractive feature of describing experimental data from the lung with a single parameter that is determined from curve-fitting experimental indicator curves.

Organ Models

Tracer outflow from an organ can be depicted by the following equation:

$$C_{out}(t) = C_{in}(t) * h_{LV}(t) * h_{mc}(t),$$
(21.13)

where C_{out} is the tracer outflow curve, C_{in} is the input tracer curve, h_{LV} is the large vessel transport function (impulse response), $h_{mc}(t)$ is the microcirculatory transport function, and * indicates mathematical convolution. A fundamental assumption of the multiple-indicator method is that C_{in} and $h_{LV}(t)$ are the same for both diffusing and reference tracers. Then, if the functions are linear, the Laplace transform (designated by a bar over the function) of Eq. (21.13) is given as follows:

$$\bar{C}_{out} = \bar{C}_{in} \cdot \bar{h}_{LV} \cdot \bar{h}_{mc}.$$
(21.14)

Several models of h_{LV} and h_{mc} have been proposed. The simplest and most often used has been the assumption of uniform capillary transport functions in which h_{mc} is considered the transport function of the capillary bed that is assumed to be identical for all parallel branches within the organ. Then

$$\bar{C}_D = \bar{C}_R \cdot \bar{h}_{capD} / \bar{h}_{capR}.$$
(21.15)

If the large vessel transport function is considered to be uniform and equal to a time delay, then the C_R function may be considered to be a measure of the capillary transit time (λ) distribution, and

$$C_D = \int_{t_a}^{t} C_R(\lambda - t_a) h_{capD}(\lambda) \, d\lambda,$$
(21.16)

where t_a is the appearance time of the indicator curve. This model is usually designated as Model II of Rose and Goresky (1976); it has been applied to the lung by Rowlett and Harris (1976). A more complex version (Model III of Rose and Goresky, 1976) has been proposed for the analysis of exchange in the coronary circulation. Model III considers that the longer large vessel transit times are connected to the longer capillary transit times, the shorter to the shorter, and so on. This model has not been reported for the analysis of lung curves. Models II and III rely on the reference curve to provide information regarding capillary transit time distributions.

Another class of heterogeneous transit time models relies on the assumption that the transit times of the microcirculatory bed are randomly coupled to the large vessel transit times (King et al., 1996). For this case, the common large vessel transit times cancel in the Laplace domain, and a form of Eq. (21.15) is obtained:

$$\bar{C}_D = \bar{C}_R \cdot \bar{h}_{mcD} / \bar{h}_{mcR} , \qquad (21.17)$$

where $\bar{h}_{mcD}$ is the transfer function of the diffusing tracer exchange in the microcirculatory network, and $\bar{h}_{mcR}$ is the transfer function of the reference tracer in that network. Kuikka et al. (1986) have applied a form of this model to indicator curves taken from the heart. The microcirculatory network transport function is considered to be a capillary network of parallel paths with differing transit times. These authors used experimental radioactive microsphere-deposition data to determine the distribution of flow in the myocardium and utilized assumptions regarding capillary density to construct a network model that incorporated the capillary transfer functions. If the flows in the various flow paths are F_i flow rates per unit weight of tissue, and the tissue weights are W_i, then Eq. (21.17) is

$$C_{out}(t) = C_{in}(t) * h_{LV}(t) * \sum_{i=1}^{n} \left(\frac{F_i W_i}{F_{tot} W_{tot}} h_{cap,i} \right), \qquad (21.18)$$

where $F_{tot} = \sum_{i=1}^{n} F_i$ and $W_{tot} = \sum_{i=1}^{n} W_i$. Then Eq. (21.17) becomes

$$\bar{C}_D = \bar{C}_R \cdot \frac{\displaystyle\sum_{i=1}^{n} F_i W_i \bar{h}_{capD,i}}{\displaystyle\sum_{i=1}^{n} F_i W_i \bar{h}_{capR,i}} . \qquad (21.19)$$

For the ED model, this equation is

$$\bar{C}_D(s) = \bar{C}_R(s) \cdot \frac{\displaystyle\sum_{i=1}^{n} F_i W_i \exp\left[-\frac{V_{ci}}{F_i}\left(s + \frac{D^{1/2}S_i}{V_{ci}} s^{1/2} \right) \right]}{\displaystyle\sum_{i=1}^{n} F_i W_i \exp\left[-\frac{V_{ci}}{F_i}s \right]}, \qquad (21.20)$$

where the i subscript on $D^{1/2}S$ and V_c refers to quantities per unit weight. For the SS model, this equation is

$$\overline{C}_D(s) = \overline{C}_R(s) \cdot \frac{\displaystyle\sum_{i=1}^{n} F_i W_i \exp\left[-\frac{V_{ci}}{F_i}\left(s + \frac{sPS_i/V_{ci}}{s + PS_i/V_{li}}\right)\right]}{\displaystyle\sum_{i=1}^{n} F_i W_i \exp\left[-\frac{V_{ci}}{F_i}s\right]}. \qquad (21.21)$$

If $D^{1/2}S_i/V_{ci}$, PS_i/V_{ci}, and V_{li}/V_{ci} are considered to be constant through the organ, and a reasonable value of V_{ci} is assumed, these equations can be fit to experimental indicator curves from the heart or lungs with accuracy. Direct experimental knowledge of flow distribution is needed. The equation predictions are relatively insensitive to the value of V_{ci}, and therefore this quantity cannot be easily identified by curve fitting. These equations have been applied to heart and lung data (Kuikka et al., 1986; Overholser et al., 1991; Caruthers et al., 1995).

King et al. (1996) have broadened the definition of h_{mci} to include a capillary segment (as used above) and small artery and vein transport functions. This addition would allow simulation of a complex network, but may offer significant problems in parameter identification.

These models offer a considerable ability to accommodate flow distribution data. Comparison of Eqs. (21.15) and (21.17) shows that the network models could be considered to be models that do not assume a uniform capillary transport function, but do assume the lung to be made up of one or more uniformly distributed networks.

A phenomenon excluded by the previous analysis is the case where capillaries are recruited or derecruited. The assumption that V_{ci} is a constant throughout the organ disallows recruitment. To address this issue, Overholser et al. (1991) developed a variable recruitment (VR) model. The VR model takes into account both flow heterogeneity and changing surface area in the heart. Caruthers et al. (1995) extended this concept to the lung, as is illustrated in Figure 21.2. The basic assumption of the VR model was that for each piece of the organ there is a self-similar relationship between $D^{1/2}S$ and flow, as illustrated in Figure 21.2. As flow within a piece increases, more capillaries are recruited, thus increasing the effective surface area. At one point, termed flow at full recruitment (F_{FR}), all of the capillaries have been recruited and $D^{1/2}S$ (or PS and V_l if the SS model is used) no longer increase with an increase in flow. The value at which $D^{1/2}S$ saturates is called $D^{1/2}S$ at full recruitment ($D^{1/2}S_{FR}$).

Each piece of the organ may not have the same flow. For each piece, however, the same $D^{1/2}S$ versus flow relationship is assumed to exist and to have the same $D^{1/2}S_{FR}$ and F_{FR} (normalized to tissue weight) as the whole organ. In Figure 21.2, one piece may have a low flow; its $D^{1/2}S_i$ is also low, as is shown by the arrows in the figure. Another piece could have a greater flow, and a greater $D^{1/2}S_i$. Still other pieces, although differing in flow, would have the same $D^{1/2}S_i$ because flow is greater than F_{FR} and all capillaries are recruited in these pieces. The $D^{1/2}S$ per unit weight for the whole organ is the weighted sum of $D^{1/2}S_i$ for all of the pieces weighted by mass $\sum_i W_i D^{1/2}S_i / \sum_i W_i$). Thus the parameters of Eq. (21.20) would be defined as

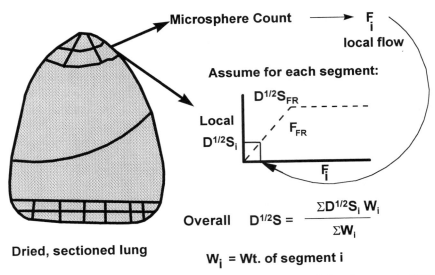

FIGURE 21.2. Diagram of variable recruitment (VR) model of Caruthers et al. (1995).

$$D^{1/2}S_i = \begin{cases} D^{1/2}S_{FR} & \text{if } F_i \geq F_{FR} \\ F_i\left(\dfrac{D^{1/2}S_{FR}}{F_{FR}}\right) & \text{if } F_i \leq F_{FR} \end{cases} \qquad (21.22)$$

Similarly, V_{ci}, the intravascular volume per unit weight of piece i, is assumed to increase linearly with flow until, when full recruitment is achieved, V_{ci} equals V_{cFR} and increases no more. Analogous relations can be derived for the SS model parameters. Then curve fitting is based on identifying the optimal value of $D^{1/2}S_{FR}$ or PS_{FR} and V_{IFR} for the SS model.

Current Issues in the Application of ID Methods to the Lung

Studies Applying Indicator Dilution in Patients

As previously discussed, ID methods may be useful in studies of ARDS patients. Such studies have been undertaken by Brigham et al. (1983), Rinaldo et al. (1986), and Harris et al. (1990). Harris et al. (1990) undertook a five-center cooperative study of ARDS patients that included baseline evaluation and progress under treatment with methylprednisolone or placebo. Hemodynamic and lung function studies including multiple-indicator dilution curves were undertaken. The results

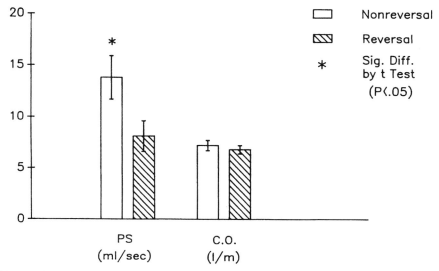

FIGURE 21.3. Mean values of *PS* for [^{14}C]-urea computed from the Crone model, Eqs. (21.8) and (21.9), and cardiac output (C.O.) for ARDS patients who reversed and those who did not reverse signs of respiratory distress (Harris et al., 1990). *PS* is significantly increased in nonreversal patients.

showed that ARDS patients could be divided into two groups: those in whom the oxygenation and radiographic signs of ARDS were reversed and those who failed to reverse these signs. This second group was presumed to have a more severe pulmonary abnormality. As is shown in Figure 21.3, values of urea *PS* computed by the Crone method for the lung were significantly higher in the nonreversal group than in the reversal group. Further, the nonreversal group showed a number of correlations between hemodynamic and transport quantities that were not seen in the reversal group. Of particular interest is the inverse relationship between *PS* and pulmonary vascular resistance, *PVR,* shown in Figure 21.4. It was speculated that the presence in the severely ill group and absence in the improving group of these correlations indicated that some compensation existed in the improving group that reduced perfusion to injured areas of these lungs. This protection seemed absent in the nonreversal patients and led to a greater variation in *PS.* This suggested that *PS* and resistance were related in severely injured lungs in a different manner than that seen in less injured lungs.

The Use of Amphipathic Tracers to Differentiate Capillary Surface Area and Permeability in the Lung

The patient research discussed above and experiments with endotoxin injury in sheep (Bradley et al., 1988) suggested that realistic endothelial injuries were a

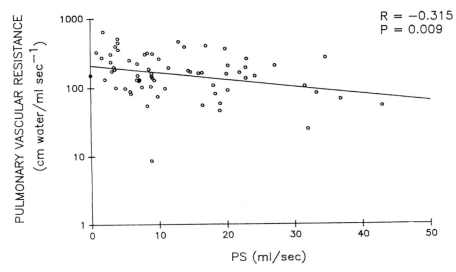

FIGURE 21.4. Log_{10} of pulmonary vascular resistance correlates with [^{14}C]-urea PS (Crone model) in ARDS patients failing to reverse the signs of respiratory failure ($R = -0.315$, $p = 0.009$). No such correlation exists for patients who reversed the signs of respiratory distress (Harris et al., 1990).

complex combination of surface area and permeability changes. Two ID methods have been studied as measures of surface area. Audi et al. (1994) have used labeled diazepam as a flow-limited marker of surface area. Olson et al. (1991) have studied amphipathic butanediol as a marker of surface area. Olson et al. (1991) undertook a series of experiments to determine whether [^{14}C]-1,4-butanediol would be more influenced by capillary surface area during lung injury than would [^{14}C]-urea. These experiments involved baseline indicator dilution curves in isolated perfused dog lungs and then studies in which the lung mass was reduced, curves repeated, and permeability was increased with alloxan. Another series was studied in which permeability was increased and lung size was reduced. In this way, sets of indicator curves were taken for surface area change, permeability change, and combined permeability and surface area change. Both urea and butanediol PS and $D^{1/2}S$ measurements were made. Then, ratios of PS and $D^{1/2}S$ were evaluated statistically to determine whether the ratio would eliminate surface effects and provide a unique measure of permeability regardless of the amount of capillary surface active in the lung. If the ratio of $D^{1/2}S$ for butanediol and urea is squared, the result is Du/Db. This ratio for lungs in which changes in both surface area and permeability were made is shown in Figure 21.5. Clearly, the ratio is statistically sensitive to permeability change even though surface area is about one-half in the reduced lung size group. Similar results were seen for the urea/butanediol ratios of PS and $D^{1/2}S$.

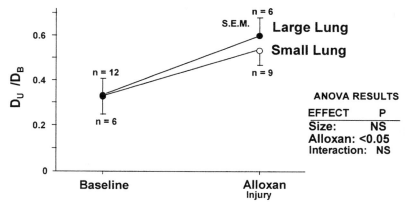

FIGURE 21.5. Comparison of the ratio of equivalent diffusivity for [^{14}C]-butanediol and [^{14}C]-urea from indicator curves taken in different sized lungs at baseline and after acute alloxan injury (Olson et al., 1991).

Adams et al. (1993) used this method to study PS and $D^{1/2}S$ in awake sheep after air embolism injury. They found that butanediol PS and $D^{1/2}S$ decreased significantly after air embolism to the lung, but urea PS and $D^{1/2}S$ did not. This suggested that the capillary surface was derecruited by the air embolism but permeability was increased since urea PS did not decrease.

Control of Capillary Surface by Flow and Pressure

The classical work on the effects of vascular pressure on capillary surface is that of West and associates (1964, 1972), who showed that alveolar pressure could affect vascular resistance and flow distribution. Mitzner and Chang (1989) have reviewed the effects of recruitment on the pulmonary vascular pressure–flow relationship. Hanson et al. (1989) have shown that increases in pulmonary arterial pressure caused by hypoxia and increased lung blood volume can recruit capillaries in Zone II lung segments viewed in vivo through a microscope over an implanted thoracic window. In addition, Dupuis et al. (1990) have shown that exercise in dogs can increase capillary permeability–surface area (PS) for norepinephrine uptake. They have interpreted this to mean that the functioning capillary surface has increased.

Some controversy exists regarding the extent to which flow can recruit capillary surface. Rickaby et al. (1987) found that airway pressure increases to Zone II conditions would derecruit capillary surface as measured from serotonin uptake in indicator dilution curves. Toivenen and Catravas (1991) found that flow increases in pump-perfused rabbit lungs would increase the surface area inferred from uptake of the angiotensin converting enzyme substrate BPAP (a polypeptide used as a diffusible indicator in indicator dilution measurements) by a factor of 3 in

Zone III. Using a method based on blue dextran efflux from the endothelium, Nelin et al. (1992) found that flow had a small effect on surface area measured by their method, but that vascular occlusion and Zone II conditions had a larger effect. In Zone II, capillary recruitment increased on the order of 30% when total flow was increased. Using microscopic tracers and morphometric techniques, Konig et al. (1993) reported that all of the capillaries in a rabbit lung were perfused under physiological conditions. Thus, investigators have observed recruitment to vary from 300 to 0%. Additional work in this area seems warranted.

Recently, Shibamoto et al. (1990) found that paraquat injury reduces the functioning capillary surface in isolated perfused dog lungs. In our laboratory, Bradley et al. (1988) found that endotoxin infusion in the awake sheep decreases *PS* for urea in Phase I (high vascular pressure phase) of the endotoxin reaction. This value increases in Phase II, the period of the reaction when protein and fluid permeability are increased. However, urea *PS* did not exceed the baseline in Phase II. Quantitative comparison of protein and urea indicator studies suggested that some capillary surface remained derecruited in Phase II, but permeability was elevated.

Overholser et al. (1994) performed a series of experiments in isolated rabbit lungs to determine the relationships among pressure, pulmonary blood flow, PVR, and capillary surface area in the uninjured lung and after hypoxia. They used

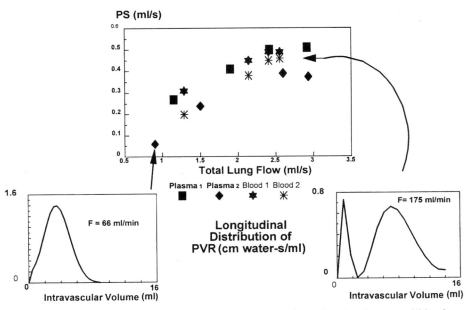

FIGURE 21.6. Crone extraction *PS* as a function of total lung flow in plasma and blood perfused rabbit lungs over Zones II and III. Note how the distribution of vascular resistance measured by the viscous bolus method changes as flow increases (Overholser et al., 1994).

labelled 1,3-propanediol for calculations of PS and $D^{1/2}S$ from indicator dilution curves, the double occlusion method to determine capillary pressure (Cope et al., 1992), and the viscous bolus method (Dawson et al., 1991) to determine PVR distribution. Flow was varied, and pressures and PVR distributions were measured at each flow. They found that PS was clearly influenced by arterial pressure. Further, hematocrit had a significant effect on this relationship. For plasma, PS increases with slight increases in pulmonary artery or capillary pressure. This suggests a mechanism where surface is recruited by a flow increase. As capillary pressure begins to rise, capillary volume increases and surface area increases. However, capillary pressure does not rise greatly because volume has increased. A related phenomenon is seen with whole blood, but surface area increase is much less sensitive to arterial pressure, which suggests a more complex situation involving red cell mechanics.

However, when PS is graphed against pulmonary flow (Figure 21.6) a close correlation is seen which is independent of hematocrit. The distribution of PVR changes as the asymptotic value of PS (full recruitment) is reached. After alveolar hypoxia under Zone II conditions, the sensitivity of PS to flow is lost and PVR distribution is altered so that resistance is concentrated in the arterial portion of the vasculature. These results suggest that, under normal circumstances, flow can control surface area precisely in Zone II, and that this control alters as full recruitment is approached. The abnormality of hypoxia curtails control of the surface area by flow in Zone II.

Effects of Increasing Total Blood Flow on Lung Capillary Recruitment and Flow Heterogeneity in Sheep

A series of experiments was performed to determine the effects of total pulmonary blood flow on two factors: the heterogeneity of blood flow as measured by the distribution of radioactive microspheres in the lung, and the $D^{1/2}S$ for ^{14}C-labeled urea and butanediol (Caruthers et al., 1995). These investigators were also interested in determining the degree to which blood flow heterogeneity could affect the computation of $D^{1/2}S$. Kuikka et al. (1986) have shown that heterogeneity measured with microspheres can affect PS calculations from coronary indicator dilution curves. Overholser et al. (1991) proposed a model of heterogeneous flow that also included recruitment of capillaries [the VR model; Eqs. (21.20) to (21.23)]. Incorporating microsphere flow data, this model allows the capillary surface area to vary directly with flow throughout the lung. The model contains two parameters, the effective diffusivity–surface product for capillary escape of small tracers when the lung is fully recruited ($D^{1/2}S_{FR}$) and the blood flow for full recruitment (F_{FR}). Data were analyzed with these two models of capillary exchange, as well as a model that assumes that flow to the lung is homogeneous, and the effects of flow on the model parameters were determined. An in situ blood perfused sheep lung with roller pump perfusion to control blood flow was used. Five flows, ranging from 1.5 to 5.0 L/min, were examined.

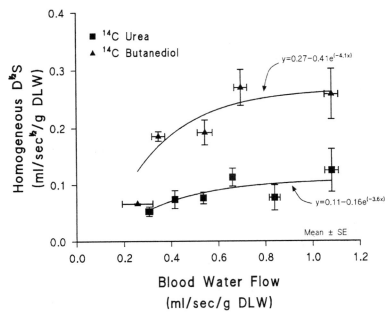

FIGURE 21.7. Effects of increasing blood flow in isolated perfused sheep lungs on $D^{1/2}S$ normalized to dry lung weight for the tracers [^{14}C]-butanediol and [^{14}C]-urea (Caruthers et al., 1995).

Pulmonary arterial and left atrial pressures were set so the lung moved from Zone II to Zone III as flow increased. At each flow, in random order, multiple-indicator dilution curves were collected using ^{14}C-labeled urea or butanediol as the diffusing tracer, and radiolabeled 15-μm microspheres were injected into the pulmonary artery. The lungs were removed, dried, cut into 8-cm³ pieces, weighed, and then counted for microsphere activity. Pieces with similar flow were then grouped to form 15 compartments of approximately equal weight. Flow heterogeneity, expressed as unitless relative dispersion (RD), decreased with increasing flow, from a high of 1.095 to a low of 0.344. The results showed that $D^{1/2}S$ increased to a maximum with increasing flow (Figure 21.7). The variable recruitment model determined that $D^{1/2}S_{FR} = 0.0119 \pm 0.013$ mls$^{-1/2}$g^{-1} and $F_{FR} = 0.640 \pm 0.025$ mls^{-1}g^{-1} for urea, and $D^{1/2}S_{FR} = 0.303 \pm 0.003$ mls$^{-1/2}$g^{-1} and $F_{FR} = 0.621 \pm 0.021$ mls^{-1}g^{-1} for butanediol (all values were normalized to dry lung weight). The ratio of $D^{1/2}S$ for urea to that for butanediol was independent of flow and averaged 0.375 ± 0.016 SEM for 21 comparable indicator curves. This provides further evidence that the ratio of these quantities is independent of surface area. Similar results were seen with PS based on the integral extraction method. As is shown in Figure 21.8, considering heterogeneous flow had little effect on computed SS PS values as compared to assuming homogeneous flow. This figure also shows that RD must be significantly larger than that seen in these uninjured lungs at full

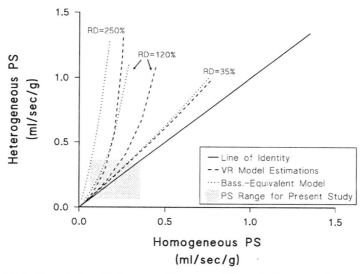

FIGURE 21.8. The relationship between *PS* computed from the Sangren–Sheppard model assuming a homogeneously perfused lung, and assuming transit time heterogeneity (Caruthers et al., 1995). The relationship varies with the relative dispersion (*RD*) of the capillary transit time distribution. Experimental *RD* for these studies at full recruitment was 35%. Results are shown for the variable recruitment model and for the model of Kuikka, Bassingthwaighte, and associates (1986), which assumes full recruitment ("Bass.-Equivalent Model").

recruitment in order for heterogeneity to affect SS *PS*. Similar results were seen for $D^{1/2}S$ (Caruthers et al., 1995).

Effects of Injury on Flow Recruitment

We also examined the effects of the infusion of phorbal myristate acetate (60 μg into the perfusate reservoir) on the Crone *PS–F* relationship (Harris et al., 1994). The *PS–F* curve is shifted below the baseline curve. These results are shown in Figure 21.9. The ratios of urea *PS* to butanediol *PS* did not change with injury (at 250, 450, and 600 ml/min the ratios were 0.56, 0.55, and 0.47 at baseline and 0.46, 0.59, and 0.58 during injury). Similar results are seen when the data are analyzed by the equivalent diffusivity and Sangren–Sheppard models. There is then some preliminary evidence that the lung is not perfusing highly permeable areas in this injury.

Summary and Conclusions

It has been established in carefully controlled experimental studies of lung vascular injury that the indicator dilution method can measure increases in lung

PS/Dry Lung Wt. (ml/s-g)

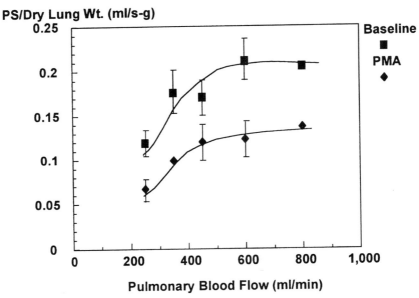

FIGURE 21.9. Effects of flow on [14C]-butanediol *PS* from the Crone model under baseline conditions and after phorbal myristate acetate infusion in isolated blood perfused dog lungs (Harris et al., 1994).

vascular *PS* and related parameters. However, further work in a wider variety of animal models and in patients has revealed significant interactions among microvascular barrier alterations, capillary derecruitment, and blood flow distribution. The complication of altered surface area joined with capillary permeability change is a significant challenge for the quantitative analysis of microvascular transport. This chapter has presented work that has been aimed at elaborating and quantifying permeability change in the presence of surface area change, as well as the effects of blood flow and capillary transit time heterogeneity on model-based interpretations of indicator dilution curves from the lung. While work is not complete, there is significant evidence that some important injuries decrease the magnitude of functioning capillary surface. Further, while heterogeneity of blood flow and capillary transit times affect the magnitude of *PS* and related parameters identified from models, the numerical alterations caused by including heterogeneity are relatively small for tracers with *PS* values in the ranges seen for [14C]-urea and [14C]-butanediol. Differences are much greater when *PS* is larger. Although it is still under development, the variable recruitment model offers the possibility of a unification of recruitment and heterogeneity, and may be a technique for accounting for both phenomena.

A number of issues remain unresolved. It is not clear what the entirely correct method for measuring the distribution of arterial, venous, and capillary transit times should be. As models increase in complexity, the sensitivity and identifiability of their parameters become increasingly problematic (Bosan and Harris, 1996). Tracers with greater sensitivities to particular forms of capillary barrier injury would be desirable. Finally, the development of methods alternative to radioisotope injection would also be desirable for clinical applications (see Olson et al., 1994, 1995, for an example of optical indicator dilution methods).

Acknowledgment. This work was supported by NIH Grants HL 07123, HL 19153, HL 39155, RR 06558, and the Martha Washington Straus–Harry H. Straus Foundation.

References

Adams, S.U., N.A. Pou, T.R. Harris, and R.J. Roselli. Indicator dilution methods using an amphipathic tracer can identify increased capillary permeability during air embolism. *FASEB J.* 7:A788, 1993.

Ashbaugh, D.G., D.B. Bigelow, T.L. Petty, B.E. Levine. Acute respiratory distress in adults. *Lancet* 2:319–323, 1967.

Audi, S.H., G.S. Krenz, J.H. Linehan, D.A. Rickaby, and C.A. Dawson. Pulmonary capillary transport function from flow-limited indicators. *J. Appl. Physiol.* 77:332–351, 1994.

Basset, G., G. Martel, F. Bouchonnet, J. Marsac, J. Sutton, J. Botter, and R. Capitini. Simultaneous detection of deuterium oxide and indocyanine green in flowing blood. *J. Appl. Physiol.* 50:1367–1371, 1981.

Bassingthwaighte, J.B. and C.A. Goresky. Modeling in the analysis of solute and water exchange in the microvasculature. In: *Handbook of Physiology—The Cardiovascular System IV,* edited by Renkin, E.M. and Michel, G.C. pp. 97–146, American Physiological Society, New York: Oxford, 1985.

Bernard, G.R. and K.L. Brigham. Adult respiratory distress syndrome. *Baylor Coll. Med. Cardiol. Series.* 7(5):5–19, 1984.

Bosan, S. and T.R. Harris. A visualization-based analysis method for multiparameter models of capillary-tissue exchange. *Ann. Biomed. Engrg.* 24:124–138, 1996.

Bradley, J.D., R.J. Roselli, R.E. Parker, and T.R. Harris. Effects of endotoxemia on the sheep lung microvascular membrane: A two-pore theory. *J. Appl. Physiol.* 64:2675–2683, 1988.

Brigham, K.L., J.D. Snell, T.R. Harris, S. Marshall, J. Haynes, R.E. Bowers, and J. Perry. Indicator dilution lung water and vascular permeability in humans: Effects of pulmonary vascular pressure. *Circ. Res.* 44:523–530, 1979.

Brigham, K.L., K. Kariman, T.R. Harris, J.R. Snapper, and S.L. Young. Correlation of oxygenation with vascular permeability-surface area but not with lung water in humans with acute respiratory failure and pulmonary edema. *J. Clin. Invest.* 72:339–349, 1983.

Byrne, K. and H.J. Sugerman. Experimental and clinical assessment of lung injury by measurement of extravascular lung water and transcapillary protein flux in ARDS: A review of current techniques. *J. Surg. Res.* 44:185–203, 1988.

Canonico, A.E. and K.L. Brigham. Biology of acute injury. In: *The Lung: Scientific Foundations,* edited by R.G. Crystal, J.B. West, Barnes, P.J., Cherniack, N.S., Weibel, E.R. pp. 2475–2498. Philadelphia: Lippincott-Raven, 1997.

Caruthers, S.D., T.R. Harris, K.A. Overholser, N.A. Pou, and R.E. Parker. The effects of flow heterogeneity on the measurement of capillary exchange in the lung. *J. Appl. Physiol.* 79:1449–1460, 1995.

Chinard, F.P. and T. Enns. Transcapillary pulmonary exchange of water in the dog. *Am. J. Physiol.* 178:197–202, 1954.

Cope, D.K., F. Grimbert, J.M. Downey, and A.E. Taylor. Pulmonary capillary pressure: A review. *Crit. Care Med.* 20:1043–1056, 1992.

Crone, C. The permeability of capillaries in various organs determined by use of the "indicator diffusion" method. *Acta. Physiol. Scand.* 58:292–305, 1963.

Dawson, C.A., J.H. Linehan, D.A. Rickaby, and G.S. Krenz. Effect of vasoconstriction on longitudinal distribution of pulmonary vascular pressure and volume. *J. Appl. Physiol.* 70:1607–1616, 1991.

Dupuis, J., C.A. Goresky, C. Juneau, A. Calderone, J.L. Rouleau, C.P. Rose, and S. Goresky. Use of norepinephrine uptake to measure lung capillary recruitment with exercise. *J. Appl. Physiol.* 68:100–113, 1990.

Gillis, C.N. and J.D. Catravas. Altered removal of vasoactive substances in the injured lung: detection of lung microvascular injury. *Ann. N.Y. Acad. Sci.* 384:458–475, 1982.

Goresky, C.A., R.F.P. Cronin, and B.E. Wangel. Indicator dilution measurements of extravascular water in the lung. *J. Clin. Invest.* 48:487–501, 1969.

Gorin, A.B., W.J. Weidner, R. Demling, and N.C. Staub. Noninvasive measurement of pulmonary transvascular protein flux in sheep. *J. Appl. Physiol.* 45:225–233, 1978.

Hanson, L., J.D. Emhardt, J.P. Bartek, L.P. Latham, L.L. Checkly, R.L. Capen, and W.W. Wagner. Site of recruitment in the pulmonary microcirculation. *J. Appl. Physiol.* 66:2079–2083, 1989.

Harris, T.R., R.D. Rowlett, and K.L. Brigham. The computation of pulmonary capillary permeability from multiple-indicator data: The effects of increased capillary pressure and alloxan treatment. *Microvasc. Res.* 12:177–196, 1976.

Harris, T.R., K.L. Brigham, and R.D. Rowlett. Pressure, serotonin and histamine effects on lung multiple-indicator curves in sheep. *J. Appl. Physiol.* 44:245–253, 1978.

Harris, T.R. and K.L. Brigham. The exchange of small molecules as a measure of normal and abnormal lung microvascular function. *Ann. N.Y. Acad. Sci.* 384:417–434, 1982.

Harris, T.R., G.R. Bernard, R.J. Roselli, C.R. Maurer, and N.A. Pou. Extravascular lung water by infrared and other measures. *Proc. Ann. Conf. Engr. Med. Biol.* 27:77, 1985.

Harris, T.R., R.J. Roselli, C.R. Maurer, R.E. Parker, and N.A. Pou. Comparison of labelled propanediol and urea as markers of lung vascular injury. *J. Appl. Physiol.* 62:1852–1859, 1987.

Harris, T.R. and R.J. Roselli. The exchange of small molecules in the normal and abnormal lung circulatory bed. In: *Respiration Physiology: A Quantitative Approach*, edited by Chang, H.K. and Paiva, M. New York: Dekker, pp. 737–791, 1989.

Harris, T.R., G.R. Bernard, K.L. Brigham, S.B. Higgins, J.E. Rinaldo, H.S. Borovetz, W.J. Sibbald, K. Kariman, and C.L. Sprung. Lung microvascular transport properties measured by multiple indicator dilution methods in ARDS patients: A comparison between patients reversing respiratory failure and those failing to reverse. *Am. Rev. Resp. Dis.* 141:272–280, 1990.

Harris, T.R., S.U. Adams, and N.A. Pou. Phorbal myristate acetate (PMA) infusion alters the recruitment of capillaries by flow in isolated perfused dog lungs. *FASEB J.* 8:A917, 1994.

Haselton, F.R., R.E. Parker, R.J. Roselli, and T.R. Harris. Lung multiple tracer analysis with an effective diffusivity model of capillary-tissue exchange. *J. Appl. Physiol.* 57:98–109, 1984a.

Haselton, F.R., R.J. Roselli, R.E. Parker, and T.R. Harris. An effective diffusivity model of pulmonary capillary exchange: General theory, limiting cases and sensitivity analysis. *Math. Biosci.* 70:237–263, 1984b.

King, R.B., G.M. Raymond, and J.B. Bassingthwaighte. Modeling blood flow heterogeneity. *Ann. Biomed. Engrg.* 24:352–372, 1996.

Konig, M.F., J.M. Lucocq, and E.R. Weibel. Demonstration of pulmonary vascular perfusion by electron and light microscopy. *J. Appl. Physiol.* 75(4):1877–1883, 1993.

Kuikka, J., M. Levin, and J.B. Bassingthwaighte. Multiple tracer dilution estimates of D- and 2-deoxy-D-glucose uptake by the heart. *Am. J. Physiol.* 250:H29–H42, 1986.

Leksell, L.G., M.S. Schreiner, A. Sjlvesto, and G.R. Neufeld. Commercial double indicator-dilution densitometer using heavy water: evaluation of oleic-acid pulmonary edema. *J. Clin. Monit.* 6:99–106, 1990.

Lewis, F.R., V.B. Elings, S.L. Hill, and J.M. Christensen. The measurement of extravascular lung water by thermal-green dye indicator dilution. *Ann. N.Y. Acad. Sci.* 384:394–410, 1982.

Mitzner, W. and H.K. Chang. Hemodynamics of the pulmonary circulation. In: *Respiration Physiology: A Quantitative Approach,* edited by Chang, H.K. and Paiva, M. New York: Dekker, pp. 561–631, 1989.

Nelin, L.D., D.L. Roerig, D.A. Rickaby, J.H. Linehan, and C.A. Dawson. Influence of flow on pulmonary vascular surface area inferred from blue dextran efflux data. *J. Appl. Physiol.* 72(3):874–880, 1992.

Olson, L.E., A. Pou, and T.R. Harris. Surface-area independent assessment of lung microvascular permeability using an amphipathic tracer. *J. Appl. Physiol.* 70:1085–1096, 1991.

Olson, L.E., D.J. Staton, M. Young, R.L. Galloway, and T.R. Harris. Sulhemoglobinated erythrocytes as an optical intravascular tracer in the lung. *Ann. Biomed. Engrg.* 22:323–331, 1994.

Olson, L.E., T.R. Harris, A. Pou, M.N. Syed-Ahmed, and R.L. Galloway. An optical multiple indicator dilution technique to measure lung permeability surface area: Calibration and baseline measurement. *IEEE Trans. BME* 42:451–463, 1995.

Overholser, K.A., M.J. Bhatte, and M.H. Laughlin. Modeling the effect of flow heterogeneity on coronary permeability-surface area. *J. Appl. Physiol.* 71:758–769, 1991.

Overholser, K.A., N.A. Lomangino, R.E. Parker, N.A. Pou, and T.R. Harris. Pulmonary vascular resistance distribution and the recruitment of microvascular area. *J. Appl. Physiol.* 77:845–855, 1994.

Pappenheimer, J.R., E.M. Renkin, and L.M. Borrero. Filtration, diffusion, and molecular sieving through peripheral capillary membranes: A contribution to the pore theory of capillary permeability. *Am. J. Physiol.* 167:13–46, 1951.

Parker, R.E., R.J. Roselli, F.R. Haselton, and T.R. Harris. Effect of perfusate hematocrit on urea permeability surface area in isolated dog lungs. *J. Appl. Physiol.* 60:1293–1299, 1986.

Renkin, E.M. Transport of potassium-42 from the blood to tissue in isolated mammalian skeletal muscles. *Am. J. Physiol.* 197:1209–1210, 1959.

Rickaby, D.A., J.H. Linehan, T.A. Bronikowski, and C.A. Dawson. Kinetics of serotonin uptake in the dog lung. *J. Appl. Physiol.* 51:405–414, 1981.

Rickaby, D.A., C.A. Dawson, J.H. Linehan, and T.A. Bronikowski. Alveolar vessel behavior in the zone 2 lung inferred from indicator-dilution data. *J. Appl. Physiol.* 63(2):778–84, 1987.

Rinaldo, J.E., H.S. Borovetz, M.C. Mancini, R.L. Hardesty, and B.P. Griffith.

Assessment of lung injury in the Adult Respiratory Distress Syndrome using multiple indicator dilution curves. *Am. Rev. Resp. Dis.* 133:1006–1010, 1986.

Rose, C.P. and C.A. Goresky. . Vasomotor control of capillary transit-time heterogeneity in the canine coronary circulation. *Circ. Res.* 39:541–544, 1976.

Roselli, R.J. and T.R. Harris. Lung fluid and macromolecular transport. In: *Respiration Physiology: A Quantitative Approach,* edited by Chang, H.K. and Paiva, M. New York: Dekker, pp. 633–735, 1989.

Roselli, R.J. and W.R. Riddle. Analysis of non-invasive microvascular macromolecular transport measurements in the lung. *J. Appl. Physiol.* 67:2343–2350, 1989.

Rowlett, R.D. and T.R. Harris. Comparative study of organ models and numerical methods for the evaluation of capillary permeability from multiple-indicator data. *Mathemat. Biosci.* 29:273–298, 1976.

Sangren, W.C. and C.W. Sheppard. Mathematical derivation of the exchange of a labelled substance between a liquid flowing in a vessel and an external compartment. *Bull. Math. Biophys.* 15:387–394, 1953.

Schuster, D.P. Positron emission tomography: Theory, and its application to the study of lung disease (State-of-the-Art). *Am. Rev. Resp. Dis.* 139:818–840, 1989.

Shibamoto, T., J.C. Parker, A.E. Taylor, and M.I. Townsley. Derecruitment of filtration surface area in paraquat-injured isolated dog lungs. *J. Appl. Physiol.* 68:1581–1589, 1990.

Syrota, A., M. Girauld, J.-J. Pocidalo, and D.L. Yudilevich. Endothelial uptake of amino acids, sugars, lipids, and prostaglandins in rat lung. *Am. J. Physiol.* 243:C20–C26, 1982.

Taylor, A.E. and J.C. Parker. Pulmonary interstitial spaces and lymphatics. In: *Handbook of Physiology—The Respiratory System I, Volume 4,* edited by A.P. Fishman, and A.B. Fisher. Bethesda, MD: American Physiology Association, pp. 167–229, 1985.

Toivonen, H.J. and J.D. Catravas. Effects of blood flow on lung ACE kinetics: Evidence for microvascular recruitment. *J. Appl. Physiol.* 71(6):2244–2254, 1991.

West, J.B., C. Dollery, and A. Naimark. Distribution of blood flow in isolated lung: Relation to vascular and alveolar pressure. *J. Appl. Physiol.* 19:713–724, 1964.

West, J.B. *Ventilation/Blood Flow and Gas Exchange.* Oxford: Blackwell, 1972.

Zelter, M., D. Lipavsky, J.M. Hoeffel, and J.F. Murray. Effect of lung injuries on ^{14}C-urea permeability surface area product in dogs. *J. Appl. Physiol.* 56:1512–1520, 1984.

22

Lipophilic Amines as Probes for Measurement of Lung Capillary Transport Function and Tissue Composition Using the Multiple-Indicator Dilution Method

Said H. Audi, John H. Linehan, Gary S. Krenz, David L. Roerig, Susan B. Ahlf, and Christopher A. Dawson

We have exploited the rapidly equilibrating interactions of nonbasic lipophilic amines such as ^{14}C-diazepam and ^{3}H-alfentanil with lung tissue to estimate the perfused lung tissue water/lipoid space ratio, and to develop a method for estimating the pulmonary capillary transit time distribution using the bolus-injection multiple-indicator dilution technique.

Estimation of extravascular lung water volume from the venous effluent concentration versus time outflow curves of a vascular reference indicator and a hydrophilic indicator such as [3]HOH depends not only on the state of tissue hydration, but also on tissue perfusion. Assuming that the extravascular volume of a rapidly equilibrating lipophilic indicator is independent of the state of tissue hydration, separation of these two effects might be facilitated if both hydrophilic and rapidly equilibrating lipophilic indicators were used. We found that the ratio of the extravascular volume accessible to [3]HOH, Q_{ew}, to the virtual extravascular volume accessible to ^{14}C-diazepam, Q_{ed}, was elevated in edematous isolated lungs even when Q_{ew} was less than normal as the result of concomitant embolization. This result suggests that Q_{ew}/Q_{ed} may be a useful index of the wet/dry weight ratio of the perfused lung tissue.

In addition, a method was developed wherein the moments of the effluent concentration curves for a reference indicator and two or more "flow-limited" indicators such as ^{14}C-diazepam and ^{3}H-alfentanil, each having a different extravascular mean residence time, can be used to estimate the mean transit time, the variance, and the third central moment of the pulmonary capillary transit time distribution. The method was implemented on data from isolated perfused dog lung lobes and rabbit lungs, and the results were consistent and encouraging; it appears to have potential for in vivo application.

A number of organic amine drugs are rapidly and extensively taken up by the lung tissue during passage through the pulmonary circulation (1, 34, 39–41).

517

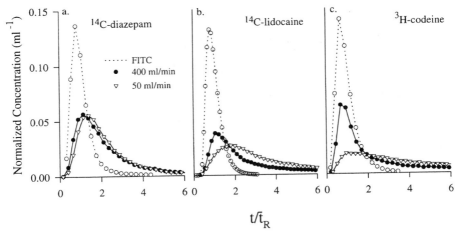

FIGURE 22.1. The venous effluent concentration versus time data for the FITC dye and labeled amine indicators ^{14}C-diazepam (a), ^{14}C-lidocaine (b), and ^{3}H-codeine (c) following the bolus injections of the indicators into the pulmonary artery of three isolated perfused rabbit lungs at two different flows, as indicated. Concentrations, C, on this and subsequent graphs are normalized to the amount of the injected indicator and are thus the fraction of the injected dose of indicator per ml of effluent perfusate. The time scale was obtained by first subtracting the tubing mean transit time, $\bar{t}_{in}$, from each sample time measured from the time the bolus was injected and then normalizing the result to the FITC dye mean transit time, $\bar{t}_R$, also obtained after subtracting $\bar{t}_{in}$ from each sample time.

Pulmonary uptake is highest with amines of moderate to high lipophilicity and pK_a greater than about 8.0 (1, 39, 40). In addition, the type and extent of association with plasma protein are important. There has been considerable interest in understanding how the lung influences the pharmacokinetics of such drugs (1, 39, 40). However, the kinetics of drug–lung interaction might also be exploited to learn about the properties of the lung itself.

Figure 22.1 shows the patterns of separation between the concentration versus time outflow curves of three lipophilic amines—^{14}C-diazepam, ^{14}C-lidocaine, and ^{3}H-codeine—and a vascular reference indicator following a single pass through an isolated perfused rabbit lung at two different flows. These three compounds represent different classes of amines with respect to physicochemical properties and types of interactions with pulmonary tissue. For ^{14}C-diazepam ($pK_a = 3.4$), a nonbasic lipophilic amine, the outflow curve peak is attenuated and shifted to longer time relative to that for the reference indicator outflow curve. In addition, on a time scale normalized to the reference indicator mean transit time, changing the flow rate had little effect on the shape of the diazepam outflow curve. These characteristics indicate that ^{14}C-diazepam equilibrates very rapidly between the perfusate and lung tissue during a single pass through the lungs, behaving as a rapidly equilibrating or a "flow-limited" indicator as defined by Goresky et al. (28).

The patterns for the basic lipophilic amines ^{14}C-lidocaine (pK_a = 7.9) and ^{3}H-codeine (pK_a = 8.4) are quite different, and the changes with flow indicate that equilibrium between the perfusate and lung tissue occurs relatively slowly on the time course of the pulmonary capillary transit time. This might be due to the existence of a barrier to diffusion into the tissue or, more likely, because the interactions of these drugs with proteins and lipoid fractions of the vascular and extravascular spaces are slow in comparison to diazepam. In either case, the outflow curves for ^{3}H-codeine and ^{14}C-lidocaine exhibit some characteristics of "barrier-limited" indicators (28). For example, the peak of the ^{3}H-codeine concentration curve was smaller than that for the reference indicator outflow curve, but it occurred at virtually the same time. Furthermore, when the perfusate flow through the lung was reduced, ^{3}H-codeine extraction increased, as indicated by the reduction in the peak of its outflow curve. The ^{14}C-lidocaine appears to be intermediate between ^{14}C-diazepam and ^{3}H-codeine.

The observations exemplified in Figure 22.1 suggest that lipophilic amines such as these may be useful as probes for studying the intact lung, the concept being that changes in the kinetics of interactions of indicators, having different physicochemical properties, with the lung tissue will reflect changes in the chemical composition and perfusion of the tissue. In this chapter, we will discuss our beginning attempts to exploit this concept using nonbasic lipophilic amines.

As one example, we have examined the potential use of diazepam for helping to interpret indicator dilution extravascular lung water measurements (22). The extravascular lung water volume, Q_{ew}, estimated from the venous effluent concentration versus time outflow curves of a vascular indicator and a hydrophilic indicator such as ^{3}HOH, is sensitive to changes in both tissue hydration and perfusion. Thus, interpretation of a change in Q_{ew} might be misleading when it is associated with a change in tissue perfusion. On the other hand, the extravascular volume accessible to a rapidly diffusing lipophilic indicator such as diazepam, Q_{ed}, which is distributed to the nonaqueous, lipoid fraction of the tissue, would also be affected by changes in tissue perfusion, but not by changes in the tissue water content per se (22).

Figure 22.2 shows the venous effluent concentration versus normalized time curves for ^{3}HOH and ^{14}C-diazepam before and after filling the alveolar space of an isolated perfused dog lung lobe with saline (22). The upper panel of Figure 22.2 shows that the mean transit time for the ^{3}HOH outflow curve increased when saline was added, whereas the lower panel shows that the outflow curves for ^{14}C-diazepam were relatively unaffected, because, according to the proposed concept, ^{14}C-diazepam disposition still involves only the nonaqueous part of the tissue.

To demonstrate how this might help in the interpretation of the extravascular water measurement, we carried out the following sequence of injections in the isolated perfused dog lung lobe (22). A bolus of ^{125}I-HSA, ^{14}C-diazepam, and ^{3}HOH was injected under control conditions. Hydrostatic edema was then induced

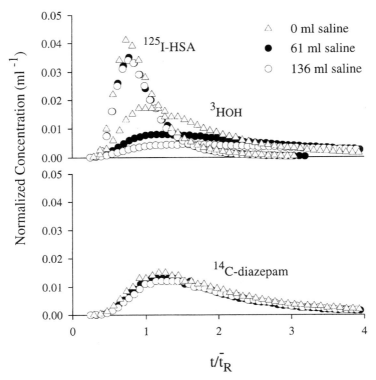

FIGURE 22.2. Lobar venous effluent concentration versus time data for ^{125}I-HSA, [3]HOH, and ^{14}C-diazepam under control conditions (0 ml saline) and after filling the alveolar space with 61 and 136 ml of saline. The time scale is the same as in Figure 22.1. [Reprinted with permission from the American Physiological Society (22).]

by raising the venous pressure to about 20 torr for 18–90 min. The venous pressure was then returned to normal and a second bolus of ^{125}I-HSA, ^{14}C-diazepam, and [3]HOH was injected. Glass beads were then injected to reduce the number of perfused vessels, and a third bolus of ^{125}I-HSA, ^{14}C-diazepam, and [3]HOH was injected. The vascular volume (Q), the extravascular water volume accessible to [3]HOH (Q_{ew}), and a virtual extravascular volume accessible to ^{14}C-diazepam (Q_{ed}) were determined from each injection. The results summarized in Figure 22.3 show that Q_{ew}, following embolism, was less than under control conditions despite the fact that tissue was still edematous. On the other hand, the ratio Q_{ew}/Q_{ed} did not decrease following embolism, reflecting the edema in the tissue surrounding the remaining perfused vessels. Thus, the Q_{ew}/Q_{ed} ratio might be thought of as an index of the wet/dry weight ratio of the perfused lung tissue (22).

We also examined the potential use of rapidly equilibrating lipophilic amines as probes of the kinematics of pulmonary capillary perfusion (2, 3). The pulmonary

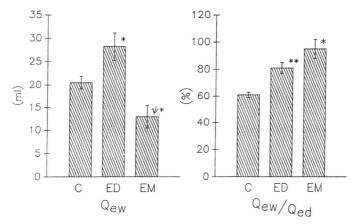

FIGURE 22.3. Comparison of changes in Q_{ew} and Q_{ew}/Q_{ed} after lung lobes were made edematous (ED) and after glass beads were introduced to occlude some of the vessels in edematous lobe (EM). Values are means ± SE; $n = 6$. Significantly different from control: $*P < 0.05$; $**P < 0.01$. ψ Significantly different from edema, $P < 0.01$. [Reprinted with permission from the American Physiological Society (22).]

capillary mean transit time and the distribution of capillary transit times about the mean are important determinants of the function of the pulmonary capillary bed. Several methods have been used to estimate the pulmonary capillary mean transit time (or volume). As exemplified in Tables 22.1 and 22.2, these methods have produced a wide range of estimates, which suggests that either the pulmonary capillary volume is quite variable, depending on the experimental conditions, or that some methods involve overly simplified assumptions that yield estimates which are not sufficiently accurate to resolve differences within the range of reported values, or both. Estimates of the contribution of the pulmonary capillary transit time distribution (transport function), $h_c(t)$, to the total transit time dispersion within the lung have also varied widely (Tables 22.1 and 22.2), and the kinematic assumptions used in transport models of the lungs have ranged from all of the pulmonary dispersion occurring in the capillaries (15, 20, 26) to none occuring in the capillaries (19). This emphasizes the need for a means of measuring $h_c(t)$.

To examine the possibility of using rapidly equilibrating nonbasic amines to provide an estimate of $h_c(t)$, we followed the approach introduced by Goresky (25, 26), which takes advantage of the separation between the outflow curves of a vascular reference indicator, $C_R(t)$, and a flow-limited indicator, $C_D(t)$, which is the result of the rapid equilibration of the flow-limited indicator between the perfusate and lung tissue within the capillary bed. As a consequence of this rapid equilibration, the concentration of the flow-limited indicator within the capillary bed behaves like a wave propagating in the flow direction at a reduced velocity relative to the vascular flow velocity. The result of this "delayed wave effect" (25)

TABLE 22.1. Examples of estimates of the pulmonary capillary blood volume for dog lungs.

Method	Q_c (ml)	Body weights (kg)	Q_c (ml/kg body weight)	Q_c as a fraction of Q (%)	Flow (L/min)	$\bar{T}_c$ (s)	RD_c (%)	Reference
D_{LCO}	21 ± 9.8	16 ± 6.5	1.3		1.9	0.7		(50)
D_{LCO}	67 ± 22	22–35	2.4					(33)
D_{LCO}	55 ± 30	24 ± 4.0	2.3					(32)
D_{LCO}	17 ± 3.3	14 ± 0.5	1.2					(9)
Ether and dye dilution+	50 ± 11*	12 ± 0.8	4.2	49 (of intrapul.)	0.7	4.6		(10)
D_{LCO}	41 ± 5.3	19 ± 2.0	2.2		1.2	1.0		(18)
O$_2$ saturation step transit time+	20 ± 3.7*	8	2.5	38 (of intrapul.)				(38)
Morphometric	119 ± 34	23 ± 1.5	5.2					(45)
D_{LCO}	48 ± 21	16–29	2.1	22 (of total)	2.7	1.1		(48)
Morphometric	84 ± 16	15–20	4.8					(17)
D_{LCO}	42 ± 6	15–20	2.4					"
Fluorescence microscopy	76 ± 23	26 ± 7.9	2.9		2.7	1.8		(14)
D_{LCO}	80 ± 27	26 ± 7.9	3.1		2.7	1.9		"
Morphometric+	89 ± 22	23 ± 3.9	3.9		1.0	5.3		(16)
D_{LCO}+	56 ± 10	23 ± 3.9	2.4		1.0	3.4		"
D_{LCO}	61 ± 12	23 ± 3.9	2.7					"
Ether and dye dilution+	110 ± 10*	23 ± 4.8	4.8	60 (of intrapul.)	1.5	4.6		(21)
Indicator dilution	62 ± 3.9	21 ± 3.5	3.0	60 (of total)	2.7	1.4	61	(24)
Fluorescence microscopy	97	> 20	< 4.9		2.9	2.0	64	(13)
Indicator dilution+ (flow-limited indicators)	78 ± 12*	21 ± 1.6	3.7	48 (of intrapul.)	1.5	3.2	75	(2)
Fluorescence microscopy	131 ± 94	19 ± 1	6.9		2.3	3.4	93	(4)

The + indicates measurements made on isolated lungs; otherwise, intact anesthtized dogs were used. The * indicates that the total pulmonary capillary blood volume, Q_c, was calculated assuming that the left lower lung lobe was 25% of total lung mass (30). The $\bar{T}_c$ is the pulmonary capillary mean transit time. The Q is the total pulmonary vascular blood volume; (of intrapul.) refers to experimental conditions under which the large pulmonary artery and vein are excluded from the measurement; (of total) refers to experimental conditions under which the large pulmonary artery and vein were included in the measurement; D_{LCO} is the pulmonary diffusing capacity for CO, and RD_c is the capillary relative dispersion. [Reprinted with permission from the American Physiological Society (2).]

TABLE 22.2. Estimates of the pulmonary capillary blood volume and mean transit time for rabbit lungs in chronological order.

Method	Q_c (ml)	Body weights (kg)	Q_c (ml/kg body weight)	Q_c as a fraction of Q (%)	Flow (ml/min)	$\overline{T}_c$ (s)	RD_c (%)	Reference
Morphometric[+]	7.2 ± 0.8 (SD)	3.6 ± 0.5 (SD)	2.0					(18)
Morphometric[+]	8.4 ± 1.3	3.1 ± 0.2	2.7		180	2.8 ± 0.4 (SD)		(2)
Fluorescence microscopy	2.7 ± 1.2	3.2 ± 0.5	0.8	10	266	0.6 ± 0.3		(17)
Fluorescence microscopy[+]	3.2 ± 1.0	3.4 ± 0.4	1.0		272	0.7 ± 0.2		(16)
Fluorescence microscopy	1.7 ± 0.6	3.2	0.5		220	0.5 ± 0.2		(11)
Indicator dilution[+] (flow-limited indicators)	4.2 ± 0.3	2.7 ± 0.2	1.6	44	198	1.3 ± 0.1	90	(2)
Fluorescence microscopy[+]	4.1 ± 2.2	3.5 ± 0.2	1.2	20	280	0.9 ± 0.5	83	(3)

The [+] indicates measurements made in isolated lungs; otherwise, intact anesthetized rabbits were used. The Q_c and $\overline{T}_c$ are the pulmonary capillary blood volume and capillary mean transit time, respectively. The RD_c is the capillary relative dispersion. The Q is the total pulmonary vascular volume. [Reprinted with permission from the American Physiological Society (3).]

is that the portion of the total dispersion of the flow-limited diffusible indicator and reference indicator that takes place within the capillary bed has the same relative dispersion even though the flow-limited diffusible indicator can have a much longer mean transit time. This results in a simple scaling relationship between the capillary transport functions for the two indicators. This scaling relationship is distorted if dispersion also takes place in the noncapillary part of the system (large vessels, tubing, and injection-sampling system) wherein both indicators are confined to the same vascular space. In order to be able to take advantage of the scaling relationship within the capillaries, one needs a means for decorticating the noncapillary dispersion from the total dispersion in $C_R(t)$ and $C_D(t)$. Goresky (25) made use of the relatively small noncapillary dispersion in the liver by assuming that all dispersion of the bolus from the injection site to the organ outflow occurs within the capillary bed. Under this assumption, the vascular reference indicator outflow curve, normalized to unit area, is a time-shifted version of the capillary transit time distribution, $h_c(t)$, and the following scaling relationship exists between $C_R(t)$ and $C_D(t)$. Let λ be the ratio of the flow-limited extravascular volume to capillary volume. If the time axis of $C_D(t)$ is shifted backward by the noncapillary transit time ($\bar{t}_n$), the concentration values and the shifted time axis of $C_D(t)$ are scaled by $(1 + \lambda)$ and $1/(1 + \lambda)$, respectively, then, shifting the scaled time axis forward by $\bar{t}_n$ should superimpose the scaled $C_D(t)$ on $C_R(t)$. Subtracting $\bar{t}_n$ from the total vascular mean transit time provides an estimate of the capillary mean transit time (25, 26, 44).

In the following we present a bolus-injection, multiple-indicator dilution method for estimating the pulmonary capillary mean transit time and transit time distribution (2, 3) that is based on the concepts introduced by Goresky (25), but relaxes the assumption of no dispersion within the noncapillary part of the system. The method is referred to as method A in references (2) and (3).

Model and Theory

We begin with the following description of the assumptions underlying the model system upon which the proposed method for estimating $h_c(t)$ is based (2, 3).

1. Each capillary element includes a capillary vessel and surrounding extravascular volume. The vascular reference indicator is confined to the vessel, but the flow-limited diffusible indicator can diffuse out of the vessel into the extravascular volume.
2. Diffusion of both vascular and diffusible indicators in the direction of flow is negligible compared to the axial convective transport; that is, indicators are transported from inflow to outflow only by convection.
3. Flow is restricted to the vascular region. Transport in the extravascular region is only by diffusion.
4. With respect to the diffusible indicator, diffusion equilibrium within the vascular and the extravascular volumes in the direction perpendicular to the flow direction is instantaneous.

5. $C_n(t)$, a function representing the transport of indicators through the noncapill-ary (arteries, veins, tubing, and injection-sampling system) part of the system, is the same for all capillary elements [random coupling conditions (8, 11, 36) between $C_n(t)$ and the capillary transit time distribution, $h_c(t)$]. To put $C_n(t)$ in perspective, $C_n(t) = C_{in}(t) * h_{av}(t)$, where $C_{in}(t)$ is a concentration function representing all of the dispersive processes occurring outside of the organ, including the dispersion caused by the injection-sampling system; $h_{av}(t)$ is the transit time distribution for both diffusible and nondiffusible indicators in the noncapillary blood vessels of the organ; and * is the convolution operator. Given the linearity, commutativity, and associativity of the model system, $C_n(t)$ could be thought of as the outlet concentration curve that would exist if all of the arteries and veins were connected directly together at a common nexus with no intervening capillaries (2, 3).

6. The A_e/A_c ratio averaged along the length of the capillary is the same for all capillaries (where A_e is the cross-sectional area perpendicular to the flow of the extravascular tissue components in which the diffusible indicators are soluble, and A_c is the capillary cross-sectional area) (2, 3).

7. The heterogeneity of the capillary transit times producing $h_c(t)$ can be thought of as being the result of having capillaries with differing lengths, flows, cross-sectional areas, or any combination thereof (2, 3).

Under the above assumptions, we have reduced the interactions between dynamics and perfusion to the following equations (2, 3), which relate the mean transit time (first moment), $\bar{t}$, the variance (second central moment), σ^2, and the third central moment, m^3, of $h_c(t)$ and $C_n(t)$ to those of the outflow curves of a vascular reference indicator, $C_R(t)$:

$$\bar{t}_R = \bar{t}_n + \bar{t}_c \tag{22.1a}$$

$$\sigma_R^2 = \sigma_c^2 + \sigma_n^2 \tag{22.1b}$$

$$m_R^3 = m_c^3 + m_n^3 \tag{22.1c}$$

and a rapidly equilibrating or flow-limited indicator, $C_D(t)$:

$$\bar{t}_D = \bar{t}_n + \left(1 + \frac{\bar{t}_e}{\bar{t}_c}\right)\bar{t}_c = \bar{t}_n + \bar{t}_c + \bar{t}_e \tag{22.2a}$$

$$\sigma_D^2 = \sigma_n^2 + \left(1 + \frac{\bar{t}_e}{\bar{t}_c}\right)^2 \sigma_c^2 \tag{22.2b}$$

$$m_D^3 = m_n^3 + \left(1 + \frac{\bar{t}_e}{\bar{t}_c}\right)^3 m_c^3 . \tag{22.2c}$$

In Eqs. (22.1a–c) and (22.2a–c), the subscripts c and n identify the moments of $h_c(t)$ and $C_n(t)$, respectively. The subscripts R and D identify the moments of $C_R(t)$ and $C_D(t)$, respectively. $\bar{t}_e$ is the extravascular mean residence time of the flow-limited indicator obtained by subtracting the mean transit time of $C_R(t)$ from that of $C_D(t)$.

Equations (22.2a–c) show that, for the flow-limited indicator, as a result of its rapid equilibration within the extravascular volume Q_e, $Q_e = \bar{t}_e \dot{Q}$, the moments of the capillary transport function are scaled by the factor $(1 + \bar{t}_e/\bar{t}_c)$. $\dot{Q}$ is the blood flow rate. Note that setting $\bar{t}_e = 0$ reduces the moments relationships for $C_D(t)$ (Eqs. 22.2a–c)) to those for $C_R(t)$ [Eqs. (22.1a–c)], as expected.

Equations (22.1a–c), and (22.2a–c) can be combined and rearranged in terms of the moments of the measured concentration curves of both $C_R(t)$ and $C_D(t)$. In this case, $\bar{t}_e$, σ_e^2, and m_e^3 are defined to be the difference between the respective moments of $C_R(t)$ and $C_D(t)$, i.e.,

$$\sigma_e^2 = \sigma_D^2 - \sigma_R^2 = \left[\left(1 + \frac{\bar{t}_e}{\bar{t}_c} \right)^2 - 1 \right] \sigma_c^2 \qquad (22.3a)$$

$$m_e^3 = m_D^3 - m_R^3 = \left[\left(1 + \frac{\bar{t}_e}{\bar{t}_c} \right)^3 - 1 \right] m_c^3 , \qquad (22.3b)$$

where

$$\bar{t}_e = \bar{t}_D - \bar{t}_R .$$

In Eqs. (22.3a–b), we see that $\bar{t}_e$, σ_e^2, and m_e^3 are known since the mean transit times, variances, and third central moments of $C_R(t)$ and $C_D(t)$ are known. The three unknowns are the moments of $h_c(t)$, namely, $\bar{t}_c$, σ_c^2, and m_c^3. These two equations with three unknowns represent an underdetermined system. In other words, we cannot specify the first three moments of $h_c(t)$ from the moments of the outflow curves of a vascular indicator and a single flow-limited indicator using Eqs. (22.3a–b).

The number of unknowns in Eqs. (22.3a–b) can be reduced from three to one, namely, $\bar{t}_c$, by imposing the assumption (25, 26) that $C_n(t)$ is an impulse function ($\sigma_n^2 = 0$ and $m_n^3 = 0$) and therefore all of the indicator dispersion is due to the distribution of capillary transit times. Under this assumption, the outflow curve for the vascular reference indicator is a time-shifted version of $h_c(t)$, and Eqs. (22.1a–c) and (22.2a–c) reduce to

$$\frac{\bar{t}_D - \bar{t}_n}{\bar{t}_R - \bar{t}_n} = \left(1 + \frac{\bar{t}_e}{\bar{t}_c} \right) \qquad (22.4a)$$

$$\frac{\sigma_D^2}{\sigma_R^2} = \left(1 + \frac{\bar{t}_e}{\bar{t}_c} \right)^2 \qquad (22.4b)$$

$$\frac{m_D^3}{m_R^3} = \left(1 + \frac{\bar{t}_e}{\bar{t}_c} \right)^3 \qquad (22.4c)$$

since $m_c^3 = m_R^3$ and $\sigma_c^2 = \sigma_R^3$ (2, 3).

Equations (22.4a–c) show that the diffusible indicator outflow curve is a scaled version of a time-shifted (time shifted by $\bar{t}_n$) reference indicator outflow curve, and they represent the linear model of Goresky et al. (25, 26) expressed in terms of moment relationships.

An analytical expression for the capillary mean transit time, $\bar{t}_c$, in terms of the measurable moments of $C_R(t)$ and $C_D(t)$ can be derived from Eqs. (22.4a–c):

$$\bar{t}_c = \frac{\bar{t}_e}{\sqrt{1 + \dfrac{\sigma_e^2}{\sigma_R^2}} - 1}. \tag{22.5}$$

Equation (22.5) should, theoretically, give the same value of $\bar{t}_c$ as that estimated using the scaling method proposed by Goresky et al. (26, 44).

Method A

For the proposed method for estimating the moments of $h_c(t)$, we relaxed the assumption that $C_n(t)$ is an impulse function to improve the estimates of the pulmonary capillary mean transit time and transit time distribution (2, 3).

To be able to take advantage of Eqs. (22.3a–b) to estimate the first three moments of $h_c(t)$, the injected bolus should include, in addition to a vascular reference indicator, at least two flow-limited indicators with different extravascular mean residence times. The moments of the outflow concentration curves for a reference indicator and each of the two flow-limited diffusible indicators result in two equations that are similar to Eqs. (22.3a–b). This will result in four equations that are similar to Eqs. (22.3a–b) with three unknowns (overdetermined system). The three unknowns, namely, the first three moments of $h_c(t)$, are then specified from the four equations by nonlinear least square optimization (2, 3).

For the purposes of parameter estimation, Eqs. (22.3a–b) were rewritten into the following forms:

$$\sigma_{ei}^2 = \sigma_{Di}^2 - \sigma_R^2 = \left[\left(1 + \frac{\bar{t}_{ei}}{\bar{t}_c}\right)^2 - 1\right]\sigma_c^2 \tag{22.6a}$$

$$m_{ei}^3 = m_{Di}^3 - m_R^3 = \left[\left(1 + \frac{\bar{t}_{ei}}{\bar{t}_c}\right)^3 - 1\right]m_c^3 \quad i = 1, 2, ..., j, \tag{22.6b}$$

where j is the number of flow-limited indicators, each with a different extravascular mean residence time, $\bar{t}_{ei}$. In Eqs. (22.6a) and (22.6b), the measurable variables are σ_{Di}^2, σ_R^2, m_{Di}^3, m_R^3, and $\bar{t}_{ei}$, and the three unknowns are $\bar{t}_c$, σ_c^2, and m_c^3.

In the estimation procedure, the moments of $h_c(t)$ were calculated by nonlinear least squares optimization using Eqs. (22.6a–b) in the following forms:

$$F(\bar{t}_{ei}, \sigma_{ei}^2, \bar{t}_c, \sigma_c^2) = 0 \tag{22.7a}$$
$$G(\bar{t}_{ei}, m_{ei}^3, \bar{t}_c, m_c^3) = 0, \tag{22.7b}$$

where

$$F(\bar{t}_{ei}, \sigma_{ei}^2, \bar{t}_c, \sigma_c^2) = \left\{ \sqrt{\left[\left(1 + \frac{\bar{t}_{ei}}{\bar{t}_c}\right)^2 - 1\right] \sigma_c^2 - \sqrt{\sigma_{ei}^2}} \right\} \bigg/ \sqrt{\sigma_{ei}^2}$$

$$G(\bar{t}_{ei}, m_{ei}^3, \bar{t}_c, m_c^3) = \left\{ \sqrt[3]{\left[\left(1 + \frac{\bar{t}_{ei}}{\bar{t}_c}\right)^3 - 1\right] m_c^3 - \sqrt[3]{m_{ei}^3}} \right\} \bigg/ \sqrt[3]{m_{ei}^3}$$

Optimization was carried out using the IMSL subroutine DBCLSF (IMSL Math/Library, version 1.0, 1987), which involves solving a nonlinear least squares problem subject to bounds on the variables using a modified Levenberg–Marquardt algorithm (37) with a finite difference Jacobian. This subroutine finds the best values of $\bar{t}_c$, σ_c^2, and m_c^3, subject to bounds, that minimize

$$\sum_{i=1}^{j} F(\bar{t}_{ei}, \sigma_{ei}^2, \bar{t}_c, \sigma_c^2)^2 + G(\bar{t}_{ei}, m_{ei}^3, \bar{t}_c, m_c^3)^2 .$$

The lower bounds on $\bar{t}_c$, σ_c^2, and m_c^3 were set at zero. The upper bound on m_c^3 was m_R^3. The upper bounds on σ_c^2, σ_{cmax}^2, and $\bar{t}_c$, $\bar{t}_{cmax}$ were obtained as follows. From Eqs. (22.3a–b), $\bar{t}_c$ and σ_c^2 are

$$\bar{t}_c = \frac{\bar{t}_e}{\sqrt{1 + \frac{\sigma_e^2}{\sigma_c^2}} - 1} \tag{22.8}$$

$$\sigma_c^2 = \frac{\sigma_e^2}{\left(1 + \frac{m_e^3}{m_c^3}\right)^{2/3} - 1} . \tag{22.9}$$

Knowing that $m_c^3 \leq m_R^3$, we see that, from Eq. (22.9),

$$\sigma_c^2 \leq \frac{\sigma_e^2}{\left(1 + \frac{m_e^3}{m_R^3}\right)^{2/3} - 1} . \tag{22.10}$$

If we denote the upper bound on σ_c^2 in Eq. (22.10) as σ_{cmax}^2, from Eq. (22.8) we obtain

$$\bar{t}_c \leq \frac{\bar{t}_e}{\sqrt{1 + \frac{\sigma_e^2}{\sigma_{c_{max}}^2}} - 1} . \tag{22.11}$$

Let the upper bound in Eq. (22.11) be denoted by $\bar{t}_{cmax}$. In addition, knowing $\bar{t}_{cmax}$, one can also set upper and lower bounds on the relative dispersion of $h_c(t)$, RD_c, using Eqs. (22.3a–b):

$$\frac{RD_e}{\sqrt{1 + 2\frac{\bar{t}_{c_{max}}}{\bar{t}_e}}} \le RD_c \le RD_e \,,$$

where

$$RD_e = \frac{\sqrt{\sigma_D^2 - \sigma_R^2}}{\bar{t}_D - \bar{t}_R} \,.$$

The mean transit time, the variance, and the third central moment of $C_R(t)$ and $C_D(t)$ can be calculated from

$$\bar{t} = \frac{\int_0^\infty tC(t)\,dt}{\int_0^\infty C(t)\,dt}\,, \qquad \sigma^2 = \frac{\int_0^\infty (t-\bar{t})^2 C(t)\,dt}{\int_0^\infty C(t)\,dt}\,, \qquad m^3 = \frac{\int_0^\infty (t-\bar{t})^3 C(t)\,dt}{\int_0^\infty C(t)\,dt}\,, \qquad (22.12)$$

where $C(t)$ is $C_R(t)$ for the reference indicator and $C_D(t)$ for the diffusible indicator. However, we have found the following approach to be useful (2–4). The first three moments of $C_D(t)$ and $C_R(t)$ curves were obtained by fitting the outflow curves to the lagged normal density function $L(t)$ as suggested by Bassingth-waighte et al. (7). The lagged normal density function $L(t)$ is defined by the following initial value problem:

$$L(t) = \frac{1}{\phi\sqrt{2\pi}} \exp\left(-\frac{1}{2}\left(\frac{t-\theta}{\phi}\right)^2\right) - \tau \frac{dL(t)}{dt}\,. \qquad (22.13)$$

The three parameters, τ, θ, and ϕ, are related to the first three moments of $L(t)$ by

$$\tau = \frac{\sqrt[3]{m^3}}{2}\,, \qquad \theta = \bar{t} - \tau, \qquad \phi = \sqrt{\sigma^2 - \tau^2}\,. \qquad (22.14)$$

We would like to point out that there is a typographical error in the definition of τ in references (2, 3). The fitting procedure consists of finding the values of the parameters τ, θ, and ϕ, for which Eq. (22.13) best fits the outflow concentration versus time data, scaled by the inverse of the flow rate. Equation (22.13) was solved numerically using the IMSL subroutine IVPAG (IMSL Math/Library, version 1.0, 1987), which solves the differential equations using an Adams–Moulton or Gear method (24). The parameter optimization was carried out using the IMSL subroutine DBCLSF.

The above algorithm of fitting the time data to $L(t)$ to obtain the moments of $C_R(t)$ and $C_D(t)$ tends to be less sensitive to noise in the tails of the concentration data than that using Eq. (22.12), as shown in reference (29). Moreover, the functional form of $L(t)$ is specified by its mean transit time, variance, and third central moment. Thus, $L(t)$ is useful when comparing the model predictions to experimental data as described in the "Experimental Application and Results" section below (2, 3).

Experimental Application and Results

To make use of method A for estimating the first three moments of $h_c(t)$, one possibility might be to find hydrophilic diffusible indicators that have access to different extravascular volumes, for example, total water volume and extracellular volume. However, the total extravascular water volume of the lung is small, and even if one could find indicators with flow-limited access to different fractions of the total extravascular water volume, the differences in moments among the indicators would be small with the potential for sensitivity to small errors. Alternatively, for a lipophilic indicator, the extravascular mean residence time is directly proportional to its tissue-to-plasma partition coefficient and is not limited by the actual extravascular volume (22). If one could find two or more lipophilic indicators that have significantly different tissue-to-plasma partition coefficients, then the use of method A to specify the first three moments of $h_c(t)$ might be adequately robust (2).

Some lipophilic indicators bind to plasma proteins in a rapidly equilibrating manner such that their tissue-to-plasma partition coefficient, and hence their apparent extravascular mean residence time, can be manipulated by altering the plasma protein concentration (2, 3, 22). For this class of indicators, assuming that the equilibrium between perfusate and tissue and between free and protein-bound indicators occurs rapidly in comparison to the pulmonary capillary mean transit time, the extravascular mean residence time, $\bar{t}_e$, is (2, 3):

$$\bar{t}_e = \frac{MQ_t}{\left(1 + \dfrac{[p]}{K}\right)\dot{Q}} \tag{22.15}$$

where M is the tissue-to-plasma partition coefficient of the diffusible indicator at equilibrium, Q_t is some tissue volume accessible to the indicator (generally not equal to the aqueous volume for lipophilic indicators), $[p]$ is the perfusate albumin concentration, K is the plasma protein binding equilibrium dissociation constant, and $\dot{Q}$ is the blood flow rate. As Eq. (22.15) shows, for constant Q_t, M, and K, $\bar{t}_e$ increases as the plasma albumin concentration, $[p]$, decreases. Thus, changing the albumin concentration is the equivalent of using a different indicator that has a different partition coefficient. By making separate injections of ^{14}C-diazepam each at a different plasma albumin concentration, data equivalent to making a single injection with several different rapidly equilibrating indicators with different extravascular mean residence times could be obtained (2, 3).

Isolated Perfused Dog Lung Lobes

Figure 22.4 shows the measured lobar venous effluent ^{125}I-HSA and ^{14}C-diazepam concentration versus time data from one lung lobe at five different plasma albumin concentrations (2, 3). The set of five ^{125}I-HSA superimpose reasonably well, suggesting that any differences in the total system transport function be-

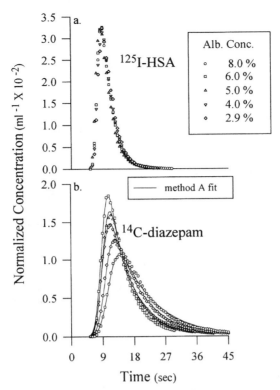

FIGURE 22.4. Venous effluent ^{125}I-HSA (a) and ^{14}C-diazepam (b) concentration versus time data at the five different plasma albumin concentrations (Alb. Conc.) following the bolus injection of the indicators upstream from the lobar artery. The solid lines on the lower panel are method A fits to the ^{14}C-diazepam data. [Reprinted with a permission from the American Physiological Society (2).]

tween albumin concentrations were small, and that the heterogeneity of the lobe perfusion over the course of the experiment was reasonably stable. On the other hand, the simultaneously obtained ^{14}C-diazepam data plotted in Figure 22.4(b) show a progressive increase in the transit time of ^{14}C-diazepam as the plasma albumin concentration was decreased, which is consistent with Eq. (22.15). The moments of $h_c(t)$ were then estimated from the data in Figure 22.4 using method A as described above by fitting $L(t)$ to the individual concentration curves and then using the moments to estimate the moments of $h_c(t)$ by regression on Eqs. (22.7a–b).

Figure 22.4(b) also shows the curves obtained by fitting the model upon which method A is based to the data. The model fits were obtained using Eqs. (22.1a–c) to obtain the moments of $C_n(t)$ for each bolus, and Eqs. (22.2a–c) to calculate the predicted moments for each diazepam outflow concentration curve. Then, the predicted moments were used to obtain the predicted $L(t)$, scaled by $1/\dot{Q}$, for each

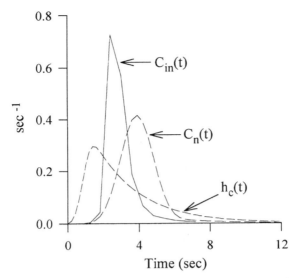

FIGURE 22.5. The measured $C_{in}(t)$, normalized to unit area, and the estimated $C_n(t)$ and $h_c(t)$ for the data shown in Figure 22.4. [Reprinted with permission from the American Physiological Society (2).]

diazepam curve. The results suggest that the model is reasonably consistent with the data from all five injections.

Figure 22.5 shows the $L(t)$ representation of $h_c(t)$, the moments of which were estimated from the moments of the data in Figure 22.4 using method A. Also shown in Figure 22.5 are the measured $C_{in}(t)$, normalized to unit area, and the estimated $C_n(t)$, obtained by numerically deconvolving (12) $h_c(t)$ and the representative $C_R(t)$ curve. The representative $C_R(t)$ curve was obtained by using the average values of the moments of the five ^{125}I-HSA outflow curves shown in Figure 22.4(a) in Eq. (22.13) (2). The results from dog lung lobes are summarized in Table 22.3.

The total vascular volume, Q, of the lung lobe can be estimated from $Q = \dot{Q} \, (\bar{t}_n - \bar{t}_{in} + \bar{t}_c)$, where $\bar{t}_{in}$ is the mean transit time for $C_{in}(t)$, and $\bar{t}_n - \bar{t}_{in}$ is the nonexchanging organ vessels (arteries and veins) mean transit time. This averaged about 41 ml, and the capillary volume, $(\dot{Q} \, \bar{t}_c)$, about 20 ml. Thus, about 48% of the total lobar vascular volume was estimated to be in the capillaries. From this study, we estimated that more than 90% of the variance of the whole organ transport function, excluding $C_{in}(t)$, was estimated to be due to the capillary bed (2). Table 22.3 shows that, on average, the pulmonary capillary relative dispersion, RD_c, was about 75%.

Isolated Perfused Rabbit Lungs

This approach of estimating the moments of $h_c(t)$ using method A by manipulating plasma albumin concentration provided estimates of the pulmonary capillary

TABLE 22.3. Summary of the moments of the capillary transit time distribution, $h_c(t)$, estimated for a dog lung lobe.

	$h_c(t)$					$\dot{Q}$	Q_c
	$\bar{t}_c$ (s)	σ_c (s)	m_c^3/σ_c^3	RD_c (%)	C.V.	(ml/s)	(ml)
Mean	3.20	2.37	1.84	75	8.7	6.13	19.6
± SD	± 0.50	± 0.31	± 0.06	± 6	± 3.3	± 0.28	± 3.0

$n = 6$. The $\bar{t}_c$, σ_c, and m_c^3/σ_c^3 are, respectively, the estimates of mean transit time, the standard deviation, and the coefficient of skewness of the capillary transport function, $h_c(t)$. The C.V. is the coefficient of variation between the data and the model prediction for the $L(t)$ corresponding to each diazepam concentration curve based on the moments of estimated $h_c(t)$ and $\bar{t}_D - \bar{t}_R$, as indicated in the text. The RD_c is the relative dispersion of $h_c(t)$, which is the ration $\sigma_c/\bar{t}_c$. The $\dot{Q}$ and Q_c are the blood flow rate and capillary volume, respectively. [Reprinted with permission from the American Physiological Society (2).]

transit time distribution of the isolated dog lung lobe and some useful information to establish the feasibility of the method. However, the method would have more practical utility if at least two indicators having the required properties were available, so that the moments of $h_c(t)$ could be estimated from a single bolus injection.

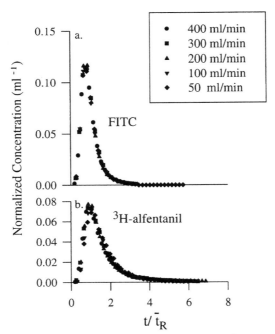

FIGURE 22.6. Venous effluent FITC dye (a) and ^{3}H-alfentanil (b) concentration versus time data at the five different flow rates following the bolus injection of the indicators upstream from the pulmonary artery of an isolated perfused rabbit lung. The time scale is the same as in Figure 22.1. [Reprinted with permission from the American Physiological Society (3).]

We found that ^{3}H-alfentanil is another rapidly equilibrating indicator whose tissue-to-plasma partition coefficient is sufficiently different from that of ^{14}C-diazepam to produce the required difference in extravascular mean residence times (3).

To demonstrate that ^{3}H-alfentanil is a rapidly diffusing (flow-limited) indicator, we examined the influence of varying flow on its effluent concentration curve. Figure 22.6 shows the measured venous effluent FITC and ^{3}H-alfentanil concentration curves from an isolated perfused rabbit lung at five flows. The venous pressure was adjusted when the flow was changed so that the sum of the arterial and venous pressures was approximately constant over the flow range. This was done to minimize the influence of flow on the total vascular volume by maintaining constant microvascular pressure (3). Both the FITC and ^{3}H-alfentanil curves are reasonably congruent when plotted against time normalized to the total pulmonary vascular mean transit time over the eightfold range of flows. The superimposition of the data in Figure 22.6 indicates that ^{3}H-alfentanil equilibrates with the tissue very rapidly on passage through the lung (3).

Having established that ^{3}H-alfentanil is a flow-limited indicator, we proceeded with estimating the first three moments of $h_c(t)$ in the isolated perfused rabbit lung using method A with ^{3}H-alfentanil and ^{14}C-diazepam as the two flow-limited

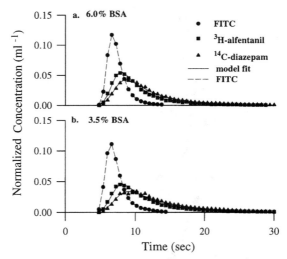

FIGURE 22.7. (a) The venous effluent concentration versus time data for the FITC dye, ^{3}H-alfentanil, and ^{14}C-diazepam following the bolus injection of these indicators into the pulmonary artery of an isolated perfused rabbit lung at 6% BSA. (b) The venous effluent concentration versus time data for the FITC dye, ^{3}H-alfentanil, and ^{14}C-diazepam following the bolus injection of these indicators into the pulmonary artery of the same isolated perfused rabbit lung at 3.5% BSA. The solid lines superimposed on the ^{3}H-alfentanil and ^{14}C-diazepam data in all three panels are method A fits to the ^{3}H-alfentanil and ^{14}C-diazepam data. Reprinted with permission from the American Physiological Society (3).

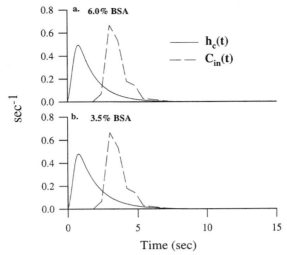

FIGURE 22.8. (a) The measured $C_{in}(t)$ normalized to unit area, and the estimated $h_c(t)$ from the outflow curves of FITC dye, ^{3}H-alfentanil, and ^{14}C-diazepam (panel a in Figure 22.7) obtained following the rapid bolus injection at 6% BSA. (b) The measured $C_{in}(t)$ normalized to unit area, and the estimated $h_c(t)$ from the outflow curves of FITC dye, ^{3}H-alfentanil, and ^{14}C-diazepam (panel b in Figure 22.7) obtained following the rapid bolus injection of these indicators at 3.5% BSA. Reprinted with permission from the American Physiological Society (3).

indicators. Figure 22.7 shows the pulmonary venous effluent concentration versus time curves for FITC dye, ^{14}C-diazepam, and ^{3}H-alfentanil from the same lung following bolus injection of these indicators at perfusate albumin concentrations of 6% (panel a) and 3.5% (panel b). The outflow concentration–time curve for ^{14}C-diazepam is more dispersed, and hence its extravascular mean residence time is longer than that for ^{3}H-alfentanil at both perfusate albumin concentrations (3).

The moments of $h_c(t)$ were estimated from each set of outflow data using method A as indicated above. Figures 22.8(a) and 22.8(b) show the $L(t)$ representation of $h_c(t)$ for each of the two sets of data in Figures 22.7(a) and 22.7(b), respectively. Also shown in Figure 22.8 are the measured $C_{in}(t)$, normalized to unit area. Neither the results of fitting the model to each of the two sets of data nor the estimated capillary transport function moments were found to be significantly dependent on the perfusate albumin concentration, as shown in Table 22.4 (combinations I and II) (3).

One important test for the hypotheses upon which method A is based was to determine if the moments of $h_c(t)$ could be estimated even if the noncapillary transit time distribution was varied. $C_n(t)$, $C_n(t = C_{in}(t)*h_{av}(t)$, can be changed without affecting $h_c(t)$ by changing $C_{in}(t)$, which can be changed by adjusting the

TABLE 22.4. Summary of the moments of the capillary transit time distribution, $h_c(t)$, estimated for a rabbit lung.

Combination (n)	$\bar{t}_c$ (s)	σ_c (s)	m_c^3/σ_c^3	RD_c (%)	C.V.	$\dot{Q}$ (ml/s)	Q_c (ml)
			$h_c(t)$				
I (7)	1.33	1.17	1.99	88	11.8	3.29	4.33
	± 0.06	± 0.21	± 0.25	± 2	± 0.9	± 0.02	± 0.20
II (6)	1.36	1.19	2.06	87	12.8	3.31	4.50
	± 0.12	± 0.30	± 0.46	± 2	± 0.7	± 0.01	± 0.40
III (6)	1.31	1.30	1.90	97	14.4	3.29	4.32
	± 0.30	± 0.40	± 0.58	± 4	± 0.9	± 0.02	± 0.67

The values are mean ± SE. (n) is the number of lungs. The $\bar{t}_c$, σ_c, and m_c^3/σ_c^3 are, respectively, the estimates of mean transit time, the standard deviation, and the coefficient of skewness of the capillary transport function, $h_c(t)$. The C.V. is the coefficient of variation between the data and the model prediction for the $L(t)$ corresponding to each diazepam and alfentanil concentration curve based on the estimated moments of $h_c(t)$ and $\bar{t}_D - \bar{t}_R$, as indicated in the text. The RD_c is the relative dispersion of $h_c(t)$, which is the ratio $\sigma_c/\bar{t}_c$. The $\dot{Q}$ and Q_c are the blood flow rate and capillary volume, respectively. The $\dot{Q}$ and Q_c were estimated from the moments of the outflow curves of: alfentanil and diazepam at 6% BSA for combination (I), and alfentanil and diazepam at 3.5% BSA for combination (II). For combination (III), the moments of $h_c(t)$ were estimated from the moments of the outflow curves of alfentanil and diazepam following the slow injection of these indicators at 6% BSA. Reprinted with permission from the American Physiological Society (3).

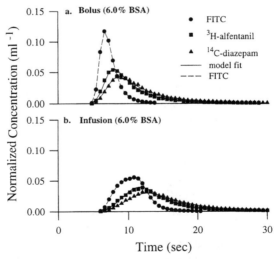

FIGURE 22.9. (a) The venous effluent concentration versus time data for the FITC dye, ^{3}H-alfentanil, and ^{14}C-diazepam following the bolus injection of these indicators into the pulmonary artery of an isolated perfused rabbit lung at 6% BSA. (b) The venous effluent concentration versus time data for the FITC dye, ^{3}H-alfentanil, and ^{14}C-diazepam following the slow injection of these indicators into the pulmonary artery of the same isolated perfused rabbit lung at 6% BSA. Reprinted with permission from the American Physiological Society (3).

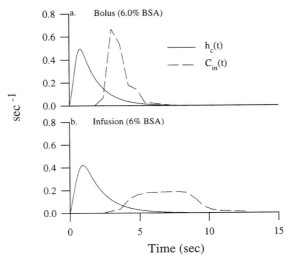

FIGURE 22.10. (a) The measured $C_{in}(t)$ normalized to unit area, and the estimated $h_c(t)$ from the outflow curves of FITC dye, ^{3}H-alfentanil, and ^{14}C-diazepam (panel a in Figure 22.9) obtained following the rapid bolus injection of these indicators at 6% BSA. (b) The measured $C_{in}(t)$ normalized to unit area, and the estimated $h_c(t)$ from the outflow curves of FITC dye, ^{3}H-alfentanil, and ^{14}C-diazepam (panel b in Figure 22.9) obtained following the slow bolus injection of these indicators at 6% BSA. Reprinted with permission from the American Physiological Society (3).

rate at which the bolus is injected. In other words, for a slow bolus injection, $C_{in}(t)$, and hence $C_n(t)$, are more dispersed than for a rapid bolus injection (2, 3).

The upper panel of Figure 22.9 shows the venous effluent concentration versus time outflow curves for FITC dye, ^{14}C-diazepam, and ^{3}H-alfentanil following a rapid bolus injection of these indicators into the pulmonary artery of an isolated perfused rabbit lung. In the same rabbit lung, the injection procedure was modified (2, 3), and a second bolus of the vascular indicator, ^{14}C-diazepam, and ^{3}H-alfentanil was infused slowly into the pulmonary artery, resulting in more dispersed venous effluent concentration versus time outflow curves, as shown in the lower panel of Figure 22.9. Therefore, Figure 22.9 shows two different sets of FITC dye, ^{14}C-diazepam, and ^{3}H-alfentanil outflow curves from the same lung with presumably the same capillary transit time distribution, but different amounts of dispersion outside the capillaries.

Figure 22.10 shows the $L(t)$ representation of $h_c(t)$ for each of the two sets of data in Figure 22.10, and the measured $C_{in}(t)$, normalized to unit area. The models fit to the two sets of data and the estimated moments of $h_c(t)$ were little affected by the dispersion of $C_n(t)$ (see combinations I and III in Table 22.4), which is consistent with the hypotheses upon which method A is based (3).

The total vascular volume in the rabbit lungs averaged about 9.5 ml, and the capillary volume, $(\dot{Q}\bar{t}_c)$, about 4.2 ml (3). Thus, about 44% of the total lobar

TABLE 22.5. Summary of relationships between the moments of the transport functions $h_c(t)$ and $C_n(t)$, and the moments of $C_R(t)$ and $C_D(t)$ after relaxing the assumptions of random coupling and constant A_e/A_c ratio for all capillary elements.

Assumptions	$C_R(t)$ moments	$C_D(t)$ moments
$C_n(t)$ and $h_c(t)$ are independent (random coupling), and the A_e/A_c ratio is distributed	$\bar{t}_R = \bar{t}_n + \bar{t}_c$ $\sigma_R^2 = \sigma_n^2 + \sigma_c^2$ $m_R^3 = m_n^3 + m_c^3$	$\bar{t}_D = \bar{t}_n + (1 + \bar{r})\bar{t}_c$ $\sigma_D^2 = \sigma_n^2 + (1 + \bar{r})^2\sigma_c^2 + (\bar{t}_c^2 + \sigma_c^2)\,\sigma_A^2$ $m_D^3 = m_n^3 + (1 + \bar{r})^3 m_c^3 + m_A^3(\bar{t}_c^2 + m_c^3 + 3\sigma_c^2\bar{t}_c)$ $\qquad + 3\sigma_A^2(1 + \bar{r})\,(m_c^3 + 2\sigma_c^2\bar{t}_c)$ For method A, $r = A_e/A_c$ ratio is same for all capillaries (i.e., $h_A(r)$ is a delta function). Hence, σ_A^2 and m_A^3 are set to zero. Note: $\bar{r} = \dfrac{\bar{t}_e}{\bar{t}_c}$
$C_n(t)$ and $h_c(t)$ are dependent (flow coupling), and the A_e/A_c ratio is distributed	$\bar{t}_R = \bar{t}_n + \bar{t}_c$ $\sigma_R^2 = \sigma_n^2 + \sigma_c^2 + 2\text{Cov}_{nc}$	$\bar{t}_D = \bar{t}_n + \bar{t}_c + \bar{t}_e$ $\sigma_D^2 = \sigma_n^2 + (1 + \bar{r})^2\sigma_c^2 + (\bar{t}_c^2 + \sigma_c^2)\,\sigma_A^2 + 2\text{Cov}_{nc}(1 + \bar{r})$
Flow coupling between $C_n(t)$ and $h_c(t)$ with a linear relationship between the capillary and noncapillary transit times (42)	$\bar{t}_R = \bar{t}_n + \bar{t}_c$ $\sigma_R^2 = \left(1 + \dfrac{\bar{t}_n}{\bar{t}_c}\right)^2 \sigma_c^2$ $m_R^3 = \left(1 + \dfrac{\bar{t}_n}{\bar{t}_c}\right)^3 m_c^3$	$\bar{t}_D = \bar{t}_n + \bar{t}_c + \bar{t}_e$ $\sigma_D^2 = \left(1 + \dfrac{\bar{t}_n}{\bar{t}_c} + \bar{r}\right)^2 \sigma_c^2$ $m_D^3 = \left(1 + \dfrac{\bar{t}_n}{\bar{t}_c} + \bar{r}\right)^3 m_c^3$

The σ_A^2 and m_A^3 are the variance and the third central moment of the frequency distribution of the ratio A_e/A_c. CoVnc = covariance of the capillary and noncapillary transit time distributions. For more details on the derivation of the above equations, refer to the appendix in reference (2). Reprinted with permission from the American Physiological Society (2).

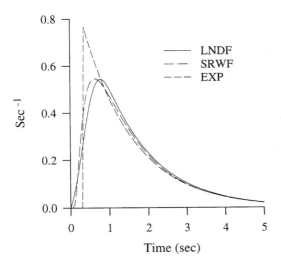

FIGURE 22.11. Approximation of the functional form of the capillary transport function using three different right-skewed functional forms, namely, single exponential (EXP), lagged normal density function (LNDF), and shifted random walk function (SRWF) (3). All three functional forms have the same first three moments estimated from the outflow data in Figure 22.8(a) using method A. Reprinted with permission from the American Physiological Society (3).

vascular volume was estimated to be in the capillaries. Most of the variance of the whole organ transit time distribution, that is, excluding $C_{in}(t)$, was due to the capillary bed (3). On average, the pulmonary capillary relative dispersion was about 90% (Table 22.4). The results from the dog and rabbit lungs can be compared with those from other studies using other methods in Tables 22.1 and 22.2.

In deriving Eqs. (22.1a–c) and (22.2a–c), two of the assumptions made were that the A_e/A_c and $C_n(t)$ are the same for all capillary elements. Table 22.5 shows the moment relationships obtained after relaxing these two assumptions (2). These moment relationships involve increased numbers of parameters, and it is not clear whether they will be useful for moment estimation in a manner analogous to method A. However, these equations can be used to put some perspective on the potential impact of distributed A_e/A_c and correlations between local capillary and conducting vessel transit times. Evaluation of some of the implications of these assumption is discussed in detail in reference (2).

We have used the lagged normal density function, $L(t)$, to approximate the functional shape of the capillary transport function from the estimated first three moments using the present method. However, it should be pointed out that this function is not the only shape that satisfies the first three moments. This is exemplified in Figure 22.11, wherein three smooth right-skewed functional forms having the same $\bar{t}$, σ^2, and m^3 as $h_c(t)$ in Figure 22.8 (panel a) are shown. In Figure 22.11, the time-shifted exponential function, which rises vertically, appears to be at one end of the range of reasonable possibilities for approximating the functional form of the capillary transport function. This functional form is similar to the one proposed by Fung and Sobin (23) for the transit time distribution for blood flow in an alveolar sheet.

Summary and Conclusion

In conclusion, we have presented two applications involving the use of nonbasic rapidly equilibrating lipophilic amines such as ^{14}C-diazepam and ^{3}H-alfentanil for estimating parameters reflecting the state of the lung. We used isolated lung preparations in which we could control variables in such a way as to test the model hypotheses. The results suggest that it may be useful to pursue the possibility that similar approaches may be applicable in vivo. In this regard it is interesting that, for alfentanil, diazepam, and other amine drugs, there are clinical chemical assays which are sensitive enough that the drugs could be used even in subpharmacological doses in nonradioactive form (34, 41).

Acknowledgment. This study was supported by the Department of Veterans Affairs, the American Heart Association of Wisconsin, the Whitaker Foundation, and the National Heart, Lung, and Blood Institute under Grant HL-24349.

References

1. Anderson, M.W., T.C. Orton, R.D. Pickett, and T.E. Eling. Accumulation of amines in the isolated perfused rabbit lungs. *J. Pharmacol. Exp. Ther.* 189:456–466, 1974.
2. Audi, S.H., G.S. Krenz, J.H. Linehan, D.A. Rickaby, and C.A. Dawson. Pulmonary capillary transport function from flow-limited indicators. *J. Appl. Physiol.* 77:332–351, 1994.
3. Audi, S.H., J.H. Linehan, G.S. Krenz, C.A. Dawson, S.B. Ahlf, and L.D. Roerig. Estimation of the pulmonary capillary transport function in isolated rabbit lungs. *J. Appl. Physiol.* 78:1004–1014, 1995.
4. Ayappa, I., L.V. Brown, P.M. Wang, and S.J. Lai-Fook. Arterial, capillary, and venous transit times and dispersion measured in isolated rabbit lungs. *J. Appl. Physiol.* 79:261–269, 1995.
5. Bachofen, H., D. Wangensteen, and E.R. Weibel. Surfaces and volumes of alveolar tissue under zone II and zone III conditions. *J. Appl. Physiol.* 53:879–885, 1982.
6. Bassingthwaighte, J.B. Plasma indicator dispersion in arteries of the human leg. *Circ. Res.* 19:332–346, 1966.
7. Bassingthwaighte, J.B., F.H. Ackerman, and E.H. Wood. Application of the Lagged Normal Density curve as a model for arterial dilution curve. *Circ. Res.* 18:398–415, 1966.
8. Bassingthwaighte, J.B. and C.A. Goresky. Modeling in the analysis of solute and water exchange in the microvasculature. In: *Handbook of Physiology, Section 2, The Cardiovascular System, Volume IV, Microcirculation, Part 1*, edited by E.M. Renkin and C.C. Michel. Bethesda, MD: Amer. Physiol. Soc., 1984, pp. 549–626.
9. Brashear, R.E., J.C. Ross, and W.J. Daly. Pulmonary diffusion and capillary blood volume in dogs at rest and with exercise. *J. Appl. Physiol.* 21:516–520, 1966.
10. Brody, J.S., E.J. Stemmler, and A.B. Dubois. Longitudinal distribution of vascular resistance in the pulmonary arteries, capillaries, and veins. *J. Clin. Invest.* 47:783–799, 1968.

11. Bronikowski, T.A., C.A. Dawson, and J.H. Linehan. On indicator dilution and perfusion heterogeneity: A stochastic model. *Math. Biosci.* 83:199–225, 1987.

12. Bronikowski, T.A., C.A. Dawson, and J.H. Linehan. Model-free deconvolution techniques for estimating vascular transport functions. *Int. J. Bio-Medical Computing.* 14:411–429, 1983.

13. Capen, R.L., W.L. Hanson, L.P. Latham, C.A. Dawson, and W.W. Wagner, Jr. Distribution of pulmonary capillary transit times in recruited networks. *J. Appl. Physiol.* 69:473–478, 1990.

14. Capen, R.L., L.P. Latham, and W.W. Wagner, Jr. Comparison of direct and indirect measurements of pulmonary capillary transit times. *J. Appl. Physiol.* 62:1150–1154, 1987.

15. Chinard, F.P. and W.O. Cua. Endothelial extraction of tracer water varies with extravascular water in dog lungs. *Am. J. Physiol.* 252(*Heart Circ. Physiol. 21*): H340–H348, 1987.

16. Crapo, J.D., R.O. Crapo, R.L. Jensen, R.R. Mercer, and E.R. Weibel. Evaluation of lung diffusing capacity by physiological and morphometric techniques. *J. Appl. Physiol.* 64:2083–2091, 1988.

17. Crapo, R.O., J.D. Crapo, and A.H. Morris. Lung tissue and capillary blood volumes by rebreathing and morphometric techniques. *Resp. Physiol.* 49:175–186, 1982.

18. Cree, E.M., D.F. Benfield, and H.K. Rasmussen. Differential lung diffusion, capillary volume and compliance in dogs. *J. Appl. Physiol.* 25:186–190, 1968.

19. Crone, C. Permeability of capillaries in various organs as determined by use of the "indicator diffusion" method. *Acta Physiol. Scand.* 48:292–305, 1963.

20. Cua, W.O., G. Basset, F. Bouchonnet, R.A. Carrick, G. Saumon, and F.P. Chinard. Endothelial and epithelial permeabilities to antipyrine in rat and dog lungs. *Am. J. Physiol.* 258(*Heart Circ. Physiol. 27*):H1321–H1333, 1990.

21. Dawson, C.A., D.A. Rickaby, J.H. Linehan, and T.A. Bronikowski. Distribution of vascular volume and compliance in the lung. *J. Appl. Physiol.* 64:266–273, 1988.

22. Dawson, C.A., D.L. Roerig, D.A. Rickaby, L.D. Nelin, J.H. Linehan, and G.S. Krenz. Use of diazepam for interpreting changes in extravascular lung water. *J. Appl. Physiol.* 72:686–693, 1992.

23. Fung, Y.C. and S.S. Sobin. Pulmonary alveolar blood flow. *Circ. Res.* 30:470–490, 1972.

24. Gear, C.W. *Numerical Initial-Value Problems in Ordinary Differential Equations.* Englewood Cliffs, NJ: Prentice-Hall, 1971.

25. Goresky, C.A. A linear method for determining liver sinusoidal and extravascular volumes. *Am. J. Physiol.* 204:626–640, 1963.

26. Goresky, C.A., R.F.P. Cronin, and B.E. Wangel. Indicator dilution measurements of extravascular water in the lungs. *J. Clin. Invest.* 48:487–501, 1969.

27. Goresky, C.A. and M. Silverman. Effect of connection of catheter distortion on calculated liver sinusoidal volumes. *Am. J. Physiol.* 207:883–892, 1964.

28. Goresky, C.A., W.H. Ziegler, and G.G. Bach. Capillary permeability, barrier-limited and flow-limited distribution. *Circ. Res.* 27:739–764, 1970.

29. Harris, T.R. and E.V. Newman. An analysis of mathematical models of circulatory indicator dilution curves. *J. Appl. Physiol.* 28:840–850, 1970.

30. Haworth, S.T., J.H. Linehan, T.A. Bronikowski, and C.A. Dawson. A hemodynamic model representation of the dog lung. *J. Appl. Physiol.* 70:15–26, 1991.

542   Said H. Audi et al.

31. Hogg, J.C., T. McLean, B.A. Martin, and B. Wiggs. Erythrocyte transit and neutrophil concentration in dog lung. *J. Appl. Physiol.* 65:1217–1225, 1988.

32. Jouasset-Strieder, D., J.M. Cahill, and J.J. Byrne. Pulmonary capillary blood volume in dogs during shock and after retransfusion. *J. Appl. Physiol.* 21:365–369, 1966.

33. Jouasset-Strieder, D., J.M. Cahill, J.J. Byrne, and E.A. Gaensler. Pulmonary diffusing capacity and capillary blood volume in normal and anemic dogs. *J. Appl. Physiol.* 20:113–116, 1965.

34. Kotrly, K.J., D.L. Roerig, S.B. Ahlf, and J.P. Kampine. First pass uptake of lidocaine, diazepam and thiopental in human lung. *Anesth. Anal.* 67:5119, 1988.

35. Kuhnle, G.E.H., F.H. Leipfinger, and A.E. Goetz. Measurement of microhemo-dynamics in the ventilated rabbit lung by intravital fluorescence microscopy. *J. Appl. Physiol.* 74:1462–1471, 1993.

36. Linehan, J.H., T.A. Bronikowski, and C.A. Dawson. Kinetics of uptake and metabo-lism by endothelial cells from indicator dilution data. *Ann. of Biomed. Eng.* 15:201–215, 1987.

37. Marquardt, D. An algorithm for least-squares estimation of nonlinear parameters, *SIAM J. Appl. Math.* 11:431–441, 1963.

38. Piiper, J. Attempts to determine volume compliance and resistance to flow of pulmo-nary vascular compartments. *Prog. Resp. Res.* 5:40–52, 1970.

39. Roerig, D.L., R.R. Dahl, C.A. Dawson, and R.I.H. Wang. Uptake of 1-α-acetyl-methadol (LAAM) and its analgesically active metabolites, nor-LAAM and dinor-LAAM in the isolated perfused rat lung. *Drug Metab. Dispos.* 11:411–416, 1983.

40. Roerig, D.L., R.R. Dahl, C.A. Dawson, and R.I.H. Wang. Effect of protein binding on the uptake and efflux of methadone and diazepam in the isolated perfused rat lung. *Drug Metab. Dispos.* 12:536–542, 1984.

41. Roerig, D.L., K.J. Kotrly, C.A. Dawson, S.B. Ahlf, J.F. Gaultieri, and J.P. Kampine. First pass uptake of verapamil, diazepam and thiopental in the human lung. *Anesth. Anal.* 69:461–466, 1989.

42. Rose, C.P. and C.A. Goresky. Vasomotor control of capillary transit time heterogeneity in the canine coronary circulation. *Circ. Res.* 39:541–554, 1976.

43. Staub, N.C. and E.L. Schultz. Pulmonary capillary length in dog, cat and rabbit. *Respi. Physiol.* 5:371–378, 1968.

44. Schwab, A.J., F. Barker III, C.A. Goresky, and K.S. Pang. Transfer of enalaprilat across rat liver cell membranes is barrier limited. *Am. J. Physiol.* 258(*Gastrointest. Liver Physiol. 21*):G461–G475, 1990.

45. Siegwart, B., P. Gehn, J. Gil, and E.R. Weibel. Morphometric estimation of pulmonary diffusing capacity. *Resp. Physiol.* 13:141–159, 1971.

46. Wang, P.M., C.D. Fike, M.R. Kaplowitz, L.V. Brown, I. Ayappa, M. Jahed, and S.J. Lai-Fook. Effects of lung inflation and blood flow on capillary transit time in isolated rabbit lungs. *J. Appl. Physiol.* 72:2420–2427, 1992.

47. Wang, P.M., Q.H. Yang, and S.J. Lai-Fook. Effect of positive airway pressure on capillary transit time in rabbit lung. *J. Appl. Physiol.* 69:2262–2268, 1990.

48. Wanner, A., R. Begin, M. Cohen, and M.A. Sackner. Vascular volumes of the pulmo-nary circulation in intact dogs. *J. Appl. Physiol.: Respirat. Environ. Exercise Physiol.* 44:956–963, 1978.

49. Weibel, E.R. and J. Gil. Structure-function relationships at the alveolar level. In: *Lung Biology in Health and Disease, Vol. 3: Bioengineering Aspects of the Lung,* edited by J.B. West. New York: Marcel Dekker, 1977, pp. 1–81.

50. Young, R.C., Jr., H. Nagano, T.R. Vaughn, Jr., and N. Staub. Pulmonary capillary blood volume in dogs; effects 5-hydroxytryptamine. *J. Appl. Physiol.* 18:264–268, 1963.

Glossary

A_c = capillary cross-sectional area

A_e = extravascular cross-sectional area

$C_D(t)$ = flow-limited indicator venous effluent concentration at time t following bolus injection

$C_{in}(t)$ = measured tubing outflow curve

$C_n(t)$ = function representing the transport of indicators through the noncapillary (arteries, veins, tubing, and injection-sampling system) part of the system

$C_R(t)$ = vascular reference indicator venous effluent concentration at time t following bolus injection

$h_{av}(t)$ = arterial–venous transit time distribution

$h_c(t)$ = capillary transit time distribution

K = plasma protein binding equilibrium dissociation constant

$L(t)$ = lagged normal density function

M = tissue-to-plasma partition coefficient of the flow-limited indicator at equilibrium

m^3 = third central moment

m^3/σ^3 = skewness coefficient

$[p]$ = perfusate albumin concentration

Q = total pulmonary vascular volume

Q_c = pulmonary capillary volume

Q_e = extravascular volume of the flow-limited indicator

Q_{ed} = virtual extravascular volume accessible to ^{14}C-diazepam

Q_{ew} = extravascular volume of ^{3}HOH

Q_t = some tissue volume accessible to the flow-limited lipophilic indicator

$\dot{Q}$ = flow

RD_c = relative dispersion of $h_c(t)$

RD_e = extravascular relative dispersion

$\overline{t}$ = mean transit time

σ^2 = variance or second central moment

σ = standard deviation

$\lambda = Q_e/Q_c$

τ, θ, and ϕ are the three parameters of the lagged normal density function $L(t)$

Publications of Carl A. Goresky

1. Goresky, C. A. A linear method for determining liver sinusoidal and extravascular volumes. *Am. J. Physiol.* 204:626–640, 1963.
2. Goresky, C. A., H. Watanabe, and D. G. Johns. The renal excretion of folic acid. *J. Clin. Invest.* 42:1841–1849, 1963.
3. Goresky, C. A. Initial distribution and rate of uptake of sulfobromophthalein in the liver. *Am. J. Physiol.* 207:13–26, 1964.
4. Goresky, C. A., and G. Kumar. Renal failure in cirrhosis of the liver. *Can. Med. Assoc. J.* 90:353–356, 1964.
5. Goresky, C. A., and M. Silverman. Effect of correction of catheter distortion on calculated liver sinusoidal volumes. *Am. J. Physiol.* 207:883–892, 1964.
6. Goresky, C. A. The nature of transcapillary exchange in the liver. *Can. Med. Assoc. J.* 92:517–522, 1965.
7. Goresky, C. A. The hepatic uptake and excretion of sulfobromophthalein and bilirubin. *Can. Med. Assoc. J.* 92:851–857, 1965.
8. Silverman, M., and C. A. Goresky. A unified kinetic hypothesis of carrier mediated transport: Its applications. *Biophys. J.* 5:487–509, 1965.
9. Chinard, F. P., T. Enns, C. A. Goresky, and M. F. Nolan. Renal transit times and distribution volumes of T-1824, creatinine, and water. *Am. J. Physiol.* 209:243–252, 1965.
10. Chinard, F. P., C. A. Goresky, T. Enns, M. F. Nolan, and R. W. House. Trapping of urea by red cells in the kidney. *Am. J. Physiol.* 209:253–263, 1965.
11. Goresky, C. A. *The nature of the exchange across sinusoids in the liver.* Ph.D. Thesis, McGill University, 1965.
12. Goresky, C. A. The distribution of substances in a flow-limited organ, the liver. In: *Compartments, Pools, and Spaces in Medical Physiology,* edited by P. E. Bergner, C. C. Lushbaugh, and E. Anderson. Oak Ridge: Division of Technical Services, U.S. Atomic Energy Commission, 1967, pp. 423–450.
13. Goresky, C. A., T. H. Holmes, and A. Sass-Kortsak. The initial uptake of copper by the liver in the dog. *Can. J. Physiol. Pharmacol.* 46:771–784, 1968.
14. Goresky, C. A., R. F. P. Cronin, and B. E. Wangel. Indicator dilution measurements of extravascular water in the lungs. *J. Clin. Invest.* 48:487–501, 1969.
15. Goresky, C. A., W. H. Ziegler, and G. G. Bach. Barrier-limited distribution of diffusible substances from the capillaries in a well-perfused organ. In: *Capillary Permeability,* edited by C. Crone and N. A. Lassen. Copenhagen: Munksgaard, 1970, pp. 171–184.

16. Goresky, C. A. The interstitial space in the liver: Its partitioning effects. In: *Capillary Permeability*, edited by C. Crone and N. A. Lassen. Copenhagen: Munksgaard, 1970, pp. 415–430.

17. Goresky, C. A., and G. G. Bach. Membrane transport and the hepatic circulation. *Ann. NY Acad. Sci.* 170:18–47, 1970.

18. Goresky, C. A., W. H. Ziegler, and G. G. Bach. Capillary exchange modeling: Barrier-limited and flow-limited distribution. *Circ. Res.* 27:739–764, 1970.

19. Ziegler, W. H., and C. A. Goresky. Transcapillary exchange in the working left ventricle of the dog. *Circ. Res.* 29:181–207, 1971.

20. Ziegler, W. H., and C. A. Goresky. Kinetics of rubidium uptake in the working dog heart. *Circ. Res.* 29:208–220, 1971.

21. Goresky, C. A., R. F. P. Cronin, L. M. Lawson, and B. E. Wangel. Extravascular lung water: Its measurements in normal and edematous lungs. In: *Central Hemodynamics and Gas Exchange*, edited by C. Giuntini. Torino: Minerva Medica, 1971, pp. 77–92.

22. Goresky, C. A., F. P. Chinard, R. F. P. Cronin, and B. E. Wangel. Extravascular water space. In: *Central Hemodynamics and Gas Exchange*, edited by C. Giuntini. Torino: Minerva Medica, 1971, pp. 449–458.

23. Goresky, C. A. The diffusional problems underlying the distribution of water in well-perfused organs: Their general implications. In: *Oxygen Supply*, edited by M. Kessler, D. F. Bruley, L. C. Clark Jr., D. W. Lubbers, I. A. Silver, and J. Strauss. Munich: Urban & Schwarzenberg, 1973, pp. 51–60.

24. Goresky, C. A., G. G. Bach, and B. E. Nadeau. On the uptake of materials by the intact liver: The concentrative transport of rubidium-86. *J. Clin. Invest.* 52:975–990, 1973.

25. Goresky, C. A., G. G. Bach, and B. E. Nadeau. On the uptake of materials by the intact liver: The transport and net removal of galactose. *J. Clin. Invest.* 52:991–1009, 1973.

26. Goresky, C. A. The transport and net removal of substances by the intact liver. In: *The Liver: Quantitative Aspects of Structure and Function*, edited by G. Paumgartner and R. Preisig. Basel: Karger, 1973, pp. 125–132.

27. Goresky, C. A., and H. L. Goldsmith. Capillary-tissue exchange kinetics: Diffusional interactions between adjacent capillaries. In: *Oxygen Transport to Tissue: Pharmacology, Mathematical Studies and Neonatology*, edited by D. F. Bruley and H. L. Bicher. New York: Plenum Press, 1973, pp. 773–781.

28. Goresky, C. A., and B. E. Nadeau. Uptake of materials by the intact liver: The exchange of glucose across the cell membranes. *J. Clin. Invest.* 53:634–646, 1974.

29. Goresky, C. A., H. H. Haddad, W. S. Kluger, B. E. Nadeau, and G. G. Bach. The enhancement of maximal bilirubin excretion with taurocholate-induced increments in bile flow. *Can. J. Physiol. Pharmacol.* 52:389–403, 1974.

30. Farack, U. M., C. A. Goresky, M. Jabbari, and D. G. Kinnear. Double pylorus: A hypothesis concerning its pathogenesis. *Gastroenterology* 66:596–600, 1974.

31. Gordon, E. R., M. Dadoun, C. A. Goresky, T. H. Chan, and A. S. Perlin. The isolation of an azobilirubin β-D-monoglucoside from dog gallbladder bile. *Biochem. J.* 143:97–105, 1974.

32. Goresky, C. A. The lobular design of the liver: its effect on uptake processes. In: *Regulation of Hepatic Metabolism (Alfred Benzon Symp. IV)*, edited by F. Lundquist and N. Tygstrup. Copenhagen: Munksgaard, 1974, pp. 808–822.

33. Goresky, C. A. The hepatic uptake process: Its implications for bilirubin transport. In:

Jaundice, edited by C. A. Goresky and M. M. Fisher. New York: Plenum Press, 1975, pp. 159–174.

34. Goresky, C. A., G. G. Bach, and B. E. Nadeau. Red cell carriage of label: Its limiting effect on the exchange of materials in the liver. *Circ. Res.* 36:328–351, 1975.

35. Schulze, K. S., C. A. Goresky, M. Jabbari, and J. O. Lough. Esophageal achalasia associated with gastric carcinoma: Lack of evidence for widespread plexus destruction. *Can. Med. Assn. J.* 112:857–864, 1975.

36. Goresky, C. A., J. W. Warnica, J. H. Burgess, and B. E. Nadeau. Effect of exercise on dilution estimates of extravascular lung water and on the carbon monoxide diffusing capacity in normal adults. *Circ. Res.* 37:379–389, 1975.

37. Goresky, C. A., and G. G. Bach. A factor in longitudinal tissue gradients: Red cell carriage. In: *Oxygen Transport to Tissue,* Vol. II, edited by J. Grote, D. D. Reneau, and G. Thews. New York: Plenum Press, 1976, pp. 505–510.

38. Goresky, C. A., and G. G. Bach. Effect of the red cell membrane on the exchange of materials in the microcirculation. In: *Microcirculation,* Vol. II, edited by J. Grayson and W. Zingg. New York: Plenum Press, 1976, pp. 63–69.

39. Goresky, C. A., and G. G. Bach. The constraining effect of limited red cell permeability on the exchange of materials in the liver. In: *Diseases of the Liver and Biliary Tract,* edited by C. M. Leevy. Basel: S. Karger AG, 1976, pp. 60–64.

40. Gordon, E. R., C. A. Goresky, T.-H. Chan, and A. S. Perlin. The isolation and characterization of bilirubin diglucuronide, the major bilirubin conjugate in dog and human bile. *Biochem. J.* 155:477–486, 1976.

41. Goresky, C. A. Uptake of materials by the intact liver: The design and analysis of experiments. In: *The Liver: Quantitative Aspects of Structure and Function,* edited by R. Preisig, J. Bircher, and G. Paumgartner. Aulendorf, Switzerland: Editio Cantor, 1976, pp. 106–125.

42. Rose, C. P., and C. A. Goresky. Vasomotor control of capillary transit time heterogeneity in the canine coronary circulation. *Circ. Res.* 39:541–554, 1976.

43. Goresky, C. A. Hepatic membrane carrier transport processes: Their involvement in bilirubin uptake. In: *Chemistry and Physiology of Bile Pigments.* Washington, D.C.: U.S. Government Printing Office, 1977, pp. 265–281.

44. Rose, C. P., C. A. Goresky, and G. G. Bach. The capillary and sarcolemmal barriers in the heart: An exploration of labeled water permeability. *Circ. Res.* 41:515–533, 1977.

45. Rose, C. P., and C. A. Goresky. Constraints on the uptake of labeled palmitate by the heart: The barriers at the capillary and sarcolemmal surfaces and the control of intracellular sequestration. *Circ. Res.* 41:534–545, 1977.

46. Livingstone, A. S., M. Potvin, C. A. Goresky, M. H. Finlayson, and E. J. Hinchey. Changes in the blood–brain barrier in hepatic coma after hepatectomy in the rat. *Gastroenterology* 73:697–704, 1977.

47. Gordon, E. R., T. H. Chan, K. Samodai, and C. A. Goresky. The isolation and further characterization of the bilirubin tetrapyrroles in bile-containing human duodenal juice and dog gallbladder bile. *Biochem. J.* 167:1–8, 1977.

48. Rasio, E. A., M. Bendayan, and C. A. Goresky. Diffusion permeability of an isolated *rete mirabile.* Circ. Res. 41:791–798, 1977.

49. Goresky, C. A., and C. P. Rose. Blood-tissue exchange in liver and heart: The influence of heterogeneity of capillary transit times. *Fed. Proc.* 36:2629–2634, 1977.

50. Chinard, F. P., C. Crone, C. A. Goresky, and N. Lassen. Memorial. William Perl, 1918–1976. *Microvasc. Res.* 13:277–281, 1977.

51. Goresky, C. A., J. W. Warnica, J. H. Burgess, and R. F. P. Cronin. Measurement of the extravascular water space in the lungs: Its dependence on alveolar blood flow. *Microvasc. Res.* 15:149–167, 1978.

52. Warnica, J. W., C. A. Goresky, and J. H. Burgess. Changes with exercise in dilution estimates of extravascular lung water in patients with mitral stenosis. *Br. Heart J.* 40:665–671, 1978.

53. Goresky, C. A., D. S. Daly, S. Mishkin, and I. M. Arias. Uptake of labeled palmitate by the intact liver: Role of intracellular binding sites. *Am. J. Physiol.* 234:E542–E553, 1978.

54. Goresky, C. A., E. R. Gordon, E. A. Shaffer, P. Paré, D. Carassavas, and A. Aronoff. Definition of a conjugation dysfunction in Gilbert's syndrome: Studies of the handling of bilirubin loads and of the pattern of bilirubin conjugates secreted in bile. *Clin. Sci. Mol. Med.* 55:63–71, 1978.

55. Alsumait, A. R., M. Jabbari, and C. A. Goresky. Pancreaticocolonic fistula: A complication of pancreatitis. *Can. Med. Assoc. J.* 119:715–719, 1978.

56. Rasio, E. A., and C. A. Goresky. Capillary limitation of oxygen distribution in the isolated *rete mirabile* of the eel (*Anguilla anguilla*). *Circ. Res.* 44:498–503, 1979.

57. Wolkoff, A. W., C. A. Goresky, J. Sellin, Z. Gatmaitan, and I. M. Arias. Role of ligandin in transfer of bilirubin from plasma into liver. *Am. J. Physiol.* 236:E638–E648, 1979.

58. Goresky, C. A. Uptake in the liver: The nature of the process. In: *Liver and Biliary Tract Physiology I. International Review of Physiology,* edited by N. B. Javitt. Baltimore: University Park Press, 1980, pp. 65–101.

59. Rose, C. P., C. A. Goresky, P. Bélanger, and M. J. Chen. Effect of vasodilation and flow rate on capillary permeability surface product and interstitial space size in the coronary circulation: A frequency domain technique for modeling multiple dilution data with Laguerre functions. *Circ. Res.* 47:312–328, 1980.

60. Cousineau, D., C. P. Rose, and C. A. Goresky. Labeled catecholamine uptake in the dog heart: Interactions between capillary wall and sympathetic nerve uptake. *Circ. Res.* 47:329–338, 1980.

61. Gordon, E. R., and C. A. Goresky. The formation of bilirubin diglucuronide by rat liver microsomal preparations. *Can. J. Biochem.* 58:1302–1310, 1980.

62. Cousineau, D., C. P. Rose, and C. A. Goresky. *In vivo* characterization of the adrenergic receptors in the working canine heart. *Circ. Res.* 49:501–510, 1981.

63. Rasio, E. A., M. Bendayan, and C. A. Goresky. The effect of hyperosmolality on the permeability and structure of the capillaries of the isolated *rete mirabile* of the eel. *Circ. Res.* 49:661–676, 1981.

64. Rose, C. P., D. Cousineau, and C. A. Goresky. The influence of capillary and tissue barriers on the extraction of metabolites and catecholamines in the coronary circulation. *Bibl. Anat.* 20:503–506, 1981.

65. Goresky, C. A. Tracer behaviour in the hepatic microcirculation. In: *Hepatic Circulation in Health and Disease,* edited by W. W. Lautt. New York: Raven Press, 1981, pp. 25–39.

66. Salcudean, S. E., P. R. Bélanger, C. A. Goresky, and C. P. Rose. The use of Laguerre functions for parameter identification in a distributed biological system. *IEEE Trans. Biomed. Eng.* 28:767–775, 1981.

67. Goresky, C. A., and E. R. Gordon. Unconjugated hyperbilirubinemia. *Med. North Am.* 19:1917–1921, 1982.

68. Cherry, R. D., D. Portnoy, M. Jabbari, D. S. Daly, D. G. Kinnear, and C. A. Goresky. Metronidazole: An alternate therapy for antibiotic-associated colitis. *Gastroenterology* 82:849–851, 1982.

69. Goresky, C. A. Cell membrane transport processes: Their role in hepatic uptake. In: *The Liver: Biology and Pathobiology,* edited by I. M. Arias, H. M. Popper, D. Schachter, and D. A. Shafritz. New York: Raven Press, 1982, pp. 581–599.

70. Goresky, C. A., P. M. Huet, and J. P. Villeneuve. Blood-tissue exchange and blood flow in the liver. In: *Hepatology, a Textbook of Liver Disease,* edited by D. Zakim and T. D. Boyer. Philadelphia: W. B. Saunders, 1982, pp. 32–63.

71. Gordon, E. R., and C. A. Goresky. A rapid and quantitative high performance liquid chromatographic method for assaying bilirubin and its conjugates in bile. *Can. J. Biochem.* 60:1050–1057, 1982.

72. Goresky, C. A. The processes of cellular uptake and exchange in the liver. *Fed. Proc.* 41:3033–3039, 1982.

73. Huet, P. M., C. A. Goresky, J. P. Villeneuve, D. Marleau, and J. O. Lough. Assessment of liver microcirculation in human cirrhosis. *J. Clin. Invest.* 70:1234–1244, 1982.

74. Goresky, C. A., E. R. Gordon, and G. G. Bach. Uptake of monohydric alcohols by liver: Demonstration of a shared enzymic space. *Am. J. Physiol.* 244 (*Gastrointest. Liver. Physiol.* 7):G198–G214, 1983.

75. Goresky, C. A., G. G. Bach, and C. P. Rose. Effects of saturating metabolic uptake on space profiles and tracer kinetics. *Am. J. Physiol.* 244 (*Gastrointest. Liver. Physiol.* 7):G215–G232, 1983.

76. Cousineau, D., C. A. Goresky, and C. P. Rose. Blood flow and norepinephrine effects on liver vascular and extravascular volumes. *Am. J. Physiol.* 244 (*Heart Circ. Physiol.* 13):H495–H504, 1983.

77. Goresky, C. A. Kinetic interpretation of hepatic multiple-indicator dilution studies. *Am. J. Physiol.* 245 (*Gastrointest. Liver. Physiol.* 8):G1–G12, 1983.

78. Cousineau, D., C. P. Rose, D. Lamoureux, and C. A. Goresky. Changes in cardiac transcapillary exchange with metabolic coronary vasodilation in the intact dog. *Circ. Res.* 53:719–730, 1983.

79. Gordon, E. R., U. Sommerer, and C. A. Goresky. The hepatic microsomal formation of bilirubin diglucuronide. *J. Biol. Chem.* 258:15028–15036, 1983.

80. Goresky, C. A. Permeability of the liver microvasculature. In: *Microcirculation of the Alimentary Tract,* edited by A. Koo, S. K. Lam, and L. H. Smaje. Singapore: World Scientific, 1983, pp. 197–208.

81. Cousineau, D., C. A. Goresky, G. G. Bach, and C. P. Rose. Effect of β-adrenergic blockade on *in vivo* norepinephrine release in canine heart. *Am. J. Physiol.* 246 (*Heart Circ. Physiol.* 15):H283–H292, 1984.

82. Goresky, C. A. The modeling of tracer exchange and sequestration in the liver. *Fed. Proc.* 43:154–160, 1984.

83. Potvin, M., M. H. Finlayson, E. J. Hinchey, J. O. Lough, and C. A. Goresky. Cerebral abnormalities in hepatectomized rats with acute hepatic coma. *Lab. Invest.* 50:560–564, 1984.

84. Gordon, E. R., P. J. Meier, C. A. Goresky, and J. L. Boyer. Mechanism and subcellular site of bilirubin diglucuronide formation in rat liver. *J. Biol. Chem.* 259:5500–5506, 1984.

85. Goresky, C. A., and A. C. Groom. Microcirculatory events in the liver and the spleen. In: *Handbook of Physiology.* Sect. 2, *The Cardiovascular System. Volume IV, The*

Microcirculation., edited by E. M. Renkin and C. C. Michel. Washington, D.C.: American Physiological Society, 1984, pp. 689–780.

86. Bassingthwaighte, J. B., and C. A. Goresky. Modeling in the analysis of solute and water exchange in the microvasculature. In: *Handbook of Physiology. Sect. 2, The Cardiovascular System.* Vol. IV, *The Microcirculation,* edited by E. M. Renkin and C. C. Michel. Bethesda, MD: American Physiological Society, 1984, pp. 549–626.

87. Rose, C. P., and C. A. Goresky. Interactions between capillary exchange, cellular entry, and metabolic sequestration processes in the heart. In: *Handbook of Physiology—The Cardiovascular System,* IV, Chap 16, edited by E. M. Renkin and C. C. Michel. Washington, D.C.: American Physiological Society, 1984, pp. 781–798.

88. Jabbari, M., R. Cherry, J. O. Lough, D. S. Daly, D. G. Kinnear, and C. A. Goresky. Gastric antral vascular ectasia: The watermelon stomach. *Gastroenterology* 87:1165–1170, 1984.

89. Rose, C. P., and C. A. Goresky. Limitations of tracer oxygen uptake in the canine coronary circulation. *Circ. Res.* 56:57–71, 1985.

90. Cousineau, D., F. Péronnet, C. P. Rose, and C. A. Goresky. Norépinéphrine plasmatique et activité sympathique cardiaque durant l'exercice dynamique chez l'homme: quelle est la relation?.' *l'Union Médicale du Canada* 114:1–5, 1985.

91. Cousineau, D., C. A. Goresky, C. P. Rose, and S. Lee. Reflex sympathetic effects on liver vascular space and liver perfusion in dogs. *Am. J. Physiol.* 248 (*Heart Circ. Physiol.* 17):H186–H192, 1985.

92. Goresky, C. A. The Landis Award Lecture for 1982—Biological barriers: Their effects on cellular entry and metabolism *in vivo. Microvasc. Res.* 29:1–17, 1985.

93. Rasio, E. A., and C. A. Goresky. Passage of ions and Dextran molecules across the *rete mirabile* of the eel: The effects of charge. *Circ. Res.* 57:74–83, 1985.

94. Luterman, L., A. R. Alsumait, D. S. Daly, and C. A. Goresky. Colonoscopic features of cecal amebomas. *Gastrointest. Endosc.* 31:204–206, 1985.

95. Jabbari, M., C. A. Goresky, J. Lough, C. Yaffe, D. Daly, and C. Coté. The inlet patch: Heterotopic gastric mucosa in the upper esophagus. *Gastroenterology* 89:352–356, 1985.

96. Goresky, C. A., G. G. Bach, A. W. Wolkoff, C. P. Rose, and D. Cousineau. Sequestered tracer outflow recovery in multiple indicator dilution experiments. *Hepatology* 5:805–814, 1985.

97. Rose, C. P., D. Cousineau, and C. A. Goresky. A clinically applicable method for the estimation of substrate transport in the coronary circulation *in vivo. Circulation* 72:81–88, 1985.

98. Goresky, C. A., G. G. Bach, E. R. Gordon, and C. P. Rose. The enzymic shared space effect: a feature of metabolic uptake by the liver. In: *Carrier-mediated Transport of Solutes from Blood to Tissue,* edited by D. L. Yudilevich and G. E. Mann. London: Longman, 1985, pp. 125–137.

99. Cousineau, D., C. A. Goresky, G. G. Bach, and C. P. Rose. *In vivo* norepinephrine dynamics in the heart: an evolving puzzle. In: *Carrier-Mediated Transport of Solutes from Blood to Tissue,* edited by D. L. Yudilevich and G. E. Mann. London: Longman, 1985, pp. 139–153.

100. Goresky, C. A. Hepatic uptake of bile pigments. In: *Bile Pigments and Jaundice: Molecular, Metabolic, and Medical Aspects,* edited by J. D. Ostrow. New York: Marcel Dekker, 1986, pp. 183–209.

101. Cousineau, D., C. P. Rose, and C. A. Goresky. Plasma expansion effect on cardiac capillary and adrenergic exchange in intact dogs. *J. Appl. Physiol.* 60:147–153, 1986.

102. Bassingthwaighte, J. B., F. P. Chinard, C. Crone, C. A. Goresky, N. A. Lassen, R. S. Reneman, and K. L. Zierler. Terminology for mass transport and exchange. *Am. J. Physiol.* 250 (*Heart Circ. Physiol.* 19):H539–H545, 1986.

103. Cousineau, D., C. A. Goresky, and C. P. Rose. Decreased basal cardiac interstitial norepinephrine release after neuronal uptake inhibition in dogs. *Circ. Res.* 58:859–866, 1986.

104. Viallet, J., J. D. MacLean, C. A. Goresky, M. Staudt, G. Routhier, and C. Law. Arctic trichinosis presenting as prolonged diarrhea. *Gastroenterology* 91:938–946, 1986.

105. Cherry, R. D., M. Jabbari, C. A. Goresky, M. Herba, D. Reich, and P. E. Blundell. Chronic mesenteric vascular insufficienty with gastric ulceration. *Gastroenterology* 91:1548–1552, 1986.

106. Goresky, C. A., D. Cousineau, C. P. Rose, and S. Lee. Lack of liver vascular response to carotid occlusion in mildly acidotic dogs. *Am. J. Physiol.* 251 (*Heart Circ. Physiol.* 20):H991–H999, 1986.

107. Lough, J., L. Rosenthall, A. Arzoumanian, and C. A. Goresky. Kupffer cell depletion associated with capillarization of liver sinusoids in carbon tetrachloride-induced rat liver cirrhosis. *J. Hepatol.* 5:190–198, 1987.

108. Goresky, C. A., M. Huet, D. Marleau, J. O. Lough, and J. P. Villeneuve. Blood-tissue exchange in cirrhosis of the liver. In: *Cirrhosis of the Liver: Methods and Fields of Research,* edited by N. Tygstrup and F. Orlandi. Oxford: Elsevier Science, 1987, pp. 143–164.

109. Goresky, C. A., and A. J. Schwab. Flow, cell entry, and metabolic disposal: Their interactions in hepatic uptake, Ch. 46. In: *The Liver: Biology and Pathobiology,* 2nd ed., edited by I. M. Arias. New York: Raven Press, 1988, pp. 807–832.

110. Sommerer, U., E. R. Gordon, and C. A. Goresky. Microsomal specificity underlying the differing hepatic formation of bilirubin glucuronide and glucose conjugates by rat and dog. *Hepatology* 8:116–124, 1988.

111. Rose, C. P., C. A. Goresky, G. G. Bach, J. B. Bassingthwaighte, and S. E. Little. In vivo comparison of non-gaseous metabolite and oxygen transport in the heart. In: *Oxygen Transport to Tissue X. Adv. Exp. Med. Biol.* 222, edited by M. Mochizuki et al. New York: Plenum Press, 1988, pp. 45–54.

112. Goresky, C. A., and G. E. Wild. Vascular disorders of the liver. *Med. North Am.* 21:4090–4101, 1988.

113. Goresky, C. A., A. J. Schwab, and C. P. Rose. Xenon handling in the liver: Red cell capacity effect. *Circ. Res.* 63:767–778, 1988.

114. Pang, K. S., W. F. Cherry, J. Accaputo, A. J. Schwab, and C. A. Goresky. Combined hepatic arterial-portal venous and hepatic arterial-hepatic venous perfusions to probe the abundance of drug metabolizing activities. *J. Pharmacol. Exp. Ther.* 247:690–700, 1988.

115. Pang, K. S., W. F. Lee, W. F. Cherry, V. Yuen, J. Accapulo, S. Fayz, A. J. Schwab, and C. A. Goresky. Effects of perfusate flow rate on measured blood volume, Disse space, intracellular water space, and drug extraction in the perfused rat liver preparations. *J. Pharmocokinet. Biopharmaceut.* 16:595–632, 1988.

116. Jabbari, M., G. Wild, C. A. Goresky, D. S. Daly, J. O. Lough, D. P. Cleland, and D. G. Kinnear. Scalloped valvulae conniventes: an endoscopic marker of celiac sprue. *Gastroenterology* 95:1518–1522, 1988.

117. Goresky, C. A., G. G. Bach, D. Cousineau, A. J. Schwab, C. Rose, S. Lee, and S. Goresky. Handling of tracer norepinephrine by the dog liver. *Am. J. Physiol.* 256 (*Gastrointest. Liver. Physiol.* 19):G107–G123, 1989.

118. Rasio, E. A., M. Bendayan, and C. A. Goresky. Effect of reduced energy metabolism and reperfusion on the permeability and morphology of the capillaries of an isolated rete mirabile. *Circ. Res.* 64:243–254, 1989.

119. St. Pierre, M. V., A. J. Schwab, C. A. Goresky, W. F. Lee, and K. S. Pang. The multiple indicator dilution technique for characterization of normal and retrograde flow in once-through rat liver perfusions. *Hepatology* 9:285–296, 1989.

120. Goresky, C. A., and C. P. Rose. The delivery of oxygen to the myocardium. In: *Analysis and Simulation of the Cardiac System—Ischemia,* edited by S. Sideman. Boca Raton, FL: CRC Press, 1989, pp. 333–353.

121. Rasio, E. A., M. Bendayan, C. A. Goresky, J. S. Alexander, and D. Shepro. Effect of phalloidin on structure and permeability of rete capillaries in the normal and hypoxic state. *Circ. Res.* 65:591–599, 1989.

122. Schwab, A. J., C. A. Goresky, and C. P. Rose. Handling of tracer bicarbonate in the liver: The relative impermeability of hepatocyte cell membranes to the ionic species. *Circ. Res.* 65:1646–1656, 1989.

123. Goresky, C. A., and E. R. Gordon. High-performance liquid chromatographic separation of bilirubin conjugates. *Anal. Biochem.* 183:269–274, 1989.

124. Dupuis, J., C. A. Goresky, C. Juneau, A. Calderone, J. L. Rouleau, C. P. Rose, and S. Goresky. Use of norepinephrine uptake to measure lung capillary recruitment with exercise. *J. Appl. Physiol.* 68:700–713, 1990.

125. Schwab, A. J., F. Barker III, C. A. Goresky, and K. S. Pang. Transfer of enalaprilat across rat liver cell membranes is barrier limited. *Am. J. Physiol.* 258 (*Gastrointest. Liver. Physiol.* 21):G461–G475, 1990.

126. Goresky, G. A., and E. R. Gordon. High-performance liquid chromatographic separation of bilirubin conjugates: The effects of change in molarity and pH. *J. Chromatog. (Biomed. Applications)* 528:123–141, 1990.

127. Pang, K. S., F. Barker III, A. J. Schwab, and C. A. Goresky. [^{14}C]urea and ^{58}Co-EDTA as reference indicators in hepatic multiple indicator dilution studies. *Am. J. Physiol.* 259 (*Gastrointest. Liver. Physiol.* 22):G32–G40, 1990.

128. Xu, N., A. Chow, C. A. Goresky, and K. S. Pang. Effects of retrograde flow on measured blood volume, Disse space, intracellular water space, and drug extraction in the perfused rat liver: Characterization by the multiple indicator dilution technique. *J. Pharmacol. Exp. Ther.* 254:914–925, 1990.

129. Cousineau, D. F., C. A. Goresky, C. P. Rose, and A. J. Schwab. Cardiac microcirculatory effects of β-adrenergic blockade during sympathetic stimulation. *Circ. Res.* 68:997–1006, 1991.

130. Pang, K. S., F. Barker III, W. F. Cherry, and C. A. Goresky. Esterases for enalapril hydrolysis are concentrated in the perihepatic venous region of the rat liver. *J. Pharmacol. Exp. Ther.* 257:294–301, 1991.

131. Pang, K. S., C. A. Goresky, and A. J. Schwab. Deterministic factors underlying drug and metabolite clearances in rat liver perfusion studies. In: *Research in perfused liver: Clinical and basic applications,* edited by F. Ballet and R. G. Thurman: Les Editions INSERM and John Libbey Eurotext, 1991, pp. 259–302.

132. Basset, G., J. B. Bassingthwaighte, F. P. Chinard, A. Gjedde, C. A. Goresky, and A. J. Hansen. Christian Crone (1926–1990): An appreciation. *Microvasc. Res.* 41:1–4, 1991.

133. Schwab, A. J., and C. A. Goresky. Free fatty acid uptake by polyethylene: What can one learn from this?. *Am. J. Physiol.* 261 (*Gastrointest. Liver Physiol.* 24):G896–G906, 1991.

134. Pang, K. S., N. Xu, and C. A. Goresky. D_2O as a substitute for ^3H_2O, as a reference indicator in liver multiple indicator dilution studies. *Am. J. Physiol. (Gastrointest. Liver Physiol.* 24) 261:G929–G936, 1991.

135. Rasio, E. A., M. Bendayan, and C. A. Goresky. Effect of temperature change on the permeability of eel rete capillaries. *Circ. Res.* 70:272–284, 1992.

136. Dupuis, J., C. A. Goresky, J. W. Ryan, J. L. Rouleau, and G. G. Bach. Pulmonary angiotensin-converting enzyme substrate hydrolysis during exercise. *J. Appl. Physiol.* 72:1868–1886, 1992.

137. Goresky, C. A., E. R. Gordon, J. R. Sanabria, S. M. Strasberg, and M. W. Flye. Changes in bilirubin pigments secreted in bile after liver transplantation. *Hepatology* 15:849–857, 1992.

138. Goresky, C. A., K. S. Pang, A. J. Schwab, F. Barker, III, W. F. Cherry, and G. G. Bach. Uptake of a protein-bound polar compound, acetaminophen sulfate, by perfused rat liver. *Hepatology* 16:173–190, 1992.

139. Kassissia, I., C. P. Rose, C. A. Goresky, A. J. Schwab, G. G. Bach, and S. Guirguis. Flow limited tracer oxygen distribution in the isolated perfused rat liver: Effects of temperature and hematocrit. *Hepatology* 16:763–775, 1992.

140. Schwab, A. J., I. A. M. de Lannoy, C. A. Goresky, K. Poon, and K. S. Pang. Enalaprilat handling by the kidney; barrier-limited cell entry. *Am. J. Physiol. 263 (Renal Fluid Electrolyte Physiol.* 32):F858–F869, 1992.

141. Barkun, J. S., A. N. Barkun, J. S. Sampalis, G. Fried, B. Taylor, M. J. Wexler, C. A. Goresky, J. L. Meakins, and the McGill Gallstone Treatment Group. Randomised controlled trial of laparoscopic versus mini cholecystectomy. *Lancet* 340:1116–1119, 1992.

142. Goresky, C. A., A. J. Schwab, and K. S. Pang. Kinetic models of hepatic transport at the organ level. In: *Hepatic transport and bile secretion: Physiology and pathophysiology,* edited by N. Tavoloni and P. D. Berk. New York: Raven Press, 1993, pp. 11–39.

143. Goresky, C. A., G. G. Bach, and A. J. Schwab. Distributed-in-space product formation *in vivo:* linear kinetics. *Am. J. Physiol.* 264 (*Heart Circ. Physiol.* 33):H2007–H2028, 1993.

144. Goresky, C. A., G. G. Bach, and A. J. Schwab. Distributed-in-space product formation *in vivo:* Enzymic kinetics. *Am. J. Physiol.* 264 (*Heart Circ. Physiol.* 33):H2029–H2050, 1993.

145. Rasio, E. A., M. Bendayan, and C. A. Goresky. Le *rete mirabile* de l'anguille: un modèle unique pour l'étude de la perméabilité capillaire. *Synthèse* 9:593–603, 1993.

146. Xu, X., A. J. Schwab, F. I. Barker, C. A. Goresky, and K. S. Pang. Salicylamide sulfate cell entry in perfused rat liver: A multiple indicator dilution study. *Hepatology* 19:229–244, 1994.

147. Dupuis, J., C. A. Goresky, and D. J. Stewart. Pulmonary removal and production of endothelin in the anesthetized dog. *J. Appl. Physiol.* 76:694–700, 1994.

148. Goresky, C. A., A. J. Schwab, and K. S. Pang. Flow, cell entry, metabolic disposal and product formation in the liver. In: *The Liver: Biology and Pathology,* 3rd ed., edited by I. M. Arias, J. L. Boyer, N. Fausto, W. B. Jakoby, D. A. Schachter and D. A. Shafritz. New York: Raven Press, Ltd., 1994, pp. 1107–1141.

149. Rose, C. P., D. Cousineau, C. A. Goresky, and J. de Champlain. Constitutive non-exocytotic norepinephrine release in sympathetic nerves of *in situ* canine heart. *Am. J. Physiol.* 266 (*Heart Circ. Physiol.* 35):H1386–H1394, 1994.

150. Goresky, C. A., W. Stremmel, C. P. Rose, S. Guirguis, A. J. Schwab, H. E. Diede, and

E. Ibrahim. The capillary transport system for free fatty acids in the heart. *Circ. Res.* 74:1015–1026, 1994.

151. Barkun, A. N., J. S. Barkun, G. M. Fried, G. Ghitulescu, O. Steinmetz, C. Pham, J. L. Meakins, C. A. Goresky, and the McGill Gallstone Treatment Group. Useful predictors of bile duct stones in patients undergoing laparoscopic cholecystectomy. *Ann. Surg.* 220:32–39, 1994.

152. Pang, K. S., F. Barker, III, A. J. Schwab, and C. A. Goresky. Demonstration of rapid entry and a cellular binding space for salicylamide in perfused rat liver: A multiple indicator dilution study. *J. Pharmacol. Exp. Ther.* 270:285–295, 1994.

153. Cousineau, D. F., C. A. Goresky, J. R. Rouleau, and C. P. Rose. Microsphere and dilution measurements of flow and interstitial space in dog heart. *J. Appl. Physiol.* 77:113–120, 1994.

154. Pang, K. S., I. A. Sherman, A. J. Schwab, W. Geng, F. Barker, III, J. A. Dlugosz, G. Cuerrier, and C. A. Goresky. Role of hepatic artery in the metabolism of phenacetin and acetaminophen: An intravital microscopic and multiple-indicator dilution study in perfused rat liver. *Hepatology* 20:672–683, 1994.

155. Pang, K. S., A. J. Schwab, C. A. Goresky, and M. Chiba. Transport, binding, and metabolism of sulfate conjugates in the liver. *Chem. Biol. Interact.* 92:179–207, 1994.

156. Rasio, E. A., M. Bendayan, and C. A. Goresky. The isolated rete. In: *Biochemistry and Molecular Biology of Fishes,* Vol. 3, edited by P. W. Hochachka and T. P. Mommsen. Amsterdam, New York: Elsevier Science, 1994.

157. Goresky, C. A., E. R. Gordon, E. J. Hinchey, and G. M. Fried. Bilirubin conjugate changes in the bile of gallbladders containing gallstones. *Hepatology* 21:373–382, 1995.

158. Rasio, E. A., M. Bendayan, and C. A. Goresky. Effects of second messengers on the permeability and morphology of eel rete capillaries. *Circ. Res.* 76:566–574, 1995.

159. Fallone, C. A., G. E. Wild, C. A. Goresky, and A. N. Barkun. Evaluation of IgA and IgG serology for the detection of *Helicobacter pylori* infection. *Can. J. Gastroenterol.* 9:105–111, 1995.

160. Cousineau, D. F., C. A. Goresky, C. P. Rose, A. Simard, and A. J. Schwab. Effects of flow, perfusion pressure, and oxygen consumption on cardiac capillary exchange. *J. Appl. Physiol.* 78:1350–1359, 1995.

161. Pang, K. S., F. Barker, A. Simard, A. J. Schwab, and C. A. Goresky. Sulfation of acetaminophen by the perfused rat liver: The effect of red blood cell carriage. *Hepatology* 22:267–282, 1995.

162. Geng, W. P., A. J. Schwab, and C. A. Goresky and K. S. Pang. Carrier-mediated uptake and excretion of bromosulfophthalein-glutathione in perfused rat liver: A multiple indicator dilution study. *Hepatology* 22:1188–1207, 1995.

163. Goresky, C. A. The 1994 G. Malcolm Brown Lecture: Biological Barriers and Medicine. *Clin Invest. Med.* 18:484–501, 1995.

164. Kassissia, I. G., C. A. Goresky, C. P. Rose, A. J. Schwab, A. Simard, P. M. Huet, and G. G. Bach. Tracer oxygen distribution is barrier-limited in the cerebral microcirculation. *Circ. Res.* 77:1201–1211, 1995.

165. Dupuis, J., C. A. Goresky, J. L. Rouleau, G. G. Bach, A. Simard, and A. J. Schwab. Kinetics of pulmonary uptake of serotonin during exercise in the dog. *J. Appl. Physiol.* 80:30–46, 1996.

166. Schwab, A. J., and C. A. Goresky. Hepatic uptake of protein-bound ligands: Effect of an unstirred Disse space. *Am. J. Physiol.* 270 (*Gastrointest. Liver. Physiol.* 33):G869–G880, 1996.

167. Dupuis, J., C. A. Goresky, and A. Fournier. Pulmonary clearance of circulating endothelin-1 in dogs in vivo: exclusive role of ETB receptors. *J. Appl. Physiol.* 81:1510–1515, 1996.

168. Sanabria, J. R., E. R. Gordon, P. R. C. Harvey, C. A. Goresky, and S. M. Straserg. Accumulation of unconjugated bilirubin in cholesterol pellets implanted in swine gallbladders. *Gastroenterology* 110:607–613, 1996.

169. Sorrentino, D., D. D. Stump, K. Van Ness, A. Simard, A. J. Schwab, S. L. Shou, C. A. Goresky, and P. D. Berk. Oleate uptake by isolated hepatocytes and the perfused rat liver is competitively inhibited by palmitate. *Am. J. Physiol.* 270 (*Gastrointest. Liver. Physiol.* 33):G385–G392, 1996.

170. Goresky, C. A., A. Simard, and A. J. Schwab. Increase hepatocyte permeability surface area product for ^{86}Rb with increase in blood flow. *Circ. Res.* 80:645–654, 1997.

171. Dupuis, J., C. A. Goresky, C. P. Rose, D. J. Stewart, P. Cernacek, A. J. Schwab, and A. Simard. Endothelin-1 myocardial clearance, production, and effect on capillary permeability in vivo. *Am. J. Physiol.* (*Heart Circ. Physiol.*) 42:H1239–H1245, 1997.

172. Pavone, E., S. N., Mehta, N. Hilzenrat, P. Bret, J. Lough, C. A. Goresky, A. N. Barkun, and M. Jabbari. Role of ERCP in the diagnosis of intraductal papillary mucinous neoplasms. *Am. J. Gastroenterol.* 92:887–890, 1997.

173. Barkun, A. N., J. S. Barkun, J. S. Sampalis, J., Caro, G. M. Fried, J. L. Meakins, L. Joseph L., C. A. Goresky. Costs and effectiveness of extracorporeal stone shock wave lithotripsy versus laparoscopic cholecystectomy: a randomized clinical trial. *Int. J. Technol. Assess. Health Care* 13:589–599, 1997.

174. Geng, W., A. J. Schwab, T. Horie, C. A. Goresky, and K. S. Pang. Hepatic uptake of bromosulfophthalein-glutathione in perfused EHBR mutant rat liver: a multiple indicator dilution study. *J. Pharmacol. Exp. Therap.* in press, 1997.

175. Chiba, M., A. J. Schwab, C. A. Goresky, and K. S. Pang. Carrier-mediated entry of 4-methylumbelliferyl sulfate: characterization by the multiple-indicator dilution technique in perfused rat liver. *Hepatology* 27:134–146, 1998.

176. Bassingthwaighte, J. B., C. A. Goresky, and J. H. Linehan (eds.). *Whole Organ Approaches to Cellular Metabolism. Capillary Permeation, Cellular Uptake, and Product Formation.* New York: Springer-Verlag, 1998.

177. Bassingthwaighte, J. B., C. A. Goresky, and J. H. Linehan. Chapter 1: Modeling in the analysis of the processes of uptake and metabolism in the whole organ. In: *Whole Organ Approaches to Cellular Metabolism,* edited by J. B. Bassingthwaighte, C. A. Goresky, and J. H. Linehan. New York: Springer-Verlag, 1988.

178. Rasio, E. A., M. Bendayan, and C. A. Goresky. Chapter 3: Study of Blood Capillary Permeability with the Rete Mirabile. In: *Whole Organ Approaches to Cellular Metabolism,* edited by J. B. Bassingthwaighte, C. A. Goresky, and J. H. Linehan. New York: Springer-Verlag, 1998.

179. Goresky, C. A., G. G. Bach, A. J. Schwab, and K. S. Pang. Chapter 13: Liver cell entry in vivo and enzymatic conversion. In: *Whole Organ Approaches to Cellular Metabolism,* edited by J. B. Bassingthwaighte, C. A. Goresky, and J. H. Linehan. New York: Springer-Verlag, 1998.

180. Pang, K. S., C. A. Goresky, A. J. Schwab, and W. Geng. Chapter 14: Multiple indicator dilution studies of drug and metabolite processing in the liver. In: *Whole Organ Approaches to Cellular Metabolism,* edited by J. B. Bassingthwaighte, C. A. Goresky, and J. H. Linehan. New York: Springer-Verlag, 1998.

Index

Page numbers for entries occurring in figures are followed by an f; those for entries occurring in tables, by a t.

557